STOCHASTIC ANALYSIS OF STRUCTURAL AND MECHANICAL VIBRATIONS

Loren D. Lutes

Texas A&M University
College Station, Texas

Shahram Sarkani

The George Washington University
Washington, D. C.

 Prentice Hall, Upper Saddle River, New Jersey 07458

Library of Congress Cataloging-in-Publication Data

Lutes, L. D. (Loren D.)
   Stochastic analysis of structural and mechanical vibrations /
Loren D. Lutes, Shahram Sarkani.
      p. cm.
   Includes bibliographical references and index.
   ISBN 0-13-490533-4
   1. Random vibration. 2. Stochastic processes. 3. Structural
dynamics. I. Sarkani, Shahram. II. Title.
TA355.L88 1997
   624.1' 76--dc20   96-33323
                                                    CIP

Acquisitions editor: **Bill Stenquist**
Editorial/production supervision: **Sharyn Vitrano**
Editor-in-chief: **Marcia Horton**
Managing editor: **Bayani Mendoza DeLeon**
Cover designer: **Bruce Kenselaar**
Director of production and manufacturing: **David W. Riccardi**
Manufacturing buyer: **Julia Meehan**
Editorial assistant: **Meg Weist**

© 1997 by Prentice-Hall, Inc.
Simon & Schuster/A Viacom Company
Upper Saddle River, New Jersey 07458

The author and publisher of this book have used their best efforts in preparing this book. These efforts include the
development, research, and testing of the theories and programs to determine their effectiveness. The author and
publisher make no warranty of any kind, expressed or implied, with regard to these programs or the documentation
contained in this book. The author and publisher shall not be liable in any event for incidental or consequential damages
in connection with, or arising out of, the furnishing, performance, or use of these programs.

Printed in the United States of America

10 9 8 7 6 5 4 3 2

ISBN 0-13-490533-4

Prentice-Hall International (UK) Limited, *London*
Prentice-Hall of Australia Pty. Limited, *Sydney*
Prentice-Hall Canada Inc., *Toronto*
Prentice-Hall Hispanoamericana, S.A., *Mexico*
Prentice-Hall of India Private Limited, *New Delhi*
Prentice-Hall of Japan, Inc., *Tokyo*
Simon & Schuster Asia Pte. Ltd., *Singapore*
Editora Prentice-Hall do Brasil, Ltda., *Rio de Janeiro*

DEDICATION

To spouses and parents:

Janis P. Stout and the memory of Louis D. Lutes and Ruth Ernestine Lutes.   *LDL*

Sepideh Farsai Sarkani, Hossein Sarkani, and Shokouh Yamin.   *SS*

CONTENTS

CHAPTER 4. FREQUENCY DOMAIN ANALYSIS

CHAPTER 5. GAUSSIAN AND NON-GAUSSIAN STOCHASTIC PROCESSES

CHAPTER 6. OCCURRENCE RATES AND DISTRIBUTIONS OF EXTREMES

PREFACE

This book is designed for use as a text for graduate courses in random vibrations or stochastic structural dynamics, such as might be offered in departments of civil engineering, mechanical engineering, aerospace engineering, ocean engineering, and applied mechanics. It is also intended for use as a reference for graduate students and practicing engineers with a similar level of preparation. The focus is on the determination of response levels for dynamical systems excited by forces that can be modeled as stochastic processes.

Since many readers will be new to the subject, our primary goal has been clarity, particularly regarding the fundamental principles and relationships. At the same time, we have sought to make the presentation sufficiently thorough and rigorous that the reader will be able to move on to more advanced work. We believe that the book can meet the needs of both those who wish to apply existing stochastic procedures to practical problems, and those who wish to prepare for research which will extend the boundaries of knowledge.

In the hopes of meeting the needs of a broad audience, we have made this book relatively self-contained. For example, all necessary definitions and analysis methods related to stochastic processes are developed within the book. Since the presentation of stochastic processes requires an understanding of random variables, though, we provide a fairly extensive appendix for the benefit of readers needing a review of that material. This appendix can be used by someone with no more than the rudimentary concepts of probability. We do presume that the reader has a background in deterministic structural dynamics or mechanical vibration, but we also give a brief review of the deterministic methods before extending them for use in stochastic problems. Some knowledge of complex functions is necessary for the understanding of important frequency domain concepts. However, we also present time domain integration techniques that provide viable alternatives to the calculus of residues. Because of this, the book should also be useful to engineers who do not have a strong background in complex analysis.

The choice of prerequisites, as well as the demands of brevity, sometimes make it necessary to omit mathematical proofs of results. We always try to give mathematically rigorous definitions and results, though, even when mathematical details are omitted. This approach is particularly important for the reader who wishes to pursue further study. We do not want to burden anyone with imprecise material that they will need to "unlearn" in proceeding to another level. An important part of the book is the inclusion, in each chapter (or major appendix), of a number of worked examples which illustrate the modeling of physical problems as well as the proper application of theoretical solutions. Similar problems are also presented as exercises to be solved by the reader.

One of the special features of the book is an early introduction of vibration problems. In our experience, many engineering students become impatient with any lengthy study of mathematical procedures for which they do not know the application. For this reason, we introduce linear vibration problems after only the introductory chapter on the modeling of stochastic problems, and a chapter on time domain methods for stochastic processes. Time domain interpretations are emphasized throughout the book, even in the presentation of important frequency domain concepts. This includes, for example, the time history implications of bandwidth, with situations varying from narrowband to white noise.

We have organized the book to give maximum flexibility to the reader or instructor. We have intentionally included more material than can be covered in the typical one-semester or one-quarter format. Different instructors will choose to include different topics within an introductory course, and this book includes enough material to accommodate a number of such possibilities. To promote this flexibility, the crucial material is concentrated in the early portions of the book. In particular, the fundamentals of stochastic modeling and analysis of vibration problems are presented by the end of Section 5.3. From this point the reader could proceed to most topics in any of the other chapters. In our own teaching we include a significant amount of the material in Chapters 6 and 7 and a selection of topics from Chapters 8 to 10, with this selection being based on the perceived needs of our students. The book, and a modest number of readings on current research, could also form the basis for a two-semester course.

It should be noted that Chapters 8 to 10 include relatively advanced topics which are the subjects of ongoing research. Thus, one must anticipate that in future years there will be significant extensions to these results on state-space analysis, nonlinear dynamics, and fatigue damage. We feel it is important, though, to provide introductions to these topics so that the reader can make use of the current research, and possibly contribute to its future.

Finally, we express our gratitude to all those from whom we have learned. This list of professors, colleagues, and students is too long to enumerate, and it continues to grow. We gladly acknowledge the benefit we have recieved from all those individuals who have read and offered comments on all or portions of the manuscript, with special thanks for the very thorough review by Professor Marc Mignolet of Arizona State University. We also particularly recognize the authors of the books from which we have taught this material over the past decades, including Y. K. Lin, S. H. Crandall, W. D. Mark, N. C. Nigam, C. Y. Yang, and D. E. Newland. The first author also acknowledges a special debt to Professor T. K. Caughey who introduced him to these concepts in the 1960s, without the benefit of a textbook.

Loren D. Lutes
Shahram Sarkani

Chapter 1

Introduction

1.1 Why Study Random Vibration?

Most structural and mechanical engineers who study probability do so specifically so that they may better estimate the likelihood that some engineering system will provide satisfactory service. This is often stated in the complementary way as estimating the probability of unsatisfactory service or failure. Thus, the study of probability generally implies that the engineer accepts the idea that it is either impossible or infeasible to devise a system which is absolutely sure to perform satisfactorily. The authors believe that this is an honest acceptance of the facts in our uncertain world, but it is somewhat of a departure from the philosophy of much of past engineering education and of the explicit form of many engineering design codes. Of course, engineers have always known that there was a possibility of failure, but they have not always made an effort to quantify the likelihood of that event and to use it in assessing the adequacy of a design. We believe that more rational design decisions will result from such explicit study of the likelihood of failure, and that is our motivation for the study of probabilistic methods.

The characterization of uncertainty in this book will always be done by methods based on probability theory. This is a purely pragmatic choice, based on the fact that these methods have been shown to be useful for a great variety of problems. Methods based on fundamentally different concepts, such as fuzzy sets, have also been demonstrated for some problems, but they will not be investigated here.

The engineering systems studied in this book are dynamical systems. Specifically, they are systems for which the dynamic motion can be modeled by a differential or integral equation or a set of such equations. Such systems usually consist of elements having mass, stiffness, and damping and exhibiting vibratory dynamic behavior. The methods presented are quite general and can be applied to a great variety of problems of structural and mechanical vibration. Examples will vary from simple mechanical oscillators to buildings or other large structures, with excitations which can be either forces or base motion. The primary emphasis will be on problems with linear models, but we will also include some study of nonlinear problems. For nonlinear problems, we will particularly emphasize methods which are direct extensions of linear methods.

The uncertainty studied will be limited to that in the excitation of the system. That is, no consideration will be given to uncertainty about the parameters of the system. Obviously there is also value in studying uncertain systems, but that is usually a rather complicated problem for a dynamical system, and this book will concentrate on the simpler problems in which the uncertainties only involve the inputs and outputs of the system. Experience has shown that there are indeed many problems in which the uncertainty about the input to the system is the key factor determining the probability of system failure. This is particularly true when the inputs are such environmental loads as earthquakes, wind, or ocean waves, but it also applies to numerous other situations such as the pressure variations in the exhaust from a jet engine.

The response of a dynamical system of the type studied here is a time history defined on a continuous set of time values. The field of probability which is applicable to such problems is called stochastic (or random) processes. Thus, the applications presented here involve the use of stochastic processes to model problems involving the behavior of dynamical systems. An individual whose background included both a course in stochastic processes and a course in either structural dynamics or mechanical vibrations might be considered to be in an ideal situation to study stochastic vibrations, but this would be an unreasonably high set of prerequisites for the beginning of such study. In particular, we will not assume prior knowledge of stochastic processes, and will develop the methods for analysis of such processes within this book. The probability background needed for the study of stochastic processes is a fairly thorough understanding of the fundamental methods for investigating random variables. This is because a stochastic process is generally viewed as a family of random variables. For the benefit of readers lacking the necessary random variable background, Appendix A gives a relatively comprehensive introduction to the topic, focusing on the aspects which are most important for the understanding of stochastic processes. Some review of this material may be appropriate for any reader who has not used these methods recently and may need a refresher.

We will expect the reader to be familiar with deterministic approaches to vibration problems by using superposition methods such as the Duhamel convolution integral and, to a lesser extent, the Fourier transform. We will present brief reviews of the principal ideas involved in these methods of vibration analysis, but the reader without a solid background in this area will probably need to do some outside reading on these topics.

1.2 Probabilistic Modeling and Terminology

Within the realm of probabilistic methods, there are several terms related to uncertainty which warrant some comment. The term "random" will be used here for any variable about which we have some uncertainty. This does not mean that no knowledge is available, but that we have less than perfect knowledge. As indicated in the previous paragraph, we will particularly use results from the area of random variables. The word "stochastic" in common usage is essentially synonymous with random, but we will use it in a somewhat more specialized way. In particular, we will use the term stochastic to imply that there is a time history involved. Thus, we will say that the dynamic response at one instant of time t is a random variable $X(t)$, but that the uncertain history of response over a range of time values is a stochastic process $\{X(t)\}$. The practice of denoting a stochastic process by putting the notation for the associated random variables in braces will be used to indicate that the stochastic process is a family of random variables — one for each t value. The term "probability," of course, will be used in the sense of fundamental probability theory. The probability of any event is a number in the range of zero to unity which models the likelihood of the event occurring. We can compute the probabilities of events which are defined in terms of random variables having certain values, or in terms of stochastic processes behaving in certain ways.

One can view the concepts of event, random variable, and stochastic process as forming a hierarchy, in order of increasing complexity. One can always give all the probabilistic information about an event by giving one number — the probability of occurrence for the event. To have all the information about a random variable generally requires knowledge of the probability of many events. In fact, we will be most concerned with so-called continuous random variables, and one must know the probabilities of infinitely many events to completely describe the probabilities of a continuous random variable. As mentioned before, a stochastic process is a family of random variables, so its probabilistic description will alway require significantly more information than does the description of any one of those random variables. We will be most concerned with the case in which the stochastic process consists of infinitely many random variables, so the additional information required will be much more than for a random variable. One can also extend this hierarchy further, with the next step being stochastic fields, which are families of stochastic processes. Within this book we will use events, random variables, and especially stochastic processes, but we will avoid stochastic fields and further generalizations.

****

Example 1.1: Let t denote time in seconds and the random variable $X(t)$, for any fixed t value, be the magnitude of the wind speed at a specified location at that time. Further, let the family of $X(t)$ random variables for all nonnegative t values be a stochastic process, $\{X(t): t \geq 0\}$, and let A be the event $A = \{X(10) \leq 5\,\mathrm{m}\,/\,\mathrm{s}\}$. Review the amount of information needed to give complete probabilistic descriptions of the event A, the random variable $X(10)$, and the stochastic process $\{X(t)\}$.

All the probabilistic information about the event A is given by one number — its probability of occurrence. Thus, we might say that $p = P(A)$ is that probability of occurrence, and the only other probabilistic statement that can be made about A is the almost trivial affirmation that $(1 - p) = P(A^c)$, in which A^c denotes the event of A not occurring, and is read as "A complement" or "not A."

We expect there to be many possible values for $X(10)$. Thus, it takes much more information to give its probabilistic description than it did to describe A. In fact, one of the simpler comprehensive ways of describing the random variable $X(10)$ is to give the probability of infinitely many events like A. That is, if we know $P[X(10) \leq u]$ for all possible u values, then we have a complete probabilistic description of the random variable $X(10)$. Thus, in going from an event to a random variable we have moved from needing one number to needing many (often infinitely many) numbers to describe the probabilities.

The stochastic process $\{X(t): t \geq 0\}$ is a family of random variables, of which $X(10)$ is one particular member. Clearly, it takes infinitely more information to give the complete probability description for this stochastic process than it does to describe any one member of the family. In particular, we would need to know the probability of events such as $[X(t_1) \leq u_1, X(t_2) \leq u_2, \cdots, X(t_j) \leq u_j]$ for all possible choices of j, $t_1, \cdots, t_j$, and $u_1, \cdots, u_j$.

If one chooses to extend this hierarchy further, then a next step could be a stochastic field giving the wind speed at many different locations, with the speed at any particular location being a stochastic process like $\{X(t)\}$.

****

It should also be noted that there exist special cases which somewhat blur the boundaries between the various levels of complexity in the common classification system based on the concepts of event, random variable, stochastic process, stochastic field, etc. In particular, there are random variables which can be described in terms of the probabilities of only a few, or even only one, event. Similarly, one can define stochastic processes which are families of only a few random variables. Within this book, we will generally use the concept of a vector random variable to describe any finite family of random variables and reserve the term stochastic process for an infinite (usually uncountable) family of random variables. Finally, we will treat a finite family of stochastic processes as a vector stochastic process, even though it could be considered a stochastic field.

****

Example 1.2: Let the random variable X denote the maintenance cost for an antenna subjected to the wind, and presume that $X = 0$ if the antenna is undamaged and $\$5,000$ (replacement cost) if it is damaged. How much information is needed to describe all probabilities of X?

Since X has only two possible values in this simplified situation, one can describe all its probabilities with only one number — $p = P(X = 5,000) \equiv P(D)$, in which D denotes the event of antenna damage occurring. The only other information that can be given about the random variable X is $P(X = 0) \equiv P(D^c) = (1 - p)$.

****

Example 1.3: Let the random variable X denote the maintenance cost for an antenna structure subjected to the wind, and presume that there are two possible types of damage. Event A denotes damage to the structure which supports the antenna dish, and it costs $\$2,000$ to repair, while event B denotes damage to the dish itself, and costs $\$3,000$ to repair. Let the random variable X denote the total maintenance cost. How much information is needed to describe all probabilities of X?

In this problem, X may take on any of four values: zero if neither the structure or the dish is damaged, $2,000$ if only the structure is damaged, $3,000$ if only the dish is damaged, and $5,000$ if both structure and dish are damaged. Thus, one can give all the probability information about X with no more than the four numbers giving the probability of X taking on each of its possible values. These are easily described by using the events A and B and the operations of complement and intersection. For example, we might write $p_1 = P(X = 5,000) = P(AB)$, $p_2 = P(X = 3,000) = P(A^cB)$, $p_3 = P(X = 2,000) = P(AB^c)$, and $p_4 = P(X = 0) = P(A^cB^c)$. Even this is somewhat redundant since we also know that $p_1 + p_2 + p_3 + p_4 = 1$, so that knowledge of only three of the probabilities, such as p_1, p_2, and p_3, would be sufficient to describe the problem.

****

Example 1.4: Consider the permanent displacement of a rigid 10 meter square foundation slab during an earthquake which causes some sliding of the underlying soil. Let X, Y, and Z denote the east-west, north-south, and vertical translations of the center of the slab, and let θ_x, θ_y, and θ_z be the rotations (in radians) about the three axes. What type of probability information is required to describe this foundation motion?

Since $\{X, Y, Z, \theta_x, \theta_y, \theta_z\}$ is a family of random variables, one could consider this to be a simple stochastic process. Since the family has only a finite number of members, though, we can equally well consider it to be a vector random variable. We will denote vectors by putting an arrow over them, and treat them as column matrices. Thus we can write $\vec{V} = [X, Y, Z, \theta_x, \theta_y, \theta_z]^T$, in which the T superscript denotes the matrix transpose operation, and this column vector $\vec{V}$ gives the permanent displacement of the foundation. Knowledge of all the probability information about $\vec{V}$ would allow us to write the probability of any event which was defined in terms of the components of $\vec{V}$. That is, we want to be able to give $P(A)$ for any event A which depends on $\vec{V}$ in the sense that we can tell whether A has or has not occurred if we know the value of the vector $\vec{V}$. Clearly we must have information such as $P(X \le 100 \, \text{mm})$ and $P(\theta_z > 0.5 \, \text{rad})$, but we must also know probabilities of intersections like $P(X \le 100 \, \text{mm}, \theta_z < 0.5 \, \text{rad}, \theta_y < 0.1 \, \text{rad})$, etc.

****

Example 1.5: Consider the permanent deformation of a system consisting of a rigid building 20 meters high resting on the foundation of Example 1.4. Let a new random variable W denote the translation to the west of a point at the top of the north face of the building, as shown in the sketch. Show the relationship between W and the vector $\vec{V}$ of Example 1.4.

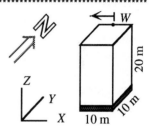

In order to describe the random variable W, we need to be able to calculate probabilities of the sort $P(W \le 200 \, \text{mm})$. We can see, though, that W is related to the components of our vector $\vec{V}$ by $W = -X + 5\theta_z - 20\theta_y$ so that $P(W \le 200 \, \text{mm}) = P(-X + 5\theta_z - 20\theta_y \le 200 \, \text{mm})$. It can be shown that one has sufficient information to compute all such terms as this if one knows $P(X \le u, Y \le v, Z \le w, \theta_x \le \alpha, \theta_y \le \beta, \theta_z \le \gamma)$ for all values of the six parameters $(u, v, w, \alpha, \beta, \gamma)$.

****

Example 1.6: For the same situation as was described in Example 1.4, consider the foundation motion at any time during the earthquake. That is, rather than simply considering permanent displacement, let

$$\vec{V}(t) = [X(t), Y(t), Z(t), \theta_x(t), \theta_y(t), \theta_z(t)]^T$$

Identify the appropriate probabilistic model for this problem.

Now the description of the motion at any one particular time t has the same complexity as the vector $\vec{V}$ in Example 1.4. A complete description of the motion at all times, though, is much more complicated, and requires information about events related to $\vec{V}(t)$ for any t value and/or to several t values. This is a problem in which we need the probabilistic description of the vector stochastic process $\{\vec{V}(t)\}$.

****

1.3 Approach to the Study of Failure Probability

Unsatisfactory performance or "failure" of a structural or mechanical dynamical system can usually be classified as being due to either "first passage" or "fatigue," and both of these failure modes will be studied within this book. Study of first-passage failure is appropriate when we consider the system to have performed unsatisfactorily if some measure of response has ever reached some critical value. Thus, for example, first-passage failure might be considered to have occurred if the stress or strain at some critical location had ever exceeded the yield level, or if the displacement had exceeded some preselected value regarded as a boundary of a region of instability due to buckling.

In order to calculate the probability of first-passage failure during a given time interval $[0, T]$, we will need terms such as $P[\max X(t) > x_{critical}]$, in which the maximum is over the set of values $0 \leq t \leq T$. Obviously, such probabilistic analysis of the maximum of a time history requires considerable knowledge about how the response $X(t_j)$ at one time relates to the response $X(t_k)$ at some other time. In fact, we will need to consider the relationship of $X(t_j)$ to X at every other time within the interval of interest. Fatigue failure differs from first-passage inasmuch as it involves an accumulation of damage over a stress or strain time history, rather than the maximum of that time history. It is similar to first passage, though, in that one cannot calculate the accumulated fatigue damage without knowing the relationship of the response X at any one time to X at every other time. As mentioned previously, analyzing how $X(t_j)$ relates to X at every other time necessitates our study of stochastic processes $\{X(t)\}$ defined on a continuous set of t values. We begin this study in Chapter 2.

Exercises

****

1.1 Assume that a given machine part may fail either due to gradual wear or due to brittle fracture. Let $X(t)$ denote the amount of gradual wear at time t. Parts (a) - (d) list various levels of detail that might be studied for this problem. For each of these, indicate whether the required model would be: an event, a scalar random variable, a vector random variable, a scalar stochastic process, a vector stochastic process, or a stochastic field. Give the answer which is the simplest adequate model.
(a) The likelihood that the part will fail during 10,000 hours of service.
(b) The likelihood that there will be brittle fracture during 10,000 hours of service.
(c) The possible time histories of $X(t)$ on the set $0 \leq t \leq 10,000$ hours.
(d) The set of possible values of $X(t)$ at $t = 10,000$ hours.

****

1.2 Let $X(t,u)$ denote the bending moment in a cantilever beam at time t and a distance u from the fixed end. Parts (a) - (h) list various levels of detail that might be studied for this problem. For each of these, indicate whether the required model would be: an event, a scalar random variable, a vector random variable, a scalar stochastic process, a vector stochastic process, or a stochastic field. Give the answer which is the simplest adequate model.

(a) The likelihood that the bending moment is exactly 3 kN·m when $t = 5$ seconds and $u = 2$ meters.

(b) The likelihood that the bending moment exceeds 2 kN·m when $t = 3$ seconds and $u = 2$ meters.

(c) The likelihoods of all the possible values of the bending moment when $t = 5$ seconds and $u = 3$ meters.

(d) The possible time histories on the set $0 \le t \le 5$ seconds of the bending moment at $u = 3$ meters.

(e) The possible values of the set of bending moments at $u = 0$ m, $u = 1$ m, $u = 2$ m, and $u = 3$ m, all observed at $t = 3$ seconds.

(f) The possible values of the set of bending moment observations at $t = 0$ seconds, $t = 1$ second, $t = 2$ seconds, $t = 3$ seconds, $t = 4$ seconds, and $t = 5$ seconds, all at location $u = 2$ meters.

(g) The possible time histories on the set $0 \le t \le 5$ seconds of the bending moments at $u = 0$ m, $u = 1$ m, $u = 2$ m, and $u = 3$ m.

(h) The possible time histories on the set $0 \le t \le 5$ seconds of the bending moments at every u value for $0 \le u \le 3$ m.

****

1.3 Let $X(t)$ and $Y(t)$ denote the tensile stress and the temperature, respectively, at time t at a given location in a critical element. Parts (a) - (j) list various levels of detail that might be studied for this problem. For each of these, indicate whether the required model would be: an event, a scalar random variable, a vector random variable, a scalar stochastic process, a vector stochastic process, or a stochastic field. Give the answer which is the simplest adequate model.

(a) The likelihood that the tensile stress exceeds 70 MPa and the temperature exceeds 500°C at time $t = 120$ seconds.

(b) The likelihood that the tensile stress exceeds 70 MPa and the temperature is less than or equal to 500°C at time $t = 120$ seconds.

(c) The likelihoods of all the possible values of the tensile stress at time $t = 120$ seconds.

(d) The likelihoods of all the possible values of the temperature at time $t = 120$ seconds.

(e) The likelihoods of all the possible combinations of tensile stress and temperature at time $t = 120$ seconds.

(f) The possible time histories of the tensile stress on the set of values $0 \le t \le 120$ seconds.

(g) The possible time histories of the temperature on the set of values $0 \le t \le 120$ seconds.

(h) The possible combinations of time histories of the tensile stress and temperature on the set of values $0 \le t \le 120$ seconds.

(i) The likelihood that $W(t) \equiv X(t) + 0.2Y(t) > 200$ at time $t = 120$ seconds.

(j) The possible time histories of $W(t)$ on the set of values $0 \le t \le 120$ seconds.

****

1.4 Let $X(t)$ denote the load applied to a crane cable at time t and $Y(t)$ denote the strength of the cable at that time. The strength is a decreasing function of time, because of wear on the cable. Parts (a) - (f) list various levels of detail that might be studied for this problem. For each of these, indicate whether the required model would be: an event, a scalar random variable, a vector random variable, a scalar stochastic process, a vector stochastic process, or a stochastic field. Give the answer which is the simplest adequate model.

(a) The likelihoods of all the possible combinations of load and strength at time $t = 15,000$ hours.
(b) The likelihoods of all possible values of the load at time $t = 15,000$ hours.
(c) The likelihood that the load will exceed the strength prior to $t = 15,000$ hours.
(d) The possible time histories of the load on the set of values $0 \le t \le 15,000$ hours.
(e) The possible time histories of the strength on the set of values $0 \le t \le 15,000$ hours.
(f) The possible combinations of time histories of the load and strength on the set of values $0 \le t \le 15,000$ hours.

****

1.5 Let $X_j(t)$ for $j = 1$ to 20 denote the time history of the shear distortion in story j of a 20 story building which is subjected to an earthquake. Let Y_j denote the maximum value of $X_j(t)$ throughout the duration of the earthquake. Parts (a) - (f) list various levels of detail that might be studied for this problem. For each of these, indicate whether the required model would be: an event, a scalar random variable, a vector random variable, a scalar stochastic process, a vector stochastic process, or a stochastic field. Give the answer which is the simplest adequate model.

(a) The likelihood that $X_5(t)$ will exceed 100 mm at any time during the earthquake.
(b) The likelihood that the translation at the top of the building will exceed 200 mm at any time during the earthquake.
(c) The likelihoods of all the possible values for Y_5.
(d) The likelihoods of all the possible values for the combination of all Y_j terms.
(e) The likelihoods of all the possible time histories of $X_5(t)$.
(f) The likelihoods of all the possible combinations of $X_j(t)$ time histories.

****

Chapter 2

Analysis of Stochastic Processes

2.1 Concept of a Stochastic Process

As noted in Chapter 1, a stochastic process can be viewed as a family of random variables. It is common practice to use braces to denote a set or collection of items, so we write $\{X(t)\}$ for a stochastic process which gives a random variable $X(t)$ for any particular value of t. The parameter t may be called the index parameter for the process, and the set of possible t values is then the index set. The basic idea is that for every possible t value there is a random variable $X(t)$. In some situations we will be precise and include the index set within the notation for the process, such as $\{X(t): t \geq 0\}$ or $\{X(t): 0 \leq t \leq 20\}$. It should be kept in mind that such a specification of the index set within the notation is nothing more than a statement of the range of t for which $X(t)$ is defined. For example, writing a process as $\{X(t): t \geq 0\}$ means that $X(t)$ is not defined for $t < 0$. We will often simplify our notation by omitting the specification of the possible values of t, unless that is apt to cause confusion. In this book, we will always treat the index set of possible t values as being continuous rather than discrete.[1] If we consider a set of $X(t)$ random variables with t chosen from a discrete set such as $\{t_1, t_2, \cdots, t_n, \cdots\}$, then we will use the nomenclature and notation of a vector random variable rather than of a stochastic process.

Another useful way to conceive of a stochastic process is in terms of its possible time histories — its variation with t for a particular observation of the process. For example, any earthquake ground acceleration record might be thought of as one of the many time histories which could have occurred for an earthquake with the same intensity at that site. In this example, as in many others, the set of possible time histories must be viewed as an infinite collection, but the concept is still useful. This idea of the set of all possible time histories may be viewed as a direct generalization of the concept of a random variable. The generalization can be illustrated by the idea of statistical sampling. A statistical sample from a random variable Y is a set of independent observations of the value of Y. Each observation is simply a number from the set of possible values of Y. After a sufficient number of such observations, one will get an idea of the likelihood of various outcomes and can estimate the probability distribution of Y and/or expected values of functions of Y. For the stochastic process $\{X(t)\}$, each observation will give an observed time history rather than simply a number. Again, a sufficient number of observations will allow us to estimate probabilities and expected values related to the process. A collection of time histories for a stochastic process is typically called an ensemble.

Figure 2.1 illustrates the idea of a statistical sample, or ensemble, from a stochastic process, using the notation $X^{(j)}(t)$ for the jth sample time history observed for the process. Of course, the

---

[1]   This is precisely described as a "continuously parameterized" or "continuously indexed" stochastic process.

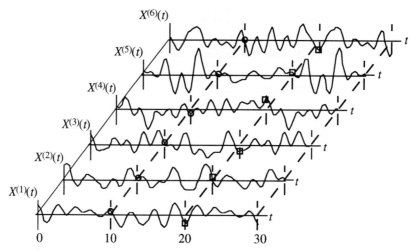

Figure 2.1 Ensemble of time histories for $\{X(t)\}$

ensemble shown in Fig. 2.1 is for illustration only and is too small (i.e., it has too few time histories) to allow one to estimate probabilities or expected values with any confidence. It shows six time histories observed in separate, independent observations of the particular $\{X(t)\}$ process. A "section" across this ensemble at any particular time gives a statistical sample for the random variable corresponding to that t value. Thus, we have observed a sample of six values for $X(10)$, a sample of six values for $X(20)$, etc. The plot illustrates the fact that there is a sort of "orthogonality" between the idea that a stochastic process is characterized by an ensemble of time histories and the idea that it is a family of random variables. A time history is a single observation including many values of t. Many observations at a single value of t give a statistical sample of the random variable $X(t)$.

In most practical problems, it is not feasible to describe a stochastic process in terms of the probability of occurrence of particular time histories (although Examples 2.1 - 2.3 will illustrate a few situations in which this is feasible). On the other hand, we can always characterize the process by using information about the joint probability distribution of the random variables which make it up. Nonetheless, it is often useful to think of the process in terms of an ensemble (usually with infinitely many members) of all its possible time histories, and to consider the characteristics of these time histories. The term "ensemble average" is often used for statistical averages across an ensemble of observed time histories. Thus, one might calculate the average value of $X(10)$ from the sample shown in Fig. 2.1 as $[X^{(1)}(10)+\cdots+X^{(6)}(10)]/6$, and this can be classified either as a statistical average of the observed values for the random variable $X(10)$ or as an ensemble average of the process $\{X(t)\}$ at time $t = 10$. It should also be noted that the term ensemble average is also sometimes used to refer to mathematical expectation. This is generally legitimate if one considers an average over an infinite ensemble. Basically, we expect a statistical average to converge to the underlying expected value if we can make our sample size infinitely large and an ensemble average is simply a statistical average for a stochastic process. One must always remember, though, that actual numerical ensemble averages are always statistical averages across a finite sample; they are not the same as expected values.

2.2   Probability Distribution

In order to have a complete probabilistic description of a stochastic process $\{X(t)\}$, one must know the probability distribution for every set of random variables belonging to that process. This can be simply stated as knowing the probability distribution for every set $\{X(t_1), X(t_2), \cdots, X(t_n)\}$, for all possible n values, and all possible choices of $\{t_1, t_2, \cdots, t_n\}$ for each n value. Since the probability distribution of a random variable or a set of random variables can always be given by a probability density function (or a cumulative distribution function), the necessary information can be written as

$$p_{X(t_1)}(u_1) \qquad\qquad \text{for all choices of } t_1 \text{ and } u_1$$
$$p_{X(t_1),X(t_2)}(u_1,u_2) \qquad \text{for all choices of } t_1,\ t_2,\ u_1, \text{ and } u_2$$
$$\vdots$$
$$p_{X(t_1)\cdots X(t_n)}(u_1,\cdots,u_n) \quad \text{for all choices of } t_1,\ \cdots,\ t_n,\ u_1,\ \cdots u_n$$
$$\vdots$$

Note, though, that this list is redundant. If one knows the joint distribution of the random variables $\{X(t_1), X(t_2), \cdots, X(t_k)\}$ corresponding to $n = k$, then the corresponding information for any n value which is less than k is simply a marginal distribution which can be easily derived from this joint distribution. Thus, the function $p_{X(t_1)\cdots X(t_k)}(u_1,\cdots,u_k)$ contains all the information for $p_{X(t_1)}(u_1)$, $p_{X(t_1),X(t_2)}(u_1,u_2)$, $\cdots$, $p_{X(t_1)\cdots X(t_{k-1})}(u_1,\cdots,u_{k-1})$. All of these statements could equally well be written in terms of cumulative distribution functions like $F_{X(t_1)\cdots X(t_n)}(u_1,\cdots,u_n)$, but there is no point in repeating them since the forms would be the same.

The probability density $p_{X(t)}(u)$ can be viewed as a function of the two variables t and u, and it is sometimes written as $p_X(u,t)$ or $p_X(u;t)$, with similar notation for the higher-order density functions. There is nothing wrong with this notation, but we choose to keep any t variable explicitly stated as an index of X to remind us that the sole function of t is to specify a particular random variable $X(t)$. Stated another way, we do not want a notation like $p_X(u,t)$ to lead the reader to consider this probability density function for the stochastic process $\{X(t)\}$ as somehow different from $p_X(u)$ for a random variable X. The probability densities for a stochastic process must be identical in concept to those for random variables, since they are simply the descriptions of the random variables which make up the process. In some cases, we will find it convenient to use simplified notations such as $X_k \equiv X(t_k)$, $\vec{X} \equiv \{X(t_1), \cdots X(t_n)\}^T$, and $p_{\vec{X}}(\vec{u}) \equiv p_{X(t_1)\cdots X(t_n)}(u_1,\cdots,u_n)$, but we will always keep index parameters [such as t in $p_{X(t)}(u)$] specifying particular random variables separate from the dummy variables denoting possible values of the random variables [such as u in $p_{X(t)}(u)$].

We noted earlier that our discussion of stochastic processes will be limited to the situation in which the index set of possible t values in $\{X(t)\}$ is continuous. This is very different from saying that the set of possible values for $X(t)$ is continuous. In fact, we will usually be considering situations in which both the set of t values and the set of u values in $p_{X(t)}(u)$ is continuous, but our definitions certainly allow the possibility of having $\{X(t)\}$ defined on a continuous index set of t values, but with each random variable $X(t)$ belonging to a discrete set. Example 2.1 is a very simple case of this situation, since in that example $X(t)$ has only three possible values for any particular t value. Recall that a random variable X may not be a continuous random variable even if it has a continuous set of possible values. If a random variable X has a continuous set of possible values but a discontinuous cumulative distribution $F_X(u)$ on those values, then X has a mixed rather than continuous distribution. Similarly, saying that the index set of possible t values is continuous does not imply that the stochastic process $\{X(t)\}$ varies continuously on that set, either in terms of time history continuity or some sort of probabilistic continuity. This issue of continuity of $\{X(t)\}$ over the continuous index set will be considered in Section 2.6.

In many problems, we will need to consider more than one stochastic process. This means that we will need information related to the joint distribution of random variables from two or more stochastic processes. For example, in order to have a complete probabilistic description of the two stochastic processes $\{X(t)\}$ and $\{Y(s)\}$, we will need to know the joint probability distribution for every set of random variables $\{X(t_1), X(t_2),\ \cdots, X(t_n), Y(s_1), Y(s_2), \cdots, Y(s_m)\}$ for all possible n and m values, and all possible choices of $\{t_1, t_2, \cdots, t_n, s_1, s_2, \cdots, s_m\}$ for each (n,m) combination. Of course this joint distribution could be described by a probability density function like $p_{X(t_1),\cdots,X(t_n),Y(s_1),\cdots,Y(s_m)}(u_1,\cdots,u_n,v_1,\cdots,v_m)$. The concept can be readily extended to more than two stochastic processes.

In summary, the fundamental definition of one or more stochastic processes is in terms of the underlying probability distribution, as given by probability density functions or cumulative distribution functions. Of course, we can never explicitly write out all of these functions since there are infinitely many of them for processes with continuous index sets. Furthermore, we will often find that we need to write out few or none of them in calculating or estimating the probabilities of interest for failure calculations. In many cases, we can gain the information that we need for stochastic processes from considering their moments and the following section considers this characterization.

2.3   Moment and Covariance Functions

One can characterize any random variable X by moments in the form of mean value, mean squared value, variance, and possibly higher moments or central moments giving information like skewness and kurtosis (see Section A.12 of Appendix A). Similarly, if we have more than one random variable then we can use cross-products, covariance, and other moments or expected values involving two or more random variables. The material presented here for a stochastic process $\{X(t)\}$ is a direct application of these concepts to the set of random variables which compose the process.

For the mean value, or expected value, of a process we will use the notation

$$\mu_X(t) \equiv \mu_{X(t)} \equiv E[X(t)] \equiv \int_{-\infty}^{\infty} u\, p_{X(t)}(u)\, du \tag{2.1}$$

That is, $\mu_X(t)$ is a function defined on the index set of possible values of t in $\{X(t)\}$, and at any particular t value, this mean value function is identical to the mean of the random variable $X(t)$. Similarly we define a function called the autocorrelation function of $\{X(t)\}$ as a cross-product of two random variables from the same process

$$\phi_{XX}(t,s) \equiv E[X(t)X(s)] \equiv \int_{-\infty}^{\infty} \int_{-\infty}^{\infty} uv\, p_{X(t),X(s)}(u,v)\, du\, dv \tag{2.2}$$

and this function is defined on a two-dimensional space with t and s each varying over all values in the index set for $\{X(t)\}$. The double subscript notation on $\phi_{XX}(t,s)$ is to distinguish autocorrelation from the corresponding concept of cross-correlation which applies when the two random variables in the cross-product are drawn from two different stochastic processes as

$$\phi_{XY}(t,s) \equiv E[X(t)Y(s)] \equiv \int_{-\infty}^{\infty} \int_{-\infty}^{\infty} uv\, p_{X(t),Y(s)}(u,v)\, du\, dv \tag{2.3}$$

The cross-correlation function $\phi_{XY}(t,s)$ is defined on the two-dimensional space with t being any value from the index set for $\{X(t)\}$ and s being any value from the index set for $\{Y(t)\}$. These two index sets may be identical or quite different.

The reader should note that the use of the term "correlation" in the expressions autocorrelation and cross-correlation is not consistent with the use of the term for random variables. For two random variables the correlation coefficient is a normalized form of the covariance, whereas autocorrelation and cross-correlation simply correspond to cross-products of two random variables. This inconsistency is unfortunate, but well established in the literature on random variables and stochastic processes.[2]

We will also use the covariances of random variables drawn from stochastic processes and refer to them as autocovariance

---

[2]   We would prefer a less confusing nomenclature, but have finally decided not to go contrary to common usage of the terms. We conclude that a change of nomenclature might make it difficult for the reader to reconcile the presentation in this book with those of other authors.

$$K_{XX}(t,s) \equiv K_{X(t),X(s)} = E\big([X(t) - \mu_X(t)][X(s) - \mu_X(s)]\big) \tag{2.4}$$

and cross-covariance

$$K_{XY}(t,s) \equiv K_{X(t),Y(s)} = E\big([X(t) - \mu_X(t)][Y(s) - \mu_Y(s)]\big) \tag{2.5}$$

Of course, the two-dimensional domain of definition of $K_{XX}(t,s)$ is identical to that for $\phi_{XX}(t,s)$, and for $K_{XY}(t,s)$ it is the same as for $\phi_{XY}(t,s)$. We can rewrite these second central moment expressions for the stochastic processes in terms of mean and cross-product values, and obtain[3]

$$K_{XX}(t,s) = \phi_{XX}(t,s) - \mu_X(t)\mu_X(s) \tag{2.6}$$

and

$$K_{XY}(t,s) = \phi_{XY}(t,s) - \mu_X(t)\mu_Y(s) \tag{2.7}$$

Note that we can also define a mean squared function for the $\{X(t)\}$ stochastic process as a special case of the autocorrelation function, and an ordinary variance function as a special case of the autocovariance function:

$$E[X^2(t)] = \phi_{XX}(t,t) \tag{2.8}$$

and

$$\sigma_X^2(t) = K_{XX}(t,t) \tag{2.9}$$

One can also extend the idea of the correlation coefficient for random variables (see Eq. A.61 of Appendix A) and write[4]

$$\rho_{XX}(t,s) \equiv \frac{K_{XX}(t,s)}{\sigma_X(t)\sigma_X(s)} = \frac{K_{XX}(t,s)}{\sqrt{K_{XX}(t,t)K_{XX}(s,s)}} \tag{2.10}$$

and

$$\rho_{XY}(t,s) \equiv \frac{K_{XY}(t,s)}{\sigma_X(t)\sigma_Y(s)} = \frac{K_{XY}(t,s)}{\sqrt{K_{XX}(t,t)K_{YY}(s,s)}} \tag{2.11}$$

Sometimes it is convenient to modify the formulation of a problem in such a way that one can analyze a mean zero process, even though the physical process requires a model with a mean value function $\mu_X(t)$ which is not zero. This is simply done by introducing a new stochastic process $\{Z(t)\}$ defined by $Z(t) = X(t) - \mu_X(t)$. This new process is mean zero [i.e., $\mu_Z(t) = 0$], and the autocovariance function for the original $\{X(t)\}$ process is given by $K_{XX}(t,s) = \phi_{ZZ}(t,s)$.

---

[3] This is the same as the operation in Eqs. A.57 and A.58 of Appendix A for random variables.

[4] It would be logical to call these functions autocorrelation coefficient and cross-correlation coefficient, but that is probably too confusing given the usage of the terms autocorrelation function and cross-correlation function.

****

Example 2.1: Consider a process $\{X(t): t \geq 0\}$ which represents the position of a body which is minimally stable against overturning. In particular, $\{X(t)\}$ has only three possible time histories:

$$X^{(1)}(t) = 0$$
$$X^{(2)}(t) = \alpha \sinh(\beta t)$$
$$X^{(3)}(t) = -\alpha \sinh(\beta t)$$

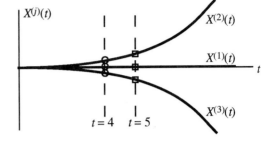

in which α and β are constants, and the three time histories represent "not falling," "falling to the right," and "falling to the left." Let the probabilities be

$$P[X(t) = X^{(1)}(t)] = 0.50$$
$$P[X(t) = X^{(2)}(t)] = 0.25$$
$$P[X(t) = X^{(3)}(t)] = 0.25$$

Find:
(a) the first-order probability distribution of any random variable $X(t)$ from $\{X(t)\}$.
(b) the joint probability distribution of any two random variables $X(t)$ and $X(s)$.
(c) the mean value function for $\{X(t)\}$.
(d) the covariance function for $\{X(t)\}$ and the variance function $\sigma_X^2(t) \equiv \sigma_{X(t)}^2$.
(e) the correlation coefficient relating the random variables $X(t)$ and $X(s)$.

(a) At any time t there are only three possible values for $X(t)$, and the probability distribution for such a simple discrete random variable can be given as the probability of each possible outcome. Thus, we have

$$P[X(t) = 0] = 0.50 \quad P[X(t) = \alpha \sinh(\beta t)] = 0.25 \quad P[X(t) = -\alpha \sinh(\beta t)] = 0.25$$

By using the Dirac delta function,[5] this can be described by a probability density function of

$$p_{X(t)}(u) = 0.50 \, \delta(u) + 0.25 \, \delta[u - \alpha \sinh(\beta t)] + 0.25 \, \delta[u + \alpha \sinh(\beta t)]$$

(b) Clearly there are three possible values for $X(t)$ and three possible values for $X(s)$, but in this case these result in only three possible values for the pair. In particular, for any t and s not equal to zero there is one point with $X(t) = X^{(1)}(t), X(s) = X^{(1)}(s)$, one with $X(t) = X^{(2)}(t), X(s) = X^{(2)}(s)$, and one with $X(t) = X^{(3)}(t)$, $X(s) = X^{(3)}(s)$. The joint probability distribution of the two random variables is completely given by

$$P[X(t) = 0, X(s) = 0] = 0.50$$
$$P[X(t) = \alpha \sinh(\beta t), X(s) = \alpha \sinh(\beta s)] = 0.25$$
$$P[X(t) = -\alpha \sinh(\beta t), X(s) = -\alpha \sinh(\beta s)] = 0.25$$

---

5   See Appendix C and Section A.2 of Appendix A.

or as a joint probability density function as

$$p_{X(t)X(s)}(u,v) = 0.50\,\delta(u)\delta(v) + 0.25\,\delta[u - \alpha\sinh(\beta t)]\delta[v - \alpha\sinh(\beta s)]$$
$$+ 0.25\,\delta[u + \alpha\sinh(\beta t)]\delta[v + \alpha\sinh(\beta s)]$$

(c) The mean of the random variable $X(t)$ is the probability weighted average over the possible values so that

$$\mu_X(t) \equiv E[X(t)] = (0.50)(0) + (0.25)[\alpha\sinh(\beta t)] + (0.25)[-\alpha\sinh(\beta t)] = 0$$

(d) Since the mean is zero, the covariance function is the same as the autocorrelation function. Thus,

$$\phi_{XX}(t,s) \equiv E[X(t)X(s)] = (0.50)(0)(0) + (0.25)[\alpha\sinh(\beta t)][\alpha\sinh(\beta s)]$$
$$+ (0.25)[-\alpha\sinh(\beta t)][-\alpha\sinh(\beta s)]$$

gives

$$K_{XX}(t,s) = \phi_{XX}(t,s) = (0.50)\alpha^2\sinh(\beta t)\sinh(\beta s)$$

Choosing $s = t$ gives the variance and mean squared values as

$$\sigma^2_{X(t)} = E[X^2(t)] = (0.50)\alpha^2\sinh^2(\beta t)$$

(e) The correlation coefficient of the two random variables is of the form

$$\rho_{XX}(t,s) = \frac{K_{XX}(t,s)}{\sigma_{X(t)}\sigma_{X(s)}} = \frac{(0.50)\alpha^2\sinh(\beta t)\sinh(\beta s)}{(0.50)\sqrt{\alpha^2\sinh^2(\beta t)\alpha^2\sinh(\beta s)}} = 1$$

Thus we see that $X(t)$ and $X(s)$ are perfectly correlated. There is a perfect linear relationship between them which can be written as

$$X(s) = \frac{\sinh(\beta s)}{\sinh(\beta t)}X(t)$$

It should be noted that such a situation with only a few possible time histories is not typical of most stochastic problems of interest.

****

Example 2.2: Consider a stochastic process $\{X(t)\}$ which may have infinitely many possible time histories, but with these time histories depending on only one random variable — a random initial condition:

$$X(t) = A\cos(\omega t)$$

in which ω is a deterministic circular frequency and A is a random variable, for which the probability distribution is known. Find expressions for the mean value function, the autocorrelation function, and the covariance function for $\{X(t)\}$.

Since only A is random in $X(t)$, we know that

$$\mu_X(t) \equiv E[X(t)] = E(A)\cos(\omega t) \equiv \mu_A \cos(\omega t)$$

$$\phi_{XX}(t,s) \equiv E[X(t)X(s)] = E(A^2)\cos(\omega t)\cos(\omega s)$$

and

$$K_{XX}(t,s) = \phi_{XX}(t,s) - \mu_X(t)\mu_X(s) = [E(A^2) - \mu_A^2]\cos(\omega t)\cos(\omega s) = \sigma_A^2 \cos(\omega t)\cos(\omega s)$$

The number of possible time histories in this example depends entirely on the number of possible values for the random variable A. If A, for example, has only five possible values, then $X(t)$ has only five possible time histories. At the other extreme, if A is a continuous random variable, then there are infinitely many (in fact uncountably many) possible time histories for $X(t)$. Even in this case, though, we know the shape of each time history since they are all cosine functions with a given frequency. Furthermore, any two time histories are either exactly in phase or 180° out of phase, but there are many possible amplitudes.
****

Example 2.3: Consider a stochastic process $\{X(t)\}$ which has time histories depending on only two random variables:

$$X(t) = A\cos(\omega t + \theta)$$

in which ω is a deterministic circular frequency and A and θ are random variables, for which the joint probability distribution is known. In particular, let A and θ be independent and θ be uniformly distributed on the range $[0, 2\pi]$. Find expressions for the mean value function, the autocorrelation function, and the covariance function for $\{X(t)\}$.

Since A and θ are independent, it can be shown that A and $\cos(\omega t + \theta)$ are also independent. The gist of this idea can be stated as follows: Independence of A and θ tells us that knowledge of the value of A will give no information about the possible values or probability of θ, and this then guarantees that knowledge of the value of A will give no information about the possible values or probability of $\cos(\omega t + \theta)$; this is the condition of independence. Using this independence, we can write

$$\mu_X(t) \equiv \mu_A E[\cos(\omega t + \theta)]$$

but we can use the given uniform distribution for θ to obtain

$$E[\cos(\omega t + \theta)] = \int_{-\infty}^{\infty} p_\theta(\eta)\,\cos(\omega t + \eta)\,d\eta = \int_0^{2\pi} \frac{1}{2\pi}\cos(\omega t + \eta)\,d\eta = 0$$

Thus $\mu_X(t) = 0$ regardless of the distribution of A, provided only that μ_A is finite. We use exactly the same ideas to find the other quantities asked for:

$$K_{XX}(t,s) = \phi_{XX}(t,s) = E(A^2)\,E[\cos(\omega t + \theta)\cos(\omega s + \theta)]$$

Using the identity $\cos(\alpha)\cos(\beta) = [\cos(\alpha+\beta) + \cos(\alpha-\beta)]/2$ allows us to obtain

$$E[\cos(\omega t + \theta)\cos(\omega s + \theta)] = \frac{1}{2}\left\{ \int_0^{2\pi} \frac{1}{2\pi}\cos[\omega(t+s)+2\eta)]d\eta + \cos[\omega(t-s)]\right\}$$

$$= \frac{1}{2}\cos[\omega(t-s)]$$

so that

$$K_{XX}(t,s) = \phi_{XX}(t,s) = \frac{E(A^2)}{2}\cos[\omega(t-s)]$$

Note that in this problem, as in Example 2.2, the shape of each time history is a cosine function with a given frequency. The distinguishing factor is that now the phase difference between any two time histories can be any angle, and all possible values are equally likely. Thus, for any distribution of A, there are infinitely many possible time histories. In fact, even if A has only one value [i.e., $p_A(u) = \delta(u - A_0)$] there are still infinitely many possible time histories but they differ only in phase angle.

****

Example 2.4: Let $\{X(t)\}$ be the stochastic process of Example 2.3, and let $\{Y(t)\}$ be a process which also has trigonometric time histories. The time histories of $\{Y(t)\}$, though, are 45° out of phase with those of $\{X(t)\}$ and they are offset by a fixed amount of 5:

$$Y(t) = 5 + A\cos(\omega t + \theta + \pi/4)$$

Note that the same two random variables A and θ define $\{X(t)\}$ and $\{Y(t)\}$. Find the cross-correlation function and the cross-covariance function for the stochastic processes.

Since the mean value function of $\{X(t)\}$ is exactly zero, Eq. 2.7 tells us that the cross-covariance function is identical to the cross-correlation function. We find this function as

$$K_{XY}(t,s) = \phi_{XY}(t,s) = E[X(t)Y(s)] = E\left([A\cos(\omega t + \theta)]\left[5 + A\cos\left(\omega s + \theta + \frac{\pi}{4}\right)\right]\right)$$

or

$$K_{XX}(t,s) = \phi_{XX}(t,s) = 5E(A)E[\cos(\omega t + \theta)] + E(A^2)E\left[\cos(\omega t + \theta)\cos\left(\omega t + \theta + \frac{\pi}{4}\right)\right]$$

or

$$K_{XY}(t,s) = \phi_{XY}(t,s) = E(A^2)E\left[\cos(\omega t + \theta)\cos\left(\omega s + \theta + \frac{\pi}{4}\right)\right]$$

Using the identity $\cos(\alpha)\cos(\beta) = [\cos(\alpha+\beta) + \cos(\alpha-\beta)]/2$ now gives

$$K_{XY}(t,s) = \phi_{XY}(t,s) = \frac{E(A^2)}{2}E\left(\cos\left[\omega(t+s)+2\theta+\frac{\pi}{4}\right] + \cos\left[\omega(t-s)-\frac{\pi}{4}\right]\right)$$

which reduces to

$$K_{XY}(t,s) = \phi_{XY}(t,s) = \frac{E(A^2)}{2}\cos\left[\omega(t-s) - \frac{\pi}{4}\right] = \frac{E(A^2)}{2\sqrt{2}}\left(\cos[\omega(t-s)] + \sin[\omega(t-s)]\right)$$

****

2.4 Stationarity of Stochastic Processes

The property of stationarity (or homogeneity) of a stochastic process $\{X(t)\}$ always refers to some aspect of the description of the process being unchanged by any arbitrary shift along the t axis. There are many types of stationarity depending on what characteristic of the process has this property of being invariant under a time shift.

The simplest type of stationarity involves only invariance of the mean value function for the process. We say that $\{X(t)\}$ is **mean-value stationary** if

$$\mu_X(t+r) = \mu_X(t) \quad \text{for any value of the shift parameter } r \tag{2.12}$$

Clearly this can only be true if $\mu_X(t)$ is the same for all t values, so that we can equally well say that $\{X(t)\}$ is mean-value stationary if

$$\mu_X(t) = \mu_X \tag{2.13}$$

in which the absence of a t argument on the right-hand side conveys the information that the mean value is independent of time. Although the notation on the right-hand side of Eq. 2.13 is the same as for the mean value of a random variable, Eq. 2.13 does refer to the mean value function of a stochastic process. Of course, having the mean value be independent of t does not imply that the $X(t)$ random variables are all the same at different values of t, or that the probability distributions of these random variables are all the same, but only that they all have the same mean value.

As a next step in specifying more rigorous stationarity, let us say that $\{X(t)\}$ is **second-moment stationary** if the second moment function (i.e., the autocorrelation function) is invariant under a time shift:

$$E[X(t+r)X(s+r)] = E[X(t)X(s)] \quad \text{for any value of the shift parameter } r$$

or

$$\phi_{XX}(t+r,s+r) = \phi_{XX}(t,s) \quad \text{for any value of the shift parameter } r \tag{2.14}$$

Since Eq. 2.14 is specified to be true for any value of r, it must be true, in particular, for $r = -s$; using this particular value gives $\phi_{XX}(t,s) = \phi_{XX}(t-s,0)$. This shows that the autocorrelation function for a second-moment stationary process is a function of only one time argument — the $(t - s)$ difference between the two time arguments in $\phi_{XX}(t,s)$. We will use an alternate notation of R in place of ϕ for such a situation in which autocorrelation can be written as a function of a single time argument. Thus, we define $R_{XX}(t-s) \equiv \phi_{XX}(t-s,0)$ so that

$$\phi_{XX}(t,s) = R_{XX}(t-s) \quad \text{for any values of } t \text{ and } s \tag{2.15}$$

or, using $\tau = t - s$

$$R_{XX}(\tau) = \phi_{XX}(t+\tau,t) \equiv E[X(t+\tau)X(t)] \quad \text{for any values of } t \text{ and } \tau \tag{2.16}$$

We similarly say that two stochastic processes $\{X(t)\}$ and $\{Y(t)\}$ are jointly second-moment stationary if

$$\phi_{XY}(t+r,s+r) = \phi_{XY}(t,s) \quad \text{for any value of the shift parameter } r \tag{2.17}$$

and we define $R_{XY}(t-s) \equiv \phi_{XY}(t-s,0)$ so that

$$\phi_{XY}(t,s) = R_{XY}(t-s) \quad \text{for any values of } t \text{ and } s \tag{2.18}$$

or

$$R_{XY}(\tau) = \phi_{XY}(t+\tau,t) \equiv E[X(t+\tau)Y(t)] \quad \text{for any values of } t \text{ and } \tau \tag{2.19}$$

A slight variation on second-moment stationarity is obtained by using the autocovariance and cross-covariance functions in place of autocorrelation and cross-correlation functions. Thus, we say that a stochastic process is **covariant stationary** if

$$K_{XX}(t,s) = G_{XX}(t-s) \quad \text{for any values of } t \text{ and } s \tag{2.20}$$

or, equivalently,

$$G_{XX}(\tau) = K_{XX}(t+\tau,t) \quad \text{for any values of } t \text{ and } \tau \tag{2.21}$$

in which Eq. 2.21 gives the definition of the new stationary autocovariance function $G_{XX}(\tau)$. Similarly, $\{X(t)\}$ and $\{Y(t)\}$ are jointly covariant stationary if

$$K_{XY}(t,s) = G_{XY}(t-s) \quad \text{for any values of } t \text{ and } s \tag{2.22}$$

and

$$G_{XY}(\tau) = K_{XY}(t+\tau,t) \quad \text{for any values of } t \text{ and } \tau \tag{2.23}$$

Note that one can also say that $G_{XX}(\tau) = R_{ZZ}(\tau)$ if $\{Z(t)\}$ is defined as the mean zero shifted version of $\{X(t)\}$: $Z(t) = X(t) - \mu_X(t)$.

Clearly one can extend this concept to definitions of third-moment stationarity, fourth-moment stationarity, skewness stationarity, kurtosis stationarity, etc. We will only explicitly look at one generic term in the sequence of moment stationary definitions. We will say that $\{X(t)\}$ is **jth-moment stationary** if

$$E[X(t_1+r)X(t_2+r)\cdots X(t_j+r)] = E[X(t_1)X(t_2)\cdots X(t_j)] \tag{2.24}$$

for all values of r and all values of $t_1, t_2, \cdots, t_j$. Using the notation $\tau_k = t_k - t_j$ for $k = 1$ to $(j - 1)$ allows this to be rewritten as

$$E[X(t_1)X(t_2)\cdots X(t_j)] = E[X(t_j + \tau_1)X(t_j + \tau_2)\cdots X(t_j + \tau_{j-1})X(t_j)] \qquad (2.25)$$

with the expression on the right-hand side being independent of t_j. Thus, the jth moment function of a jth-moment stationary process is a function of only the $(j - 1)$ time arguments $\tau_1, \tau_2, \cdots, \tau_{j-1}$ giving increments between the original time values of $t_1, t_2, \cdots, t_j$. Stationarity always reduces the number of necessary time arguments by one. Of course we had already explicitly demonstrated this fact for first and second moments, showing that $\mu_X(t)$ depends on zero time arguments (is independent of time), and $\phi_{XX}(t, s)$ depends on the one time argument $\tau = t - s$.

Although, a given stochastic process may simultaneously have various types of moment stationarity, this is not necessary. One of the most commonly considered combinations of moment stationarity involves the first and second moments. If a process $\{X(t)\}$ does have both mean-value and second-moment stationarity, then it is easily shown from Eq. 2.6 that it is also covariant stationary. Often such a process is described in the alternate way of saying that it is mean and covariant stationary, and Eq. 2.6 then shows that it is also second-moment stationary.

There are also forms of stationarity which are not defined in terms of moment functions. Rather, they are defined in terms of probability distributions being invariant under a time shift. The general relationship is that $\{X(t)\}$ is **jth-order stationary** if

$$p_{X(t_1+r)\cdots X(t_j+r)}(u_1, \cdots, u_j) = p_{X(t_1)\cdots X(t_j)}(u_1, \cdots, u_j) \qquad (2.26)$$

for all values of $t_1, \cdots, t_j, u_1, \cdots, u_j$ and the shift parameter r. This includes, as special cases, $\{X(t)\}$ being **first-order stationary** if

$$p_{X(t+r)}(u) = p_{X(t)}(u) \qquad (2.27)$$

for all values of t and u and the shift parameter r, and **second-order stationary** if

$$p_{X(t_1+r)X(t_2+r)}(u_1, u_2) = p_{X(t_1)X(t_2)}(u_1, u_2) \qquad (2.28)$$

for all values of t_1, t_2, u_1, u_2 and the shift parameter r.

There are strong interrelationships among the various sorts of stationarity. One of the simplest of these involves the stationarities defined in terms of the probability distribution, as in Eqs. 2.26 - 2.28. Specifically, from consideration of marginal distributions it is easy to show that if $\{X(t)\}$ is jth-order stationary, then it is also first-order stationary, second-order stationary, and up to $(j-1)$-order stationary. Note that the same hierarchy does not apply to moment stationarity. It is quite possible to define a process which is second-moment stationary but not mean-value stationary, although there may be little practical usefulness for such a process. A slight variation on this is to have a process which is covariant stationary, but not mean-value or second-moment stationary; this process certainly does approximate various physical phenomena for which the mean value changes with time, but the variance is constant.

It is also important to note that jth-order stationarity always implies jth-moment stationarity, since a jth moment function can be calculated by using a jth-order probability distribution. Thus, for example, second-order stationarity implies second-moment stationarity. However, second-order stationarity also implies first-order stationarity, and this then implies first-moment (mean-value) stationarity. In general, we can say that jth-order stationarity implies stationarity of all moments up to and including the jth.

When comparing moment and order stationarity definitions, it is not always possible to say which is stronger. For example, consider the question of whether second-moment or first-order stationarity is stronger. Since first-order stationarity states that $p_{X(t)}(u)$ has time shift invariance, it implies that any moment which can be calculated from that probability density function also has time shift invariance. Thus, first-order stationarity implies that the jth moment $E[X^j(t)]$ for the random variable $X(t)$ at one instant of time has time shift invariance for any value of j. This seems to be a form of jth-moment stationarity which is certainly not implied by second-moment stationarity. On the other hand, first-order stationarity says nothing about the relationship of $X(t)$ to $X(s)$ for $t \neq s$, whereas second-moment stationarity does. Thus, there is no answer as to which condition is stronger. In general, jth-order stationarity implies time shift invariance of moments of any order, so long as they only depend on the values of $X(t)$ at no more than j different times. Similarly kth-moment stationarity implies time shift invariance only of kth moment functions, which may involve values of $X(t)$ at up to k different time values. For $k > j$ one cannot say whether jth-order or kth-moment stationarity is stronger, since each implies certain behavior which the other does not.

The most restrictive type of stationarity is called strict stationarity. We say that $\{X(t)\}$ is **strictly stationary** if it is jth-order stationary for any value of j. This implies that any order probability density function has time shift invariance, and any order moment function has time shift invariance. In the stochastic process literature, one also frequently encounters the terms weakly stationary and/or wide-sense stationary. It appears that both of these terms are usually used to mean a process which is both mean-value and covariant stationary, but caution is advised since there is some variation in usage, with meanings ranging up to second-order stationary. When we refer to a stochastic process as being stationary, and give no qualification as to type of stationarity, we will generally mean strictly stationary. In some situations, though, the reader may find that some weaker form of stationarity is adequate for the calculations being performed. For example, if the analytical development or problem solution only involves second-moment calculations, then strict stationarity will be no more useful than second-moment stationarity. Thus, one can also say that the word stationary without qualifier simply means that all moments and/or probability distributions being used in the given problem are invariant under a time shift.

****

Example 2.5:   Identify applicable types of stationarity for the stochastic process with $X^{(1)}(t) = 0$, $X^{(2)}(t) = \alpha \sinh(\beta t)$, $X^{(3)}(t) = -\alpha \sinh(\beta t)$, and $P[X(t) = X^{(1)}(t)] = 0.50$, $P[X(t) = X^{(2)}(t)] = 0.25$, $P[X(t) = X^{(3)}(t)] = 0.25$.

In Example 2.1 we found the first-order probability density function to be

$$p_{X(t)}(u) = 0.50\,\delta(u) + 0.25\,\delta[u - \alpha\sinh(\beta t)] + 0.25\,\delta[u + \alpha\sinh(\beta t)]$$

Clearly this function is not invariant under a time shift. Thus, $\{X(t)\}$ is not first-order stationary, and this precludes the possibility of its having any jth-order stationarity. We did find, though, that the mean value function for this process is a constant $[\mu_X(t) = 0]$, so it is time shift invariant. Thus, the process is mean-value stationary. The autocorrelation function $\phi_{XX}(t,s) = (0.50)\alpha^2\sinh(\beta t)\sinh(\beta s)$ determined in Example 2.1 clearly demonstrates that the process is not second-moment stationary, since its dependence on t and s is not of the form $(t - s)$.

****

Example 2.6: Identify applicable types of stationarity for the stochastic process with $X(t) = A\cos(\omega t)$ in which A is a random variable.

The mean value function and autocorrelation function of this process were found in Example 2.2 as $\mu_X(t) = \mu_A\cos(\omega t)$ and $\phi_{XX}(t,s) = E(A^2)\cos(\omega t)\cos(\omega s)$. These functions clearly show that this $\{X(t)\}$ is neither mean-value or second-moment stationary.

We can investigate first-order stationarity by deriving the probability density function for the random variable $X(t)$. For any fixed t, this is a special case of $X = cA$, in which c is a deterministic constant. This form gives $F_X(u) = F_A(u/c)$ for $c > 0$ and $F_X(u) = 1 - F_A(u/c)$ for $c < 0$. Taking a derivative with respect to u then gives $p_X(u) = |c|^{-1}p_A(u/c)$. Thus, for $c = \cos(\omega t)$ we have

$$p_X(u) = \frac{1}{|\cos(\omega t)|}p_A\left(\frac{u}{\cos(\omega t)}\right)$$

and this does not appear to be invariant under a time shift. Thus, the process is not generally first-order stationary, or jth-order stationary for any j.

****

Example 2.7: Investigate mean-value, second-moment, and covariance stationarity for the stochastic process with $X(t) = A\cos(\omega t + \theta)$ in which A and θ are independent random variables with θ uniformly distributed on the range $[0, 2\pi]$.

In Example 2.3 we found the mean value, covariance, and autocorrelation functions for this stochastic process to be $\mu_X(t) = 0$ and $K_{XX}(t,s) = \phi_{XX}(t,s) = E(A^2)\cos[\omega(t - s)]/2$. These functions are all invariant under a time shift (remembering that a time shift will change t and s by an equal amount in the K

and ϕ functions. Thus, $\{X(t)\}$ does have stationarity of the mean value, the second moment, and the covariance functions. We can rewrite the autocorrelation and autocovariance functions as

$$G_{XX}(\tau) = R_{XX}(\tau) = \frac{E(A^2)}{2}\cos(\omega\tau)$$

****

Example 2.8: Consider the two stochastic processes $\{X(t)\}$ and $\{Y(t)\}$ defined in Example 2.4 with

$$K_{XY}(t,s) = \phi_{XY}(t,s) = \frac{E(A^2)}{2}\cos\left[\omega(t-s) - \frac{\pi}{4}\right] = \frac{E(A^2)}{2\sqrt{2}}\left(\cos[\omega(t-s)] + \sin[\omega(t-s)]\right)$$

Are $\{X(t)\}$ and $\{Y(t)\}$ jointly covariant stationary?

The covariance function does have the time shift property, so $\{X(t)\}$ and $\{Y(t)\}$ are jointly covariant stationary. We can rewrite the cross-correlation and cross-covariance functions as

$$G_{XY}(\tau) = R_{XY}(\tau) = \frac{E(A^2)}{2}\cos\left(\omega\tau - \frac{\pi}{4}\right) = \frac{E(A^2)}{2\sqrt{2}}[\cos(\omega\tau) + \sin(\omega\tau)]$$

****

2.5 Properties of Autocorrelation and Autocovariance

There are a number of properties of the autocorrelation and autocovariance functions which apply for any stochastic process in any problem. We will list some of those here, giving both general versions and the forms that apply for stationary processes. Probably one of the more obvious, but nonetheless significant, properties is **symmetry**:

$$\phi_{XX}(s,t) = \phi_{XX}(t,s) \quad \text{and} \quad K_{XX}(s,t) = K_{XX}(t,s) \tag{2.29}$$

Rewriting these equations for a process with the appropriate stationarity (second-moment and covariant stationarity, respectively) gives

$$R_{XX}(-\tau) = R_{XX}(\tau) \quad \text{and} \quad G_{XX}(-\tau) = G_{XX}(\tau) \tag{2.30}$$

Next we note that knowledge of an autocorrelation (autocovariance) function for certain time arguments also **bounds** the magnitude of the function for other values of the arguments. In particular, application of the Schwarz inequality (see Eq. A.64 of Appendix A) shows that

$$|\phi_{XX}(t,s)| \le \sqrt{\phi_{XX}(t,t)\phi_{XX}(s,s)} = \sqrt{E[X^2(t)]E[X^2(s)]} \tag{2.31}$$

and

$$|K_{XX}(t,s)| \le \sqrt{K_{XX}(t,t)K_{XX}(s,s)} = \sigma_X(t)\sigma_X(s) \tag{2.32}$$

These relationships provide bounds on the value of $\phi_{XX}(t,s)$ anywhere off the 45° line of the (t,s) plane in terms of the values on that diagonal line. One particular implication of this is the fact that $|\phi_{XX}(t,s)| < \infty$ everywhere if $E[X^2(t)]$ is bounded for all values of t, and $|K_{XX}(t,s)| < \infty$ everywhere if $\sigma_X(t) < \infty$ for all t. The standard nomenclature is that $\{X(t)\}$ is a second-order process if its autocorrelation function is always finite.[6] If $\{X(t)\}$ has the appropriate stationarity, then the bounding relationships become

$$|R_{XX}(\tau)| \le R_{XX}(0) = E[X^2] \quad \text{and} \quad |G_{XX}(\tau)| \le G_{XX}(0) = \sigma_X^2 \tag{2.33}$$

We will now investigate certain **continuity** properties of the autocorrelation and autocovariance functions. It is not necessary that these functions be continuous everywhere, but there are some limitations on the types of discontinuities that they can have. The basic result is that if either of the functions is continuous in the neighborhood of a point (t,t) and in the neighborhood of a point (s,s), then it must also be continuous in the neighborhood of the point (t,s). The first step in clarifying this result is a statement of precisely what we mean by continuity for these functions of two arguments. We say that $\phi(t,s)$ is continuous at the point (t,s) if

$$\lim_{\substack{\varepsilon_1 \to 0 \\ \varepsilon_2 \to 0}} \phi(t+\varepsilon_1, s+\varepsilon_2) = \phi(t,s) \tag{2.34}$$

meaning that we obtain the same limit $\phi(t,s)$ as ε_1 and ε_2 both tend to zero along any route in the two-dimensional space. A special case of this relationship is that $\phi(t,s)$ is continuous at the point (t,t) if

$$\lim_{\substack{\varepsilon_1 \to 0 \\ \varepsilon_2 \to 0}} \phi(t+\varepsilon_1, t+\varepsilon_2) = \phi(t,t)$$

The reader should carefully consider the distinction between this last statement and the much weaker condition of $\phi(t+\varepsilon, t+\varepsilon)$ tending to $\phi(t,t)$ as ε tends to zero, which corresponds only to approaching the point (t,t) along a 45° diagonal in the (t,s) space. This seemingly minor issue has confused many students in the past.

In order to derive conditions for the continuity of $\phi_{XX}(t,s)$ at the point (t,s), we write the autocorrelation function as an expected value and look at the difference between the value at (t,s) and at $(t+\varepsilon_1, s+\varepsilon_2)$:

$$\phi_{XX}(t+\varepsilon_1, s+\varepsilon_2) - \phi_{XX}(t,s) = E[X(t+\varepsilon_1)X(s+\varepsilon_2) - X(t)X(s)]$$

---

6   The reader is cautioned that the word "order" in this term relates only to the order of the moment being considered, not to the order of a probability distribution, as in our definition of jth-order stationarity.

which can be rewritten as

$$\phi_{XX}(t+\varepsilon_1,s+\varepsilon_2) - \phi_{XX}(t,s) = E\big([X(t+\varepsilon_1) - X(t)][X(s+\varepsilon_2) - X(s)] +$$
$$[X(t+\varepsilon_1) - X(t)]X(s) + X(t)[X(s+\varepsilon_2) - X(s)]\big) \qquad (2.35)$$

Thus, we are assured of continuity of ϕ_{XX} at the point (t,s) if the expected value of each of the three terms in the parentheses is zero. We can easily find sufficient conditions for this to be true by using the Schwarz inequality. For example,

$$E\big([X(t+\varepsilon_1) - X(t)]X(s)\big) \le \sqrt{E\big([X(t+\varepsilon_1) - X(t)]^2\big)E\big([X(s)]^2\big)}$$

and converting the right-hand side back into autocorrelation functions gives

$$E\big([X(t+\varepsilon_1) - X(t)]X(s)\big) \le \sqrt{[\phi_{XX}(t+\varepsilon_1,t+\varepsilon_1) - 2\phi_{XX}(t+\varepsilon_1,t) + \phi_{XX}(t,t)]}\sqrt{\phi_{XX}(s,s)}$$

Now we can see that the first of these two square root terms tends to zero as $\varepsilon_1 \to 0$ if ϕ_{XX} is continuous at the point (t,t), regardless of whether it is continuous anywhere else in the (t,s) plane. Thus, continuity of ϕ_{XX} at the point (t,t) will assure that the second term in Eq. 2.35 goes to zero as $\varepsilon_1 \to 0$. One can show the same result for the first term in Eq. 2.35 by exactly the same set of manipulations. Similarly, the third term in Eq. 2.35 goes to zero as $\varepsilon_2 \to 0$ if ϕ_{XX} is continuous at the point (s,s). The sum of these three results shows that Eq. 2.34 holds for ϕ_{XX} if ϕ_{XX} is continuous at both the point (t,t) and the point (s,s). That is, if ϕ_{XX} is continuous at both the point (t,t) and the point (s,s), then it is also continuous at the point (t,s). Exactly the same result applies for the autocovariance function: If K_{XX} is continuous at both the point (t,t) and the point (s,s), then it is also continuous at the point (t,s).

We again emphasize that there is no requirement that ϕ_{XX} or K_{XX} be continuous. The requirement is that if either function is discontinuous at a point (t,s), then it must also be discontinuous at point (t,t) and/or point (s,s). One implication of this result is that if ϕ_{XX} or K_{XX} is continuous everywhere along the (t,t) diagonal of the (t,s) plane, then it must be continuous everywhere in the plane.

The general continuity relationship just derived can be applied to the special case of a stationary process by simply replacing ϕ_{XX} and K_{XX} by R_{XX} and G_{XX} and noting that the points (t,t) and (s,s) both correspond to $\tau = 0$. The result is that continuity of $R_{XX}(\tau)$ or $G_{XX}(\tau)$ at $\tau = 0$ is sufficient to prove that the function is continuous everywhere. Alternatively, we can say that if $R_{XX}(\tau)$ or $G_{XX}(\tau)$ is discontinuous for any value of the argument τ, then it must also be discontinuous at $\tau = 0$.
****

Example 2.9: Consider a very erratic stochastic process $\{X(t)\}$ which has discontinuities in its time histories at discrete values of t:

$$X(t) = A_j \qquad \text{for} \qquad j\Delta \leq t < (j+1)\Delta$$

or

$$X(t) = \sum_j A_j \big(U[t-j\Delta]-U[t-(j+1)\Delta]\big)$$

in which the A_j random variables are independent and identically distributed with mean zero and variance $E(A^2)$, and $U(\cdot)$ is the unit step function (see Section A.2 of Appendix A). The time histories of this process have discontinuities at the times $t = j\Delta$ as the process jumps from the random variable A_{j-1} to A_j.

Find the $\phi_{XX}(t,s)$ autocorrelation function and identify the (t,s) points at which the function is discontinuous.

Using the definition $\phi_{XX}(t,s) = E[X(t)X(s)]$, we find that

$$\phi_{XX}(t,s) = E(A^2) \quad \text{if} \quad t \text{ and } s \text{ are in the same time interval (of length } \Delta)$$
$$= 0 \qquad \text{otherwise}$$

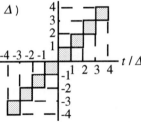

This process is uncorrelated with itself at times t and s unless t and s are in the same Δ time interval, varying from $j\Delta$ to $(j+1)\Delta$.

The sketch indicates the region in which $\phi_{XX}(t,s) \neq 0$.

Clearly the $\phi_{XX}(t,s)$ function is discontinuous along the boundaries of each of the size Δ squares shown. Any one of these boundaries corresponds to either t being an integer times Δ or s being an integer times Δ (with the corners of the squares having both t and s being integers times Δ). The location of these discontinuities is consistent with the fact that along the $45°$ diagonal of the plane we have discontinuities of $\phi_{XX}(t,s)$ at (t,t) whenever t is an integer times Δ.

This example illustrates a situation in which $\phi_{XX}(t,s)$ is discontinuous at some points (t,t), even though $\phi_{XX}(t,t) = E(A^2)$ is continuous everywhere along the line of possible t values. The apparent discrepancy between these two statements is simply due to the difference in definition of one-dimensional and two-dimensional continuity. Looking at one-dimensional continuity of $\phi_{XX}(t,t)$ at $t = 2\Delta$, for example, corresponds to approaching $(2\Delta,2\Delta)$ along the $45°$ diagonal of the (t,s) plane, whereas the discontinuity of $\phi_{XX}(t,s)$ at the point $(2\Delta,2\Delta)$ is due to the fact that a different limit is achieved if the point is approached along some other paths, such as along a line with a negative $45°$ slope.

****

Example 2.10: Consider an even more erratic stochastic process $\{X(t)\}$ for which

$$E[X^2(t)] = E(A^2)$$

but

$$E[X(t)X(s)] = 0 \quad \text{if} \quad t \neq s$$

This process is the limit of the process in Example 2.9 as Δ goes to zero. It is second-moment stationary and is uncorrelated with itself at any two distinct times, no matter how close together they are. Find the $R_{XX}(\tau)$ autocorrelation function and identify points at which it is discontinuous.

Using the definition that $R_{XX}(\tau) = E[X(t + \tau)X(t)]$, we have the stationary autocorrelation function as

$$R_{XX}(\tau) = E(A^2) \quad \text{if} \quad \tau = 0$$

and

$$R_{XX}(\tau) = 0 \quad \text{if} \quad \tau \neq 0$$

Clearly this function is discontinuous at the origin, at $\tau = 0$. It is continuous everywhere else. Note that this is the only possible type of $R_{XX}(\tau)$ function with discontinuity at only one point. Discontinuity at any point other than $\tau = 0$ would imply that there was also discontinuity at $\tau = 0$. Discontinuity at $\tau = 0$, though, only implies the possibility (not the fact) of discontinuity at other values of τ.

Stating our result for this problem in terms of the general (rather than stationary) form of autocorrelation function, we have the rather odd situation of a $\phi_{XX}(t,s)$ function which is discontinuous everywhere along the major diagonal $s = t$, and is continuous everywhere else; this is consistent with Example 2.9 as $\Delta \to 0$.
****

Example 2.11: Give a reason why each of the following $\phi(t,s)$ and $R(\tau)$ functions could not be the autocorrelation function for any process. That is, there could be no $\{X(t)\}$ process with $\phi_{XX}(t,s) = \phi(t,s)$ or $R_{XX}(\tau) = R(\tau)$:

(a) $\phi(t,s) = (t^2 - s)^{-2}$
(b) $\phi(t,s) = 2 - \cos(t - s)$
(c) $\phi(t,s) = U\left[4 - (t^2 + s^2)\right]$
(d) $R(\tau) = \tau^2 e^{-\tau^2}$
(e) $R(\tau) = e^{-\tau(\tau-1)}$
(f) $R(\tau) = U(1 - |\tau|)$

There might be various possible problems with a particular candidate function, and it only takes one violation of a necessary condition to prove that a function cannot be an autocorrelation function. Nonetheless, we will investigate the matter of symmetry for each of the six functions. If the function is symmetric, we will also investigate the matter of boundedness. We will investigate continuity only for (c) and (f), since the other four are clearly continuous everywhere.

(a) Checking symmetry: We have $\phi(s,t) = (s^2 - t)^{-2} \neq \phi(t,s) = (t^2 - s)^{-2}$, therefore this function violates the symmetry required of an autocorrelation function.

(b) Checking symmetry: Yes, $\phi(s,t) = 2 - \cos(s - t) = \phi(t,s)$, so it does have the necessary symmetry.

Checking boundedness: We have $\phi(t,t) = \phi(s,s) = 1$, but $\phi(t,s) = 2 - \cos(t - s) > 1$ for some (t,s)

values. Thus, this function violates Eq. 2.31 bounding $\phi(t,s)$ by $\sqrt{\phi(t,t)\phi(s,s)}$.

(c) Since this function has discontinuities, it is reasonable to start by checking to see whether those discontinuities satisfy the necessary conditions. Clearly $U[4-(t^2+s^2)]$ is discontinuous on the circle $t^2+s^2=4$. This gives discontinuity along the diagonal of the (t,s) plane only at the points $\left(-\sqrt{2},-\sqrt{2}\right)$ and $\left(\sqrt{2},\sqrt{2}\right)$. Clearly there are many (t,s) points for which $\phi(t,s)$ is discontinuous even though the function is continuous at (t,t) and (s,s). For example, $t=2$ and $s=0$ gives such a point. Thus, the discontinuity of this function is not of the type allowed by the Schwarz inequality.

Since we have already shown a reason why $\phi(t,s)$ cannot be an autocorrelation function, we need not check the other conditions, but we will do so anyway.
Checking symmetry: Yes, $\phi(s,t)=U[4-(s^2+t^2)]=\phi(t,s)$, so it does have the necessary symmetry.

Checking boundedness: We have $\phi(t,t)=U(4-2t^2)=U(2-t^2)$ and $\phi(s,s)=U(2-s^2)$ so that $\sqrt{\phi(t,t)\phi(s,s)}=\sqrt{U(2-t^2)U(2-s^2)}$, which is unity on the square shown in the sketch. However, $\phi(t,s)=U[4-(t^2+s^2)]$ is unity on the circle in the sketch. Since there are portions of the circle which are not included in the square, we have a situation with $\phi(t,s)>\sqrt{\phi(t,t)\phi(s,s)}$ whenever (t,s) is in one of the portions of the circle which is outside the square. Thus, the necessary boundedness condition is violated also.

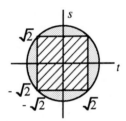

(d) Checking symmetry: Now we have a stationary form of $R(\tau)$, so the necessary symmetry condition is that $R(-\tau)=R(\tau)$. Clearly this is satisfied.

Checking boundedness condition: We need $|R(\tau)|\le R(0)$. Clearly this is violated since $R(0)=0$, but $R(\tau)$ is not zero everywhere.

(e) Checking symmetry: $R(-\tau)=e^{\tau(-\tau-1)}=e^{-\tau^2-\tau}\ne R(\tau)=e^{-\tau^2+\tau}$ so symmetry is violated.

(f) Checking continuity: $R(\tau)$ is continuous at $\tau=0$, so the necessary condition is that it be continuous everywhere. Clearly this is violated, so $R(\tau)$ cannot be an autocorrelation function.

Checking symmetry: $R(-\tau)=R(\tau)$ so symmetry is satisfied.

Checking boundedness condition: We do have $|R(\tau)|\le R(0)$, so there is no problem with the boundedness condition.
****

2.6 Limits of Stochastic Processes

The next two major issues with which we wish to deal are (1) ergodicity and (2) differentiability of a process $\{X(t)\}$. Both of these involve taking a limit of a stochastic process as the index set approaches some value, and this is closely related to the simpler idea of continuity of a process. In particular, continuity involves the behavior of $\{X(t)\}$ as t approaches some particular value t_1, ergodicity involves evaluating an expected value of a function of the process from a limit as $T \to \infty$ of a time average over a time history of length T, and finding the derivative $\dot{X}(t)$ involves taking a limit involving of $X(t + h) - X(t)$ as $h \to 0$. Since continuity is probably the easiest to visualize, we will use it in this section to illustrate the idea of the limit of a stochastic process, even though the other two properties have more overall importance to the application of stochastic process methods to vibration problems. It should also be noted, though, that some sort of continuity of a process must be a condition for differentiability.

The concept of the limit of a deterministic function must be generalized somewhat to handle situations in which one wishes to find a limit of a stochastic process. The difficulty, of course, is that a stochastic process is a different random variable at each value of the index argument, so that the limit of a stochastic process is like the limit of a sequence of random variables. There are a number of definitions of convergence of a stochastic process, in all of which one expects the limit to be some random variable. Rather than give a general discussion of this topic, we will focus on the particular sort of convergence to a limit which we wish to use in the three situations cited in the preceding paragraph.

We say that $\{X(t)\}$ **converges in probability** to the random variable Y as $t \to t_1$ if

$$\lim_{t \to t_1} P[|\, X(t) - Y \,| \geq \varepsilon] = 0 \quad \text{for any } \varepsilon > 0 \tag{2.36}$$

Note that the limit in Eq. 2.36 is on the probability of an event, so that this is an unambiguous deterministic limit. The well-known Chebyshev inequality provides a very useful tool for any problem that requires bounding of the probability of a random variable exceeding some particular value. This can be written in a quite general form as

$$P(|Z| \geq b) \leq \frac{E(|Z|^c)}{b^c} \tag{2.37}$$

for any random variable Z and any nonnegative numbers b and c. Thus, condition 2.36 will be satisfied if

$$\lim_{t \to t_1} E\left(|\, X(t) - Y \,|^c\right) = 0 \quad \text{for some } c > 0 \tag{2.38}$$

Note that Eq. 2.38 is very general, with convergence in probability assured if Eq. 2.38 holds for any positive real exponent c. Unfortunately it is difficult to use the relationship in Eq. 2.38 unless the exponent is an even integer such that the power can be expanded in terms of moments of the random variables. Any even integer will serve this purpose, but choosing $c = 2$ allows us to use the lowest possible moments. Thus, we will generally choose a much more restrictive condition than Eq. 2.38, and say that we are assured of convergence in probability to Y if we have **mean-square convergence**:

$$\lim_{t \to t_1} E\big([X(t) - Y]^2\big) = 0 \qquad (2.39)$$

Turning now to the issue of continuity of $\{X(t)\}$ at $t = t_1$, we simply need to replace the variable Y in the preceding equations and say that $\{X(t)\}$ is **continuous in probability** at $t = t_1$ if

$$\lim_{t \to t_1} E\big(\mid X(t) - X(t_1) \mid^c\big) = 0 \qquad \text{for some } c > 0 \qquad (2.40)$$

and, in particular, if it is **mean-square continuous**:

$$\lim_{t \to t_1} E\big([X(t) - X(t_1)]^2\big) = 0$$

Expanding the quadratic expression inside the expected value allows us to rewrite this condition for mean-square continuity as

$$\lim_{t \to t_1} \phi_{XX}(t,t) - 2\phi_{XX}(t,t_1) + \phi_{XX}(t_1,t_1) = 0 \qquad (2.41)$$

Clearly Eq. 2.41 is satisfied if $\phi_{XX}(t,s)$ is continuous at the point (t_1,t_1). Thus, we see that mean-square continuity of a process depends only on the continuity of the autocorrelation function of the process. The process $\{X(t)\}$ is mean-square continuous at time $t = t_1$ if and only if $\phi_{XX}(t,s)$ is continuous at (t_1,t_1).

It is easy to show that any process for which all time histories are continuous will have mean-square continuity. However, the mean-square continuity condition is considerably weaker than a requirement that all possible time histories of the process be continuous. This will be illustrated in Example 2.13, but first we will consider a simpler situation.

****

Example 2.12: Identify the t values corresponding to mean-square continuity of the process $\{X(t)\}$ with $X(t) = A_j$ for $j\Delta \le t < (j+1)\Delta$ with the A_j random variables being independent, and identically distributed with mean zero and variance $E(A^2)$.

We have already shown in Example 2.9 that $\phi_{XX}(t,s)$ for this process is $E(A^2)$ on the squares with dimension Δ along the 45° line, and is zero outside those squares. Thus, this autocorrelation is discontinuous along the boundaries of those squares, and continuous elsewhere. This means that $\phi_{XX}(t,s)$ is continuous at (t_1,t_1) if $t_1 \neq j\Delta$ for any j value. The process $\{X(t)\}$ is mean-square continuous except at the times $t = j\Delta$. Note that $t = j\Delta$ describes the instants of time when the time histories of $\{X(t)\}$ are almost sure to have discontinuities. Thus, in this example, $\{X(t)\}$ is mean-square continuous when the time histories are continuous, and it is mean-square discontinuous when the time histories are discontinuous.

****

Example 2.13: Let $\{N(t)\}$ be what is called a Poisson process defined by

$$P[N(t) = k] = \frac{e^{-\lambda t}(\lambda t)^k}{k!} \quad \text{for} \quad k = 1, 2, \cdots, \infty$$

Note that $N(t)$ is always integer valued and $N(0) = 0$. This distribution arises in many areas of applied probability, with $\{N(t)\}$ representing the number of occurrences (the count) of some event during the time interval $[0,t]$. It corresponds to having the number of occurrences in two non-overlapping (i.e., disjoint) time intervals be independent of each other, and with their being identically distributed if the time intervals are of the same length. The parameter λ represents the mean rate of occurrence. Consider the continuity of $\{N(t)\}$.

Note that any time history of this process is a "stair-step" function. It is zero until the time of first occurrence, stays at unity until the time of second occurrence, etc., as shown in the sketch of the kth sample time history. We could write

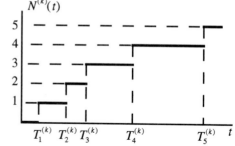

$$N(t) = \sum_{j=1}^{\infty} U[t - T_j]$$

with the random variable T_j denoting the time of jth occurrence. Clearly these time histories are not continuous at the times of occurrence. Furthermore, any time t is a time of occurrence for some of the possible time histories, so there is no time at which we can say that all time histories are continuous.

Let us now find the autocorrelation function for $\{N(t)\}$. For $t \geq s$ we can define $\Delta N = N(t) - N(s)$ and write

$$\phi_{NN}(t,s) \equiv E[N(t)N(s)] = E([N(s) + \Delta N]N(s)) = E[N^2(s)] + \mu_{\Delta N}\mu_N(s) \quad \text{for} \quad t \geq s$$

in which the independence of $N(s)$ and ΔN has been used. To proceed further we need the mean value function of $\{N(t)\}$:

$$\mu_N(t) = \sum_{k=0}^{\infty} kP[N(t) = k] = \sum_{k=1}^{\infty} \frac{e^{-\lambda t}(\lambda t)^k}{(k-1)!} = \lambda t$$

Now we use the fact that the distribution of ΔN is the same as that of $N(t - s)$ to obtain

$$\phi_{NN}(t,s) = \sum_{k=0}^{\infty} k^2 P[N(s)=k] + \lambda^2(t-s)s = \sum_{k=2}^{\infty} \frac{e^{-\lambda t}(\lambda s)^k}{(k-2)!} + \sum_{k=1}^{\infty} \frac{e^{-\lambda t}(\lambda s)^k}{(k-1)!} + \lambda^2(ts-s^2)$$

or

$$\phi_{NN}(t,s) = \lambda s(\lambda t + 1) \quad \text{for} \quad t \ge s \ge 0$$

The general relationship can be written as

$$\phi_{NN}(t,s) = \lambda^2 ts + \lambda \min(t,s) \quad \text{for} \quad t \ge 0, s \ge 0$$

It is easily verified that this function is continuous everywhere on the plane of (t,s) values, so $\{N(t)\}$ is mean-square continuous everywhere. This is in spite of the fact that there is nowhere at which all its time histories are continuous, nor does it have any continuous time histories, except the trivial one corresponding to no arrivals.
****

Example 2.14: Identify the t values giving mean-square continuity of the process $\{X(t)\}$ of Example 2.10 with the autocorrelation $\phi_{XX}(t,s)$ of $E(A^2)$ for $t = s$ and zero off this diagonal line.

Since $\phi_{XX}(t,s)$ is discontinuous at every point (t_1,t_1) on the 45° line, this $\{X(t)\}$ process is not mean-square continuous anywhere.
****

2.7 Ergodicity of a Stochastic Process

The concept of ergodicity has to do with using a time average obtained from one time history of a stochastic process as a substitute for a mathematical expectation. The stochastic process is said to be ergodic (in some sense) if the two are the same. Recall the idea of an ensemble of possible time histories of the process, as illustrated in Figure 2.1. As noted before, an expected value can be thought of as a statistical average across an infinite ensemble of time histories, and this is an average on a section orthogonal to a particular time history. Thus, there is no obvious reason why the two should be the same, even though they are both averages across infinitely many values of $\{X(t)\}$. To illustrate the idea we will begin with consideration of the simplest expected value related to a stochastic process $\{X(t)\}$ — the mean value. A truncated time average over sample j corresponding to the mean of the process can be written as

$$\frac{1}{T} \int_{-T/2}^{T/2} X^{(j)}(t)\, dt \tag{2.42}$$

We say that $\{X(t)\}$ is **ergodic in mean value** if it is mean-value stationary and the time average of expression 2.42 tends to $\mu_X = E[X(t)]$ as T tends to infinity, regardless of the value of j. We imposed the stationarity condition, since taking a time average like that in expression 2.42 could not possibly approximate the mean unless that mean was independent of time. Similarly we say that

$\{X(t)\}$ is **ergodic in second moment** if it is second-moment stationary and

$$R_{XX}(\tau) \equiv E[X(t+\tau)X(t)] = \lim_{T\to\infty} \frac{1}{T} \int_{-T/2}^{T/2} X^{(j)}(t+\tau)X^{(j)}(t)\,dt \tag{2.43}$$

There are as many types of ergodicity as there are stationarity. For any stationary characteristic of the stochastic process, one can define a corresponding time average, and ergodicity of the proper type will assure that the two are the same. The only additional type of ergodicity that we will list is the one related to the first-order probability distribution of the process. We will say that $\{X(t)\}$ is **first-order ergodic** if it is first-order stationary and

$$F_{X(t)}(u) = \lim_{T\to\infty} \frac{1}{T} \int_{-T/2}^{T/2} U[u - X^{(j)}(t)]\,dt \tag{2.44}$$

Note that the unit step function in the integrand is always zero or unity, such that the integral gives exactly the amount of time for which $X^{(j)}(t) \leq u$ within the interval $[-T/2, T/2]$. Thus, this type of ergodicity is a condition that the probability that $X(t) \leq u$ at any time t is the same as the fraction of the time that any time history $X^{(j)}(t)$ is less than or equal to u. We mention Eq. 2.44, both because it illustrates an idea which is somewhat different than the moment ergodicity of Eqs. 2.42 and 2.43, and because it gives another property that one often wishes to determine in practical problems.

The reason that people usually want to consider stochastic processes to be ergodic really has to do with the issue of statistics rather than probability theory. That is, they wish to determine information about the moments and/or probability distribution of a process $\{X(t)\}$ from observed values from the process. If a sufficiently large ensemble of observed time histories is available, then one can use ensemble averages and there is no need to invoke a condition of ergodicity. In most situations, though, it is impossible or impractical to obtain a very large ensemble of observed time histories from a given physical process. Thus, for example, one may wish to determine moments and/or the probability distribution of the wind speed at Easterwood Airport on April 27 from one long time history obtained on that date in 1997, rather than wait many years to obtain an ensemble of time histories, all recorded on April 27. Furthermore, even if it is possible to wait to obtain the many-year sample, one may suspect that the process may not be stationary over such a long time, so that it may be necessary to consider each annual sample as being from a different process or a process with different parameter values. Thus, the use of the ergodicity property is almost essential in many situations involving observed data from physical processes.

At least two cautions should be noted regarding ergodicity. The first is that there may be some difficulty in proving that a physical process is ergodic. Once we have chosen a mathematical model, we can usually show that it has a particular form of ergodicity, but this avoids the more fundamental issue of whether this ergodic model is the appropriate model for the physical problem. Furthermore, even if we know that Eq. 2.44, for example, is satisfied, we will not have an infinite time history available to use in evaluating the cumulative distribution function $F_{X(t)}(u)$ from the time average.

The best that we can ever hope for in a physical problem is a long finite sample time history. If we were sure that a process was ergodic in the proper way, then we would probably feel more confident in using a time average such as in Eq. 2.42, 2.43, or 2.44 to estimate the corresponding expected value, but it would still only be an estimate.

To illustrate the idea of ergodicity theorems, consider the proof of convergence of expression 2.42 to μ_X. What we wish to prove is convergence in probability:

$$\lim_{T \to \infty} P\left(|Q_T| \geq \varepsilon\right) = 0 \quad \text{for any } \varepsilon > 0 \tag{2.45}$$

in which

$$Q_T = \mu_X - \frac{1}{T} \int_{-T/2}^{T/2} X(t)\, dt \tag{2.46}$$

As noted in the preceding section, this is assured if $\lim_{T \to \infty} E\left[|Q_T|^c\right] = 0$ for any $c > 0$, and in particular if we have mean-square convergence:

$$\lim_{T \to \infty} E\left[|Q_T|^2\right] = 0 \tag{2.47}$$

One can find conditions to assure that Eq. 2.47 is true by rewriting Q_T^2 as

$$Q_T^2 = \left(\frac{1}{T} \int_{-T/2}^{T/2} [X(t) - \mu_X]\, dt\right)^2 = \frac{1}{T^2} \int_{-T/2}^{T/2} \int_{-T/2}^{T/2} [X(t_1) - \mu_X][X(t_2) - \mu_X]\, dt_1\, dt_2$$

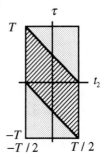

Taking the expected value of this quantity, rewriting the integrand as $G_{XX}(t_1 - t_2)$, and eliminating t_1 by the change of variables $\tau = t_1 - t_2$ gives

$$E\left(Q_T^2\right) = \frac{1}{T^2} \int_{-T/2}^{T/2} \int_{-T/2}^{T/2} G_{XX}(t_1 - t_2)\, dt_1\, dt_2 = \frac{1}{T^2} \int_{-T/2}^{T/2} \int_{-T/2-t_2}^{T/2-t_2} G_{XX}(\tau)\, d\tau\, dt_2$$

The parallelogram in the sketch shows the domain of integration in the (t_2, τ) plane. We now replace this integral by the one over the rectangle shown, minus the integrals over the two triangular corners. Using the symmetry of $G_{XX}(\tau)$, it is easily shown that the two triangles give the same integral, so that the result can be written as

$$E\left(Q_T^2\right) = \frac{1}{T^2}\left[\int_{-T/2}^{T/2} \int_{-T}^{T} G_{XX}(\tau)\, d\tau\, dt_2 - 2\int_{-T/2}^{T/2} \int_{T/2-t_2}^{T} G_{XX}(\tau)\, d\tau\, dt_2\right] \tag{2.48}$$

We now easily perform the integration with respect to t_2 in the first term. We can also reverse the order of integration in the second term and then perform its integration with respect to t_2 to obtain a sufficient condition for ergodicity in mean value as

$$\lim_{T \to \infty} \frac{1}{T} \int_{-T}^{T} \left(1 - \frac{|\tau|}{T}\right) G_{XX}(\tau) d\tau = 0 \qquad (2.49)$$

Note that the development of Eq. 2.49 from 2.47 is exact. Thus, Eq. 2.49 is necessary as well as sufficient for ergodicity in mean value to hold in the sense of convergence in mean square. This does not imply, however, that Eq. 2.49 is necessary for ergodicity in mean value to hold in the sense of convergence in probability.

Rather than performing the integration with respect to t_2 in the second term in Eq. 2.48, we can introduce the new variable $t = T/2 - t_2$ to obtain a result which can be written as

$$E\left(Q_T^2\right) = \frac{1}{T} \int_{-T}^{T} G_{XX}(\tau) d\tau - \frac{1}{T} \int_{0}^{T} \frac{2}{T} \int_{t}^{T} G_{XX}(\tau) d\tau \, dt$$

From this form one can show that

$$\lim_{T \to \infty} E\left(Q_T^2\right) = 0 \quad \text{if} \quad \lim_{T \to \infty} \frac{1}{T} \int_{-T}^{T} G_{XX}(\tau) d\tau = 0 \qquad (2.50)$$

so that this gives a sufficient condition for ergodicity in mean value. The condition of Eq. 2.50 is certainly satisfied if $G_{XX}(\tau)$ is integrable over the entire real line, so that $\int_{-\infty}^{\infty} G_{XX}(\tau) d\tau < \infty$, but this is a somewhat more restrictive condition than Eq. 2.50. For example, the integral is finite if $G_{XX}(\tau)$ for large values of τ tends to zero like $|\tau|^{-(1+\varepsilon)}$ for some small $\varepsilon > 0$, but Eq. 2.50 is satisfied provided only that $G_{XX}(\tau)$ tends to zero in any fashion as $|\tau|$ tends to infinity.

It may also be useful to note a situation which does not meet the conditions of ergodicity in mean value. If $G_{XX}(\tau)$ is a constant (other than zero) for all τ values, then the process violates the condition in Eq. 2.49 which is necessary for convergence in mean square. Consider what it means, though, for a process to be covariant stationary with a constant autocovariance function $G_{XX}(\tau)$. This would give $X(t + \tau)$ and $X(t)$ as being perfectly correlated, for any value of τ, and it can be shown (see Section A.8 of Appendix A) that this implies that there is a linear relationship of the form $X(t + \tau) = aX(t) + b$ for some constants a and b. Combining this condition with covariant stationarity requires that each time history of $\{X(t)\}$ be simply a constant value. This is an almost trivial case of a stochastic process which is always equal to a random variable. Each possible time history $X^{(j)}(t)$ is a constant $X^{(j)}$ independent of time, but each random sample may give a different value for that constant. It should be no surprise that this process is not ergodic in mean value. The time average in Eq. 2.42 is simply the constant $X^{(j)}$ and there is no convergence to μ_X as T tends to infinity. A necessary condition for ergodicity in mean value is that the random variables $X(t + \tau)$ and $X(t)$ not be perfectly correlated as $\tau \to \infty$.

The condition for ergodicity in mean value can be extended to the ergodicity in second moment of Eq. 2.43 by defining a new random process $\{Y(t)\}$ as

$$Y(t) = X(t + \tau)X(t)$$

for a particular τ value. Then Eq. 2.43 holds for that particular τ value if $\{Y(t)\}$ is ergodic in mean value. The autocovariance for this $\{Y(t)\}$ process can be written in terms of the second and fourth moment functions of the $\{X(t)\}$ process as

$$G_{YY}(s) = E[X(t + \tau + s)X(t + s)X(t + \tau)X(t)] - R_{XX}^2(\tau) \qquad (2.51)$$

Thus we can say that the $\{X(t)\}$ process is ergodic in second moment if it is fourth-moment stationary and the $G_{YY}(s)$ function satisfies a condition like Eq. 2.49, with the integration being over the variable s. Similarly we can define a $\{Z(t)\}$ process such that

$$Z(t) = U[u - X(t)]$$

with autocovariance function of

$$G_{ZZ}(\tau) = E(U[u - X(t + \tau)]U[u - X(t)]) - F_X^2(u) = F_{X(t+\tau),X(t)}(u,u) - F_X^2(u) \qquad (2.52)$$

for a given u value. The process $\{X(t)\}$ is first-order ergodic (Eq. 2.44) if it is second-order stationary and the $G_{ZZ}(\tau)$ function of Eq. 2.52 satisfies a condition like Eq. 2.49.

We will not proceed further with proofs of ergodicity. Equations 2.47 - 2.50 illustrate that it is not necessarily difficult to perform the manipulations to obtain conditions for convergence in mean square, and that those conditions are not generally very restrictive. Note that applying Eqs. 2.49 - 2.51 to show ergodicity in second moment requires conditions on the fourth moment function of $\{X(t)\}$, and the corresponding relationship to show jth-moment ergodicity in general would involve conditions on the $(2j)$th moment function of $\{X(t)\}$. This need for consideration of higher moments can be seen to follow directly from the use of mean-square convergence. If one could derive simple conditions to show $E(|Q_T|^c)$ tending to zero for an appropriate Q_T error term and some c which is only slightly greater than zero, then one could also prove ergodicity without consideration of such high moment functions. Unfortunately, such simple conditions are not available.

It is our observation that in common practice people assume that processes are ergodic unless there is some obvious physical reason why that would be inappropriate. For example, if $\{X(t)\}$ denotes the position of a body which has a finite probability of instability, then no one time history will be representative of the entire process since it will only represent one observation of either stable or unstable behavior of the system. Example 2.1 is of this type. We previously saw that the process used in that example is mean-value stationary, but it is clearly not mean-value ergodic. In terms of checking the ergodicity conditions presented here, one obvious difficulty is that the process is not covariant stationary.

Even for clearly nonstationary processes, it is sometimes necessary to use time averages as estimates of expected values. For example, one might modify Eq. 2.42 and use

$$\frac{1}{T}\int_{t-T/2}^{t+T/2}X^{(j)}(s)\,ds$$

as an estimate of $\mu_X(t)$ for a process which was known not to be mean-value stationary. Intuitively, this seems reasonable if $\mu_X(t)$ varies relatively slowly, but we will not go into any analysis of the error of such an approximation.

2.8 Stochastic Derivative

Most deterministic dynamics problems are governed by differential equations. Our goal is to analyze such systems when the deterministic excitation and response time histories are replaced by stochastic processes. This means that we want to analyze stochastic differential equations, but the first step in doing this must be the identification of an acceptable concept of the derivative of a stochastic process $\{X(t)\}$ with respect to its index parameter t:

$$\dot{X}(t) = \frac{dX(t)}{dt}$$

At each t value $X(t)$ is a different random variable, so that the idea of a Riemann definition of the derivative gives $\dot{X}(t)$ as being like a limit as h goes to zero of

$$Y(t,h) \equiv \frac{X(t+h) - X(t)}{h} \tag{2.53}$$

with the numerator of the expression being a different random variable for each h value. We will take a very simple approach and say that if any property of $Y(t,h)$ exists and has a limit as $h \to 0$, then that property also describes $\dot{X}(t)$. That is, if $E[Y(t,h)]$ exists and has a limit as $h \to 0$, then that limit is $\mu_{\dot{X}}(t) = E[\dot{X}(t)]$; if $E[Y(t,h)Y(s,h)]$ exists and has a limit as $h \to 0$, then that limit is $\phi_{\dot{X}\dot{X}}(t,s)$; if the probability distribution of $Y(t,h)$ exists and has a limit as $h \to 0$, then that limit is the probability distribution of $\dot{X}(t)$; etc.

Let us now investigate the moments of $\{\dot{X}(t)\}$ by evaluating those for $Y(t,h)$. For the mean value function we have

$$\mu_{\dot{X}}(t) = \lim_{h\to 0} E[Y(t,h)] = \lim_{h\to 0} E\left(\frac{X(t+h) - X(t)}{h}\right) = \lim_{h\to 0} \frac{\mu_X(t+h) - \mu_X(t)}{h}$$

or

$$\mu_{\dot{X}}(t) = \frac{d\mu_X(t)}{dt} \tag{2.54}$$

This important result can be stated in words as "the mean of the derivative is the derivative of the mean." This certainly sounds reasonable since expectation is a linear operation and we want stochastic differentiation also to be linear, so as to be consistent with the deterministic definition of the derivative. Since we are defining the stochastic derivative by a limit of the linear operation in Eq. 2.53, though, we are assured of this linearity.

Before proceeding to find the autocorrelation function for the derivative, let us consider the intermediate step of finding the cross-correlation function between $\{X(t)\}$ and $\{\dot{X}(t)\}$. We say that

$$\phi_{X\dot{X}}(t,s) = \lim_{h \to 0} E[X(t)Y(s,h)] = \lim_{h \to 0} E\left(X(t)\frac{X(s+h)-X(s)}{h}\right)$$

which gives

$$\phi_{X\dot{X}}(t,s) = \lim_{h \to 0}\left[\frac{\phi_{XX}(t,s+h)-\phi_{XX}(t,s)}{h}\right] \equiv \frac{\partial\,\phi_{XX}(t,s)}{\partial\,s} \tag{2.55}$$

showing that this cross-correlation function is obtained by taking a partial derivative of the autocorrelation function. Similarly, the autocorrelation function for the derivative process

$$\phi_{\dot{X}\dot{X}}(t,s) = \lim_{\substack{h_1 \to 0 \\ h_2 \to 0}} E[Y(t,h_1)Y(t,h_2)] = \lim_{\substack{h_1 \to 0 \\ h_2 \to 0}} E\left(\frac{X(t+h_1)-X(t)}{h_1}\frac{X(s+h_2)-X(s)}{h_2}\right)$$

gives

$$\phi_{\dot{X}\dot{X}}(t,s) = \lim_{\substack{h_1 \to 0 \\ h_2 \to 0}} \frac{\phi_{XX}(t+h_1,s+h_2)-\phi_{XX}(t,s+h_2)-\phi_{XX}(t+h_1,s)+\phi_{XX}(t,s)}{h_1 h_2}$$

or

$$\phi_{\dot{X}\dot{X}}(t,s) = \frac{\partial^2 \phi_{XX}(t,s)}{\partial t\,\partial s} \tag{2.56}$$

Conditions for higher-order moment functions could be derived in the same way. Also, a first-order probability distribution for $\dot{X}(t)$ could be derived from the second-order joint probability distribution for $X(t)$ and $X(t+h)$, which governs the probability distribution of $Y(t,h)$.

Note that our approach implies that $\mu_{\dot{X}}(t)$ exists if and only if $\mu_X(t)$ exists and has a derivative at the point t. Furthermore, $\mu_X(t)$ cannot be differentiable at t unless it is continuous at that point. Similarly, existence of the cross-correlation function $\phi_{X\dot{X}}(t,s)$ requires that $\phi_{XX}(t,s)$ be continuous at (t,s) and that the partial derivative exist at that point. The corresponding condition that $\phi_{XX}(t,s)$ be twice differentiable in order that the autocorrelation function $\phi_{\dot{X}\dot{X}}(t,s)$ exist, requires that the first partial derivative of $\phi_{XX}(t,s)$ be continuous at the point of interest. This final point is sometimes slightly confusing and will be illustrated in Examples 2.16 and 2.17.

If the process $\{X(t)\}$ is second-moment stationary, then one can rewrite the conditions of Eqs. 2.55 and 2.56 in stationary notation. First note that using $\phi_{XX}(t,s) = R_{XX}(t-s)$ gives

$$\phi_{X\dot{X}}(t,s) = \frac{\partial R_{XX}(t-s)}{\partial s} = \left[-\frac{dR_{XX}(\tau)}{d\tau} \right]_{\tau=t-s}$$

and

$$\phi_{\dot{X}\dot{X}}(t,s) = \frac{\partial^2 R_{XX}(t-s)}{\partial t \partial s} = \left[-\frac{d^2 R_{XX}(\tau)}{d\tau^2} \right]_{\tau=t-s}$$

These two expressions show that $\phi_{X\dot{X}}(t,s)$ and $\phi_{\dot{X}\dot{X}}(t,s)$, if they exist, are functions only of the time difference $\tau = t-s$. Thus, the derivative process is also second-moment stationary and $\{X(t)\}$ and $\{\dot{X}(t)\}$ are jointly second-moment stationary so that one can also write the results as

$$R_{X\dot{X}}(\tau) = -\frac{dR_{XX}(\tau)}{d\tau} \tag{2.57}$$

and

$$R_{\dot{X}\dot{X}}(\tau) = -\frac{d^2 R_{XX}(\tau)}{d\tau^2} \tag{2.58}$$

Using the results given for the mean value function and correlation functions involving the derivative of $\{X(t)\}$, one can easily show that corresponding results hold for the covariance functions describing $\{\dot{X}(t)\}$:

$$K_{X\dot{X}}(t,s) = \frac{\partial K_{XX}(t,s)}{\partial s}, \qquad\qquad G_{X\dot{X}}(\tau) = -\frac{dG_{XX}(\tau)}{d\tau} \tag{2.59}$$

$$K_{\dot{X}\dot{X}}(t,s) = \frac{\partial^2 K_{XX}(t,s)}{\partial t \partial s}, \qquad\qquad G_{\dot{X}\dot{X}}(\tau) = -\frac{d^2 G_{XX}(\tau)}{d\tau^2} \tag{2.60}$$

Let us now consider the derivative of a function of a stochastic process. Specifically, we will define a new stochastic process $\{Z(t)\}$ by $Z(t) = g[X(t)]$, and investigate the derivative process $\{\dot{Z}(t)\}$. By the procedure of Eq. 2.53, the behavior of $\dot{Z}(t)$ must be the same as that of

$$\frac{Z(t+h) - Z(t)}{h} = \frac{g[X(t+h)] - g[X(t)]}{h}$$

in the limit as h goes to zero. In particular,

$$E[\dot{Z}(t)] = \lim_{h \to 0} E\left(\frac{Z(t+h) - Z(t)}{h} \right) = \lim_{h \to 0} \frac{E(g[X(t+h)]) - E(g[X(t)])}{h} \equiv \frac{d}{dt} E(g[X(t)])$$

provided that $E(g[X(t)])$ is a differentiable function of t. This can be rewritten as

$$E\left(\frac{d}{dt}g[X(t)]\right) = \frac{d}{dt}E(g[X(t)]) \tag{2.61}$$

to emphasize the very important result that the order of differentiation and expectation can generally be reversed. The results presented in Eqs. 2.54 - 2.60 can be considered as special cases of this general relationship.

The result in Eq. 2.61 will be very useful to us in our study of stochastic vibrations, particularly in the method of state space analysis. The derivative of $g[X(t)]$ can be written out in the same way as for a deterministic function giving

$$\frac{d}{dt}E(g[X(t)]) = E\left(\dot{X}(t)g'[X(t)]\right) \tag{2.62}$$

in which $g'(\cdot)$ denotes the derivative of $g(\cdot)$ with respect to its total argument. For example,

$$\frac{d}{dt}E[X^2(t)] = E[2X(t)\dot{X}(t)] = 2\phi_{X\dot{X}}(t,t) \tag{2.63}$$

and

$$\frac{d}{dt}E[X^j(t)] = j\,E[X^{j-1}(t)\dot{X}(t)] \tag{2.64}$$

One situation in which these relationships become particularly simple is when the processes involved are stationary. Thus, if $\{X(t)\}$ is second-moment stationary we can say that $E[X^2(t)]$ is a constant and Eq. 2.63 then tells us that $E[X(t)\dot{X}(t)] = 0$. This particular result could also be seen by noting that $E[X(t)\dot{X}(t)] = R_{X\dot{X}}(0)$, using Eq. 2.57 to show that this is the negative of the slope of the $R_{XX}(\tau)$ function at $\tau = 0$, and arguing that this slope at the origin must be zero (if it exists) since $R_{XX}(\tau)$ is a symmetric function. Equation 2.64 shows the somewhat less obvious fact that $E[X^{j-1}(t)\dot{X}(t)] = 0$ for any process $\{X(t)\}$ which is jth-moment stationary. If $\{X(t)\}$ is strictly stationary, then one can argue that $E(g[X(t)])$ must be a constant for any function $g(\cdot)$ so that Eq. 2.62 gives $E\left(\dot{X}(t)g'[X(t)]\right) = 0$ for any function $g'(\cdot)$.

The procedure can also be extended to expected values involving more than one stochastic process, giving expressions such as

$$\frac{d}{dt}E[X^j(t)Z^k(t)] = j\,E[X^{j-1}(t)\dot{X}(t)Z^k(t)] + k\,E[X^j(t)Z^{k-1}(t)\dot{Z}(t)] \tag{2.65}$$

which includes such special cases as

$$\frac{d}{dt}E[X(t)\dot{X}(t)] = E[\dot{X}^2(t)] + E[X(t)\ddot{X}(t)]$$

and

$$\frac{d}{dt}E[X(t)\ddot{X}(t)] = E[\dot{X}(t)\ddot{X}(t)] + E[X(t)\dddot{X}(t)]$$

If the processes are second-moment stationary, then the derivatives of expected values must be zero and these two equations give us $E[X(t)\ddot{X}(t)] = -E[\dot{X}^2(t)]$ and $E[X(t)\dddot{X}(t)] = -E[\dot{X}(t)\ddot{X}(t)]$, but this latter equation gives $E[X(t)\dddot{X}(t)] = 0$ since $E[\dot{X}(t)\ddot{X}(t)]$ is one half of the derivative of $E[\dot{X}^2(t)]$. Clearly one can derive many such relationships between different expected values, and those given are only illustrative examples.

****

Example 2.15: Consider the differentiability of a stochastic process $\{X(t)\}$ with mean value and autocovariance functions of

$$\mu_X(t) = \exp(3t) \qquad G_{XX}(\tau) = K_{XX}(t+\tau, t) = 2\exp(-5\tau^2)$$

Both of these functions are differentiable everywhere so there is no problem with the existence of the mean and covariance of the derivative process $\{\dot{X}(t)\}$. The mean value function is given by

$$\mu_{\dot{X}}(t) = \frac{d\mu_X(t)}{dt} = 3\exp(3t)$$

Since $\{X(t)\}$ is covariant stationary, we can use

$$K_{X\dot{X}}(t+\tau, t) = G_{X\dot{X}}(\tau) = -\frac{dG_{XX}(\tau)}{d\tau} = 20\tau\exp(-5\tau^2)$$

and

$$K_{\dot{X}\dot{X}}(t+\tau, t) = G_{\dot{X}\dot{X}}(\tau) = -\frac{d^2 G_{XX}(\tau)}{d\tau^2} = 20(1-10\tau^2)\exp(-5\tau^2)$$

Alternatively, we could have found the nonstationary autocorrelation function for $\{X(t)\}$ as

$$\phi_{XX}(t,s) = K_{XX}(t,s) + \mu_X(t)\mu_X(s) = 2\exp[-5(t-s)^2] + \exp[3(t+s)]$$

and used

$$\phi_{X\dot{X}}(t,s) = \frac{\partial \phi_{XX}(t,s)}{\partial s} = 20(t-s)\exp[-5(t-s)^2] + 3\exp[3(t+s)]$$

and

$$\phi_{\dot{X}\dot{X}}(t,s) = \frac{\partial^2 \phi_{XX}(t,s)}{\partial t \partial s} = 20[1-10(t-s)^2]\exp[-5(t-s)^2] + 9\exp[3(t+s)]$$

It is easy to verify that this gives

$$\phi_{X\dot{X}}(t,s) = K_{X\dot{X}}(t,s) + \mu_X(t)\mu_{\dot{X}}(s)$$

and

$$\phi_{\dot{X}\dot{X}}(t,s) = K_{\dot{X}\dot{X}}(t,s) + \mu_{\dot{X}}(t)\mu_{\dot{X}}(s)$$

This is simply a confirmation of something which we know must be true, since these relationships hold for any stochastic processes, as given in Eqs. 2.6 and 2.7.

****

Example 2.16: Consider the differentiability of a mean zero stationary stochastic process $\{X(t)\}$ with autocorrelation function

$$R_{XX}(\tau) = e^{-a|\tau|}$$

in which a is a positive constant.

Since $\mu_X(t) = 0$ everywhere, its derivative is also zero and we can say that $\mu_{\dot{X}}(t) = 0$. Next, we can take the derivative of Eq. 2.57 and obtain

$$R_{X\dot{X}}(\tau) = e^{-a|\tau|}\frac{d(a|\tau|)}{d\tau} = ae^{-a|\tau|}\mathrm{sgn}(\tau) = ae^{-a|\tau|}[2U(\tau)-1]$$

This shows that the cross-correlation of $\{X(t)\}$ and $\{\dot{X}(t)\}$ exists (i.e., is finite) for all τ values. However, this first derivative is discontinuous at $\tau = 0$, so that the autocorrelation for $\{\dot{X}(t)\}$:

$$R_{\dot{X}\dot{X}}(\tau) = e^{-a|\tau|}[2a\delta(\tau) - a^2]$$

is infinite for $\tau = 0$. Note that the derivative has an infinite mean squared value. This example illustrates that caution must be used in evaluating the second partial derivative of Eq. 2.56. If we had simply investigated the situations with $\tau < 0$ and $\tau > 0$ we would have found $R_{\dot{X}\dot{X}}(\tau) = -a^2 e^{-a|\tau|}$ in both of those regions. Thus, we find that $R_{\dot{X}\dot{X}}(\tau)$ approaches the same limit as we approach from either side toward the condition which should give mean squared value. The fact that this limit is negative (namely, $-a^2$), though, is a sure indication that this could not be the mean squared value. The reader is urged to use considerable caution in investigating the second moment properties of the derivatives of a stochastic process, unless the autocorrelation function for the original process is an analytic function, assuring that all its derivatives exist.

****

Example 2.17: Consider the differentiability of the Poisson process $\{N(t):t \geq 0\}$ of Example 2.13, for which

$$\mu_N(t) = \lambda t$$

and

$$\phi_{NN}(t,s) = \lambda^2 ts + \lambda \min(t,s) = \lambda^2 ts + \lambda sU(t-s) + \lambda tU(s-t)$$

Based on the mean value function of $\{N(t)\}$ we can say that the mean of the derivative process exists and is equal to

$$\mu_{\dot{N}}(t) = \frac{d\mu_N(t)}{dt} = \lambda$$

Similarly, the cross-correlation of $\{N(t)\}$ and $\{\dot{N}(t)\}$ should be given by

$$\phi_{N\dot{N}}(t,s) = \frac{\partial \phi_{NN}(t,s)}{\partial s} = \lambda^2 t + \lambda\, U(t-s) + \lambda\,(t-s)\delta(s-t)$$

in which $\delta(\cdot)$ denotes the Dirac delta function (see Section A.2 of Appendix A). The last term of this expression is always zero, though, so that $\phi_{N\dot{N}}(t,s)$ always exists (i.e., is always finite) and is given by

$$\phi_{N\dot{N}}(t,s) = \lambda^2 t + \lambda\, U(t-s)$$

Note that this first derivative function is discontinuous along the line $t=s$, so that the second partial derivative of Eq. 2.56 does not exist. We can write it formally as

$$\phi_{\dot{N}\dot{N}}(t,s) = \frac{\partial^2 \phi_{NN}(t,s)}{\partial t \partial s} = \lambda^2 + \lambda\,\delta(t-s)$$

which again emphasizes that the autocorrelation of $\{\dot{N}(t)\}$ is finite only for $t \neq s$. The mean squared value of the derivative process, which is the $\phi_{\dot{N}\dot{N}}(t,s)$ autocorrelation function for $t=s$, is infinite.

This example illustrates a situation in which the time histories of $\{N(t)\}$ are not differentiable, but one can still consider a derivative process $\{\dot{N}(t)\}$, with the limitation that it has an infinite mean squared value. As in Example 2.16, the second derivative exists both for $t < s$ and for $t > s$, and the limit is the same (i.e., λ^2) as one approaches $t=s$ from either side. Nonetheless, the second derivative does not exist for $t=s$. A discontinuity in a function can always be expected to contribute such a Dirac delta function term to the derivative of the function.

****

Example 2.18: Consider the differentiability of the stochastic process $\{X(t)\}$ of Example 2.9, which has $X(t)$ equal to some A_j in each finite time interval, and with the A_j identically distributed, mean zero, and independent.

Taking the expected value gives $\mu_X(t) = \mu_A$, so that we can say that

$$\mu_{\dot{X}}(t) = 0$$

In Example 2.9 we found the autocorrelation function for $\{X(t)\}$ as

$$\phi_{XX}(t,s) = E(A^2)\sum_j \left(U[t - j\Delta] - U[t - (j+1)\Delta]\right)\left(U[s - j\Delta] - U[s - (j+1)\Delta]\right)$$

This function is discontinuous on the boundaries of the dimension Δ squares along the diagonal. Thus, the cross-correlation function $\phi_{X\dot{X}}(t,s)$ is infinite for some (t,s) values:

$$\phi_{X\dot{X}}(t,s) = E(A^2)\sum_j (U[t-j\Delta]-U[t-(j+1)\Delta])(\delta[s-j\Delta]-\delta[s-(j+1)\Delta])$$

In particular, it is infinite for $s = k\Delta$ for any integer k, if $(k-1)\Delta < t < (k+1)\Delta$ (i.e., along the s boundaries of the dimension Δ squares along the diagonal). Similarly, $\phi_{\dot{X}X}(t,s)$ is infinite along all the t boundaries of the same squares along the diagonal. Everywhere except on the boundaries of these squares, one obtains $\phi_{X\dot{X}}(t,s) = 0$ and $\phi_{\dot{X}X}(t,s) = 0$, reflecting the fact that this process has a zero derivative except at certain discrete times at which the derivative is infinite. This agrees with the behavior of the time histories of the process.

****

Example 2.19: Consider the differentiability of the stochastic process $\{X(t)\}$ with $X^{(1)}(t) = 0$, $X^{(2)}(t) = \alpha\sinh(\beta t)$, $X^{(3)}(t) = -\alpha\sinh(\beta t)$, and $P[X(t) = X^{(1)}(t)] = 0.50$, $P[X(t) = X^{(2)}(t)] = 0.25$, $P[X(t) = X^{(3)}(t)] = 0.25$.

In Example 2.1 we showed that $\mu_X(t) = 0$ and $\phi_{XX}(t,s) = (0.50)\,\alpha^2\sinh(\beta t)\sinh(\beta s)$. Thus, we can now take derivatives of these functions to obtain

$$\mu_{\dot{X}}(t) = 0$$

$$\phi_{X\dot{X}}(t,s) = (0.50)\,\alpha^2\beta\,\sinh(\beta t)\cosh(\beta s)$$

and

$$\phi_{\dot{X}\dot{X}}(t,s) = (0.50)\,\alpha^2\beta^2\cosh(\beta t)\cosh(\beta s)$$

Since the functions are analytic everywhere on the domain $t \geq 0$ for which $\{X(t)\}$ is defined, there is no problem of existence of the moments of the derivative.

Since this $\{X(t)\}$ was defined in terms of its time histories, one can also differentiate those time histories and obtain corresponding time history relationships for the derivative process $\{\dot{X}(t)\}$. We anticipate that this should give the same results as shown for the moments of $\{\dot{X}(t)\}$. If the results are not consistent, then it must mean that our definition of the stochastic derivative is defective. We will now check this. We find that

$$\dot{X}^{(1)}(t) = 0$$
$$\dot{X}^{(2)}(t) = \alpha\beta\cosh(\beta t)$$
$$\dot{X}^{(3)}(t) = -\alpha\beta\cosh(\beta t)$$

which gives

$$\mu_{\dot{X}}(t) \equiv E[\dot{X}(t)] = (0.50)(0) + (0.25)[\alpha\beta\cosh(\beta t)] + (0.25)[-\alpha\beta\cosh(\beta t)] = 0$$

$$\phi_{\dot{X}\dot{X}}(t,s) = (0.25)[\alpha\sinh(\beta t)][\alpha\beta\cosh(\beta s)] + (0.25)[-\alpha\sinh(\beta t)][-\alpha\beta\cosh(\beta s)]$$

$$\phi_{\dot{X}\dot{X}}(t,s) = (0.25)[\alpha\beta\cosh(\beta t)][\alpha\beta\cosh(\beta s)] + (0.25)[-\alpha\beta\cosh(\beta t)][-\alpha\beta\cosh(\beta s)]$$

These results agree exactly with those obtained from differentiating the moment functions of $\{X(t)\}$, confirming that our definition of the stochastic derivative is consistent.

****

Example 2.20: For the stochastic process $\{X(t)\}$ with $X(t) = A\cos(\omega t)$ in which A is a random variable, find $\mu_{\dot{X}}(t)$, $\phi_{X\dot{X}}(t,s)$, and $\phi_{\dot{X}\dot{X}}(t,s)$. Confirm that differentiating the moment functions for $\{X(t)\}$ and analyzing the time histories of the derivative process $\{\dot{X}(t)\}$ give the same results.

From Example 2.2, we have $\mu_X(t) = \mu_A\cos(\omega t)$ and $\phi_{XX}(t,s) = E(A^2)\cos(\omega t)\cos(\omega s)$. Taking derivatives of these functions gives

$$\mu_{\dot{X}}(t) = -\mu_A\omega\sin(\omega t)$$

$$\phi_{X\dot{X}}(t,s) = -E(A^2)\omega\cos(\omega t)\sin(\omega s)$$

and

$$\phi_{\dot{X}\dot{X}}(t,s) = E(A^2)\omega^2\sin(\omega t)\sin(\omega s)$$

The relationship for the time histories of $\{\dot{X}(t)\}$ is

$$\dot{X}(t) = -A\omega\sin(\omega t)$$

and it is obvious that this relationship gives the same moment functions for $\{\dot{X}(t)\}$ as were already obtained by differentiating the moment functions for $\{X(t)\}$.

****

In many cases, analysts choose to use a definition of stochastic derivative which is more precise but also more restrictive than what we have used here. In particular, if

$$\lim_{h\to 0} E\left(\left[\dot{X}(t) - Y(t,h)\right]^2\right) = 0 \qquad (2.66)$$

then $\{\dot{X}(t)\}$ is said to be the **mean-square derivative** of $\{X(t)\}$. Expanding this expression, one can show that the mean-square derivative exists if Eqs. 2.55 and 2.56 are both satisfied. In general, one must say that the mean-square derivative does not exist if $E[\dot{X}^2(t)]$ is infinite. This situation is illustrated in Examples 2.16, 2.17, and 2.18, in which the mean-square derivative does not exist in Examples 2.16 and 2.17 for any t value, and in 2.18 for certain t values. These examples demonstrate that it may be overly restrictive to limit attention only to the mean-square derivative.

Note that we have only considered the first derivative $\{\dot{X}(t)\}$ of a stochastic process $\{X(t)\}$, whereas our dynamics equations will usually require at least two derivatives. In particular, we will usually need to include displacement, velocity, and acceleration terms in our differential equations, and these can be modeled as three stochastic processes $\{X(t)\}$, $\{\dot{X}(t)\}$, and $\{\ddot{X}(t)\}$. This presents no difficulty, though, since we can simply say that $\{\ddot{X}(t)\}$ is the derivative of $\{\dot{X}(t)\}$, reusing the concept of the first derivative. In this way, one can define the general jth-order derivative of $\{X(t)\}$ by j applications of the derivative procedure.

Finally, it should also be noted that the various expressions given here for auto- or cross-correlation functions could equally well be written in terms of auto or cross-covariance functions. This follows directly from the fact that an auto- or cross-covariance function is identical to an auto- or cross-correlation function for the special situation in which the mean value functions are zero. One cross-covariance result of particular significance is the fact that the random variables $X(t)$ and $\dot{X}(t)$ are uncorrelated at any time t for a covariant stationary $\{X(t)\}$ process, which is the generalization of $E[X(t)\dot{X}(t)]$ being zero for a second-moment stationary process.

2.9 Stochastic Integral

In order to complete our idea of stochastic calculus, we need to define stochastic integrals. In fact, we will consider three slightly different types of stochastic integrals which we will use in applications. The conditions for existence of a stochastic integral are usually easier to visualize than for a stochastic derivative, which is basically the same as for deterministic functions. Whereas existence of a derivative depends on smoothness conditions on the original function, an integral generally exists if the integrand is bounded and tends to zero sufficiently rapidly at any infinite limits of integration.

The simplest integral of a stochastic process $\{X(t)\}$ is the simple definite integral of the form

$$Z = \int_a^b X(t)\,dt \tag{2.67}$$

For constant limits a and b, the quantity Z must be a random variable. We can follow exactly the same approach that we used for derivatives and define a Riemann sum which approximates Z, then say that the moments and/or other properties of Z are the limits of the corresponding quantities for the Riemann sum. For example, we can say that

$$Y_n = \Delta t \sum_{j=1}^{n} X(j\,\Delta t)$$

with $\Delta t = (b-a)/n$. Then,

$$\mu_Z \equiv E(Z) = \lim_{n \to \infty} E(Y_n) = \lim_{n \to \infty} \Delta t \sum_{j=1}^{n} \mu_X(j\,\Delta t) = \int_a^b \mu_X(t)\,dt \tag{2.68}$$

$$E(Z^2) = \lim_{n\to\infty} E(Y_n^2) = \lim_{n\to\infty} (\Delta t)^2 \sum_{j=1}^{n}\sum_{k=1}^{n} \phi_{XX}(j\,\Delta t, k\Delta t) = \int_a^b\int_a^b \phi_{XX}(t,s)\,dt\,ds \qquad (2.69)$$

etc. Thus, we will say that $E(Z)$ exists if the integral of $\mu_X(t)$ from a to b exists and that $E(Z^2)$ exists if the integral of $\phi_{XX}(t,s)$ over the square domain exists. Clearly, these integrals will exist for finite values of a and b if the first and second moments of $\{X(t)\}$ are finite.

The next integral that we will consider is the stochastic process which can be considered the antiderivative:

$$Z(t) = \int_a^t X(s)\,ds \qquad (2.70)$$

with a being any constant. Equations 2.68 and 2.69 now generalize to give the mean value function and autocorrelation function for the new process $\{Z(t)\}$ as

$$\mu_Z(t) \equiv E[Z(t)] = \int_a^t \mu_X(s)\,ds \qquad (2.71)$$

$$\phi_{XZ}(t_1,t_2) \equiv E[X(t_1)Z(t_2)] = \int_a^{t_2} \phi_{XX}(t_1,s_2)\,ds_2 \qquad (2.72)$$

and

$$\phi_{ZZ}(t_1,t_2) \equiv E[Z(t_1)Z(t_2)] = \int_a^{t_2}\int_a^{t_1} \phi_{XX}(s_1,s_2)\,ds_1\,ds_2 \qquad (2.73)$$

Again, we will generally have no problem with the existence of these moments. Also note that Eqs. 2.71 - 2.73 are exactly the inverse forms of the relationships in Eqs. 2.54 - 2.56, confirming that $\{\dot{Z}(t)\} = \{X(t)\}$, which is the desired inverse of Eq. 2.70. The idea of the compatibility of our concepts of stochastic derivative and stochastic integral will be essential in our study of dynamical systems. For example, we will need to be able to use the usual deterministic ideas that velocity is the derivative of displacement and displacement is the integral of velocity when both displacement and velocity are stochastic processes.

There is no assurance that a stationary process $\{X(t)\}$ will give a stationary integral process $\{Z(t)\}$. For example, $\{X(t)\}$ being mean-value stationary only assures that μ_X is independent of t, and thus Eq. 2.71 then gives the nonstationary mean value $\mu_Z(t) = \mu_X t$ for $\{Z(t)\}$. Similarly, a stationary autocorrelation function for $\{X(t)\}$ may not give a stationary autocorrelation function for $\{Z(t)\}$, as will be illustrated in Example 2.21.

The third type of stochastic integral that we will use is a generalization of Eq. 2.67 in a slightly different way. This time we will keep the limits of integration as constants, but the integral will be a stochastic process because of the presence of a kernel function in the integrand:

$$Z(\eta) = \int_a^b X(t) g(t,\eta) dt \qquad (2.74)$$

For any particular value of the new variable η this integral is a random variable exactly like Eq. 2.67, and for any reasonably smooth kernel function $g(t,\eta)$ this new family of random variables can be considered to be a stochastic process $\{Z(\eta)\}$, with η as the index parameter. From Eqs. 2.68 and 2.69, we have

$$\mu_Z(\eta) = \int_a^b \mu_X(t) g(t,\eta) dt \qquad (2.75)$$

and

$$\phi_{ZZ}(\eta_1,\eta_2) = \int_a^b \int_a^b \phi_{XX}(t,s) g(t,\eta_1) g(s,\eta_2) dt\,ds \qquad (2.76)$$

and other properties can be derived as needed. This type of stochastic process is needed for the study of dynamics. In the time domain analysis of Chapter 3, η represents another time variable, and η is a frequency parameter in the Fourier analysis of Chapter 4.

****

Example 2.21: Consider the integral process $\{Z(t): t \geq 0\}$ defined by

$$Z(t) = \int_0^t X(s)\,ds$$

for $\{X(t)\}$ being a mean zero stationary stochastic process with autocorrelation function $R_{XX}(\tau) = e^{-\alpha|\tau|}$ in which α is a positive constant. Find the cross-correlation function of $\{X(t)\}$ and $\{Z(t)\}$ and the autocorrelation function for $\{Z(t)\}$.

From Eq. 2.72 we have

$$\phi_{XZ}(t_1,t_2) = \int_0^{t_2} \phi_{XX}(t_1,s_2)\,ds_2 = \int_0^{t_2} R_{XX}(t_1 - s_2)\,ds_2 = \int_0^{t_2} \exp(-\alpha|t_1 - s_2|)\,ds_2$$

For $t_2 \leq t_1$ this gives

$$\phi_{XZ}(t_1,t_2) = \int_0^{t_2} \exp(-\alpha t_1 + \alpha s_2)ds_2 = e^{-\alpha t_1}\frac{e^{\alpha t_2}-1}{\alpha} = \frac{[e^{-\alpha(t_1-t_2)} - e^{-\alpha t_1}]}{\alpha}$$

and for $t_2 > t_1$ the integral must be split into two parts, giving

$$\phi_{XZ}(t_1,t_2) = \int_0^{t_1} \exp(-\alpha t_1 + \alpha s_2)ds_2 + \int_{t_1}^{t_2} \exp(\alpha t_1 - \alpha s_2)\,ds_2 = e^{-\alpha t_1}\frac{e^{\alpha t_1}-1}{\alpha} + e^{\alpha t_1}\frac{e^{-\alpha t_2} - e^{-\alpha t_1}}{-\alpha}$$

or

$$\phi_{XZ}(t_1,t_2) = \frac{2 - e^{-\alpha t_1} - e^{-\alpha(t_2-t_1)}}{\alpha}$$

The autocorrelation of $\{Z(t)\}$ can now be found as

$$\phi_{ZZ}(t_1,t_2) = \int_0^{t_1} \phi_{XZ}(s_1,t_2)\,ds_1$$

For $t_2 \leq t_1$ we find that

$$\phi_{ZZ}(t_1,t_2) = \int_0^{t_2} \frac{2 - e^{-\alpha s_1} - e^{-\alpha(t_2-s_1)}}{\alpha}\,ds_1 + \frac{e^{\alpha t_2}-1}{\alpha}\int_{t_2}^{t_1} e^{-\alpha s_1}\,ds_1$$

$$= \frac{1}{\alpha}\left[2t_2 - \frac{(e^{-\alpha t_2}+1)(e^{\alpha t_2}-1)}{\alpha} + \frac{(e^{\alpha t_2}-1)(e^{-\alpha t_1}-e^{-\alpha t_2})}{-\alpha} \right]$$

$$= [2\alpha t_2 - e^{-\alpha(t_1-t_2)} + e^{-\alpha t_1} + e^{-\alpha t_2} - 1]/\alpha^2$$

and for $t_2 > t_1$

$$\phi_{ZZ}(t_1,t_2) = \int_0^{t_1} \frac{2 - e^{-\alpha s_1} - e^{-\alpha(t_2-s_1)}}{\alpha}\,ds_1 = \frac{1}{\alpha}\left[2t_1 - \frac{e^{-\alpha t_1}-1}{-\alpha} - \frac{e^{-\alpha t_2}(e^{\alpha t_1}-1)}{\alpha} \right]$$

$$= [2\alpha t_1 - e^{-\alpha(t_2-t_1)} + e^{-\alpha t_1} + e^{-\alpha t_2} - 1]/\alpha^2$$

Note that this $\phi_{ZZ}(t_1,t_2)$ function can be combined as

$$\phi_{ZZ}(t_1,t_2) = [2\alpha \min(t_1,t_2) - e^{-\alpha|t_2-t_1|} + e^{-\alpha t_1} + e^{-\alpha t_2} - 1]/\alpha^2$$

for any values of t_1 and t_2. In this form it is obvious that the function does have the necessary symmetry, $\phi_{ZZ}(t_1,t_2) = \phi_{ZZ}(t_2,t_1)$. Note that $\{Z(t)\}$ is not second-moment stationary and it is not jointly second-moment stationary with $\{X(t)\}$, even though $\{X(t)\}$ is second-moment stationary.
****
Example 2.22: Consider the integral process $\{Z(t): t \geq 0\}$ defined by

$$Z(t) = \int_0^t X(s)\,ds$$

for $X(t)$ equal to some random variable A_j in each time increment $j\Delta \leq t \leq (j+1)\Delta$, and with the A_j being identically distributed, mean zero, and independent. Find the mean value function of $\{Z(t)\}$, the cross-correlation function of $\{X(t)\}$ and $\{Z(t)\}$, and the autocorrelation function for $\{Z(t)\}$.

Since $\mu_X(t) = \mu_A$,we have

$$\mu_Z(t) = \mu_A t$$

In Example 2.9 we found that the autocorrelation function $\phi_{XX}(t,s)$ for $\{X(t)\}$ is equal to $E(A^2)$ in squares of dimension Δ along the 45° line of the (t,s) plane and zero elsewhere.

In order to simplify the presentation of the results, we will define a new integer time function $k(t)$ as the number of full time increments included in $[0,t]$:

$$k(t) = \operatorname{int}(t/\Delta) = j \qquad \text{if} \qquad j\Delta \leq t < (j+1)\Delta$$

This allows us to classify more simply any (t_1, t_2) point into one of three sets: points within the squares of dimension Δ along the 45° line of the (t_1, t_2) plane are described by $k(t_1) = k(t_2)$, by $k(t_1)\Delta \leq t_2 < [k(t_1)+1]\Delta$, or by $k(t_2)\Delta \leq t_1 < [k(t_2)+1]\Delta$; points below the squares are described by $k(t_1) > k(t_2)$, by $t_2 < k(t_1)\Delta$, or by $t_1 \geq [k(t_2)+1]\Delta$; and points above the squares are described by $k(t_1) < k(t_2)$, by $t_2 \geq [k(t_1)+1]\Delta$, or by $t_1 < k(t_2)\Delta$.

Then using $\phi_{XZ}(t_1, t_2) = \int_0^{t_2} \phi_{XX}(t_1, s_2)\, ds_2$ gives

$$
\begin{aligned}
\phi_{XZ}(t_1, t_2) &= 0 && \text{if} && k(t_1) > k(t_2) \\
&= E(A^2)[t_2 - k(t_1)\Delta] && \text{if} && k(t_1) = k(t_2) \\
&= E(A^2)\Delta && \text{if} && k(t_1) < k(t_2)
\end{aligned}
$$

The function $\phi_{XZ}(t_1, t_2)$ versus t_2 grows linearly from zero to $E(A^2)\Delta$ as t_2 varies from $k(t_1)\Delta$ to $[k(t_1)+1]\Delta$. The same function versus t_1 is a stair-step which is constant at the level $E(A^2)\Delta$ for $t_1 < k(t_2)\Delta$, at the level $E(A^2)[t_2 - k(t_1)\Delta]$ for $k(t_2)\Delta \leq t_1 < [k(t_2)+1]\Delta$, and at zero for $t_1 \geq [k(t_2)+1]\Delta$.

Now $\phi_{ZZ}(t_1, t_2) = \int_0^{t_1} \phi_{XZ}(s_1, t_2)\, ds_1$ gives

$$
\begin{aligned}
\phi_{ZZ}(t_1, t_2) &= E(A^2)\Delta\, t_1 && \text{if} && k(t_1) < k(t_2) \\
&= E(A^2)\big(k(t_1)\Delta^2 + [t_2 - k(t_1)\Delta][t_1 - k(t_1)\Delta]\big) && \text{if} && k(t_1) = k(t_2) \\
&= E(A^2)\Delta\, t_2 && \text{if} && k(t_1) > k(t_2)
\end{aligned}
$$

Note that if Δ is small compared to t_1 and t_2 then one can say that $\phi_{ZZ}(t_1, t_2) \approx E(A^2)\Delta\,\min(t_1, t_2)$. This expression is exact for $k(t_1) \neq k(t_2)$ and has an error of order Δ^2 for $k(t_1) = k(t_2)$.

****

Example 2.23: Consider the integral process $\{Z(t)\}$ defined by

$$Z(t) = \int_{-\infty}^{t} X(s)\, ds$$

for $X(t)$ equal to some random variable A in the time increment $0 \leq t \leq \Delta$, and equal to zero elsewhere. Find the mean value function of $\{Z(t)\}$, the cross-correlation function of $\{X(t)\}$ and $\{Z(t)\}$, and the autocorrelation function for $\{Z(t)\}$.

First we note that the process $\{X(t)\}$ has mean value function and autocorrelation function given by

$$\mu_X(t) = \mu_A[U(t) - U(t - \Delta)]$$

and

$$\phi_{XX}(t_1, t_2) = E(A^2)[U(t_1) - U(t_1 - \Delta)][U(t_2) - U(t_2 - \Delta)]$$

Now $\mu_Z(t) = \int_{-\infty}^{t} \mu_X(s)\, ds$ gives

$$\mu_Z(t) = \mu_A[t\, U(t) - (t - \Delta)U(t - \Delta)]$$

which grows linearly from zero to $\mu_A \Delta$ on the interval $0 \le t \le \Delta$. Similarly, $\phi_{XZ}(t_1, t_2) = \int_{-\infty}^{t_2} \phi_{XX}(t_1, s_2)\, ds_2$ gives

$$\phi_{XZ}(t_1, t_2) = E(A^2)[U(t_1) - U(t_1 - \Delta)][t_2 U(t_2) - (t_2 - \Delta)U(t_2 - \Delta)]$$

which is $E(A^2) t_1 t_2$ within the square $0 \le t_1 < \Delta$, $0 \le t_2 < \Delta$, is $E(A^2)\Delta$ for $0 \le t_1 < \Delta$ and $t_2 \ge \Delta$ and is zero elsewhere. Finally $\phi_{ZZ}(t_1, t_2) = \int_{-\infty}^{t_1} \phi_{XZ}(s_1, t_2)\, ds_1$ gives

$$\phi_{ZZ}(t_1, t_2) = E(A^2)[t_1 U(t_1) - (t_1 - \Delta)U(t_1 - \Delta)][t_2 U(t_2) - (t_2 - \Delta)U(t_2 - \Delta)]$$

which is $E(A^2) t_1 t_2$ within the square $0 \le t_1 < \Delta$, $0 \le t_2 < \Delta$, is $E(A^2) t_1 \Delta$ for $0 \le t_1 < \Delta$ and $t_2 \ge \Delta$, is $E(A^2) t_2 \Delta$ for $t_1 \ge \Delta$, $0 \le t_2 < \Delta$, is $E(A^2)\Delta^2$ for $t_1 \ge \Delta$, $t_2 \ge \Delta$, and is zero elsewhere.
****

Exercises
****

Time Histories of Stochastic Processes
****

2.1 Let a stochastic process $\{X(t)\}$ be defined by

$X(t) = A \exp(-Bt)$ on $t \ge 0$

where A and B are independent random variables with μ_A and σ_A^2 known for A and the probability density function of B being

$p_B(u) = \lambda \exp(-\lambda u)U(u)$

in which $\lambda > 0$ is a constant.
Find the mean value function $\mu_X(t)$ and the autocorrelation function $\phi_{XX}(t, s)$.
****

2.2 Let a stochastic process $\{X(t)\}$ be defined by

$X(t) = A \exp(-t) + B \exp(-3t)$ on $t \ge 0$

in which A and B are random variables with the joint probability density function

$p_{AB}(u, v) = [U(u) - U(u - 1)][U(v) - U(v - 1)]$

Find the mean value function $\mu_X(t)$ and the autocorrelation function $\phi_{XX}(t, s)$.
****

2.3 Let $\{X(t)\}$ be a stochastic process which depends on three random variables $\{A,B,C\}$:
$$X(t) = A + Bt + Ct^2$$
In order to simplify the description of the random variables let them be written as a vector $\vec{V}$ so that
the mean values are also a vector $\vec{\mu}_V \equiv E(\vec{V})$, and the mean squared values and cross-products can be
arranged in a matrix $E(\vec{V}\vec{V}^T)$. Specifically, let

$$\vec{V} = \begin{Bmatrix} A \\ B \\ C \end{Bmatrix} \quad \vec{\mu}_V \equiv E(\vec{V}) = \begin{Bmatrix} 1 \\ 2 \\ 3 \end{Bmatrix} \quad \text{and} \quad E(\vec{V}\vec{V}^T) = \begin{bmatrix} 4 & -1 & 6 \\ -1 & 9 & 0 \\ 6 & 0 & 19 \end{bmatrix}$$

(a) Find the mean function $\mu_X(t)$.
(b) Find the autocorrelation function $\phi_{XX}(t,s)$.
(c) Find the mean squared value $E[X^2(t)]$.
(d) Find the variance $\sigma_X^2(t)$.
****

2.4 Let the second-moment stationary processes $\{X_1(t)\}$ and $\{X_2(t)\}$ represent the motions at two
different points in a complex structural system. Let the correlation matrix for $\{X_1(t)\}$ and $\{X_2(t)\}$
be given by

$$\mathbf{R}_{\vec{X}\vec{X}}(\tau) \equiv E[\vec{X}(t+\tau)\,\vec{X}^T(t)] = \begin{bmatrix} g(\tau) + g(2\tau) & 2g(\tau) - g(2\tau) \\ 2g(\tau) - g(2\tau) & 4g(\tau) + g(2\tau) \end{bmatrix}$$

in which $g(\tau) = e^{-b|\tau|}[\cos(\omega_0\tau) + (b/\omega_0)\sin(\omega_0|\tau|)]$ for constants b and ω_0. Let $\{Z(t)\}$ denote
the relative motion between the two points: $Z(t) = X_2(t) - X_1(t)$
(a) Find the cross-correlation function $R_{X_1Z}(\tau) \equiv E[X_1(t+\tau)Z(t)]$.
(b) Find the autocorrelation function value of Z: $R_{ZZ}(\tau) \equiv E[Z(t+\tau)Z(t)]$.
****

Stationarity
****

2.5 For each of the following $\phi_{XX}(t,s)$ autocorrelation functions, tell whether the stochastic process
$\{X(t)\}$ is second-moment stationary.

(a) $\phi_{XX}(t,s) = \dfrac{1}{1+(t-s)^2}$

(b) $\phi_{XX}(t,s) = \dfrac{1}{1+(t-s)^2} U(1-|t|)U(1-|s|)$

(c) $\phi_{XX}(t,s) = \dfrac{1}{1+t^2+s^2}$

(d) $\phi_{XX}(t,s) = e^{-t^2-s^2}$

(e) $\phi_{XX}(t,s) = \cos(t)\cos(s) + \sin(t)\sin(s)$

(f) $\phi_{XX}(t,s) = \cos(t-s)/[1+(t-s)^2]$

(g) $\phi_{XX}(t,s) = (t-1)(s-1)$

(h) $\phi_{XX}(t,s) = (1-t^2)(1-s^2)[U(t+1)-U(t-1)][U(s+1)-U(s-1)]$.
****

2.6 Consider a stochastic process $\{X(t)\}$ with autocorrelation function

$$\phi_{XX}(t,s) = \dfrac{ts}{1+ts(1+s^2-2ts+t^2)} U(t)U(s)$$

(a) Find $E[X^2(t)]$ for $t \geq 0$ and sketch it versus t.

(b) Use the limit as $t \to \infty$ of $\phi_{XX}(t+\tau, t)$ to show that $\{X(t)\}$ tends to become second-moment stationary in that limiting situation.

****

Properties of Mean and Covariance Functions

****

2.7 It is asserted that none of the following $\phi(t,s)$ functions could be the autocorrelation function for any stochastic process. That is, there does not exist any stochastic process $\{X(t)\}$ such that $E[X(t)X(s)] = \phi(t,s)$. For each of the ϕ functions, give a reason why the assertion must be true.

(a) $\phi(t,s) = \dfrac{1}{(s^2+1)(t^2+2)}$

(b) $\phi(t,s) = \dfrac{(s-t)^2}{(s-t)^2+1}$

(c) $\phi(t,s) = \exp[-(s-t)^2]$ for $s \leq t$

$\qquad\quad = \exp[-2(s-t)^2]$ for $s > t$

(d) $\phi(t,s) = \exp[-(s-t)^2]$ for $s+t > 0$

$\qquad\quad = \exp[-2(s-t)^2]$ for $s+t \leq 0$

(e) $\phi(t,s) = (s-t)^2 \exp[-(s-t)^2]$

****

2.8 It is asserted that none of the following $\phi(t,s)$ functions could be the autocorrelation function for any stochastic process. That is, there does not exist any stochastic process $\{X(t)\}$ such that $E[X(t)X(s)] = \phi(t,s)$. For each of the ϕ functions, give a reason why the assertion must be true.

(a) $\phi(t,s) = \dfrac{s}{(s^2+1)(t^2+1)}$

(b) $\phi(t,s) = 1 - \exp\left[-(s-t)^2\right]$

(c) $\phi(t,s) = 1$ for $t^2 + s^2 \leq 1$

$\qquad\quad = e^{-(s-t)^2}$ for $t^2 + s^2 > 1$

(d) $\phi(t,s) = \dfrac{1}{1+t^2 s^4}$

(e) $\phi(t,s) = (s-t)^2 \cos(s-t)$

****

2.9 It is asserted that none of the following $R(\tau)$ functions could be the autocorrelation function for any stochastic process. That is, there does not exist any $\{X(t)\}$ such that $E[X(t)X(s)] = R(t-s)$. For each of the R functions, give a reason why the assertion must be true:

(a) $R(\tau) = \exp(-\tau^2)U(\tau)$

(b) $R(\tau) = \dfrac{\tau^2}{1+\tau^4}$

(c) $R(\tau) = \exp(-2\tau^2)U(1-\tau^2)$

(d) $R(\tau) = \exp(-2\tau^2)\sin(5\tau)$

****

2.10 Assume that for a stationary process $\{X(t)\}$ you know a conditional probability density function of the form $p_{X(t+\tau)}[v|X(t)=u]$. Further, assume that $\lim_{\tau\to\infty} p_{X(t+\tau)}[v|X(t)=u] = p_{X(t)}(v)$ for any u and t values.

Give an integral expression for the autocorrelation function $R_{XX}(\tau)$ in terms of the conditional probability density function.

[It will be shown in Chapter 8 that a conditional probability density function of this type can sometimes be derived from a Fokker-Planck equation.]

****

Continuity

****

2.11 For each of the following $\phi_{XX}(t,s)$ autocorrelation functions, identify any t values at which $\{X(t)\}$ is not mean-square continuous.

(a) $\phi_{XX}(t,s) = \dfrac{1}{1+(t-s)^2}U(1-|t|)U(1-|s|)$

(b) $\phi_{XX}(t,s) = \dfrac{1}{1+(t^2+s^2)}$

(c) $\phi_{XX}(t,s) = (1-t^2)(1-s^2)U(1-|t|)U(1-|s|)$

(d) $\phi_{XX}(t,s) = \cos(t)\cos(s) + \sin(t)\sin(s)$

(e) $\phi_{XX}(t,s) = \cos(t)\cos(s)U\left(\dfrac{\pi}{2}-|t|\right)U\left(\dfrac{\pi}{2}-|s|\right)$

(f) $\phi_{XX}(t,s) = \cos(t)\cos(s)U(\pi-|t|)U(\pi-|s|)$

****

Derivative

****

2.12 For each of the following $\phi_{XX}(t,s)$ autocorrelation functions, determine whether there exist $\phi_{X\dot{X}}(t,s)$ and $\phi_{\dot{X}\dot{X}}(t,s)$ functions which are finite for all (t,s) values. Identify the (t,s) values for which a finite value of either of the ϕ functions does not exist. Also give the value of $E[\dot{X}^2(t)]$ for all t values for which it is finite. In each case, a and b are real constants and $a > 0$.

(a) $\phi_{XX}(t,s) = \exp(-|t-s|)$

(b) $\phi_{XX}(t,s) = \exp[-(s-t)^2]U(ts)$

(c) $\phi_{XX}(t,s) = (1-t^2)(1-s^2)U(1-t^2)U(1-s^2)$

(d) $\phi_{XX}(t,s) = e^{-(s-t)^2}U(t)U(s) + \dfrac{[1-U(t)][1-U(s)]}{1+(s-t)^2}$

(e) $\phi_{XX}(t,s) = ts$ for $0 \le t \le 1,\ 0 \le s \le 1$

            $= e^{-|t-s|}$ for $t > 1,\ s > 1$

            $= 0$ otherwise

****

2.13 For the derivative of the $\{X(t)\}$ stochastic process of Exercise 2. 6, find the limit of the mean squared value: $\lim_{t\to\infty} E[\dot{X}^2(t)]$.

****

2.14 For each of the following $R_{XX}(\tau)$ stationary autocorrelation functions, determine whether there exist $R_{X\dot{X}}(\tau)$, $R_{\dot{X}\dot{X}}(\tau)$, $R_{\dot{X}\ddot{X}}(\tau)$, and $R_{\ddot{X}\ddot{X}}(\tau)$ functions which are finite for all τ values. Identify the τ values for which a finite value of any of the R functions does not exist. Also give the values of $E[\dot{X}^2(t)]$ and $E[\ddot{X}^2(t)]$ if they are finite. In each case, a and b are real constants and $a > 0$.

(a) $R_{XX}(\tau) = \exp(-a\tau^2)$

(b) $R_{XX}(\tau) = \exp(-a|\tau|)\cos(b\tau)$

(c) $R_{XX}(\tau) = (1 + a|\tau|)\exp(-a|\tau|)$

****

2.15 For each of the following $R_{XX}(\tau)$ stationary autocorrelation functions, determine whether there exist $R_{X\dot{X}}(\tau)$, $R_{\dot{X}\dot{X}}(\tau)$, $R_{\dot{X}\ddot{X}}(\tau)$, and $R_{\ddot{X}\ddot{X}}(\tau)$ functions which are finite for all τ values. Identify the τ values for which a finite value of any of the R functions does not exist. Also give the values of $E[\dot{X}^2(t)]$ and $E[\ddot{X}^2(t)]$ if they are finite. In each case, a and b are real constants and $a > 0$.

(a) $R_{XX}(\tau) = \dfrac{b^2}{b^2 + \tau^2}$

(b) $R_{XX}(\tau) = \left(1 - \dfrac{1}{2}|\tau|\right)\exp(-a|\tau|)$

(c) $R_{XX}(\tau) = \dfrac{\sin(a\tau)}{\tau}$

****

2.16 Consider a second-order stochastic process $\{X(t)\}$ with $\mu_X(t) \equiv 0$, and autocorrelation function $\phi_{XX}(t,s) = (1-t^2)(1-s^2)U(1-t^2)U(1-s^2)$

(a) Find $\phi_{X\dot{X}}(t,s)$ for all (t,s) values for which it exists and identify any (t,s) values for which it does not exist.

(b) Find $\phi_{\dot{X}\dot{X}}(t,s)$ for all (t,s) values for which it exists and identify any (t,s) values for which it does not exist.

****

2.17 For the stochastic process $\{X(t)\}$ defined in Exercise 2.1:

(a) Find the mean value function $\mu_{\dot{X}}(t)$ of the derivative process $\{\dot{X}(t)\}$.

(b) Find the cross-correlation function $\phi_{X\dot{X}}(t,s)$ between $\{X(t)\}$ and the derivative process.

(c) Find the autocorrelation function $\phi_{\dot{X}\dot{X}}(t,s)$ of the derivative process.

(d) Confirm that $\mu_{\dot{X}}(t)$, $\phi_{X\dot{X}}(t,s)$, and $\phi_{\dot{X}\dot{X}}(t,s)$ are the same whether determined from derivatives of the moment functions for $\{X(t)\}$ or by analyzing the time histories of the derivative process $\{\dot{X}(t)\}$.

****

2.18 For the stochastic process $\{X(t)\}$ defined in Exercise 2.2:

(a) Find the mean value function $\mu_{\dot{X}}(t)$ of the derivative process $\{\dot{X}(t)\}$.

(b) Find the cross-correlation function $\phi_{X\dot{X}}(t,s)$ between $\{X(t)\}$ and the derivative process.

(c) Find the autocorrelation function $\phi_{\dot{X}\dot{X}}(t,s)$ of the derivative process.

(d) Confirm that $\mu_{\dot{X}}(t)$, $\phi_{X\dot{X}}(t,s)$, and $\phi_{\dot{X}\dot{X}}(t,s)$ are the same whether determined from derivatives of the moment functions for $\{X(t)\}$ or by analyzing the time histories of the derivative process $\{\dot{X}(t)\}$.

****

2.19 For the stochastic process $\{X(t)\}$ defined in Exercise 2.3:

(a) Find the mean value function $\mu_{\dot{X}}(t)$ of the derivative process $\{\dot{X}(t)\}$.
(b) Find the cross-correlation function $\phi_{X\dot{X}}(t,s)$ between $\{X(t)\}$ and the derivative process.
(c) Find the autocorrelation function $\phi_{\dot{X}\dot{X}}(t,s)$ of the derivative process.
(d) Confirm that $\mu_{\dot{X}}(t)$, $\phi_{X\dot{X}}(t,s)$, and $\phi_{\dot{X}\dot{X}}(t,s)$ are the same whether determined from derivatives of the moment functions for $\{X(t)\}$ or by analyzing the time histories of the derivative process $\{\dot{X}(t)\}$.

****

2.20 Consider a stochastic process $\{Y(t)\}$ used to model an earthquake ground acceleration. This $\{Y(t)\}$ is formed by multiplying a stationary process $\{X(t)\}$ by a specified time function:

$$Y(t) = X(t)[at\exp(-bt)]U(t)$$

in which a and b are positive constants. Presume that the standard deviations σ_X and $\sigma_{\dot{X}}$ are known positive constants and that $\mu_X = 0$.
(a) Find $E[Y^2(t)]$.
(b) Find $E[\dot{Y}^2(t)]$.
(c) Find $E[Y(t)\dot{Y}(t)]$.

****

2.21 Consider a stochastic process $\{X(t)\}$ with mean $\mu_X(t)$ and variance $\sigma_X^2(t)$ at any time t. Let the conditional distribution of the derivative of the process be given by

$$p_{\dot{X}(t)}[v\,|\,X(t) = u] = \delta(v - bu)$$

in which b is some positive constant and $\delta(\cdot)$ is the Dirac delta function.
(a) Find $E[\dot{X}(t)|X(t) = u]$, $E[X(t)\dot{X}(t)|X(t) = u]$, and $E[\dot{X}^2(t)|X(t) = u]$.
(b) Find $E[\dot{X}(t)]$, $E[X(t)\dot{X}(t)]$, and $E[\dot{X}^2(t)]$ as functions of $\mu_X(t)$ and $\sigma_X^2(t)$.
(c) Is it possible for $\{X(t)\}$ to be mean-value stationary? That is, is the given information consistent with $\mu_X(t)$ being constant and $\mu_{\dot{X}}(t)$ being zero?
(d) Show that it is not possible for $\{X(t)\}$ to be second-moment stationary, by showing that the information found in part (c) is inconsistent with $E[X^2(t)]$ and $E[\dot{X}^2(t)]$ being constant.

****

2.22 Let $\{X(t)\}$ be a stationary stochastic process with known values of $E[X^2(t)]$, $E[\dot{X}^2(t)]$, and $E[\ddot{X}^2(t)]$. Find $E[\ddot{X}(t)\dot{X}(t)]$ and $E[\dot{X}(t)\dddot{X}(t)]$ and $E[X(t)\dddot{X}(t)]$ in terms of the known quantities. [Each dot denotes a derivative with respect to t.]

****

2.23 Let $\{X(t)\}$ be a stationary stochastic process with known values of $E[X^2(t)]$, $E[\dot{X}^2(t)]$, $E[\ddot{X}^2(t)]$ and $E[\dddot{X}^2(t)]$. Find $E[\dddot{X}(t)\ddot{X}(t)]$, $E[\ddot{X}(t)\dddot{X}(t)]$, $E[\dot{X}(t)\dddot{X}(t)]$, and $E[X(t)\dddot{X}(t)]$ in terms of the known quantities. [Each dot denotes a derivative with respect to t.]

****

Integral
****

2.24 Find the $\phi_{XZ}(t_1,t_2)$ and $\phi_{ZZ}(t_1,t_2)$ functions for the stochastic process $\{Z(t): t \geq 0\}$ defined as the stochastic integral $Z(t) = \int_0^t X(s)\,ds$, in which $\{X(t)\}$ has the autocorrelation function $\phi_{XX}(t_1,t_2) = b\exp(-ct_1 - ct_2)$, with b and c being positive constants.

****

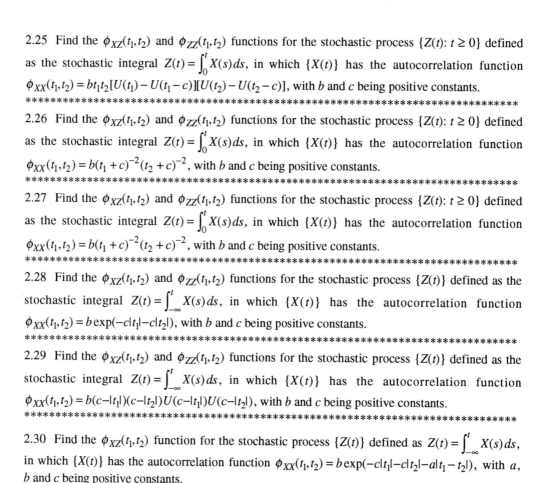

2.25 Find the $\phi_{XZ}(t_1,t_2)$ and $\phi_{ZZ}(t_1,t_2)$ functions for the stochastic process $\{Z(t): t \geq 0\}$ defined as the stochastic integral $Z(t) = \int_0^t X(s)\,ds$, in which $\{X(t)\}$ has the autocorrelation function $\phi_{XX}(t_1,t_2) = bt_1t_2[U(t_1) - U(t_1 - c)][U(t_2) - U(t_2 - c)]$, with b and c being positive constants.

****

2.26 Find the $\phi_{XZ}(t_1,t_2)$ and $\phi_{ZZ}(t_1,t_2)$ functions for the stochastic process $\{Z(t): t \geq 0\}$ defined as the stochastic integral $Z(t) = \int_0^t X(s)\,ds$, in which $\{X(t)\}$ has the autocorrelation function $\phi_{XX}(t_1,t_2) = b(t_1 + c)^{-2}(t_2 + c)^{-2}$, with b and c being positive constants.

****

2.27 Find the $\phi_{XZ}(t_1,t_2)$ and $\phi_{ZZ}(t_1,t_2)$ functions for the stochastic process $\{Z(t): t \geq 0\}$ defined as the stochastic integral $Z(t) = \int_0^t X(s)\,ds$, in which $\{X(t)\}$ has the autocorrelation function $\phi_{XX}(t_1,t_2) = b(t_1 + c)^{-2}(t_2 + c)^{-2}$, with b and c being positive constants.

****

2.28 Find the $\phi_{XZ}(t_1,t_2)$ and $\phi_{ZZ}(t_1,t_2)$ functions for the stochastic process $\{Z(t)\}$ defined as the stochastic integral $Z(t) = \int_{-\infty}^t X(s)\,ds$, in which $\{X(t)\}$ has the autocorrelation function $\phi_{XX}(t_1,t_2) = b\exp(-c|t_1| - c|t_2|)$, with b and c being positive constants.

****

2.29 Find the $\phi_{XZ}(t_1,t_2)$ and $\phi_{ZZ}(t_1,t_2)$ functions for the stochastic process $\{Z(t)\}$ defined as the stochastic integral $Z(t) = \int_{-\infty}^t X(s)\,ds$, in which $\{X(t)\}$ has the autocorrelation function $\phi_{XX}(t_1,t_2) = b(c-|t_1|)(c-|t_2|)U(c-|t_1|)U(c-|t_2|)$, with b and c being positive constants.

****

2.30 Find the $\phi_{XZ}(t_1,t_2)$ function for the stochastic process $\{Z(t)\}$ defined as $Z(t) = \int_{-\infty}^t X(s)\,ds$, in which $\{X(t)\}$ has the autocorrelation function $\phi_{XX}(t_1,t_2) = b\exp(-c|t_1| - c|t_2| - a|t_1 - t_2|)$, with a, b and c being positive constants.

****

Chapter 3

Time Domain Linear Vibration Analysis

3.1 Deterministic Dynamics

Before beginning our study of stochastic dynamics, we will review some of the fundamental ideas of deterministic linear time domain dynamic analysis. For the moment, we will consider our linear time-invariant system to have one scalar excitation $f(t)$ and one scalar response $x(t)$. In Chapter 7, we will use the idea of superposition to extend this analysis to any number of excitation and response components. Treating the excitation as the input to our linear system, and the response as its output, gives a situation which can be represented by the schematic diagrams in Fig. 3.1.

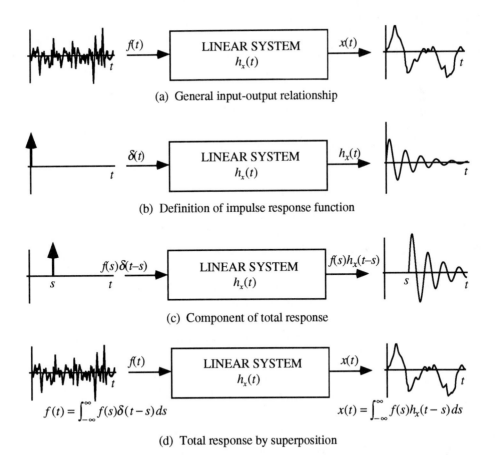

(a) General input-output relationship

(b) Definition of impulse response function

(c) Component of total response

(d) Total response by superposition

Figure 3.1 Schematic of general linear system

The function $h_x(t)$ in Fig. 3.1 is defined as the response of the system to one particular excitation. In particular, if $f(t) = \delta(t)$ then $x(t) = h_x(t)$, in which $\delta(\cdot)$ is the Dirac delta function.[1] We will investigate how to find the $h_x(t)$ function in the following section, but first we want to emphasize that this single function is adequate to characterize the response of the system to any excitation $f(t)$. The demonstration of this fact is illustrated by the steps shown in Fig. 3.1. We derive the result by superposition, first writing $f(t)$ as a superposition of Dirac delta functions. In particular, we write $f(t)$ as the convolution integral (see Eq. C.2 of Appendix C):

$$f(t) = \int_{-\infty}^{\infty} f(s)\delta(t-s)\,ds \equiv \int_{-\infty}^{\infty} f(t-r)\delta(r)\,dr \tag{3.1}$$

in which the first integral is illustrated in Fig. 3.1(c), and the final form is obtained by the change of variables $r = t - s$. (Both of the two equivalent forms are given since either may be more convenient for use in a particular situation.) Now we note that our definition of the $h_x(t)$ function assures us that the excitation component $\delta(t-s)$ will induce a response $h_x(t-s)$ or, equivalently, that $\delta(r)$ will induce $h_x(r)$. Multiplying these delta function pulses of excitation by their amplitudes from Eq. 3.1 and superimposing the responses gives

$$x(t) = \int_{-\infty}^{\infty} f(s)h_x(t-s)\,ds \equiv \int_{-\infty}^{\infty} f(t-r)h_x(r)\,dr \tag{3.2}$$

Either of the integrals in Eq. 3.2 is called the **Duhamel convolution integral** for the linear system, and the function $h_x(t)$ is called the **impulse response function** for response x of the system.

It should be noted that the definition of $h_x(t)$ as the $x(t)$ response to $f(t) = \delta(t)$ means that $h_x(t)$ is the response when there has never been any other excitation of the system except the pulse at time $t = 0$. In particular, $h_x(t)$ is the response when the initial conditions on the system are such that $x(t)$ and all of its derivatives are zero at time $t \to -\infty$. A system is said to be causal if $h_x(t) \equiv 0$ for $t < 0$. This simply means that the response to an excitation $f(t) = \delta(t)$ does not begin to appear until time $t = 0$, which we would certainly expect to be true for a physical system. We do not expect the response to precede the excitation, but one should keep in mind that this is a condition of physics and cannot be proved mathematically. For a causal system note that the limits of integration in Eq. 3.2 can be modified as

$$x(t) = \int_{-\infty}^{t} f(s)h_x(t-s)\,ds \equiv \int_{0}^{\infty} f(t-r)h_x(r)\,dr \tag{3.3}$$

The term $h_x(t-s)$ in the first integral of Eq. 3.3 gives the response at time t due to an excitation pulse at time s for $s \leq t$. Similarly, $h_x(r)$ in the second integral gives the response at time t to an excitation pulse at time $(t-r)$ for $r \geq 0$. For reasons of convenience we will use the $-\infty$ to ∞

[1]   See Appendix C for more information on the Dirac delta function.

limits of Eq. 3.2 in general, with the causal nature of $h_x(r)$ being taken into account as needed in specific examples.

It is also instructive to consider the steady state response to a static excitation $f(t)$, which has a constant value f_0 for all t. Equation 3.2 gives this steady state response as $x(t) = f_0 h_{x,static}$, with

$$h_{x,static} \equiv \int_{-\infty}^{\infty} h_x(r)\,dr \qquad (3.4)$$

Thus, the system has a static steady state x response to this static excitation if and only if $h_{x,static}$ is finite. Of course, the condition for $h_x(r)$ to be integrable so that $|h_{x,static}| < \infty$ is that $h_x(r)$ tends to zero faster than r^{-1} as r goes to infinity. We will find it convenient in the following discussions to classify any linear system as having a bounded or infinite static response depending on whether $|h_{x,static}| < \infty$ or $|h_{x,static}| = \infty$.

Finally we note that it is sometimes more convenient to consider the response of the linear system to be a superposition of the effects of some initial conditions at a fixed time t_0 and the effects of the part of the excitation occurring after time t_0. That is, if a sufficient set of initial conditions is known at time t_0, then we do not need to know the excitation prior to time t_0 in order to find the dynamic response after time t_0. The response of the system caused by excitation after time t_0 is found by limiting the range of integration in the convolution integrals of Eqs. 3.2 or 3.3 to include only $s \geq 0$ or $r \leq t - t_0$, and the additional term giving the response due to the initial conditions is found by considering the system with $f(t) = 0$.

3.2   Evaluation of Impulse Response Functions

We will be primarily concerned with analysis of the dynamics of systems governed by differential equations. A general form of such an nth-order differential equation can be written as

$$\sum_{j=0}^{n} a_j \frac{d^j x(t)}{dt^j} = f(t) \qquad (3.5)$$

By its definition, the impulse response function for such a system must satisfy the corresponding differential equation

$$\sum_{j=0}^{n} a_j \frac{d^j h_x(t)}{dt^j} = \delta(t) \qquad (3.6)$$

subject to the boundary condition that $h_x(t)$ and all of its derivatives tend to zero as $t \to -\infty$. An obvious difficulty with Eq. 3.6 is the fact that the right-hand side is not truly a function. The nature of the Dirac delta function, though, turns out to be a definite advantage. In particular, we know that $h_x(t)$ is identically zero for $t < 0$, so that we only need to be concerned with values of t which are greater than or equal to zero. Clearly, we must be very careful at $t = 0$, but for $t > 0$ the differential equation is homogeneous:

$$\sum_{j=0}^{n} a_j \frac{d^j h_x(t)}{dt^j} = 0 \quad \text{for} \quad t > 0 \tag{3.7}$$

In order to have a unique solution to this nth-order differential equation, we must have n initial conditions or boundary conditions. We will use initial values of $h_x(t)$ and its first $(n - 1)$ derivatives immediately after time zero. In particular, we will write $h_x(0^+)$ for the limit of $h_x(t)$ as t tends to zero from the positive side, so that the initial information we need can be described as the values of $h_x(t)$ and its first $(j - 1)$ derivatives at time $t = 0^+$. We must derive these initial conditions from the behavior of Eq. 3.6 in the neighborhood of $t = 0$.

Since $\delta(t)$ is infinite for $t = 0$, Eq. 3.6 tells us that at least one of the terms on the left-hand side of the equation must also be infinite at $t = 0$. Let us presume for the moment that it is the jth-order term which is infinite. In particular, we will assume that

$$\frac{d^j h_x(t)}{dt^j} = b\delta(t)$$

for some b value. This causes difficulties, though, unless $j = n$. In particular, we now have

$$\frac{d^{j+1} h_x(t)}{dt^{j+1}} = b\frac{d\delta(t)}{dt} \equiv b\delta'(t)$$

As explained in Appendix C, the precise definition of $\delta(t)$ is in terms of the limit of a sequence involving bounded functions. For example, if we consider $\delta(t)$ to be like the limit as $\Delta \to 0$ of $[(1-|t/\Delta|)/\Delta]U(\Delta-|t|)$ as shown in Fig. 3.2, then we have $\delta'(t)$ as a limit of $[-\text{sgn}(t)/\Delta^2]U(\Delta-|t|)$ so that $\delta'(t)/\delta(t) \to \infty$ as $\Delta \to 0$. This same result holds true for any sequence which we might consider as tending to $\delta'(t)$. Thus, we must consider the magnitude of $\delta'(t)$ to be infinitely larger than that of $\delta(t)$, so that we cannot satisfy Eq. 3.6 in the neighborhood of the origin if the jth derivative of $h_x(t)$ is like $\delta(t)$ and the $(j - 1)$st derivative also appears in the

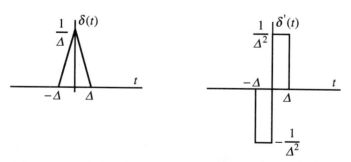

Figure 3.2 Possible approximations of the Dirac delta function and its derivative

equation. On the other hand, if we say that the nth derivative of $h_x(t)$ in the neighborhood of the origin is $b\delta(t)$, then all the other terms on the left-hand side of Eq. 3.6 are finite so that the equation can be considered to be satisfied at the origin if $b = a_n^{-1}$. Using

$$\frac{d^n h_x(t)}{dt^n} = a_n^{-1}\delta(t) \quad \text{for very small } |t| \tag{3.8}$$

now gives us the initial conditions for Eq. 3.7 at time $t = 0^+$. In particular, the integral of Eq. 3.8 gives

$$\frac{d^{n-1}h_x(t)}{dt^{n-1}} = a_n^{-1}U(t) \quad \text{for very small } |t|$$

in which $U(t)$ is the unit step function (see Eqs. A.2 and A.9 of Appendix A). Thus,

$$\left[\frac{d^{n-1}h_x(t)}{dt^{n-1}}\right]_{t=0^+} = a_n^{-1} \tag{3.9}$$

Further integrations give continuous functions so that

$$\left[\frac{d^j h_x(t)}{dt^j}\right]_{t=0^+} = 0 \quad \text{for} \quad j \le n - 2 \tag{3.10}$$

This now provides adequate conditions to assure that we can find $h_x(t)$ as a unique solution of Eq. 3.7.

Since determination of the impulse response function involves finding the solution to an initial condition problem, it is closely related to the general problem mentioned at the end of Sec. 3.1, regarding finding the response of a system with given initial conditions. In general we expect initial conditions to be given as values of $x(t_0)$ and the derivatives $x^{(j)}(t_0)$ for $j = 1, \cdots, n - 1$. We have already found that $h_x(t)$ is exactly the solution to the problem with an initial condition given in Eq. 3.9. Thus, the response resulting from having the $(n - 1)$st-order derivative be $x^{(n-1)}(t_0)$ at time t_0, instead of a_n^{-1} at time 0^+, is simply $[x^{(n-1)}(0^+)/a_n^{-1}]\ h_x(t - t_0)$. The responses to other initial conditions at time t_0 can also be found by first placing the initial value at time 0^+, then shifting the time axis of the response by the amount t_0.

We will now derive the impulse response functions for some example problems, each of which is described by a differential equation. We will also list one physical system which is described by each of the differential equations. It should be kept in mind, though, that the impulse response function characterizes the dynamic behavior of any system governed by the given differential equation, and not only the particular physical example that is given. Since the class of systems governed by any given differential equation is much broader than any particular example that we might give, it is useful to consider the examples to be defined by the differential equations, not by the physical examples cited.

****

Example 3.1: Consider a linear system governed by the first-order differential equation

$$c\dot{x}(t) + kx(t) = f(t)$$

The sketch shows one physical system which is governed by
this differential equation, with k being a spring stiffness, c a
dashpot value, and $f(t)$ an applied force.

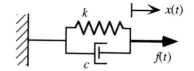

Find the impulse function $h_x(t)$ such that Eq. 3.1 describes the solution of the problem.

Rewriting Eqs. 3.7 and 3.9 gives

$$c\frac{dh_x(t)}{dt} + kh_x(t) = 0 \quad \text{for} \quad t > 0 \quad \text{with} \quad h_x(0^+) = c^{-1}$$

Taking the usual approach of trying an exponential form for the solution of a linear differential equation, we
try $h_x(t) = ae^{\lambda t}$. This has the proper initial condition if $a = c^{-1}$, and it satisfies the equation if $c\lambda + k = 0$, or
$\lambda = -k/c$, so that the answer is

$$h_x(t) = c^{-1}e^{-kt/c}U(t)$$

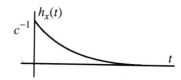

Taking the derivative of this function gives

$$h_x'(t) = -kc^{-2}e^{-kt/c}U(t) - c^{-1}e^{-kt/c}\delta(t) = -kc^{-2}e^{-kt/c}U(t) + c^{-1}\delta(t)$$

in which the last term has been simplified by taking advantage of the fact that $e^{-kt/c} = 1$ at the one t
value for which $\delta(t)$ is not zero. Substitution of $h_x(t)$ and $h_x'(t)$ confirms that we have, indeed, found
the solution of the differential equation $ch_x'(t) + kh_x(t) = \delta(t)$.

Note also that

$$h_{x,static} \equiv \int_{-\infty}^{\infty} h_x(r)\,dr = k^{-1} < \infty$$

so that the $x(t)$ response for this system has a bounded static value. For the physical model shown, it is
fairly obvious that the static response to force f_0 should be f_0/k, so that it should be no surprise that
$h_{x,static} = k^{-1}$

****

Example 3.2: Consider a linear system governed by the differential equation

$$m\ddot{x}(t) + c\dot{x}(t) = f(t)$$

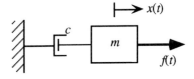

The sketch shows one physical system which is governed by this differential equation, with c being a dashpot value and $f(t)$ the force applied to the mass.

Find the impulse function $h_x(t)$ such that Eq. 3.1 describes the solution of the problem.

This time we obtain

$$m\frac{d^2h_x(t)}{dt^2} + c\frac{dh_x(t)}{dt} = 0 \quad \text{for} \quad t > 0 \quad \text{with} \quad h'_x(0^+) = m^{-1} \quad \text{and} \quad h_x(0^+) = 0$$

Proceeding as before we find that the homogeneous equation is satisfied by $ae^{\lambda t}$ with $m\lambda^2 + c\lambda = 0$. Thus, $\lambda = 0$ and $\lambda = -c/m$ are both possible and we must consider the general solution to be the linear combination of these two terms: $h_x(t) = a_1 + a_2 e^{-ct/m}$. Meeting the initial conditions at $t = 0^+$ then gives $-a_2 c/m = m^{-1}$ and $a_1 + a_2 = 0$, so that $a_1 = -a_2 = c^{-1}$ and the impulse response function is

$$h_x(t) = c^{-1}[1 - e^{-ct/m}]U(t)$$

Note that $h_{x,static} = \infty$ for this system, so that the static response of $x(t)$ is infinite, which agrees with what one would anticipate for the physical model shown.

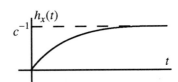

One may also note a strong similarity between Examples 3.1 and 3.2. In particular, if we rewrite Example 3.2 in terms of $y(t) = \dot{x}(t)$, then the equation is

$$m\dot{y}(t) + cy(t) = f(t)$$

which is identical to the equation for $x(t)$ in Example 3.1 if c is replaced by m and k is replaced by c, showing that $h_y(t) = m^{-1}e^{-ct/m}U(t)$. However, the impulse response for $y(t)$ should be exactly the derivative of that for $x(t)$, so that the solution to Example 3.1 gives us

$$h'_x(t) = h_y(t) = m^{-1}e^{-ct/m}U(t)$$

which does agree with the $h_x(t)$ that we derived for this problem. Also, this shows that the static response of $\dot{x}(t)$ for this system is finite, even though that of $x(t)$ is infinite.
****

Example 3.3:   Consider a linear system governed by the
differential equation

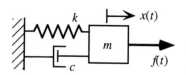

$$m\ddot{x}(t) + c\dot{x}(t) + kx(t) = f(t)$$

The sketch shows one physical system which is governed by this differential equation, with m being a mass, k being a spring stiffness and c being a dashpot value. This system is called the single-degree-of-freedom system, and will play a key role in our study of stochastic dynamics, but at this point it is simply one more example of a relatively simple dynamical system. Find the impulse function $h_x(t)$ such that Eq. 3.1 describes the solution of the problem.

Provided that $m \neq 0$ we can divide the equation by m to obtain another common form of the governing equation as

$$\ddot{x} + 2\zeta\omega_0\dot{x} + \omega_0^2 x = f(t)/m = -\ddot{z}(t)$$

in which $\omega_0 \equiv \sqrt{k/m}$ and $\zeta \equiv c/(2\sqrt{km})$ are called the undamped natural circular frequency and the fraction of critical damping, respectively, of the system. In the same way as we analyzed Examples 3.1 and 3.2, we find that the response $h_x(t)$ to $f(t) = \delta(t)$ must satisfy

$$\frac{d^2 h_x(t)}{dt^2} + 2\zeta\omega_0 \frac{dh_x(t)}{dt} + \omega_0^2 h_x(t) = 0 \quad \text{for} \quad t > 0$$

with $h_x'(0^+) = m^{-1}$ and $h_x(0^+) = 0$. We will present results only for the situation with $|\zeta| < 1$, since that is the situation which is usually of most practical interest.[2] For $|\zeta| < 1$ the general solution to the homogenous equation can be written as $h(t) = e^{-\zeta\omega_0 t}[A\cos(\omega_d t) + B\sin(\omega_d t)]$, in which $\omega_d = \omega_0(1 - \zeta^2)^{1/2}$ is called the damped natural circular frequency. Applying the initial conditions at time $t = 0^+$ then gives

$$h_x(t) = \frac{e^{-\zeta\omega_0 t}}{m\omega_d}\sin(\omega_d t)U(t)$$

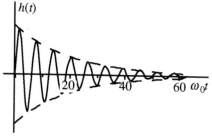

The shape of this function is shown in the sketch for the special case of $\zeta = 0.05$. The exponential envelope shown dashed in the figure simply omits the sinusoidal term of $h_x(t)$.

Based on the physical model shown, we can predict that the static response of this equation will be bounded with $h_{x,static} = k^{-1}$, as in Example 3.1, and integration of $h_x(t)$ confirms that this is true.
****

---

2   The reader is asked to investigate a case with $|\zeta| > 1$ in Exercise 3.1.

3.3 Stochastic Dynamics

The simplest formulation of stochastic linear dynamics is simply to replace the deterministic functions $f(t)$ and $x(t)$ in Eq. 3.2 by stochastic processes $\{F(t)\}$ and $\{X(t)\}$, giving

$$X(t) = \int_{-\infty}^{\infty} F(s) h_x(t-s) \, ds \equiv \int_{-\infty}^{\infty} F(t-r) h_x(r) \, dr \tag{3.11}$$

which can be considered as a special case of the stochastic convolution integral defined in Eq. 2.74. Note, in particular, that Eq. 3.11 must be true for every time history of excitation and response. We can now learn much about the characteristics of the $\{X(t)\}$ process by studying expectations obtained from Eq. 3.11. The first such result is that

$$\mu_X(t) \equiv E[X(t)] = \int_{-\infty}^{\infty} \mu_F(s) h_x(t-s) \, ds \equiv \int_{-\infty}^{\infty} \mu_F(t-r) h_x(r) \, dr \tag{3.12}$$

which follows directly from reversing the order of integration and expectation, as in Chapter 2. Note that Eq. 3.12 is of exactly the same form as Eq. 3.2, which describes deterministic dynamics. This means that finding the mean value of the stochastic response is always just the same as solving a deterministic problem with excitation $\mu_F(t)$.

In the same way we can write the autocorrelation function for the response as

$$\phi_{XX}(t_1, t_2) \equiv E[X(t_1)X(t_2)] = E\left[\int_{-\infty}^{\infty} \int_{-\infty}^{\infty} F(s_1) h_x(t_1-s_1) F(s_2) h_x(t_2-s_2) \, ds_1 \, ds_2 \right]$$

and taking the expectation inside the integrals gives

$$\phi_{XX}(t_1, t_2) = \int_{-\infty}^{\infty} \int_{-\infty}^{\infty} \phi_{FF}(s_1, s_2) h_x(t_1-s_1) h_x(t_2-s_2) \, ds_1 \, ds_2 \tag{3.13}$$

or

$$\phi_{XX}(t_1, t_2) = \int_{-\infty}^{\infty} \int_{-\infty}^{\infty} \phi_{FF}(t_1-r_1, t_2-r_2) h_x(r_1) h_x(r_2) \, dr_1 \, dr_2 \tag{3.14}$$

Clearly, this idea can be expanded to any moment function. The general jth moment function of the response can be written as

$$E[X(t_1)X(t_2)\cdots X(t_j)] = \int_{-\infty}^{\infty} \cdots \int_{-\infty}^{\infty} E[F(s_1)\cdots F(s_j)] h_x(t_1-s_1)\cdots h_x(t_j-s_j) \, ds_1 \cdots ds_j \tag{3.15}$$

Again, all of these results follow directly from using the Duhamel convolution integral and reversing the order of integration and expectation.

Similarly, one can consider more than one dynamic response. Thus, if some other response of interest is given by

$$y(t) = \int_{-\infty}^{\infty} f(s)h_y(t-s)\,ds$$

then one can derive cross-product terms such as the cross-correlation function:

$$\phi_{XY}(t_1,t_2) \equiv E[X(t_1)Y(t_2)] = \int_{-\infty}^{\infty}\int_{-\infty}^{\infty} \phi_{FF}(s_1,s_2)h_x(t_1-s_1)h_y(t_2-s_2)\,ds_1\,ds_2 \qquad (3.16)$$

for the stochastic responses $\{X(t)\}$ and $\{Y(t)\}$ caused by a stochastic excitation $\{F(t)\}$.

Note that the presentation to this point has been very general, with no limitation on the type of response which the stochastic processes $\{X(t)\}$ and/or $\{Y(t)\}$ represent. Usually we will use $\{X(t)\}$ to represent a displacement, but the preceding mathematical expressions certainly do not require that. If $\{X(t)\}$ is a displacement, then $\{\dot{X}(t)\}$ and $\{\ddot{X}(t)\}$, respectively, will denote the velocity and acceleration of response, and we will often be interested in these quantities as well.

Recall our definition of $h_x(t)$ as being the response $x(t) = h_x(t)$ when $f(t) = \delta(t)$. For $x(t)$ being response displacement, this gives us $\dot{x}(t) = h_x'(t)$ and $\ddot{x}(t) = h_x''(t)$ when $f(t) = \delta(t)$. Thus, the impulse response functions giving the velocity and acceleration responses, respectively, to the Dirac delta function pulse of excitation are $h_{\dot{x}}(t) = h_x'(t)$ and $h_{\ddot{x}}(t) = h_x''(t)$. These expressions can then be used to obtain moments of $\{\dot{X}(t)\}$ and $\{\ddot{X}(t)\}$ in forms which are exactly parallel to Eqs. 3.12 - 3.15. Two of the more important of these expressions are

$$\phi_{X\dot{X}}(t_1,t_2) = \frac{\partial}{\partial t_2}\phi_{XX}(t_1,t_2) = \int_{-\infty}^{\infty}\int_{-\infty}^{\infty} \phi_{FF}(s_1,s_2)h_x(t_1-s_1)h_x'(t_2-s_2)\,ds_1\,ds_2 \qquad (3.17)$$

and

$$\phi_{\dot{X}\dot{X}}(t_1,t_2) = \frac{\partial^2\phi_{XX}(t_1,t_2)}{\partial t_1 \partial t_2} = \int_{-\infty}^{\infty}\int_{-\infty}^{\infty} \phi_{FF}(s_1,s_2)h_x'(t_1-s_1)h_x'(t_2-s_2)\,ds_1\,ds_2 \qquad (3.18)$$

In addition, one can obtain an integral expression for any desired higher-order cross-product term, such as

$$E[X(t_1)\dot{X}(t_2)\ddot{X}(t_3)] = \int_{-\infty}^{\infty}\int_{-\infty}^{\infty}\int_{-\infty}^{\infty} E[F(s_1)F(s_2)F(s_3)]h_x(t_1-s_1)h_x'(t_2-s_2)h_x''(t_3-s_3)\,ds_1\,ds_2\,ds_3$$

Other forms of moment information, such as central moments or cumulants, can also be obtained from the moments given here. For example, using $K_{XX}(t_1,t_2) = \phi_{XX}(t_1,t_2) - \mu_X(t_1)\mu_X(t_2)$ gives the autocovariance function for $\{X(t)\}$ as

$$K_{XX}(t_1,t_2) = \int_{-\infty}^{\infty}\int_{-\infty}^{\infty} \phi_{FF}(s_1,s_2)h_x(t_1-s_1)h_x(t_2-s_2)\,ds_1\,ds_2$$
$$- \int_{-\infty}^{\infty}\mu_F(t_1)h_x(t_1-s_1)\,ds_1\int_{-\infty}^{\infty}\mu_F(t_2)h_x(t_2-s_2)\,ds_2$$

or

$$K_{XX}(t_1,t_2) = \int_{-\infty}^{\infty}\int_{-\infty}^{\infty} K_{FF}(s_1,s_2)h_x(t_1-s_1)h_x(t_2-s_2)\,ds_1\,ds_2 \qquad (3.19)$$

One important aspect of the dynamics of linear systems is the "uncoupling" of the moments of response. That is, the jth moment function of the response depends only on the jth moment function of the excitation. Thus, one can compute the mean value function of the stochastic response even if the only information given about the excitation is its mean value function. Similarly, knowledge of only the autocorrelation function for the excitation is sufficient to allow calculation of the autocorrelation function for the response, or calculation of the cross-product of two response measures such as $\{X(t)\}$ and $\{\dot{X}(t)\}$. Also, covariance functions for the response depend only on the autocovariance function for the excitation.

If we are given a complete set of initial conditions at some time t_0, then we will find it convenient to modify the convolution integral of Eq. 3.11 to include only excitation after time t_0 and to add terms representing the response to the initial conditions.[3] Since these initial conditions may be random, we will write the results as

$$X(t) = \sum_j Y_j g_j(t-t_0) + \int_{t_0}^{\infty} F(s)h_x(t-s)\,ds$$

in which the Y_j random variable denotes the jth initial condition value and the $g_j(t-t_0)$ term represents the response due to a unit value of that one initial condition. The expressions for the mean value and autocovariance functions of the response can then be written as

$$\mu_X(t) = \sum_j \mu_{Y_j} g_j(t-t_0) + \int_{t_0}^{\infty} \mu_F(s)h_x(t-s)\,ds \qquad (3.20)$$

and

$$K_{XX}(t_1,t_2) = \sum_{j_1}\sum_{j_2} K_{Y_{j_1}Y_{j_2}} g_{j_1}(t_1-t_0) g_{j_2}(t_2-t_0) + \sum_j \int_{t_0}^{\infty} K_{Y_j,F(s)} g_j(t_1-t_0)h_x(t_2-s)\,ds$$

$$+ \sum_j \int_{t_0}^{\infty} K_{Y_j,F(s)} g_j(t_2-t_0)h_x(t_1-s)\,ds + \int_{t_0}^{\infty}\int_{t_0}^{\infty} K_{FF}(s_1,s_2)h_x(t_1-s_1)h_x(t_2-s_2)\,ds_1\,ds_2 \qquad (3.21)$$

A similar expression can be written out for the response autocorrelation function. In many situations, of course, the excitation after time t_0 may be independent of the initial conditions at time t_0. If this is true, then the $K_{XX}(t_1,t_2)$ result is simplified since the $K_{Y_j,F(s)}$ cross-covariance in the second expression is zero, eliminating this term. Another simplified situation which may arise is when there is no stochastic excitation, so that the dynamic response is due only to the random initial conditions. Finally, in the special case where the initial conditions are deterministic, all the covariance terms involving the initial conditions are zero, but the mean value and autocorrelation

---

[3] By a complete set of initial conditions we mean that enough conditions are given at time t_0 to assure that the solution for $t > t_0$ is unique.

functions are still affected by the initial conditions.

One can also use a conditional probability distribution for the $\{X(t)\}$ process in Eq. 3.11 in order to obtain conditional versions of the mean value, autocovariance function, and autocorrelation function. This is possible for any type of conditioning event, but we will be particularly interested in the situation in which the given event, which will be denoted by A, includes a complete set of deterministic initial conditions at time t_0. We can then write the conditional mean and covariance in the same form as Eqs. 3.20 and 3.21 with the random initial condition Y_j replaced by a deterministic y_j whose value is known when A is known. The conditional mean value of the response then depends on the initial conditions and the conditional mean of the excitation as

$$E[X(t)|A] = \sum_j y_j\, g_j(t-t_0) + \int_{t_0}^{\infty} E[F(s)|A]h_x(t-s)\,ds \tag{3.22}$$

The conditional covariance of the response is

$$Cov[X(t_1),X(t_2)|A] = \int_{t_0}^{\infty}\int_{t_0}^{\infty} Cov[F(s_1),F(s_2)|A]h_x(t_1-s_1)h_x(t_2-s_2)\,ds_1\,ds_2 \tag{3.23}$$

which depends only on the conditional covariance of the excitation.

3.4   Response to Stationary Excitation

Consider now the special case in which the excitation process $\{F(t)\}$ has existed since time $t = -\infty$ and is stationary. First, if $\{F(t)\}$ is mean-value stationary, then $\mu_F(t-r) = \mu_F$ is a constant in Eq. 3.12 so that

$$\mu_X(t) \equiv E[X(t)] = \mu_F \int_{-\infty}^{\infty} h_x(r)\,dr = \mu_F h_{x,static} \tag{3.24}$$

This shows that $\mu_X(t)$ is not a function of t in this case. There are, however, two distinct possibilities. Either the integral is infinite so that $\mu_X(t)$ does not exist, or $\mu_X(t) = \mu_X$ is a constant. In particular, if a system has a finite static response and the excitation is mean-value stationary, then the stochastic response is mean-value stationary. If the system has an infinite static response and $\mu_F \neq 0$, then the stochastic response has an infinite mean value.

Similarly, if the excitation is second-moment stationary then $\phi_{FF}(t_1-r_1,t_2-r_2) = R_{FF}(t_1-r_1-t_2+r_2)$ in Eq. 3.14 and this gives

$$\phi_{XX}(t_1,t_2) = \int_{-\infty}^{\infty}\int_{-\infty}^{\infty} R_{FF}(t_1-t_2-r_1+r_2)h_x(r_1)h_x(r_2)\,dr_1\,dr_2$$

If this second moment of the response $\phi_{XX}(t_1,t_2)$ is finite, then it is a function only of the difference between the two time arguments (t_1-t_2), showing that $\{X(t)\}$ is second-moment stationary.

Thus, we can write $\tau = t_1 - t_2$ to get the stationary autocorrelation function

$$R_{XX}(\tau) = \int_{-\infty}^{\infty}\int_{-\infty}^{\infty} R_{FF}(\tau - r_1 + r_2)h_x(r_1)h_x(r_2)\,dr_1\,dr_2 \qquad (3.25)$$

Second-moment stationarity of $\{F(t)\}$ is also sufficient to assure that the cross-correlation of two responses $\{X(t)\}$ and $\{Y(t)\}$ has the stationary form of

$$R_{XY}(\tau) = \int_{-\infty}^{\infty}\int_{-\infty}^{\infty} R_{FF}(\tau - r_1 + r_2)h_x(r_1)h_y(r_2)\,dr_1\,dr_2 \qquad (3.26)$$

including as a special case

$$R_{X\dot{X}}(\tau) = \int_{-\infty}^{\infty}\int_{-\infty}^{\infty} R_{FF}(\tau - r_1 + r_2)h_x(r_1)h_x'(r_2)\,dr_1\,dr_2 \qquad (3.27)$$

Also, Eqs. 3.25 - 3.27 are equally valid if the stationary R functions are all replaced by stationary autocovariance and cross-covariance G functions.

Let us now consider the issue of whether $R_{XX}(\tau)$ and $G_{XX}(\tau)$ will be bounded for a system with a stationary excitation. From the Schwarz inequality we know that $R_{XX}(\tau)$ is bounded by the mean squared value and $G_{XX}(\tau)$ is bounded by the variance: $|R_{XX}(\tau)| \leq R_{XX}(0) \equiv E[X^2]$ and $|G_{XX}(\tau)| \leq G_{XX}(0) \equiv \sigma_X^2$. Thus, $R_{XX}(\tau)$ and $G_{XX}(\tau)$ are bounded for all τ if and only if $E[X^2]$ and σ_X^2, respectively, are bounded. As already noted, μ_X will be infinite if $\mu_F \neq 0$ and the system is such that it does not have a finite static response (i.e., if $h_{x,static} = \infty$). The fact that $E[X^2] = \mu_X^2 + \sigma_X^2$ tells us, then, that $E[X^2]$ is also infinite in that situation, even if the variance is finite.

It is possible for the variance σ_X^2 to be infinite even if $h_x(t)$ or μ_F is such that μ_X is finite, and it is also possible for the variance to be finite when the mean is infinite. In order to investigate the boundedness of the variance, we will rewrite Eq. 3.25 for the stationary covariance of the response and set $\tau = 0$ to obtain the response variance as

$$\sigma_X^2 = \int_{-\infty}^{\infty}\int_{-\infty}^{\infty} G_{FF}(r_2 - r_1)h_x(r_1)h_x(r_2)\,dr_1\,dr_2 = \int_{-\infty}^{\infty} G_{FF}(r_3)\left[\int_{-\infty}^{\infty} h_x(r_2 - r_3)h_x(r_2)\,dr_2\right]dr_3 \qquad (3.28)$$

in which we have used the change of variables $r_3 = r_2 - r_1$. The absolute value of the inner integral in this expression can be bounded as

$$\left| \int_{-\infty}^{\infty} h_x(r_2 - r_3)h_x(r_2)\,dr_2 \right| \leq \int_{-\infty}^{\infty} h_x^2(r)\,dr$$

and this integral will be finite if $h_x(t)$ is bounded and the system has a finite $x(t)$ static response $(h_{x,static} < \infty)$. Thus, we can bound the variance of the response as

$$\sigma_X^2 \le h_{x,variance} \int_{-\infty}^{\infty} |G_{FF}(r_3)| dr_3 \qquad (3.29)$$

in which

$$h_{x,variance} \equiv \int_{-\infty}^{\infty} h_x^2(r)\, dr \qquad (3.30)$$

is a characteristic of the linear system. We see that the system must have a finite variance of stationary response if it has a finite static response and the excitation is such that its autocorrelation function is absolute value integrable. It should be noted that, although these conditions are sufficient to assure a finite stationary response covariance, they do not preclude the possibility of covariant stationary response under different conditions.

****

Example 3.4: Investigate the boundedness of the first and second moments of stationary stochastic response of the system $\ddot{X}(t) + c\dot{X}(t) = F(t)$, for which the impulse response function was derived in Example 3.2 as $h_x(t) = c^{-1}[1 - e^{-ct}]U(t)$.

Since the static response value for this system is infinite ($h_{x,static} = \infty$), we know that the mean value and autocorrelation function of the $\{X(t)\}$ response will generally not exist for an $\{F(t)\}$ process which is mean-value and second-moment stationary. In particular, $\mu_X(t) = \mu_F h_{x,static}$ is finite only if $\mu_F = 0$, in which case $\mu_X = 0$ also.

Since μ_X is infinite if $\mu_F \ne 0$, we also know that $\phi_{XX}(t_1,t_2)$ is infinite for this situation. It turns out, though, that $\phi_{XX}(t_1,t_2)$ is infinite for $\phi_{FF}(s_1,s_2) = R_{FF}(s_1 - s_2)$, even if $\mu_F = 0$. In particular, the response variance is infinite. To demonstrate this, substitute the given $h_x(t)$ into Eq. 3.28 to obtain

$$\sigma_X^2 = c^{-2} \int_{-\infty}^{\infty} G_{FF}(r_3) \left[\int_{max(0,r_3)}^{\infty} [1 - e^{-c(r_2 - r_3)} - e^{-cr_2} + e^{-c(2r_2 - r_3)}] dr_2 \right] dr_3$$

The unity term in the integrand of the integral with respect to r_2, combined with the infinite range of integration, shows that this integral is infinite. Thus, σ_X^2 is infinite for any $R_{FF}(\tau)$ function, as is $E[X^2(t)]$.

Recall that the derivative response for this problem does have a finite static value, since its impulse response function of $h_{\dot{x}}(t) = h_x'(t) = e^{-ct}U(t)$ gives $h_{\dot{x},static} = c^{-1}$. Thus, the mean value and autocorrelation of stationary $\{\dot{X}(t)\}$ response generally do exist for this system, even though those for $\{X(t)\}$ do not.
****

Example 3.5: Investigate the behavior of the first and second moments of stochastic response of the system $c\dot{X}(t) + kX(t) = 0$, for which the impulse response function was derived in Example 3.1 as $h_x(t) = c^{-1}e^{-kt/c}U(t)$. In particular, investigate $\mu_X(t)$, $K_{XX}(t_1,t_2)$, and $\phi_{XX}(t_1,t_2)$ for this situation with no excitation process, but with a random variable initial condition of $X(t_0) = Y$ with μ_Y and σ_Y known.

We can use Eqs. 3.20 and 3.21 if we know the value of the $g(t)$ function. Note that one needs only one initial condition for this first-order system, so that the given value of $X(t_0)$ is sufficient to describe a unique solution. Furthermore, Example 3.1 shows that the impulse response function $h_x(t) = c^{-1}e^{-kt/c}U(t)$ is the response of the system to an initial condition of $X(0) = c^{-1}$, so that we have $g(t) = ch_x(t-t_0) = e^{-k(t-t_0)/c}U(t-t_0)$. Thus, for $t > t_0$ we have

$$\mu_X(t) = \mu_Y g(t-t_0) = \mu_Y e^{-k(t-t_0)/c}$$

and

$$K_{XX}(t_1,t_2) = \sigma_Y^2 g(t_1-t_0)g(t_2-t_0) = \sigma_Y^2 e^{-k(t_1+t_2-2t_0)/c}$$

Putting these two together gives the autocorrelation function of the response as

$$\phi_{XX}(t_1,t_2) = E(Y^2)g(t_1-t_0)g(t_2-t_0) = E(Y^2)e^{-k(t_1+t_2-2t_0)/c}$$

By letting $t_1 = t_2 = t$, one can also find the response variance as $\sigma_X^2(t) = \sigma_Y^2 e^{-2k(t-t_0)/c}$ and the mean squared value is given by a similar expression.

****

3.5   Delta-Correlated Excitations

There are a number of important physical problems in which the excitation process $\{F(t)\}$ is so erratic that $F(t)$ and $F(s)$ are almost independent unless t and s are almost equal. For a physical process there generally is some dependence between $F(t)$ and $F(s)$ for t and s sufficiently close, but this dependence may decay very rapidly as the separation between t and s grows. To illustrate this idea, let T_c denote the time difference $t - s$ over which $F(t)$ and $F(s)$ are significantly dependent, so that $F(t)$ and $F(s)$ are essentially independent if $|t - s| > T_c$. Then if T_c is sufficiently small compared to other characteristic times of the problem being considered, it may be possible to approximate $\{F(t)\}$ by the limiting process for which $F(t)$ and $F(s)$ are independent for $t \neq s$. This limiting process is called a delta-correlated process. The motivation for using a delta-correlated process is strictly convenience. Computing response statistics for a delta-correlated excitation is often much easier than for a nearly delta-correlated one.

We now wish to investigate the details of how a delta-correlated $\{F(t)\}$ excitation process must be defined in order that it will cause a response $\{X(t)\}$ which approximates that for a physical problem with a nearly delta-correlated excitation. We will do this by focusing on the covariance functions of $\{F(t)\}$ and $\{X(t)\}$, starting with the covariant stationary situation. Independence of $F(t)$ and $F(s)$, of course, implies that the covariance $K_{FF}(t,s)$ is zero, as well as corresponding relationships involving higher moment functions of the process.[4] Thus, a covariant stationary delta-correlated $\{F(t)\}$ would have $G_{FF}(\tau) = 0$ for $\tau \neq 0$. Recall that Eq. 3.29 gave a bound on

---

[4]   Precisely, one can say that $F(t)$ and $F(s)$ are independent if and only if all the cumulant functions are zero, whereas covariance is only the second cumulant function. Some discussion of cumulants is given in Appendix A.

the response variance as $h_{x,variance}$ times the integral of the absolute value of $G_{FF}(\tau)$, in which $h_{x,variance}$ is a characteristic of the linear system. This integral of the absolute value of $G_{FF}(\tau)$, though, will be zero if we choose our delta-correlated process to have a finite value at $\tau = 0$ and to be zero for $\tau \neq 0$. Thus, our delta-correlated excitation process will not give any response unless we allow $G_{FF}(0)$ to be infinite. This leads us to the standard form for the covariance function of a stationary delta-correlated process:

$$G_{FF}(\tau) = G_0 \delta(\tau) \tag{3.31}$$

which is a special case of the covariance function for a nonstationary delta-correlated process:

$$K_{FF}(t,s) = G_0(t)\delta(t-s) \tag{3.32}$$

One should note that Eqs. 3.31 and 3.32 only assure that $F(t)$ and $F(s)$ are uncorrelated for $t \neq s$, whereas we defined the delta-correlated property to mean that $F(t)$ and $F(s)$ are independent for $t \neq s$. That is, the lack of correlation given in Eqs. 3.31 and 3.32 is necessary for a delta-correlated process, but it is not the definition of the process.[5]

The convenience of using a delta-correlated excitation in studies of dynamic response can be illustrated by substituting Eq. 3.31 for the stationary covariance function of the excitation into Eq. 3.25, giving the response covariance as

$$G_{XX}(\tau) = \int_{-\infty}^{\infty}\int_{-\infty}^{\infty} G_0 \delta(\tau - r_1 + r_2)h_x(r_1)h_x(r_2)\,dr_1\,dr_2$$

The presence of the Dirac delta function now allows easy evaluation of one of the integrals, with the result that the response covariance $G_{XX}(\tau)$ is given by a single (rather than a double) integral

$$G_{XX}(\tau) = G_0 \int_{-\infty}^{\infty} h_x(\tau + r)h_x(r)\,dr \tag{3.33}$$

A special case of this formula is the variance of the response given by

$$\sigma_X^2 = G_{XX}(0) = G_0 \int_{-\infty}^{\infty} h_x^2(r)\,dr \equiv G_0 h_{x,variance} \tag{3.34}$$

Note that this shows that the variance bound in Eq. 3.29 is exactly the variance when the excitation is delta-correlated. This also shows that the response to delta-correlated excitation is always unbounded if the $h_{x,variance}$ integral of the square of the impulse response function is infinite.

In the same way, we can substitute Eq. 3.32 into Eq. 3.19 to obtain the covariance of response

---

[5]   The term "white noise" is commonly used to refer to processes of the general type of our delta-correlated process. Sometimes, though, white noise implies only Eq. 3.32 rather than the more general independence property of a delta-correlated process.

to a nonstationary delta-correlated excitation as

$$K_{XX}(t_1, t_2) = \int_{-\infty}^{\infty} \int_{-\infty}^{\infty} G_0(s_1) \, \delta(s_1 - s_2) h_x(t_1 - s_1) h_x(t_2 - s_2) \, ds_1 \, ds_2$$

which simplifies to

$$K_{XX}(t_1, t_2) = \int_{-\infty}^{\infty} G_0(s_1) h_x(t_1 - s_1) h_x(t_2 - s_1) \, ds_1 \tag{3.35}$$

The variance in this situation is given by

$$\sigma_X^2(t) = K_{XX}(t, t) = \int_{-\infty}^{\infty} G_0(s) h_x^2(t - s) \, ds = \int_{-\infty}^{\infty} G_0(t - r) h_x^2(r) \, dr \tag{3.36}$$

which is seen to be exactly in the form of a convolution of the function $G_0(t)$ giving the intensity of the excitation and the square of the impulse response function for the dynamic system.

Delta-correlated excitations have another convenient feature when one considers the conditional distribution of the response, as in Eqs. 3.22 and 3.23. For a causal system, one can say that the response at any time t_0 is a function of the excitation $F(t)$ only for $t \leq t_0$. For a delta-correlated excitation we can say that $F(t)$ for $t > t_0$ is independent of $F(t)$ for $t \leq t_0$, and this implies that $F(t)$ for $t > t_0$ is independent of the response of the system at time t_0. Thus, if the conditioning event A only involves the value of the excitation and the response at time t_0 then we can say that $F(t)$ for $t > t_0$ is independent of A. This makes Eqs. 3.22 and 3.23 depend on the unconditional mean and covariance of the excitation, and gives

$$E[X(t)|A] = \sum_j y_j \, g_j(t - t_0) + \int_{t_0}^{\infty} \mu_F(s) \, h_x(t - s) \, ds \tag{3.37}$$

and

$$\text{Cov}[X(t_1), X(t_2)|A] = \int_{t_0}^{\infty} G_0(s) h_x(t_1 - s) h_x(t_2 - s) \, ds \tag{3.38}$$

Note that the final equation is of exactly the same form as the unconditional variance in Eq. 3.35, except that it has a different limit of integration.

∗∗∗

Example 3.6: Show that one particular delta-correlated process is the so-called "shot noise" defined by

$$F(t) = \sum_j F_j \, \delta(t - T_j)$$

in which $\{T_1, T_2, \cdots, T_j, \cdots\}$ are the arrival times for a Poisson process $\{N(t)\}$, and $\{F_1, F_2, \cdots, F_j, \cdots\}$ is a sequence of identically distributed random variables which are independent of each other and of the arrival times.

The nature of the Poisson process is such that knowledge of past arrival times gives no information about future arrival times. Specifically, the inter-arrival time $T_{j+1} - T_j$ is independent of $\{T_1, \cdots, T_j\}$. Combined with the independence of the F_j pulse magnitudes, this assures us that $F(t)$ must be independent of $F(s)$ for $t \neq s$. For $s < t$, knowledge of $F(s)$ may give some information about past arrival times and pulse magnitudes, but it gives no information about the likelihood that a pulse will arrive at future time t or about the likely magnitude of such a pulse if it does arrive at time t. Thus, $\{F(t)\}$ is delta-correlated.

One can also investigate the covariance function for the shot noise process, to confirm that it agrees with Eq. 3.32. We will only do this for the situation with $\lambda \equiv E[\dot{N}(t)]$ as a constant expected rate of arrivals in the Poisson process. We first define a process $\{Q(t)\}$ which is the integral of $\{F(t)\}$:

$$Q(t) = \sum_j F_j U(t - T_j)$$

We wish to use the covariance of $\{Q(t)\}$ in finding the covariance function of $\{F(t)\}$. Note that the conditional mean of $\{Q(t)\}$ given that $N(t) = m$ is

$$E[Q(t) \mid N(t) = m] = \sum_{k=1}^{m} E[F_k] = \mu_F m$$

and the conditional autocorrelation function for $\{Q(t)\}$ given that $N(t) = m$ and $N(s) = n$ is

$$E[Q(t)Q(s) \mid N(t) = m, N(s) = n] = \sum_{j=1}^{m} \sum_{k=1}^{n} E[F_j F_k] = \mu_F^2 mn + \sigma_F^2 \min(m,n)$$

in which we have presumed that $0 \leq T_1 \leq T_2 \leq \cdots \leq T_j$. Taking the unconditional expectation of these quantities then gives

$$E[Q(t)] = \sum_{k=1}^{E[N(t)]} E[F_k] = \mu_F E[N(t)] = \mu_F \lambda t$$

and

$$E[Q(t)Q(s)] = \sum_{j=1}^{E[N(t)]} \sum_{k=1}^{E[N(s)]} E[F_j F_k] = \mu_F^2 E[N(t)N(s)] + \sigma_F^2 E(N[\min(s,t)])$$

In Example 2.13, we found that the Poisson process has an autocorrelation function of

$$E[N(t)N(s)] = \lambda^2 ts + \lambda \min(t,s)$$

Thus, we obtain

$$E[Q(t)Q(s)] = \mu_F^2 \left(\lambda^2 ts + \lambda \min(t,s) \right) + \sigma_F^2 \lambda \min(s,t)] = \mu_F^2 \lambda^2 ts + E(F^2) \lambda \min(t,s)$$

These expressions demonstrate that the autocovariance function of $\{Q(t)\}$ is given by

$$K_{QQ}(t,s) = E(F^2)\lambda \min(t,s) = E(F^2)\lambda\, t\, U(s-t) + E(F^2)\lambda\, s\, U(t-s)$$

Since $F(t) = \dot{Q}(t)$, taking the partial derivative of this expression with respect to s gives

$$K_{QF}(t,s) = E(F^2)\lambda\, t\, \delta(s-t) - E(F^2)\lambda\, s\, \delta(t-s) + E(F^2)\lambda\, U(t-s)$$

but the first two terms on the right-hand side of this expression exactly cancel each other, so that the partial derivative gives $K_{FF}(t,s) = E(F^2)\lambda\delta(t-s)$. This is exactly of the form of Eq. 3.32, with $G_0(t) = E(F^2)\lambda$. We have shown that the shot noise $\{F(t)\}$ is covariant stationary when the arrival rate λ is a constant. Nonstationary shot noise delta-correlated processes can also be generated in this way by taking the arrival rate to be nonstationary.

****

3.6 Response of Linear Single-Degree-of-Freedom Oscillator

Most of the remainder of this chapter will be devoted to the study of one particular linear system subjected to stationary delta-correlated excitation. In a sense this constitutes one extended example problem, but it warrants special attention because it will form the basis for much of what we will do in analyzing the stochastic response of other dynamical systems. The system to be studied is the so-called "single-degree-of-freedom" (SDF) oscillator introduced in Example 3.3.

The SDF oscillator may be regarded as the prototypical vibratory system, and it has probably been studied more thoroughly than any other dynamical system. Essentially, it is the simplest differential equation which exhibits fundamentally oscillatory behavior. In particular, as we saw in Example 3.3, its impulse response function is oscillatory if damping is small, and this is a basic characteristic of vibratory systems. For both deterministic and stochastic situations, the dynamic behavior of the linear SDF system can be regarded as providing the fundamental basis for much of both the analysis and the interpretation of results for more complicated vibratory systems. In particular, the response of more complicated (multi-degree-of-freedom) linear systems is typically found by superimposing modal responses, each of which represents the response of an SDF system, and the dynamic behavior of a vibratory nonlinear system is usually interpreted in terms of how it resembles and how it differs from that of a linear system.

In a somewhat similar way, the delta-correlated excitation can be considered as prototypical, inasmuch as it is simple and the nature of any other excitation process is often interpreted in terms of how it differs from a delta-correlated process. This idea will be investigated in more detail in Chapter 4.

The differential equation for stochastic motion of the SDF system is commonly written as either

$$m\ddot{X}(t) + c\dot{X}(t) + kX(t) = F(t) \tag{3.39}$$

or

$$\ddot{X}(t) + 2\zeta\omega_0\dot{X}(t) + \omega_0^2 X(t) = F(t)/m = -\ddot{Z}(t) \tag{3.40}$$

in which ω_0 and ζ are called the undamped natural circular frequency and the fraction of critical damping, respectively, of the system. The usual interpretation of this equation for mechanical vibration problems is that m, k, and c denote the magnitudes of a mass, spring, and dashpot, respectively, attached as shown in Fig. 3.3. For the fixed base system of Fig. 3.3(a), the term $f(t)$ is the force externally applied to the mass, and $x(t)$ is the location of the mass relative to its position when the system is at rest. When the system has an imposed motion at the base, as in Fig. 3.3(b), $x(t)$ denotes the change in the location of the mass relative to the moving base and $f(t)$ is the product of the mass and the negative of the base acceleration.[6] In this physical model we expect m, k, and c to all be nonnegative.

First we will consider the nonstationary stochastic response which results when the SDF oscillator is initially at rest and the delta-correlated excitation is suddenly applied to the system at time $t = 0$ and then is stationary for all later time. Mathematically this condition can be written as

$$F(t) = W(t)U(t)$$

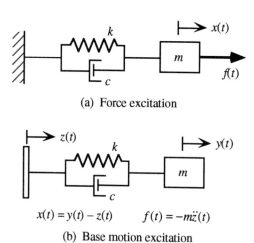

(a) Force excitation

$$x(t) = y(t) - z(t) \qquad f(t) = -m\ddot{z}(t)$$

(b) Base motion excitation

Figure 3.3 Single-degree-of-freedom oscillator

---

[6] Alternatively, one can apply the term SDF to any system which is governed by a single second-order differential equation like Eq. 3.39, provided that the m and k coefficients have the same sign.

in which $W(t)$ is a stationary delta-correlated process and the unit step $U(t)$ is included to eliminate excitation prior to $t = 0$, and thereby assure that the system is at rest at that time. Using $K_{WW}(t,s) = G_0 \delta(t-s)$ to denote the covariance function of $W(t)$ gives

$$K_{FF}(t,s) = G_0 \delta(t-s) U(t) \tag{3.41}$$

with G_0 being a constant. We will first investigate the general nonstationary response of this oscillator, then consider the limiting behavior for t tending to infinity and for t nearly zero.

Recall that the impulse function of Eq. 3.39 was found in Example 3.3 to be $h_x(t) = (m\omega_d)^{-1}$ $e^{-\zeta\omega_0 t} \sin(\omega_d t) U(t)$. We will find the nonstationary covariance of the response of Eq. 3.39 by substitution of this $h_x(t)$ and $K_{FF}(t,s)$ from Eq. 3.41 into Eq. 3.19, giving

$$K_{XX}(t_1,t_2) = \frac{G_0}{m^2 \omega_d^2} \int_{-\infty}^{\infty} \int_{-\infty}^{\infty} \delta(s_1-s_2) U(s_1) e^{-\zeta\omega_0(t_1-s_1)} \sin[\omega_d(t_1-s_1)] U(t_1-s_1)$$

$$e^{-\zeta\omega_0(t_2-s_2)} \sin[\omega_d(t_2-s_2)] U(t_2-s_2) ds_1 ds_2$$

Taking advantage of the unit step functions and the Dirac delta function in the integrand, one can rewrite this expression as

$$K_{XX}(t_1,t_2) = \frac{G_0}{m^2 \omega_d^2} \int_0^{\min(t_1,t_2)} e^{-\zeta\omega_0(t_1+t_2-2s)} \sin[\omega_d(t_1-s)] \sin[\omega_d(t_2-s)] ds\, U(t_1) U(t_2)$$

One simple way to evaluate this integral is to use the identity $\sin(\alpha) = (e^{i\alpha} - e^{-i\alpha})/(2i)$ for each of the sine terms, giving the integrand as the sum of four different exponential functions of s, each of which can be easily integrated. After substitution of the fact that $t_1 + t_2 - 2\min(t_1,t_2) = |t_1 - t_2|$ and performing considerable algebraic simplification, the result can be written as

$$K_{XX}(t_1,t_2) = \frac{G_0}{4m^2 \zeta\omega_0^3} U(t_1) U(t_2) \left\{ e^{-\zeta\omega_0|t_1-t_2|} \left[\cos[\omega_d(t_1-t_2)] + \frac{\zeta\omega_0}{\omega_d} \sin[\omega_d|t_1-t_2|] \right] - \right.$$
$$\left. e^{-\zeta\omega_0(t_1+t_2)} \left[\frac{\omega_0^2}{\omega_d^2} \cos[\omega_d(t_1-t_2)] + \frac{\zeta\omega_0}{\omega_d} \sin[\omega_d(t_1+t_2)] - \frac{\zeta^2\omega_0^2}{\omega_d^2} \cos[\omega_d(t_1+t_2)] \right] \right\} \tag{3.42}$$

The rather lengthy expression in Eq. 3.42 contains a great deal of information about the stochastic dynamic behavior of the SDF system. We will use it to investigate a number of matters of interest.

In many situations it is also important to study the response levels for derivatives of X. The covariance functions for such derivatives could be obtained in exactly the same way as Eq. 3.42 was obtained. For example, we could find $h_{\dot{x}}(t)$ and substitute it instead of $h_x(t)$ into Eq. 3.19 to obtain a covariance function for the $\{\dot{X}(t)\}$ process. Rather than performing more integration, though, we can use expressions from Chapter 2 to derive variance properties of derivatives of $\{X(t)\}$ directly from Eq. 3.42. For example, Eq. 2.59 gives the cross-covariance of $\{X(t)\}$ and $\{\dot{X}(t)\}$ as

$$K_{X\dot{X}}(t_1,t_2) = \frac{\partial K_{XX}(t_1,t_2)}{\partial t_2}$$

and Eq. 2.60 gives the covariance function for $\{\dot{X}(t)\}$ as

$$K_{\dot{X}\dot{X}}(t_1,t_2) = \frac{\partial^2 K_{XX}(t_1,t_2)}{\partial t_1 \partial t_2}$$

Substituting $K_{XX}(t_1,t_2)$ from Eq. 3.42 into these expressions gives

$$K_{X\dot{X}}(t_1,t_2) = \frac{G_0}{4m^2 \zeta \omega_0 \omega_d} U(t_1)U(t_2)\left\{ e^{-\zeta\omega_0|t_1-t_2|}\sin[\omega_d(t_1-t_2)]- \right.$$
$$\left. e^{-\zeta\omega_0(t_1+t_2)}\left[\sin[\omega_d(t_1-t_2)] - \frac{\zeta\omega_0}{\omega_d}\cos[\omega_d(t_1-t_2)] + \frac{\zeta\omega_0}{\omega_d}\cos[\omega_d(t_1+t_2)] \right] \right\}$$
$$(3.43)$$

and

$$K_{\dot{X}\dot{X}}(t_1,t_2) = \frac{G_0}{4m^2 \zeta \omega_0} U(t_1)U(t_2)\left\{ e^{-\zeta\omega_0|t_1-t_2|}\left[\cos[\omega_d(t_1-t_2)] - \frac{\zeta\omega_0}{\omega_d}\sin[\omega_d|t_1-t_2|] \right] - \right.$$
$$\left. e^{-\zeta\omega_0(t_1+t_2)}\left[\frac{\omega_0^2}{\omega_d^2}\cos[\omega_d(t_1-t_2)] - \frac{\zeta\omega_0}{\omega_d}\sin[\omega_d(t_1+t_2)] - \frac{\zeta^2\omega_0^2}{\omega_d^2}\cos[\omega_d(t_1+t_2)] \right] \right\} \quad (3.44)$$

Next let us consider the variance of the response by letting $t_2 = t_1 = t$ in Eq. 3.42, giving

$$\sigma_X^2(t) \equiv K_{XX}(t,t) = \frac{G_0}{4m^2 \zeta \omega_0^3}\left\{ 1 - e^{-2\zeta\omega_0 t}\left[\frac{\omega_0^2}{\omega_d^2} + \frac{\zeta\omega_0}{\omega_d}\sin(2\omega_d t) - \frac{\zeta^2\omega_0^2}{\omega_d^2}\cos(2\omega_d t) \right] \right\}U(t)$$
$$(3.45)$$

This important expression was apparently first published by Caughey and Stumpf in 1961. It is plotted in Fig. 3.4 for several values of the damping parameter ζ. Figure 3.4 also includes a curve for $\zeta = 0$, which cannot be obtained directly from Eq. 3.45. Rather, one must take the limit of the equation as ζ approaches zero, obtaining

$$\sigma_X^2(t) = \frac{G_0}{4m^2 \omega_0^3}\{2\omega_0 t - \sin(2\omega_0 t)\}U(t) \quad \text{for} \quad \zeta = 0. \quad (3.46)$$

Note that for any damping value other than zero, $\sigma_X^2(t)$ tends to an asymptote as t tends to

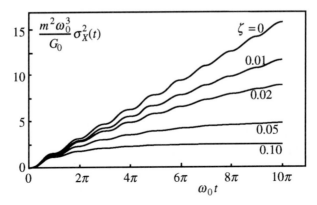

Figure 3.4 Nonstationary growth of variance

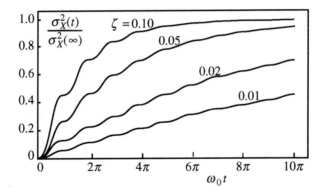

Figure 3.5 Approach of nonstationary variance to asymptote

infinity. In fact, Eq. 3.45 gives $\sigma_X^2(t) \to G_0 / (4m^2 \zeta \omega_0^3)$ for $t \to \infty$, which agrees exactly with the bound of Eq. 3.29 for any excitation of this system. This confirms the general result derived in Eq. 3.34, that the variance bound of Eq. 3.29 is identical to the response to a stationary delta-correlated excitation process.

The form of Fig. 3.4 emphasizes the important fact that the response variance grows more rapidly when damping is small. On the other hand, it appears that the initial rate of growth is approximately independent of damping. We will examine this aspect in more detail in Example 3.7, when we consider the limiting behavior for t near zero. The form of normalization used in Fig. 3.4 gives the asymptote for each curve as $(4\zeta)^{-1}$ for $t \to \infty$. The fact that the asymptote is different for each curve rather obscures the information about the rate at which each curve approaches its asymptote, and how this rate depends on damping. In order to illustrate this aspect more clearly, Fig. 3.5 shows plots of $\sigma_X^2(t)$ normalized by the asymptote for several nonzero values of damping.

In this form of the plots each curve tends to unity as t becomes large, and it is clear that the asymptotic level is reached much more quickly in a system with larger damping.

Since response variance is a very important quantity in practical applications of stochastic dynamics, we will also consider some simple approximations and bounds of Eq. 3.45. One possible approximation is to note that the oscillatory terms in Figs. 3.4 or 3.5 seem to be relatively unimportant compared to the exponential approach of the variance to its asymptotic value. This leads to the idea of replacing the oscillatory sine and cosine terms by their average value of zero in Eq. 3.45 and obtaining the averaged expression of

$$\sigma_X^2(t) \approx \frac{G_0}{4m^2 \zeta \omega_0^3}\left(1 - \frac{\omega_0^2}{\omega_d^2}e^{-2\zeta\omega_0 t}\right)U(t) = \frac{G_0}{4m^2 \zeta \omega_0^3}\left(1 - \frac{e^{-2\zeta\omega_0 t}}{1 - \zeta^2}\right)U(t)$$

The actual variance value oscillates about this smooth curve, with the amplitude of the oscillation decaying exponentially. Alternatively, one can obtain a simpler result which is almost identical to this for small damping values by neglecting the ζ^2 term and writing

$$\sigma_X^2(t) \approx \frac{G_0}{4m^2 \zeta \omega_0^3}\left(1 - e^{-2\zeta\omega_0 t}\right)U(t) \tag{3.47}$$

In some situations one may wish to obtain a rigorous upper bound on the nonstationary variance expression of Eq. 3.45. This can be obtained by using the amplitude of the sum of the two oscillatory terms:

$$Ampl.\left(\frac{\zeta\omega_0}{\omega_d}\sin(2\omega_d t) - \frac{\zeta^2\omega_0^2}{\omega_d^2}\cos(2\omega_d t)\right) = \frac{\zeta\omega_0}{\omega_d}\sqrt{1 + \frac{\zeta^2\omega_0^2}{\omega_d^2}} = \frac{\zeta\omega_0^2}{\omega_d^2}$$

Replacing the oscillatory terms by the negative of the amplitude gives the bound

$$\sigma_X^2(t) \leq \frac{G_0}{4m^2 \zeta \omega_0^3}\left\{1 - \frac{e^{-2\zeta\omega_0 t}}{1 + \zeta}\right\}U(t) \tag{3.48}$$

Figure 3.6 compares the exact response variance of Eq. 3.45 with the smooth approximation of Eq. 3.47 and the bound of Eq. 3.48 for the special case of $\zeta = 0.10$. Even for this case with rather large damping it is seen that the curves are all relatively near each other, and for smaller damping they are much closer. It should be noted, though, that the smooth exponential approximation never adequately describes the response growth for very small values of time, such as $\omega_0 t < \pi / 4$. As previously indicated, this small time situation will be examined in more detail in Example 3.7.

The exponential forms of Eqs. 3.47 and 3.48 are very valuable if one needs an approximation

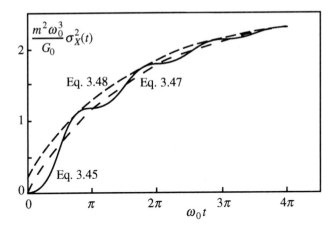

Figure 3.6 Approximation and bound of nonstationary variance for $\zeta = 0.10$

of the time that it takes for the response variance to reach some specified level. Solving Eq. 3.47 for t shows that $\sigma_X^2(t)$ reaches a specified level B when

$$t \approx \frac{-1}{2\zeta\omega_0} \ln\left[1 - \frac{4m^2\zeta\omega_0^3}{G_0}B\right]$$

Similarly, the bound gives $\sigma_X^2(t) \leq B$ for

$$t = \frac{-1}{2\zeta\omega_0} \ln\left[1 - (1+\zeta)\frac{4m^2\zeta\omega_0^3}{G_0}B\right]$$

Such an approximation or bound of the response variance is an important part of the estimation or bounding of the reliability for structural systems which are susceptible to failure due to excessive deformation or stress.

In the same way, we can investigate the variance of $\{\dot{X}(t)\}$ by taking $t_2 = t_1 = t$ in Eq. 3.44. The result can be written as

$$\sigma_{\dot{X}}^2(t) = \frac{G_0}{4m^2\zeta\omega_0}\left\{1 - e^{-2\zeta\omega_0 t}\left[\frac{\omega_0^2}{\omega_d^2} - \frac{\zeta\omega_0}{\omega_d}\sin(2\omega_d t) - \frac{\zeta^2\omega_0^2}{\omega_d^2}\cos(2\omega_d t)\right]\right\}U(t) \qquad (3.49)$$

The similarity between this equation and that for $\sigma_X^2(t)$ in Eq. 3.45 is striking. The only differences are a multiplication of Eq. 3.45 by ω_0^2 and a change of sign of one term. Given this similarity in the equations, it is not surprising to find that $\sigma_{\dot{X}}^2(t)$ and $\sigma_X^2(t)$ grow in a very similar manner. This is shown in Fig. 3.7, wherein the result of Eq. 3.49 is normalized and superimposed upon the results

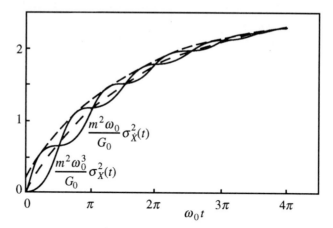

Figure 3.7 Variance of $\{\dot{X}(t)\}$ compared with results of Fig. 3.6 for $\zeta = 0.10$

from Fig. 3.6. This figure shows that the primary difference between $\sigma_{\dot{X}}^2(t)$ and $\sigma_X^2(t)$ is the phase difference caused by the change of sign on the $\sin(2\omega_d t)$ term. This difference has little significance except for small values of t. Figure 3.8 also shows that Eqs. 3.47 and 3.48, when multiplied by ω_0^2, provide a smooth approximation and a bound, respectively, on $\sigma_{\dot{X}}^2(t)$. Note that for smaller values of damping, the growth of $\sigma_{\dot{X}}^2(t)$ and $\sigma_X^2(t)$ will be even more similar than for the case shown in Fig. 3.7.

Finally, we note that by taking $t_2 = t_1 = t$ in Eq. 3.43 we can find the cross-covariance of the two random variables $X(t)$ and $\dot{X}(t)$ denoting the response and its derivative at the same instant of time. The result can be written as

$$K_{X\dot{X}}(t,t) = \frac{G_0}{4m^2\omega_d^2}e^{-2\zeta\omega_0 t}\left[1 - \cos(2\omega_d t)\right]U(t) = \frac{G_0}{2m^2\omega_d^2}e^{-2\zeta\omega_0 t}\sin^2(2\omega_d t)U(t) \quad (3.50)$$

Figure 3.8 shows a plot of this function for $\zeta = 0.10$. Two key features of this cross-covariance are that it tends to zero as t becomes large and it is exactly zero at times t which are an integer multiple of $\pi/(2\omega_d)$. Thus, $X(t)$ and $\dot{X}(t)$ are uncorrelated twice during each period of response of the oscillator. Furthermore, $K_{X\dot{X}}(t,t)$ is always relatively small. For a lightly damped system (i.e., for $\zeta \ll 1$) we can say that the maximum cross-covariance is approximately equal to the bound of $G_0/(2m^2\omega_d^2)$. Note that this bound does not depend on ζ, whereas the covariances of $\{X(t)\}$ and $\{\dot{X}(t)\}$ both vary like ζ^{-1}, so become very large when $\zeta \ll 1$. This, then, implies that the correlation coefficient

$$\rho_{X\dot{X}}(t,t) \equiv \frac{K_{X\dot{X}}(t,t)}{\sqrt{K_{XX}(t,t)K_{\dot{X}\dot{X}}(t,t)}}$$

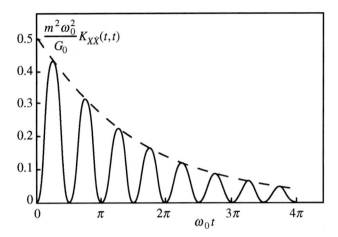

Figure 3.8 Covariance of $\{X(t)\}$ and $\{\dot{X}(t)\}$ for $\zeta = 0.10$

is generally small for the response of the single-degree-of-freedom system. The exception is when t is small, since in that situation $K_{XX}(t,t)$ and $K_{X\dot{X}}(t,t)$ in the denominator of the correlation coefficient are also small.

It may be instructive to note that one can also obtain Eq. 3.50 from Eq. 3.45, rather than using Eq. 3.42. In particular, Eq. 2.63 states that

$$\frac{d}{dt}E[X^2(t)] = E[2X(t)\dot{X}(t)]$$

and this gives

$$\phi_{X\dot{X}}(t,t) = \frac{1}{2}\frac{d\phi_{XX}(t,t)}{dt}$$

Using this along with the fact that $E[\dot{X}(t)]$ is the derivative of $E[X(t)]$ gives the covariance relationship as

$$K_{X\dot{X}}(t,t) = \frac{1}{2}\frac{d\sigma_X^2(t)}{dt} \tag{3.51}$$

Substituting $\sigma_X^2(t)$ from Eq. 3.45 into this expression gives exactly Eq. 3.50.

It should be noted that the equations developed in this section may also be used to describe the conditional covariance of the response of the SDF system when one is given the values of $X(t_0)$ and $\dot{X}(t_0)$ at some specified time t_0. In particular, the response at time t due to the portion of a delta-correlated excitation occurring after time t_0 is exactly the same as the response at time $t - t_0$ of a system at rest at time $t = 0$. Thus, Eqs. 3.38 and 3.42 give the conditional covariance of the response as

$$Cov[X(t_1), X(t_2) | X(t_0) = u, \dot{X}(t_0) = v] = \frac{G_0}{4m^2 \zeta \omega_0^3} \left\{ e^{-\zeta \omega_0 |t_1 - t_2|} \left[\cos[\omega_d(t_1 - t_2)] + \frac{\zeta \omega_0}{\omega_d} \sin[\omega_d |t_1 - t_2|] \right] \right.$$
$$-e^{-\zeta \omega_0 (t_1 + t_2 - 2t_0)} \left[\frac{\omega_0^2}{\omega_d^2} \cos[\omega_d(t_1 - t_2)] + \frac{\zeta \omega_0}{\omega_d} \sin[\omega_d(t_1 + t_2)] - \frac{\zeta^2 \omega_0^2}{\omega_d^2} \cos[\omega_d(t_1 + t_2 - 2t_0)] \right] \right\}$$

$$(3.52)$$

provided that t_1 and t_2 are both greater than t_0. From this relationship one can find the conditional variance of $X(t)$ and $\dot{X}(t)$, their cross-covariance, etc. Similarly, the conditional mean of Eq. 3.37 can be written as

$$E[X(t) | X(t_0) = u, \dot{X}(t_0) = v] = u g(t - t_0) + v mh_x(t - t_0) + \int_{t_0}^{\infty} \mu_F(s) h_x(t - s) ds \qquad (3.53)$$

in which

$$g(t) = e^{-\zeta \omega_0 t} \left[\cos(\omega_d t) + \frac{\zeta \omega_0}{\omega_d} \sin(\omega_d t) \right]$$

is the deterministic response to a unit displacement initial condition at time zero, and $mh_x(t)$ is the response to a unit velocity initial condition at time zero.

3.7   Stationary SDF Response to Delta-Correlated Excitation

In the preceding section we analyzed the response of the single-degree-of-freedom system to a delta-correlated excitation which was identically zero for $t < 0$, then was the same as a stationary process for $t \geq 0$. Note that a simple shift along the time axis will give us corresponding results for a problem in which the excitation begins to become effective at some time t_0 instead of at zero. That is, replacing t with $t - t_0$ in any of the response equations in the preceding section gives the corresponding response measure for this new loading situation with $F(t) = W(t)U(t - t_0)$. One particular situation which we might consider is when $t_0 = -\infty$, so that the excitation has always been effective. In this situation we expect the response to be stationary, and the fact that $t - t_0 = \infty$ shows that the stationary response levels are the same as asymptotes for $t \rightarrow \infty$ of the nonstationary response from the previous section. Stated differently, the nonstationary response from the previous section tends to stationary response at times long after the instant when the excitation first became active. This simple result will hold true for any linear system with a stationary excitation, if a stationary response exists for the problem.

If we had not already investigated nonstationary levels of $K_{XX}(t_1, t_2)$ for the SDF oscillator, then the easiest way to find the covariance function for stationary response would be to use Eq. 3.25 written for covariance with $G_{FF}(\tau) = G_0 \delta(\tau)$ to obtain

$$G_{XX}(\tau) = \int_{-\infty}^{\infty}\int_{-\infty}^{\infty} G_{FF}(\tau - r_1 + r_2)h_x(r_1)h_x(r_2)\,dr_1\,dr_2 = G_0\int_{-\infty}^{\infty} h_x(r)h_x(r-\tau)\,dr$$

Evaluation of this integral is somewhat simpler than the determination of Eq. 3.42, since the limits of integration are simpler. However, since we already have Eq. 3.42 we can obtain exactly the same stationary response covariance by letting t_1 and t_2 both go to infinity in that equation while holding $\tau = t_1 - t_2$ to a finite level. The result is

$$G_{XX}(\tau) = \frac{G_0}{4m^2\zeta\omega_0^3}e^{-\zeta\omega_0|\tau|}\left[\cos(\omega_d\tau) + \frac{\zeta\omega_0}{\omega_d}\sin(\omega_d|\tau|)\right] \tag{3.54}$$

Similarly, the cross-covariance of $\{X(t)\}$ and $\{\dot{X}(t)\}$ and the covariance of $\{\dot{X}(t)\}$ can be obtained either from covariance equations of the form of Eqs. 3.25 and 3.26 with $G_{FF}(\tau) = G_0\delta(\tau)$, or from the large t limits of Eqs. 3.43 and 3.44; the results are

$$G_{X\dot{X}}(\tau) = \frac{G_0}{4m^2\zeta\omega_0\omega_d}e^{-\zeta\omega_0|\tau|}\sin(\omega_d\tau) \tag{3.55}$$

and

$$G_{\dot{X}\dot{X}}(\tau) = \frac{G_0}{4m^2\zeta\omega_0}e^{-\zeta\omega_0|\tau|}\left[\cos(\omega_d\tau) - \frac{\zeta\omega_0}{\omega_d}\sin(\omega_d|\tau|)\right] \tag{3.56}$$

Figure 3.9 shows normalized versions of these three stationary covariance results for the situation with $\zeta = 0.10$. Note that Eqs. 3.54 and 3.56 are even functions of τ, thereby satisfying the necessary symmetry condition for an autocovariance function. The cross-covariance of Eq. 3.55, however, is an odd function of τ. Comparing Eqs. 3.54 and 3.56 shows that the covariance of $\{\dot{X}(t)\}$ differs from that of $\{X(t)\}$ only by the inclusion of an ω_0^2 factor in the multiplier in front, and by the sign of the $\sin(\omega_d|\tau|)$ term. Since this latter term is multiplied by the small parameter ζ, it is not surprising that the normalized plots of $G_{XX}(\tau)$ and $G_{\dot{X}\dot{X}}(\tau)$ in parts (a) and (c) of the figure are almost the same. The difference between the shapes of the $G_{XX}(\tau)$ and $G_{\dot{X}\dot{X}}(\tau)$ plots is significant only when τ is very small. At $\tau = 0$ the slope of the $G_{\dot{X}\dot{X}}(\tau)$ plot is discontinuous, while $G_{XX}(\tau)$ is smooth.

The stationary response variance values can now be obtained either by choosing $\tau = 0$ in Eqs. 3.54 and 3.56 or by letting t tend to infinity in the nonstationary response expressions in Eqs. 3.47 and 3.49, giving

$$\sigma_X^2 = \frac{G_0}{4m^2\zeta\omega_0^3} \quad \text{and} \quad \sigma_{\dot{X}}^2 = \frac{G_0}{4m^2\zeta\omega_0} \tag{3.57}$$

Similarly, either Eq. 3.55 or 3.50 shows that the covariance of the two random variables $X(t)$ and $\dot{X}(t)$ is zero at any instant of time during stationary response. Note that the two stationary variance values given in Eq. 3.57 are exactly the multipliers which have appeared on all the nonstationary and stationary variance values derived for the $\{X(t)\}$ and $\{\dot{X}(t)\}$ processes. One can also rewrite these

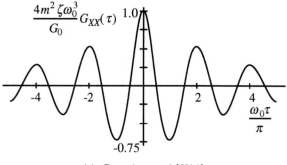

(a) Covariance of $\{X(t)\}$

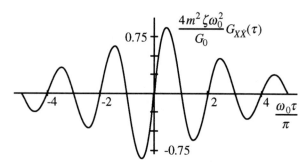

(b) Cross-covariance of $\{X(t)\}$ and $\{\dot{X}(t)\}$

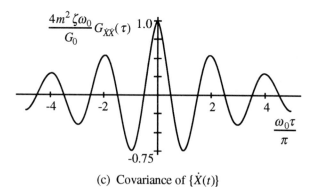

(c) Covariance of $\{\dot{X}(t)\}$

Figure 3.9 Stationary covariance for SDF with $\zeta = 0.10$

important values in terms of the original parameters of the physical model shown in Fig. 3.3, obtaining

$$\sigma_X^2 = \frac{G_0}{2kc} \quad \text{and} \quad \sigma_{\dot{X}}^2 = \frac{G_0}{2mc} \tag{3.58}$$

These expressions reveal the possibly surprising facts that the stationary variance of the displacement $\{X(t)\}$ is unaffected by the mass of the system and the stationary variance of the velocity $\{\dot{X}(t)\}$ is unaffected by the stiffness of the system.

∗∗∗

Example 3.7: Analyze the response covariance for the SDF of Eq. 3.39 with $c = 0$ and $k = 0$. Rather than looking for limits of the general expressions as c and k tend to zero, let us directly analyze the response of the simplified equation

$$m\ddot{X}(t) = F(t)$$

Since $\ddot{X}(t) = F(t)/m$, we can directly write the expression for the covariance function of the acceleration as

$$K_{\ddot{X}\ddot{X}}(t,s) = \frac{1}{m^2} K_{FF}(t,s) = \frac{G_0}{m^2} \delta(t-s) U(t) U(s)$$

Repeatedly integrating this expression gives

$$K_{\dot{X}\ddot{X}}(t,s) = \frac{G_0}{m^2} U(t-s) U(t) U(s)$$

$$K_{\dot{X}\dot{X}}(t,s) = \frac{G_0}{m^2} \min(t,s) U(t) U(s) = \frac{G_0}{m^2}[t+(s-t)U(t-s)] U(t) U(s)$$

$$K_{X\dot{X}}(t,s) = \frac{G_0}{2m^2}\left[t^2 - (t-s)^2 U(t-s)\right] U(t) U(s)$$

and

$$K_{XX}(t,s) = \frac{G_0}{6m^2}\left[s^2(3t-s)U(t-s)+t^2(3s-t)[1-U(t-s)]\right] U(t) U(s)$$

$$= \frac{G_0}{6m^2}[\max(t,s)]^2[3\max(t,s)-\min(t,s)] U(t) U(s)$$

By letting $s = t$ in the $K_{\dot{X}\dot{X}}(t,s)$ and $K_{XX}(t,s)$ expressions we can find the response variances as

$$\sigma_{\dot{X}}^2(t) = \frac{G_0}{m^2} t\, U(t) \quad \text{and} \quad \sigma_X^2(t) = \frac{G_0}{3m^2} t^3 U(t)$$

showing that application of the delta-correlated excitation to a mass with no restoring force gives a velocity variance which grows linearly with time, and a displacement variance which grows cubicly. Similarly, the cross-covariance of displacement and velocity at time t is

$$K_{X\dot{X}}(t,t) = \frac{G_0}{2m^2} t^2 U(t)$$

which gives their correlation coefficient as $\rho_{X\dot{X}} \equiv K_{X\dot{X}}(t,t) / \sigma_X \sigma_{\dot{X}} = \sqrt{3}/2$. One can go one step further with this type of analysis and notice that the cross-covariance function for velocity and acceleration is finite but has a discontinuity at $s = t$. We have $K_{\dot{X}\ddot{X}}(t,s) = G_0/m^2$ for $s < t$, and this covariance is zero for $s > t$.

Although the mass with no restoring force may seem like a very impractical system, its response actually does describe an important situation. In particular, it is the same as the limit of the response of the general SDF system of Eq. 3.39 immediately after the application of the delta-correlated excitation. The key reason for this is the fact that at $t = 0$, both $X(t)$ and $\dot{X}(t)$ in Eq. 3.39 are zero. Thus, we know that the restoring force $c\dot{X} + kX$ starts at zero when $t = 0$. Furthermore, we have found that the variance of $X(t)$ and $\dot{X}(t)$ grow continuously, so that there must be a small range of t values for which $c\dot{X} + kX$ remains small. During this range of t values, one can approximate Eq. 3.39 by the simplified expression of this example. In fact, it can be shown that all the variance and covariance functions we have derived here agree exactly with the limits of power series expansions of the corresponding expressions for the SDF system.

One implication of these results is that the variance of velocity and displacement of the SDF system initially grow like $G_0 t / m^2$ and $G_0 t^3 / (3m^2)$. These rates of growth assure that the initial values of displacement are relatively much smaller than those of velocity, and one can see that this initial behavior shows quite clearly for the SDF response in Fig. 3.7. For the simple mass, the linear and cubic growth, respectively, of velocity and displacement variance will eventually lead to displacement levels being relatively large compared to velocity, as both grow toward infinity. For the SDF system the large time behavior is quite different, though, since the spring and dashpot become effective. In particular, both displacement and velocity variance grow in the same manner toward stationary values, as illustrated in Fig. 3.7. For both the simple mass and the SDF system, the variance of acceleration is infinite for all time during which a delta-correlated excitation is applied.

****

3.8 Nearly Delta-Correlated Processes

As stated previously, no physical process is truly delta-correlated, even though delta-correlated processes are often used in practical problems to approximate excitations which are nearly independent of themselves at distinct times t and s. Equations 3.31 and 3.32 illustrate that one must somehow choose G_0 (either as a constant or a function of time) in order to define the covariance function of the approximating delta-correlated process. For a covariant stationary process, one logical way to do this is on the basis of the variance bound in Eq. 3.29. In particular, if one chooses G_0 for the delta-correlated process to be the same as the integral of the absolute value of $G_{FF}(\tau)$ for the physical process:

$$G_0 = \int_{-\infty}^{\infty} |G_{FF}(r_3)| dr_3 \tag{3.59}$$

then Eq. 3.29 will give the same bound on the response variance for the idealized approximation of the excitation as for the physical process excitation.

In order to generalize Eq. 3.59 to a situation with a nonstationary $K_{FF}(s_1, s_2)$, let us introduce new time variables as $\tau = s_1 - s_2$ and $t = (s_1 + s_2)/2$. The first of these variables is the time difference, and the second is a symmetric function of s_1 and s_2 which becomes time s in the limit of $s_1 = s_2 = s$. We then have

$$K_{FF}(s_1, s_2) = K_{FF}\left(t + \frac{\tau}{2}, t - \frac{\tau}{2}\right)$$

For a nearly delta-correlated process it is clear that this covariance function must decay rapidly as τ increases. The generalization of Eq. 3.59 is then to choose

$$G_0(t) = \int_{-\infty}^{\infty} K_{FF}\left(t + \frac{\tau}{2}, t - \frac{\tau}{2}\right) d\tau \tag{3.60}$$

One may note that Eq. 3.59 is the stationary special case of Eq. 3.60.

The choice of G_0 will be given more consideration in Chapter 4, when we consider frequency decompositions of processes. It will be found that, in some applications, we can identify choices for G_0 which are better than Eq. 3.60.

****

Example 3.8: Choose G_0 for a delta-correlated approximation of a stochastic process $\{F(t)\}$ with autocovariance function

$$G_{FF}(\tau) = A e^{-c|\tau|}$$

From Eqs. 3.59 and 3.34 we see that choosing $G_0 = 2A/c$ will give the response variance of any linear oscillator with the delta-correlated excitation to be the same as the bound of Eq. 3.29 for the response variance for $\{F(t)\}$ excitation. Of course $\{F(t)\}$ is nearly delta-correlated only if c is large, and we can also note that $G_{FF}(\tau)$ can be considered to tend to $(2A/c)\delta(\tau)$ in the limit as $c \to \infty$.
****

Exercises
****
Impulse Response Functions
****
3.1 Consider a linear system for which the response $x(t)$ to an excitation $f(t)$ is governed by the differential equation
$$\ddot{x}(t) + 5\dot{x}(t) + 6x(t) = f(t)$$

Find the impulse response function $h(t)$ by considering $f(t) = \delta(t)$. Sketch your answer.

****

3.2 Consider a linear system for which the response $x(t)$ to an excitation $f(t)$ is governed by the third-order differential equation

$$\frac{d^3x(t)}{dt^3} + 2\zeta\omega_0 \frac{d^2x(t)}{dt^2} + \omega_0^2 \frac{dx(t)}{dt} = f(t)$$

with $0 < \zeta < 1$. Find the impulse response function $h(t)$ by considering $f(t) = \delta(t)$. Specifically:

(a) Find three different functions which each solve the homogeneous third-order equation.

(b) Identify the three initial conditions (at $t = 0^+$) which must be satisfied by $h(t)$.

(c) Solve for the constants in the homogeneous solution.

****

First-Order System

****

3.3 Consider the response $\{X(t)\}$ of a linear system described by

$$c\dot{X}(t) + kX(t) = F(t)$$

for which the impulse response function was derived in Example 3.1 as $h_x(t) = c^{-1}e^{-kt/c}U(t)$.

(a) Find expressions for $\mu_X(t)$ and $\mu_{\dot{X}}(t)$ for response to a general $\{F(t)\}$ excitation.

(b) Find expressions for $\phi_{XX}(t_1,t_2)$, $\phi_{X\dot{X}}(t_1,t_2)$ and $\phi_{\dot{X}\dot{X}}(t_1,t_2)$ for a general $\{F(t)\}$ excitation.

(c) Find simplified expressions for μ_X and $R_{XX}(\tau)$ for an $\{F(t)\}$ process which is mean-value and second-moment stationary.

****

3.4 Let A be a random variable with $\mu_A = 0$ and $\sigma_A^2 = 1$. Let this random variable be taken as the initial condition at time zero $[X(0) = A]$ of a linear system governed by

$$\dot{X}(t) = bX(t) \quad \text{in which } b \text{ is a constant.}$$

(a) Find the following autocorrelation and crosscorrelation functions of the response:

$$\phi_{XX}(t_1,t_2), \ \phi_{X\dot{X}}(t_1,t_2), \text{ and } \phi_{\dot{X}\dot{X}}(t_1,t_2)$$

(b) Are $E[X^2(t)]$ and $E[\dot{X}^2(t)]$ finite for all $t > 0$? What, if any, restrictions on the value of b are required?

(c) Can $\{X(t)\}$ tend to a covariant stationary process as t goes to infinity?

****

3.5 Consider the response $\{X(t)\}$ of a linear system described by

$$\ddot{X}(t) + c\dot{X}(t) = F(t) \quad \text{in which } c > 0 \text{ is a constant, and } \{F(t)\} \text{ is a stationary stochastic}$$

process. Under what conditions will either of the mean squared responses $E[X^2(t)]$ and $E[\dot{X}^2(t)]$ be stationary and finite?

****

3.6 Consider the response $\{X(t)\}$ of a linear system described by

$$\dot{X}(t) + aX(t) = W(t)U(t)$$

in which $\{W(t)\}$ is delta-correlated process with $E[W(t)] = 0$ and $E[W(t_1)W(t_2)] = G_0\delta(t_1 - t_2)$. Note that the unit steps give $\{W(t)\}$ as being applied to the oscillator only for $t \geq 0$. The system is initially at rest: $X(0) = 0$.

(a) Find $\mu_X(t)$ and $\phi_{XX}(t_1, t_2)$ for all times.

(b) Find $E[X^2(t)]$ for all $t > 0$.

(c) Find the stationary autocorrelation function: $R_{XX}(\tau) = \lim_{t \to \infty} \phi_{XX}(t + \tau, t)$.

****

3.7 Consider the response $\{X(t)\}$ of a linear system described by

$$c\dot{X}(t) + kX(t) = F(t)$$

with the deterministic initial condition $X(0) = x_0$. For $t > 0$, $\{F(t)\}$ is a mean zero, nonstationary delta-correlated process with

$$E[F(t)] = 0 \text{ and } E[F(t_1)F(t_2)] = G_0[1 - e^{-\alpha t_1}]\delta(t_1 - t_2).$$

(a) Find the mean value function of $\{X(t)\}$: $\mu_X(t)$.

(b) Find the covariance of $\{X(t)\}$: $K_{XX}(t_1, t_2)$.

(c) Find $E[X^2(t)]$.

****

3.8 Consider a linear system whose response $\{X(t)\}$ to an excitation $\{F(t)\}$ is governed by

$$\ddot{X}(t) + c\dot{X}(t) = F(t) \qquad \text{in which } c > 0 \text{ is a constant.}$$

The excitation $\{F(t)\}$ is a delta correlated process modulated by the unit step function. That is,

$$K_{FF}(t_1, t_2) = G_0\delta(t_1 - t_2)U(t_1)U(t_2)$$

(a) Find the variance of $\{X(t)\}$: $\sigma_X^2(t)$.

(b) Find the variance of $\{\dot{X}(t)\}$: $\sigma_{\dot{X}}^2(t)$.

(c) Discuss the behavior of the response variances as t tends to $+\infty$.

****

3.9 Consider the response $\{X(t)\}$ of a linear system described by

$$\dot{X}(t) + aX(t) = W(t)[U(t) - U(t - T)]$$

in which $\{W(t)\}$ is stationary delta-correlated process with

$$E[W(t)] = 0 \quad \text{and} \quad E[W(t_1)W(t_2)] = G_0\delta(t_1 - t_2)$$

Note that the unit steps give $\{W(t)\}$ as being applied to the oscillator only for $0 \leq t \leq T$. The system is initially at rest: $X(0) = 0$.

(a) Find $\mu_X(t)$ and $\phi_{XX}(t_1, t_2)$ for all times.

(b) Sketch $E[X^2(t)]$ versus t (for all $t > 0$).

****

3.10 Consider a linear system governed by the first-order differential equation

$$\dot{X}(t) + aX(t) = F(t) \qquad \text{for} \qquad t \geq 0$$

in which a is a constant, with $0 < a < 1$. Let $\{X(t)\}$ have the random initial condition $X(0) = Y$ and let the forcing function consist of one random impulse at time $t = (4a)^{-1}$:

$F(t) = Z\delta[t - 1/(4a)]$, in which Y and Z are random variables with

$\quad E(Y) = 1 \qquad E(Y^2) = 2 \qquad E(Z) = 0 \qquad E(Z^2) = 5 \qquad E(YZ) = 1.5$

(a) Find $E[X(t)]$ for all $t \geq 0$.

(b) Find $E[X^2(t)]$ for all $t \geq 0$.

(c) Sketch your answers to (a) and (b) versus t.

****

3.11 Consider a linear system governed by the first-order differential equation

$\quad \dot{X}(t) + aX(t) = F(t)$

The excitation $\{F(t)\}$ is a stationary random process with a nonzero mean value, $\mu_F = b$, and an autocovariance function of $K_{FF}(t + \tau, t) = G_0 e^{-c|\tau|}$ for all t and τ. The terms a, b, and c are constants with $a > 0$ and $c > 0$.

(a) Find the mean value of the response $\{X(t)\}$: μ_X.

(b) Find the mean squared value of the response $\{X(t)\}$: $E(X^2)$.

****

3.12 Let the random process $\{Z(t)\}$ denote the ground acceleration during an earthquake. One possible model is a segment from a stationary process described by

$\quad \dot{Z}(t) + bZ(t) = bW(t)$

in which $\{W(t)\}$ is a mean zero delta-correlated process. Let $K_{WW}(t, s) = G_0 \delta(s - t)$.

(a) Find the stationary value of $E(Z^2)$.

(b) Data for North American earthquakes indicate that $b = 6\pi$ rad/sec (i.e., 3 Hz) is a reasonable choice. Using this b value, find G_0 such that the stationary standard deviation of $\{Z(t)\}$ is 0.981 m/s^2 (i.e., 0.1 g). Give your units for G_0 also.

****

3.13 Consider a causal linear system for which the impulse response function $h_x(t)$ is known. That is, if the input is $F(t)$ then the response is

$\quad X(t) = \int_{-\infty}^{\infty} F(s) h_x(t - s) \, ds$

(a) Find the general integral expression relating $E[X(t_1)X(t_2)X(t_3)]$ to the corresponding input expression $E[F(t_1)F(t_2)F(t_3)]$.

(b) Simplify this integral for the special case of a stationary delta-correlated excitation with

$\quad E[F(t_1)F(t_2)F(t_3)] = Q_0 \delta(t_1 - t_3)\delta(t_2 - t_3)$

(c) For the special case of $h_x(t) = e^{-at} U(t)$ and the delta-correlated excitation of part (b), find the stationary value of $E(X^3)$.

****

SDF System

****

3.14 Consider the response $\{X(t)\}$ of a linear SDF system described by

$$\ddot{X}(t) + 2\zeta\omega_0\dot{X}(t) + \omega_0^2 X(t) = F(t)$$

with the deterministic initial conditions $X(0) = 0$, $\dot{X}(0) = v_0$. For $t > 0$, $\{F(t)\}$ is a stationary delta-correlated process with $E[F(t)] = 0$ and $E[F(t_1)F(t_2)] = G_0\delta(t_1 - t_2)$.

(a) Find the mean value function of the response: $\mu_X(t)$ for $t \geq 0$.

(b) Find the variance of $\{X(t)\}$: $\sigma_X^2(t)$ for $t \geq 0$.

(c) Find $E[X^2(t)]$ for $t \geq 0$.

Give all answers in terms of the parameters ω_0, ζ_0, G_0, v_0, and $\omega_d = \omega_0(1 - \zeta_0^2)^{1/2}$.

****

3.15 Consider the response $\{X(t)\}$ of a linear SDF system described by

$$\ddot{X}(t) + 2\zeta\omega_0\dot{X}(t) + \omega_0^2 X(t) = F(t)$$

with a random initial condition: $X(0) = Y$, $\dot{X}(0) = 0$. The forcing function consists of one random impulse at time T: $F(t) = Z\delta(t - T)$ in which $T > 0$ is a constant and Y and Z are random variables with $E(Y) = 1$ $E(Y^2) = 2$ $E(Z) = 0$ $E(Z^2) = 5$ $E(YZ) = 1.5$

(a) Find $E[X(t)]$ for all $t \geq 0$.

(b) Find $E[X^2(t)]$ for all $t \geq 0$.

(c) Sketch your answers to (a) and (b) versus t.

****

3.16 Assume that you are concerned about the possibility of collision between two buildings during wind excited vibration. The east-west clearance between the buildings is 35 mm. Model each building as an SDF system:

$$\ddot{X}_j(t) + 2\zeta_j\omega_j\dot{X}_j(t) + \omega_j^2 X_j(t) = F_j(t)$$

in which X_j = displacement (to the east) of the top of building j. The building parameters are

Building A: $\omega_A = 2\pi$ rad/sec $\zeta_A = 0.01$

Building B: $\omega_B = \pi$ rad/sec $\zeta_B = 0.01$

The critical condition is with the wind blowing from the north, so you can neglect any static deflection of the buildings. Model the east-west excitation of each building (due to vortex shedding, etc.) as a delta-correlated process: $E[F_j(t)F_j(s)] = G_j\delta(t - s)$. Noting that $F_j(t)$ is the force per unit mass in each SDF system, consider the excitation intensities to be

$$G_A = 40 \ (\text{mm}^2/\text{s}^3) \quad \text{and} \quad G_B = 80 \ (\text{mm}^2/\text{s}^3)$$

(a) Find the standard deviation σ_{X_j} for the stationary displacement response at the top of each building.

(b) Let Y_B denote the displacement of building B at the level of the top of building A, and assume that $Y_B = X_B/2$. Assume that $\{X_A(t)\}$ and $\{X_B(t)\}$ are independent random processes, and find the standard deviation σ_Z of the relative displacement at the top of building A: $Z(t) = X_A(t) - Y_B(t)$.

(c) Presume that you wish to reduce the probability of collision by adding damping to one of the structures. To which structure would you add the damping? Briefly explain your answer.

****

3.17 Consider approximating an earthquake ground acceleration of 30-second duration by using a mean zero, delta-correlated process with

$$E[\ddot{Y}(t)\ddot{Y}(s)] = (1.5 \text{ m}^2 / \sec^3)\,\delta\,(t - s)[U(t) - U(t - 30)]$$

in which t is time in seconds. Let the ground velocity have an initial value of zero: $\dot{Y}(0) = 0$. Evaluate $E[\dot{Y}^2(t)]$ and sketch this mean squared value versus t over the range of $0 \le t \le 50$.
****

3.18 Let the random process $\{Z(t)\}$ denote the ground acceleration during an earthquake. One model which has been studied is described by

$$a\dot{Z}(t) + Z(t) = W(t)$$

in which $\{W(t)\}$ is a delta-correlated process. (Using $a = 0.05$ sec/radian gives a fairly reasonable model.) Let $\{X(t)\}$ represent the response of a single-degree-of-freedom structural model excited by this ground motion:

$$m\ddot{X}(t) + c\dot{X}(t) + kX(t) = -mZ(t)$$

Substitute $Z(t)$ from the second equation into the first equation in order to obtain an ordinary differential equation relating $\{X(t)\}$ and $\{W(t)\}$.
****

Chapter 4

Frequency Domain Analysis

4.1 Frequency Content Of A Stochastic Process

The Fourier transform provides the classical method for decomposing a time history into its frequency components (see Appendix D for background information on Fourier analysis). For any time history $f(t)$, we will denote the Fourier transform by $\tilde{f}(\omega)$ and define it as

$$\tilde{f}(\omega) = \frac{1}{2\pi} \int_{-\infty}^{\infty} f(t) e^{-i\omega t} dt \tag{4.1}$$

The inverse relationship is then

$$f(t) = \int_{-\infty}^{\infty} \tilde{f}(\omega) e^{i\omega t} d\omega \tag{4.2}$$

The interpretation of the Fourier transform as a frequency decomposition is based on Eq. 4.2. Since an integral may be viewed as the limit of a summation, Eq. 4.2 shows that the original time history $f(t)$ is essentially a summation of harmonic terms, with $\tilde{f}(\omega)$ being the amplitude of the $e^{i\omega t}$ component having frequency ω. This amplitude is generally complex, as shown by Eq. 4.1, so one must consider its absolute value in order to determine how much of the total $f(t)$ time history is contributed by frequency ω. A condition assuring the existence of the Fourier transform $\tilde{f}(\omega)$ is that $f(t)$ be absolute-value integrable

$$\int_{-\infty}^{\infty} |f(t)| dt < \infty \tag{4.3}$$

and Eq. 4.2 produces exactly the original $f(t)$ function at all points of continuity of $f(t)$. At a point of finite discontinuity of $f(t)$, Eq. 4.2 gives a value midway between the left-hand and right-hand limits at the point.

We now wish to apply the Fourier transform idea to a stochastic process $\{X(t)\}$. Simply using a stochastic integrand in Eq. 4.1 gives

$$\tilde{X}(\omega) = \frac{1}{2\pi} \int_{-\infty}^{\infty} X(t) e^{-i\omega t} dt \tag{4.4}$$

which is a stochastic integral of the form of Eq. 2.74. This defines a new stochastic process $\{\tilde{X}(\omega)\}$ on the set of all possible ω values. Taking the expectation of Eq. 4.4 shows that the mean value function for $\{\tilde{X}(\omega)\}$, if it exists, can be written as

$$\mu_{\tilde{X}}(\omega) = \frac{1}{2\pi} \int_{-\infty}^{\infty} \mu_X(t) e^{-i\omega t} dt = \tilde{\mu}_X(\omega) \tag{4.5}$$

The purpose of the last term in Eq. 4.5 is to emphasize that the mean value function for the Fourier transform of X is exactly the Fourier transform of the mean value function for X.

Next, we consider the second moments of the $\{\tilde{X}(\omega)\}$ process, but we slightly modify the definition of the second moment function since $\{\tilde{X}(\omega)\}$ is not real. For a complex stochastic process, we will use the standard procedure of defining the second moment function with a complex conjugate (denoted by the superscript *) on the second term:

$$\phi_{\tilde{X}\tilde{X}}(\omega_1, \omega_2) = E[\tilde{X}(\omega_1)\tilde{X}^*(\omega_2)]$$

Note that this is identical to Eq. 2.2 for a real process. One reason for the inclusion of the complex conjugate is so that $\phi_{\tilde{X}\tilde{X}}(\omega_1, \omega_2)$ will be real along the line $\omega_2 = \omega_1$. In particular, we have $\phi_{\tilde{X}\tilde{X}}(\omega, \omega) = E[|\tilde{X}^2(\omega)|]$. Using this definition and Eq. 4.4 gives

$$\phi_{\tilde{X}\tilde{X}}(\omega_1, \omega_2) = \frac{1}{(2\pi)^2} \int_{-\infty}^{\infty} \int_{-\infty}^{\infty} \phi_{XX}(t, s) e^{-i(\omega_1 t - \omega_2 s)} dt\, ds \tag{4.6}$$

The corresponding definition of autocovariance function is

$$K_{\tilde{X}\tilde{X}}(\omega_1, \omega_2) = E\left([\tilde{X}(\omega_1) - \mu_{\tilde{X}}(\omega_1)][\tilde{X}(\omega_2) - \mu_{\tilde{X}}(\omega_2)]^*\right)$$

and this is easily shown to be

$$K_{\tilde{X}\tilde{X}}(\omega_1, \omega_2) = \frac{1}{(2\pi)^2} \int_{-\infty}^{\infty} \int_{-\infty}^{\infty} K_{XX}(t, s) e^{-i(\omega_1 t - \omega_2 s)} dt\, ds \tag{4.7}$$

Expressions for higher moment functions can, obviously, also be derived.

One major difficulty in applying the Fourier transform procedure to many problems of interest is the fact that the expressions just written will not exist if $\{X(t)\}$ is a stationary process. In particular, Eqs. 4.3 and 4.5 show that $\mu_{\tilde{X}}(\omega)$ may not exist unless $\mu_X(t)$ is absolute value integrable, and this requires that $\mu_X(t)$ tend to zero as $|t|$ tends to infinity. In general, $\mu_{\tilde{X}}(\omega)$ exists for a mean-value stationary process only if the process is mean zero, in which case $\mu_{\tilde{X}}(\omega)$ is also zero. Similarly, Eqs. 4.6 and 4.7 show that $\phi_{\tilde{X}\tilde{X}}(\omega_1, \omega_2)$ and $K_{\tilde{X}\tilde{X}}(\omega_1, \omega_2)$ may not exist unless $\phi_{XX}(t, s)$ and $K_{XX}(t, s)$, respectively, tend to zero as t and s tend to infinity. These conditions cannot be uniformly met for a stochastic process which is second-moment stationary or covariant stationary, respectively. In particular, we have $\phi_{XX}(t, s) = R_{XX}(t - s)$ and $K_{XX}(t, s) = G_{XX}(t - s)$ for a stationary process so that t and s tending to infinity with a finite value of $|t - s|$ would not give $\phi_{XX}(t,s)$ or $K_{XX}(t,s)$ tending to zero, and the integrals might not exist. Thus, we will modify the Fourier transform procedure in order to obtain a form which will apply to stationary processes.

4.2 Spectral Density Functions

To avoid the problem of existence of the Fourier transforms, consider a new stochastic process $\{X_T(t)\}$ which is a truncated version of our original stationary process $\{X(t)\}$:

$$X_T(t) = X(t)[U(t + T / 2) - U(t - T / 2)] \qquad (4.8)$$

in which $U(\cdot)$ denotes the unit step function (see Sec. A.2 of Appendix A). The Fourier transform of $X_T(t)$ is sure to exist since the process is defined to be zero if $|t| > T/2$, as illustrated in Fig. 4.1. In particular, the moment integrals in Eqs. 4.5 - 4.7 will exist, since the limits of integration will now be bounded. Of course, we must consider the behavior as T goes to infinity or we will not have a description of the complete $\{X(t)\}$ process, and we know that the integrals in Eqs. 4.5 - 4.7 may diverge in that situation. Thus, we must find a way of normalizing our results in such a way that a limit exists as T goes to infinity.

We really need not give any more attention to the mean value function of the Fourier transform for a stationary process. If $\{X(t)\}$ is mean-value stationary and $\mu_X \neq 0$, then the Fourier transform of $\mu_X(t)$ contains only an infinite spike at the origin: $\mu_{\tilde{X}}(\omega) = \mu_X \delta(\omega)$. The more interesting situation involves the second moments of the process. Specifically, we will consider $\{X(t)\}$ to be stationary, so that

$$\phi_{\tilde{X}_T \tilde{X}_T}(\omega_1, \omega_2) = \frac{1}{(2\pi)^2} \int_{-T/2}^{T/2} \int_{-T/2}^{T/2} R_{XX}(t - s) e^{-i(\omega_1 t - \omega_2 s)} \, dt \, ds \qquad (4.9)$$

and

$$K_{\tilde{X}_T \tilde{X}_T}(\omega_1, \omega_2) = \frac{1}{(2\pi)^2} \int_{-T/2}^{T/2} \int_{-T/2}^{T/2} G_{XX}(t - s) e^{-i(\omega_1 t - \omega_2 s)} \, dt \, ds \qquad (4.10)$$

In most applications, we will find that two of the random variables which make up a stochastic process are independent if they correspond to observation times which are infinitely far apart. That

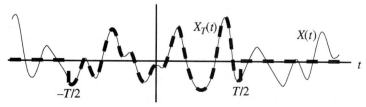

Figure 4.1 Truncated time history of $X(t)$

is, $X(t + \tau)$ becomes uncorrelated with $X(t)$ as τ becomes very large. This causes $G_{XX}(\tau)$ to tend to zero and $R_{XX}(\tau)$ to tend to $\mu_X(t + \tau)\mu_X(t)$ as $|\tau|$ tends to infinity. One consequence of this behavior is the fact that as T tends to infinity the integral in Eq. 4.9 may grow much more rapidly than the one in Eq. 4.10. Thus, we will focus on the better behaved case and restrict our attention to covariance only. If the mean is stationary and zero, then this is the same as the second moment of the process. If $\mu_X(t)$ does not equal zero, then analysis of covariance gives us information about how $X(t)$ varies from this deterministic mean value.

We will now analyze the covariance function of the Fourier transform of the truncated stochastic process $\{X_T(t)\}$ when $\{X(t)\}$ is a covariant stationary process. The integral in Eq. 4.10 can be rearranged by using exactly the same changes of variables and other manipulations as were used in deriving the ergodicity condition in Eq. 2.49, although the presence of the exponential term makes things a little more complicated this time. The result can be written as

$$K_{\tilde{X}_T\tilde{X}_T}(\omega_1,\omega_2) = \frac{1}{(2\pi)^2}\left(\frac{2\sin[(\omega_1-\omega_2)T/2]}{(\omega_1-\omega_2)}\int_{-T}^{T}G_{XX}(\tau)e^{-i\omega_1\tau}\,d\tau\right.$$

$$-2\cos[(\omega_1-\omega_2)T/2]\int_0^T G_{XX}(\tau)\left[\frac{\sin(\omega_1\tau)-\sin(\omega_2\tau)}{(\omega_1-\omega_2)}\right]d\tau$$

$$\left.+\frac{2\sin[(\omega_1-\omega_2)T/2]}{(\omega_1-\omega_2)}\int_0^T G_{XX}(\tau)[\cos(\omega_1\tau)-\cos(\omega_2\tau)]d\tau\right) \qquad (4.11)$$

Note that this expression is quite well behaved if ω_1 and ω_2 are not nearly equal. In particular, if we impose the condition that the autocovariance function is absolutely integrable:

$$\int_{-\infty}^{\infty}|G_{XX}(\tau)|d\tau < \infty \qquad (4.12)$$

then it is clear that we have $K_{\tilde{X}_T\tilde{X}_T}(\omega_1,\omega_2)$ bounded for any T value, including $T \to \infty$, for $\omega_1 \neq \omega_2$. The situation is not so good, though, when ω_1 and ω_2 are nearly equal. Letting $\omega_2 = \omega$ and $\omega_1 = \omega + \Delta\omega$, and then taking the limit as $\Delta\omega \to 0$ so that ω_1 and ω_2 approach a common value, gives $\sin(\omega_1\tau) \approx \sin(\omega\tau) + \Delta\omega\,\tau\cos(\omega\tau)$ and $\cos(\omega_1\tau) \approx \cos(\omega\tau) - \Delta\omega\,\tau\sin(\omega\tau)$. Substituting these expressions into Eq. 4.11, along with $\sin(\Delta\omega\,\tau) \approx \Delta\omega\,\tau$, shows that the final term in Eq. 4.11 drops out and the other two terms give

$$K_{\tilde{X}_T\tilde{X}_T}(\omega,\omega) = \frac{1}{(2\pi)^2}\left(T\int_{-T}^{T}G_{XX}(\tau)e^{-i\omega\tau}\,d\tau - 2\int_0^T G_{XX}(\tau)\,\tau\cos(\omega\tau)d\tau\right)$$

Clearly, the first term in this expression will grow proportionally with T as T goes to infinity, under the condition of Eq. 4.12. It is not immediately obvious how the second term will behave, but it can be shown that it grows less rapidly than the first, so that the first term always dominates in the limit.[1] Thus, we can say that

$$K_{\tilde{X}_T \tilde{X}_T}(\omega,\omega) \rightarrow \frac{T}{(2\pi)^2} \int_{-T}^{T} G_{XX}(\tau) e^{-i\omega\tau} d\tau \quad \text{as} \quad T \rightarrow \infty \qquad (4.13)$$

    Since Eq. 4.13 grows proportionally with T for large T values, we can define a normalized form which will exist for $T = \infty$ by dividing $K_{\tilde{X}_T \tilde{X}_T}(\omega_1,\omega_2)$ by T. Note, though, that this normalized autocovariance of the Fourier transform will go to zero as T goes to infinity for $\omega_1 \neq \omega_2$ since the unnormalized form was finite in that situation. Thus, all the useful information which this normalized form can give us about the stationary process is included in the special case $K_{\tilde{X}_T \tilde{X}_T}(\omega,\omega)$, corresponding to $\omega_1 = \omega_2$. Therefore, we define the new function of frequency as

$$S_{XX}(\omega) = \lim_{T\to\infty} \frac{2\pi}{T} K_{\tilde{X}_T \tilde{X}_T}(\omega,\omega) \equiv \lim_{T\to\infty} \frac{2\pi}{T} E\left([\tilde{X}_T(\omega) - \mu_{\tilde{X}_T(\omega)}][\tilde{X}_T(\omega) - \mu_{\tilde{X}_T(\omega)}]^* \right) \qquad (4.14)$$

in which the inclusion of the factor of 2π is arbitrary, but traditional. An equivalent form is

$$S_{XX}(\omega) = \lim_{T\to\infty} \frac{2\pi}{T} E[|\tilde{X}_T(\omega) - \tilde{\mu}_{X_T}(\omega)|^2] \qquad (4.15)$$

Noting that the fact that $X(t)$ is real gives $\tilde{X}_T(-\omega) = \tilde{X}_T^*(\omega)$, we can also rewrite this expression as

$$S_{XX}(\omega) = \lim_{T\to\infty} \frac{2\pi}{T} E\left([\tilde{X}_T(\omega) - \tilde{\mu}_{X_T}(\omega)][\tilde{X}_T(-\omega) - \tilde{\mu}_{X_T}(-\omega)] \right) \qquad (4.16)$$

We will call this function $S_{XX}(\omega)$ the **autospectral density function** of $\{X(t)\}$. Other terms that are commonly used for it are "power spectral density" or simply "spectral density."[2]

    Comparing Eqs. 4.13 and 4.14 shows a very important result:

$$S_{XX}(\omega) = \frac{1}{2\pi} \int_{-\infty}^{\infty} G_{XX}(\tau) e^{-i\omega\tau} d\tau \qquad (4.17)$$

---

[1] We will not include a proof of the general case, but it is very easy to verify the special case in which $|G_{XX}(\tau)| \leq c\tau^{-1-\varepsilon}$ for some positive constants c and ε. Then, the absolute value of the second integral in the equation is bounded by $cT^{1-\varepsilon}$ as T goes to infinity, so that the second term becomes less significant than the first term.

[2] Many authors define the spectral density slightly differently, using the autocorrelation function rather than the autocovariance function. The difficulty with that approach is that it gives a spectral density which generally does not exist if the Fourier transform of $\mu_X(t)$ does not exist, as when $\{X(t)\}$ is mean-value stationary and $\mu_X \neq 0$. This problem is then circumvented by looking at the spectral density of a mean zero process defined as $Y(t) = X(t) - \mu_X(t)$. The definition used here gives exactly the same result in a more direct manner.

The autospectral density function is exactly the Fourier transform of the autocovariance function. Furthermore, this implies that the inverse must also be true:

$$G_{XX}(\tau) = \int_{-\infty}^{\infty} S_{XX}(\omega) e^{i\omega\tau} d\omega \qquad (4.18)$$

We are assured that this Fourier transform pair will exist for a process $\{X(t)\}$ satisfying Eq. 4.12. Note, also, the similarity of Eqs. 4.17 and 4.7. Equation 4.7 defines $K_{\tilde{X}\tilde{X}}(\omega_1, \omega_2)$ as the double Fourier transform of the general $K_{XX}(t, s)$ covariance function, one transform with respect to t and one with respect to s. For the covariant stationary process we know that $K_{XX}(t, s)$ can be replaced by a function $G_{XX}(\tau)$ of the one time argument $\tau = t - s$, and Eq. 4.17 shows that the autospectral density function is the Fourier transform of this function with respect to its single argument.

Similar to Eq. 4.14, one can define a cross-spectral density function for two jointly covariant stationary stochastic processes $\{X(t)\}$ and $\{Y(t)\}$ as

$$S_{XY}(\omega) = \lim_{T \to \infty} \frac{2\pi}{T} K_{\tilde{X}_T \tilde{Y}_T}(\omega, \omega) = \lim_{T \to \infty} \frac{2\pi}{T} E\big([\tilde{X}_T(\omega) - \tilde{\mu}_{X_T}(\omega)][\tilde{Y}_T(-\omega) - \tilde{\mu}_{Y_T}(-\omega)]\big) \quad (4.19)$$

and show that it is the Fourier transform of the cross-correlation function of the processes:

$$S_{XY}(\omega) = \frac{1}{2\pi} \int_{-\infty}^{\infty} G_{XY}(\tau) e^{-i\omega\tau} d\tau \qquad (4.20)$$

One can also define higher-order spectral density functions in terms of Fourier transforms of higher-order moment or cumulant functions of a stochastic process. In practice, though, these are much less frequently used, and we will not discuss them here. A brief introduction to the so-called "bispectrum" and "trispectrum" will be given in Section 5.5. It should also be noted that we have only considered stationary processes in our discussion of spectral densities, although this limitation is not essential. The most common nonstationary generalization of the frequency decomposition concept makes use of the Stieltjes integral in defining the evolutionary spectral density. This topic is discussed by Nigam (1983) and by Lin and Cai (1995).

4.3 Properties of Spectral Density Functions

First we note three physical features of the autospectral density function. Directly from the definition in Eq. 4.14, we see that for any stochastic process $\{X(t)\}$ we must have

$$S_{XX}(\omega) \text{ is always real for all values of } \omega \qquad (4.21)$$

$$S_{XX}(\omega) \geq 0 \quad \text{for all values of } \omega \qquad (4.22)$$

and

$$S_{XX}(-\omega) = S_{XX}(\omega) \quad \text{for all values of } \omega \qquad (4.23)$$

The fact that $S_{XX}(\omega)$ is real and symmetric can also be immediately verified from Eq. 4.17, keeping in mind that $G_{XX}(\tau)$ is real and symmetric. Confirming that Eq. 4.17 also implies that $S_{XX}(\omega)$ is nonnegative requires use of the fact that $K_{XX}(t,s)$ is a nonnegative definite function. This does not mean that $K_{XX}(t,s)$ is never less than zero, but rather that $\sum_{j,k} a_j a_k K_{XX}(t_j,t_k) \geq 0$ for any real numbers a_j. This property follows from noting that this double summation over j and k is exactly the variance of a random variable which is defined as $\sum_j a_j X(t_j)$.

Possibly the most important feature of the various spectral density functions is that each of them gives all the information about a corresponding covariance (or cumulant) function. In particular, the usual autospectral density function of Eqs. 4.14 - 4.18 gives all the information about the autocovariance function of the $\{X(t)\}$ process and the cross-spectral density function of Eqs. 4.19 and 4.20 gives all the information about the cross-covariance of $\{X(t)\}$ and $\{Y(t)\}$. We know that this must be true since the inverse Fourier transform formula (as in Eq. 4.18) allows us to compute the covariance function from knowledge of the spectral density function. Combined with the fundamental idea of the Fourier transform, this shows that a spectral density function is a complete frequency decomposition of a stationary variance function. For example, $S_{XX}(\omega)$ can definitely be considered as a single frequency component since its definition depends only on the frequency ω part of the Fourier transform (i.e., the frequency decomposition) of $\{X(t)\}$, and knowledge of all the $S_{XX}(\omega)$ frequency components is sufficient information to recreate the variance function. These are precisely the characteristics of a frequency decomposition of any quantity.

Another important property of autospectral density functions is obtained from the special case corresponding to setting $\tau = 0$ in equation 4.18, giving

$$\sigma_X^2 = G_{XX}(0) = \int_{-\infty}^{\infty} S_{XX}(\omega)\, d\omega \qquad (4.24)$$

This shows that the variance of a covariant stationary process can always be found from the area under its autospectral density function. This property is frequently used for calculation of variance values in stochastic dynamic analysis.

The symmetry of autospectral density has also led to the use of an alternate form called the single-sided autospectral density. One argument for using this representation is that all the information is contained within the half of the function corresponding to $0 \leq \omega < \infty$, so why bother with the other half. In order to retain the property that variance is given by an integral of autospectral density, as in Eq. 4.24, the single-sided autospectral density is typically taken to be $2S_{XX}(\omega)$. It should also be noted that there are other normalizations of the autospectral density in common usage, so that one must be quite careful in interpreting the meaning of any quoted value for the quantity. In particular, some prefer to replace Eq. 4.24 with an integral over frequency in Hz (cycles per second) rather than radians per second, and this requires a modification of autospectral density by a factor of 2π. Furthermore, this variation may be found in conjunction with either the two-sided $S_{XX}(\omega)$ that we use or the single sided version, giving four different possibilities. Throughout this book we will use only the two-sided autospectral density function defined by Eq. 4.14, since use of various forms seems to add unnecessary possibilities for confusion.

Next we will use Eq. 4.19 to evaluate the derivatives of the autocovariance function, as

$$\frac{d^j G_{XX}(\tau)}{d\tau^j} = (i)^j \int_{-\infty}^{\infty} \omega^j S_{XX}(\omega) e^{i\omega\tau} d\omega$$

and substitute these expressions into relationships between derivatives of autocovariance functions and covariance functions involving the stochastic derivative of a process, as given in Eqs. 2.59 and 2.60. Thus, the cross-covariance of $\{X(t)\}$ and $\{\dot{X}(t)\}$ is given by

$$G_{X\dot{X}}(\tau) = -\frac{dG_{XX}(\tau)}{d\tau} = \int_{-\infty}^{\infty} (-i\omega) S_{XX}(\omega) e^{i\omega\tau} d\omega$$

and the autocovariance of $\{\dot{X}(t)\}$ is

$$G_{\dot{X}\dot{X}}(\tau) = -\frac{d^2 G_{XX}(\tau)}{d\tau^2} = \int_{-\infty}^{\infty} \omega^2 S_{XX}(\omega) e^{i\omega\tau} d\omega$$

Comparing these two results with Eq. 4.18 shows that the spectral densities involving $\{\dot{X}(t)\}$ are simply related to those for $\{X(t)\}$. For example, the expression for $G_{X\dot{X}}(\tau)$ is precisely in the form of an inverse Fourier transform of $(-i\omega) S_{XX}(\omega)$. The uniqueness of the Fourier transform then tells us that

$$S_{X\dot{X}}(\omega) = -i\omega \, S_{XX}(\omega) \tag{4.25}$$

Similarly, the expression for $G_{\dot{X}\dot{X}}(\tau)$ tells us that

$$S_{\dot{X}\dot{X}}(\omega) = \omega^2 S_{XX}(\omega) \tag{4.26}$$

and if we write $X^{(j)}$ for the jth-order stochastic derivative with respect to t of $\{X(t)\}$, then we can obtain the general result that

$$S_{X^{(j)}X^{(k)}}(\omega) = (-1)^k (i)^{j+k} \omega^{j+k} S_{XX}(\omega) \tag{4.27}$$

4.4   Narrowband Processes

We will now investigate the relationship between the shape of an autospectral density function and the nature of the possible time histories for the stochastic process which it describes. We will begin this investigation with consideration of the so-called "narrowband processes." A process $\{X(t)\}$ is said to be narrowband if the autospectral density $S_{XX}(\omega)$ is very small except within a narrow band of frequencies. Since $S_{XX}(\omega)$ is an even function (Eq. 4.23), this really means that the band of significant frequencies appears both for positive and negative ω values. We can state this as $S_{XX}(\omega) \approx 0$ unless $|\omega| \approx \omega_c$ for some given characteristic frequency ω_c. To illustrate this idea more clearly, let us choose the example with $S_{XX}(\omega)$ equal to a constant value S_0 if ω is within a distance b of either $+\omega_c$ or $-\omega_c$, and identically zero for all other frequencies, as shown in Fig. 4.2. Mathematically, this can be written as

$$S_{XX}(\omega) = S_0 \big(U[|\omega| - (\omega_c - b)] - U[|\omega| - (\omega_c + b)] \big)$$

We can now use the inverse Fourier transform of Eq. 4.18 to find the corresponding autocovariance function

$$G_{XX}(\tau) = S_0 \int_{-\omega_c - b}^{-\omega_c + b} e^{i\omega\tau} \, d\omega + S_0 \int_{\omega_c - b}^{\omega_c + b} e^{i\omega\tau} \, d\omega = 2S_0 \int_{\omega_c - b}^{\omega_c + b} \cos(\omega\tau) \, d\omega$$

or

$$G_{XX}(\tau) = 2S_0 \frac{\sin[(\omega_c + b)\tau] - \sin[(\omega_c - b)\tau]}{\tau} = 4S_0 \frac{\cos(\omega_c\tau)\sin(b\tau)}{\tau}$$

Setting $\tau = 0$ in this expression gives the variance of the process as $\sigma_X^2 = 4bS_0$, which could also have been found from the area under the autospectral density curve, as given in Eq. 4.24. We can use this variance as a normalization factor and write the autocovariance function in a convenient form as

$$G_{XX}(\tau) = \sigma_X^2 \cos(\omega_c\tau) \frac{\sin(b\tau)}{b\tau}$$

In order for this process to be classified as narrowband, it is necessary that the parameter b be small in some sense. In particular, if $b \ll \omega_c$ then the oscillations of the $\sin(b\tau)$ term are very slow compared to those at frequency ω_c. In this case, one can consider $\sin(b\tau)/(b\tau)$ to provide a slowly varying envelope for the $\sigma_X^2 \cos(\omega_c\tau)$ term. Figure 4.3 shows a sketch of this behavior for $b = \omega_c/10$. For small values of τ it is seen that $G_{XX}(\tau)$ is well approximated by $\sigma_X^2 \cos(\omega_c\tau)$, but the amplitude of this cosine function decays as τ increases. This is a fundamental characteristic of any

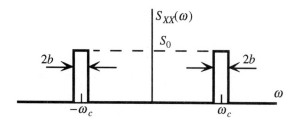

Figure 4.2 Ideal narrowband process

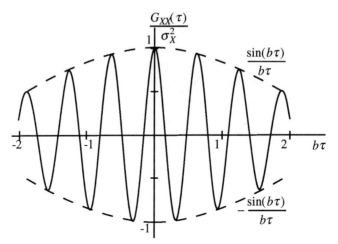

Figure 4.3 Autocovariance function for a narrowband process

narrowband process, and is the feature by which a process can be identified as being narrowbanded based only on knowledge of the autocovariance function.

   The ultimate narrowband process can be considered to be the process considered in Example 2.3, for which the autocovariance function was exactly given as $G_{XX}(\tau) = \sigma_X^2 \cos(\omega_c \tau)$, with no decay of the amplitude of $G_{XX}(\tau)$. The autospectral density for this example is $S_{XX}(\omega) = (\sigma_X^2/2)\delta(\omega + \omega_c) + (\sigma_X^2/2)\delta(\omega - \omega_c)$, showing that there are variance contributions only from components at the discrete frequencies $\omega = \pm\omega_c$. This is not too surprising, though, since the time histories of this process are exactly cosine waves, with amplitude and phase which are random variables. That is, any particular time history is a simple harmonic function with frequency ω_c and fixed amplitude and phase, although different time histories generally have different values for amplitude and phase. Thus, we see that a more general narrowband process, with an autocovariance function similar to that shown in Fig. 4.3, must be somewhat similar to the process having harmonic time histories. We will examine this idea more carefully since it provides a way to relate the frequency domain concept of a narrowband process directly to the characteristics of the time histories of the process.

   The key idea that we want to emphasize is that if any process has an autocovariance function which approximates $\sigma_X^2 \cos(\omega_c \tau)$, then the quantity $Y(t) \equiv [X(t) - \mu_X(t)]$ for most of its time histories must closely approximate a harmonic function with frequency ω_c; thus, one can write

$$X(t) = \mu_X(t) + A(t)\cos[\omega_c t + \theta(t)] \qquad (4.28)$$

in which the amplitude $A(t)$ and phase $\theta(t)$ are slowly varying. More precisely, if $G_{XX}(\tau) \approx \sigma_X^2 \cos(\omega_c \tau)$ over the range $|\tau| \leq T_0$, then a segment of length T_0 of a time history of $Y(t)$ will

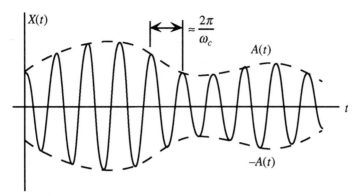

Figure 4.4 Typical time history of a narrowband process

almost surely be well approximated by the harmonic function with frequency ω_c. The reasoning is simply that if $E[Y(t+\tau)Y(t)] \approx E(Y^2)\cos(\omega_c\tau)$, then with probability one $Y(t+\tau) \approx Y(t)\cos(\omega_c\tau)$, and this is true for a range of t and τ values only if $Y(t)$ has the prescribed harmonic form. Thus, we see that for any narrowband process we should expect a typical segment of a time history of $Y(t)$ to approximate a segment from a harmonic function, as illustrated in Fig. 4.4. Conversely, if the time histories of $Y(t) = [X(t) - \mu_X(t)]$ do have the form of harmonic functions with a fixed frequency ω_c and slowly varying amplitude and phase, then we know that $\{X(t)\}$ will have a narrowband autospectral density function. Furthermore, neglecting $\dot{A}(t)$ and $\dot{\theta}(t)$ for this nearly harmonic time history gives $\sigma_X^2 \approx E[A^2(t)]/2$ and $\sigma_{\dot{X}}^2 \approx \omega_c^2 E[A^2(t)]/2$, which implies that the characteristic frequency ω_c for any narrowband process must be approximated by

$$\omega_c \approx \frac{\sigma_{\dot{X}}}{\sigma_X} \qquad (4.29)$$

The ideas of amplitude and phase of a stochastic process are investigated in much more detail in Section 4.9, but without the requirement that $\{X(t)\}$ be narrowbanded in nature.

4.5   Broadband Processes and White Noise

As noted in the preceding section, a narrowband process is one for which the variance is all contributed by components in the vicinity of a single frequency ω_c and its reflection $-\omega_c$. At the opposite extreme is a process for which all components contribute equally, such that the spectral density is the same for all ω values. Let us investigate a process $\{X(t)\}$ of this type with $S_{XX}(\omega) = S_0$. The first thing that one may notice about this autospectral density function is that it is not integrable. Thus, Eq. 4.24 gives $\sigma_X^2 = \infty$, showing that no physically meaningful process has precisely this autospectral density. Nonetheless, let us investigate the process further, since we will find that it can be used to approximate meaningful processes. Since $S_{XX}(\omega)$ is not integrable, the inverse Fourier transform takes the degenerate form of a Dirac delta function, giving the autocovariance function as $G_{XX}(\tau) = 2\pi S_0\, \delta(\tau)$, which does satisfy the forward Fourier transform of Eq. 4.17. Thus, we see that the autocovariance function for this process is the same as for the stationary delta-correlated process introduced in Section 3.5. Conversely, we can say that a delta-correlated process will always give an autospectral density which is a constant. Clearly, the relationship between the autospectral density level S_0 and the covariance parameter G_0 used in Eq. 3.31 is simply $G_0 = 2\pi S_0$.

A process with a constant value of $S_{XX}(\omega)$ is commonly referred to as white noise, by analogy with white light, which supposedly contains equal contributions from all relevant frequency components. The basic difference between the terms "delta-correlated" and "white noise" in defining a stochastic process is that the former term focuses attention on the time domain characterization of the process, while the latter focuses attention on the frequency domain. A process $\{X(t)\}$ which is so erratic that $X(t)$ and $X(s)$ are independent of each other for any two distinct times ($t \neq s$), also contains all frequencies equally in the frequency decomposition of its covariance.[3] The time histories of any delta-correlated process, of course, must be extremely erratic. In fact, if the time histories had any finite probability of being continuous at some value of t, then $X(t)$ and $X(s)$ could not be independent if s were near t. Thus, a truly delta-correlated process is so erratic as to preclude the drawing of sample time histories. Since a truly constant autospectral density gives the unbounded Dirac delta function as the autocovariance function, it is useful also to examine physically realizable processes which approach this delta-correlated process in the limit. Examples 4.1 and 4.2 present two such situations.

****

Example 4.1: Let $\{X(t)\}$ be a covariant stationary process with autospectral density function given by

$$S_{XX}(\omega) = S_0 e^{-\gamma|\omega|}$$

Find the autocovariance function for the process, and verify that it tends to that of a delta-correlated process in the limit as $\gamma \to 0$.

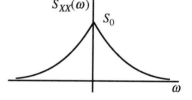

The inverse Fourier transform of $S_{XX}(\omega)$, as in Eq. 4.18, gives

$$G_{XX}(\tau) = S_0 \int_{-\infty}^{\infty} e^{-\gamma|\omega|}\, e^{i\omega\tau}\, d\omega$$

$$= S_0 \int_{-\infty}^{0} e^{(\gamma+i\tau)\omega}\, d\omega + S_0 \int_{0}^{\infty} e^{(-\gamma+i\tau)\omega}\, d\omega$$

or

$$G_{XX}(\tau) = S_0 \left(\frac{1}{\gamma+i\tau} + \frac{1}{\gamma-i\tau} \right) = \frac{2\gamma S_0}{\gamma^2 + \tau^2}$$

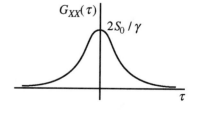

[3] Also the frequency decompositions of higher-order cumulants, as represented by bispectra, trispectra, etc., are made up of equal contributions from all frequencies as well.

If we let γ tend to zero for any $\tau \neq 0$, we see that $G_{XX}(\tau)$ tends to zero, as it should for a delta-correlated process. On the other hand, for $\tau = 0$ we have $\sigma_X^2 = G_{XX}(0) = 2S_0 / \gamma$, which tends to infinity as γ tends to zero, which again is characteristic of a delta-correlated process. The only remaining issue to check, in verifying that this autocovariance is that of a delta-correlated process, regards its integrability with respect to τ. However, we know, without any further explicit integration, that the Fourier transform of this autocovariance function is $S_0 e^{-\gamma|\omega|}$, as written in Eq. 4.17. Setting $\omega = 0$ in this Fourier transform shows that the autocovariance function is integrable, and confirms that setting $G_0 = 2\pi S_0$ does make $G_{XX}(\tau)$ tend to the form $G_0 \delta(\tau)$ of Eq. 3.31 as $\gamma \to 0$.

****

Example 4.2: Let $\{X(t)\}$ be a covariant stationary process with autospectral density function given by

$$S_{XX}(\omega) = S_0 \frac{\alpha^2}{\alpha^2 + \omega^2}$$

Find the autocovariance function for the process, and verify that it tends to that of a delta-correlated process in the limit as $\alpha \to \infty$.

One must be slightly more creative in finding the inverse Fourier transform of $S_{XX}(\omega)$ in this example:

$$G_{XX}(\tau) = S_0 \alpha^2 \int_{-\infty}^{\infty} \frac{e^{i\omega\tau}}{\alpha^2 + \omega^2} d\omega$$

For someone familiar with the theory of complex variables, the most straightforward approach is to note that the integrand has poles at $\omega = \pm i\alpha$ and use the calculus of residues to evaluate the integral. Alternatively, one might look in a table of Fourier transforms to find the same result. In the present situation, we can avoid doing either of these things by comparing this integral with the results in Example 4.1. In particular, the Fourier transform giving the autospectral density as a function of the autocovariance in that example is

$$S_{XX}(\omega) = S_0 e^{-\gamma|\omega|} = \frac{1}{2\pi}\int_{-\infty}^{\infty} G_{XX}(\tau)e^{-i\omega\tau}\,d\tau = \frac{1}{2\pi}\int_{-\infty}^{\infty}\frac{2\gamma S_0}{\gamma^2 + \tau^2}e^{-i\omega\tau}\,d\tau$$

and simplification of this integral and changing variables gives

$$\int_{-\infty}^{\infty}\frac{e^{i\omega\tau}}{\alpha^2 + \omega^2}\,d\omega = \frac{\pi}{\alpha}e^{-\alpha|\tau|}$$

Applying this result to the current problem gives $G_{XX}(\tau) = S_0 \pi \alpha\, e^{-\alpha|\tau|}$. Clearly, this satisfies the conditions of $G_{XX}(\tau) \to 0$ as $\alpha \to \infty$ for $\tau \neq 0$, and $G_{XX}(\tau) \to \infty$ as $\alpha \to \infty$ for $\tau = 0$. Furthermore, the integral of $e^{-\alpha|\tau|}$ from $-\infty$ to ∞ is $1/(2\alpha)$, confirming that the limit of $G_{XX}(\tau)$ can be written as $2\pi S_0 \delta(\tau)$.

****

4.6   Linear Dynamics and Harmonic Transfer Functions

In Chapter 3, we formulated time domain expressions for stochastic dynamics of linear systems directly from the deterministic Duhamel convolution integral. Now we will pursue a parallel development, using the common deterministic frequency domain formulation of linear dynamics to give us information about stochastic response. Using the Fourier transform, one can describe the input and output, respectively, of a time-invariant linear system either as $f(t)$ and $x(t)$ or as $\tilde{f}(\omega)$ and $\tilde{x}(\omega)$, as shown in Fig. 4.5.

The function $H_x(\omega)$ in Fig. 4.5 is called the harmonic transfer function, and it is defined to be the ratio between $x(t)$ and $f(t)$ when $f(t)$ is the pure harmonic $e^{i\omega t}$. That is, if $f(t) = e^{i\omega t}$, then $x(t) = H_x(\omega) \, e^{i\omega t}$. Using superposition, then, one can say that $\tilde{f}(\omega)e^{i\omega t}$ induces a response of $H_x(\omega)\tilde{f}(\omega)e^{i\omega t}$, so that a time history of input of

$$f(t) = \int_{-\infty}^{\infty} \tilde{f}(\omega)e^{i\omega t} \, d\omega$$

causes a time history of response of

$$x(t) = \int_{-\infty}^{\infty} H_x(\omega)\tilde{f}(\omega)e^{i\omega t} \, d\omega \qquad (4.30)$$

Comparing this equation with the standard inverse Fourier transform shows that

$$\tilde{x}(\omega) = H_x(\omega)\tilde{f}(\omega) \qquad (4.31)$$

Equation 4.31 is generally regarded as the standard form of the frequency domain input-output relationship for linear dynamics, but it should be remembered that it also implies the time domain relationship of Eq. 4.30.

Before proceeding to stochastic structural dynamics, it is useful to note the relationship between the harmonic transfer function $H_x(\omega)$ and the impulse response function $h_x(t)$ used in the time domain formulation. One easy way of identifying this relationship is to use the harmonic function $f(t) = e^{i\omega t}$ as the input in the convolution integral of Eq. 3.2. This gives

$$x(t) = \int_{-\infty}^{\infty} f(t-r)\, h_x(r)\, dr = \int_{-\infty}^{\infty} e^{i\omega(t-r)} h_x(r)\, dr = e^{i\omega t} \int_{-\infty}^{\infty} e^{-i\omega r} h_x(r)\, dr$$

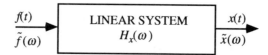

Figure 4.5 Schematic of general linear system

The fact that this response must be $x(t) = H_x(\omega)\, e^{i\omega t}$, from the definition of the harmonic transfer function, shows that $H_x(\omega)$ is exactly 2π times the Fourier transform of $h_x(t)$:

$$H_x(\omega) = 2\pi \tilde{h}_x(\omega) = \int_{-\infty}^{\infty} e^{-i\omega r} h_x(r)\, dr \tag{4.32}$$

Switching to stochastic processes $\{F(t)\}$ and $\{X(t)\}$ for input and output of the linear system now gives us the relationship of $\tilde{X}(\omega) = H_x(\omega)\tilde{F}(\omega)$ for the stochastic Fourier transform processes $\{\tilde{F}(\omega)\}$ and $\{\tilde{X}(\omega)\}$, just as Eq. 4.31 described the deterministic situation. Taking the expectation of this relationship shows that the Fourier transform of the mean value function of the output can be written as $\tilde{\mu}_X(\omega) = H_x(\omega)\tilde{\mu}_F(\omega)$. Furthermore, if the processes are covariant stationary then the definition of autospectral density in Eqs. 4.14 - 4.16 gives

$$S_{XX}(\omega) = H_x(\omega)H_x(-\omega)S_{FF}(\omega) = |H_x(\omega)|^2 S_{FF}(\omega) \tag{4.33}$$

which is the fundamental frequency domain relationship regarding the autocovariance of the response of a linear dynamic system. It can be viewed as a frequency domain form of Eq. 3.25, which gave the second moment function of the stationary response. A major difference is that finding the response second moment or covariance function from Eq. 3.25 involves a double integral in the time domain, while Eq. 4.33 gives us the response autospectral density from simply a multiplication of functions in the frequency domain.

After the response autospectral density is found, one can obtain the response variance from a single frequency domain integration, as in Eq. 4.24, as an alternative to the double convolution integral of Eq. 3.25. Thus, it is sometimes easier to use frequency domain manipulations to find the response variance rather than using the time domain relationships. There is one complication with this approach, though. For many problems, the form of the spectral densities is such that one must use the theory of residues from complex variable analysis in order to evaluate conveniently the integral over ω values. The reader who is not proficient with complex analysis should keep in mind that the time domain integration will always give the same results, and may actually be easier to perform. The choice between time domain and frequency domain calculations of variance is often primarily a matter of personal preference of the analyst.

Similar to Eq. 4.33 one can write the cross-spectral density for two response processes $\{X(t)\}$ and $\{Y(t)\}$ as

$$S_{XY}(\omega) = H_x(\omega)H_y(-\omega)S_{FF}(\omega) = H_x(\omega)H_y^*(\omega)S_{FF}(\omega) \tag{4.34}$$

and the cross-spectral density between the excitation and the response is

$$S_{XF}(\omega) = H_x(\omega)S_{FF}(\omega) \tag{4.35}$$

We can also confirm the corresponding relationships involving derivatives, as given in Eqs. 4.25 - 4.27, by noting that if $x(t) = H_x(\omega)e^{i\omega t}$ then $\dot{x}(t) = i\omega H_x(\omega)e^{i\omega t}$, so that the harmonic transfer function for the derivative response is simply $H_{\dot{x}}(\omega) = i\omega H_x(\omega)$. Since an integral of a cross-spectral density function always gives a cross-covariance term, these cross-covariances can also be evaluated from the frequency domain analysis.

One other advantage of frequency domain analysis involves the ease with which the harmonic transfer functions can be evaluated for a system governed by a linear differential equation. For example, if we consider the general nth-order system (as in Eq. 3.5):

$$\sum_{j=0}^{n} a_j \frac{d^j x(t)}{dt^j} = f(t)$$

then substitution of $f(t) = e^{i\omega t}$ and $x(t) = H_x(\omega)\,e^{i\omega t}$ gives

$$H_x(\omega) = \left(\sum_{j=0}^{n} a_j (i\omega)^j \right)^{-1} \tag{4.36}$$

Thus, one needs use only algebra to find the harmonic transfer function for the system, whereas finding the corresponding time domain impulse response function involves solution of an initial value problem for the differential equation.

The input-output autospectral density relationship in Eq. 4.33 shows that any linear system has the effect of shaping autospectral density, in the sense that the autospectral density of the output is $|H_x(\omega)|^2$ times that of the input. Systems specifically designed to produce a particular shape of output autospectral density are commonly called filters, and Eq. 4.33 shows that any linear system can be regarded as a linear filter. Three common categories of such filters are called low-pass, high-pass, and band-pass filters, depending on whether the output autospectral density is dominated by low-frequency components, high-frequency components, or some band of frequency components. It should be noted, though, that not all linear systems fall into these three categories since one can design a linear system to approximate any desired harmonic transfer function.

The frequency domain analysis procedures for dynamic response are applied to two simple nonoscillatory systems in the following two examples, then the much more important situation with oscillatory response is investigated in Section 4.7.

****

Example 4.3: Consider a linear system governed by the differential equation

$$c\dot{x}(t) + kx(t) = f(t)$$

for which the impulse response function was found in Example 3.1 as $h_x(t) = c^{-1}e^{-kt/c}U(t)$. Find the harmonic transfer function from the Fourier transform of $h_x(t)$, as in Eq. 4.32, and compare with the result from Eq. 4.36. Also find the autospectral density and the variance of the response $\{X(t)\}$ for the situation when $f(t)$ is replaced by a white noise (or delta-correlated) process $\{F(t)\}$ with $S_{FF}(\omega) = S_0$.

From Eq. 4.32 we have

$$H_x(\omega) = \int_{-\infty}^{\infty} e^{-i\omega r} h_x(r)\,dr = c^{-1}\int_0^{\infty} e^{-i\omega r} e^{-kr/c}\,dr = c^{-1}(i\omega + k/c)^{-1} = (k + i\omega c)^{-1}$$

which is obviously identical to what Eq. 4.36 gives.

The autospectral density of the response is now given by Eq. 4.33 as

$$S_{XX}(\omega) = |H_x(\omega)|^2 S_{FF}(\omega) = \frac{S_0}{k^2 + \omega^2 c^2}$$

The most direct method to evaluate the response variance from this autospectral density is to note that $S_{XX}(\omega)$ has poles at $\omega = \pm ik/c$ and use the calculus of residues.[4]

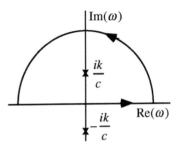

Thus, if we evaluate the integral along the real axis of the complex ω space by closing the contour in the upper half-space then we get the integral as being equal to $2\pi i$ times the residue at $\omega = ik/c$:

$$\sigma_X^2 = \int_{-\infty}^{\infty} \frac{S_0}{k^2 + \omega^2 c^2}\,d\omega = (2\pi i)\lim_{\omega \to ik/c} \frac{S_0(\omega - ik/c)}{c^2(\omega - ik/c)(\omega + ik/c)} = \frac{\pi S_0}{ck}$$

For this situation with a delta-correlated excitation, we can also obtain the response variance from a single integral in the time domain rather than the usual double integral. In particular, rather than using Eq. 3.28, we can use Eq. 3.36 to give $\sigma_X^2 = G_0 h_{x,variance}$ for $G_{FF}(\tau) = G_0\delta(\tau)$ and $h_{x,variance}$ being the integral of the square of the impulse response function, as given in Eq. 3.30. Thus,

$$h_{x,variance} = \int_{-\infty}^{\infty} h_x^2(r)\,dr = c^{-2}\int_0^{\infty} e^{-2kr/c}\,dr = c^{-2}(2k/c)^{-1} = (2ck)^{-1}$$

Also, our autospectral density function agrees with $G_{FF}(\tau) = G_0\delta(\tau)$ only if $G_0 = 2\pi S_0$, confirming that $\sigma_X^2 = \pi S_0/(ck)$.

****

Example 4.4: Find the harmonic transfer function for the linear system governed by the differential equation

$$m\ddot{x}(t) + c\dot{x}(t) = f(t)$$

---

[4] The details of this technique are included in any introductory textbook on complex analysis.

for which the impulse response function was found in Example 3.2 as $h_x(t) = c^{-1}[1 - e^{-ct/m}]U(t)$. Also consider the autospectral density and variance of the stochastic response of this system for a stochastic excitation with autospectral density $S_{FF}(\omega)$.

Strictly speaking, the Fourier transform of $h_x(t)$ does not exist for this problem since $h_x(t)$ is not integrable, which is the same condition as $h_{x,static} = \infty$ (as was pointed out in Example 3.2). Thus, one cannot directly apply Eq. 4.32. There is no difficulty, though, in using Eq. 4.36 to write

$$H_x(\omega) = \frac{1}{i\omega c - \omega^2 m}$$

Thus, the autospectral density of covariant stationary stochastic response will be

$$H_x(\omega) = \frac{S_{FF}(\omega)}{\omega^2 c^2 + \omega^4 m^2}$$

Note that the denominator of this expression tends to zero as ω tends to zero. In particular, it behaves like ω^2 for small values of ω, and this assures that the integral of the autospectral density will not exist if $S_{FF}(0) \neq 0$. Thus, this system has infinite response covariance for most stationary stochastic excitations, which agrees with our earlier observation that $h_{x,static} = \infty$. We do see, though, that if $S_{FF}(\omega)$ is zero or behaves like $|\omega|^b$ with $b > 1$, then the response variance will be finite. This shows that, for some excitations, a system can have a finite response variance even though it has $h_{x,static} = \infty$.
****

Example 4.5: Estimate the autospectral density $S_{XX}(\omega_c)$ based only on a single long time history of the process $\{X(t)\}$ and the results obtained from an ideal band-pass filter with a harmonic transfer function of $H(\omega) = 1$ if $|\omega \pm \omega_c| \leq \varepsilon$, and $H(\omega) = 0$ otherwise.

Let $\{Z(t)\}$ be the output from the filter when $\{X(t)\}$ is the input. This then gives us $S_{ZZ}(\omega) = S_{XX}(\omega)$ within the pass-band and $S_{ZZ}(\omega) = 0$ otherwise, as shown in the sketch.

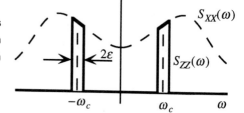

The $\{Z(t)\}$ process will be mean zero since the stationary mean value of $\{X(t)\}$ is a component at frequency zero, and this is blocked by the filter. Thus, the variance and mean squared value of $\{Z(t)\}$ are both given by

$$E[Z^2(t)] = \int_{-\infty}^{\infty} S_{ZZ}(\omega)\,d\omega = 2\int_0^{\infty} S_{ZZ}(\omega)\,d\omega = 2\int_{\omega_c - \varepsilon}^{\omega_c + \varepsilon} S_{XX}(\omega)\,d\omega$$

and if ε is small enough that $S_{XX}(\omega)$ is nearly linear across the range of integration, then we have

$$E[Z^2(t)] \approx 4\varepsilon\, S_{XX}(\omega_c)$$

Thus, $E[Z^2(t)]/4\varepsilon$ can be used as an estimate of the unknown $S_{XX}(\omega_c)$. Assuming covariance ergodicity of $\{Z(t)\}$, the value of $E[Z^2(t)]$ can be estimated from a time average, giving

$$S_{XX}(\omega_c) \approx \frac{1}{4\varepsilon T}\int_0^T Z^2(s)\,ds$$

Note that the finite bandwidth ε tends to smooth the measured autospectral density, and a peak in $S_{XX}(\omega)$ may be almost lost if it is not wider than 2ε. Based on this fact it may appear that one would want to choose ε as small as possible, so as to minimize this error due to variation of $S_{XX}(\omega)$ across the pass band. On the other hand, it can be shown that the convergence of the time average of $Z^2(t)$ to its expected value is slower when the bandwidth is smaller. Thus, if one uses a very small ε value, then it is necessary also to have a very large T value, and this may not be feasible in some situations.

In order to estimate the entire $S_{XX}(\omega)$ autospectral density function in this way, one must have a filter with a variable center frequency ω_c (and possibly a variable ε), and repeat the procedure for different ω_c values until the function is defined with sufficient resolution. This technique for estimating $S_{XX}(\omega)$ has been quite important in the past, but it would not be commonly used now, in an age of inexpensive digital computation. Section 4.10 discusses the approach that is usually used now to estimate $S_{XX}(\omega)$ from recorded data.
****

4.7 Response of Linear SDF Oscillator

We will now present an extended example, applying our frequency domain analysis procedures to investigate the response of the linear single-degree-of-freedom system described by Eqs. 3.39 and 3.40

$$m\ddot{X}(t) + c\dot{X}(t) + kX(t) = F(t)$$

or

$$\ddot{X}(t) + 2\zeta\omega_0\dot{X}(t) + \omega_0^2 X(t) = F(t)/m$$

As pointed out in Chapter 3, this system plays a key role in our analysis of oscillatory systems, as distinguished from the simpler systems of Examples 4.3 and 4.4 which have oscillatory response only if they have oscillatory excitation.

First we can find the harmonic transfer function from Eq. 4.36 as

$$H_x(\omega) = \frac{1}{(k + i\omega c - \omega^2 m)} = \frac{1}{m(\omega_0^2 + 2i\zeta\omega_0\omega - \omega^2)} \tag{4.37}$$

We could also check this expression by finding the Fourier transform of the impulse response function, which was derived in Example 3.3 as $h_x(t) = (m\omega_d)^{-1}e^{-\zeta\omega_0 t}\sin(\omega_d t)$ in which $\omega_d = \omega_0(1-\zeta^2)^{1/2}$.

From Eq. 4.37, we can write the autospectral density of the response of the SDF system as

$$S_{XX}(\omega) = S_{FF}(\omega)|H_x(\omega)|^2 = \frac{S_{FF}(\omega)}{m^2[(\omega_0^2 - \omega^2)^2 + (2\zeta\omega_0\omega)^2]} \qquad (4.38)$$

Figure 4.6 shows a plot of $m^2\omega_0^4|H_x(\omega)|^2 \equiv k^2|H_x(\omega)|^2$ versus the normalized frequency ω/ω_0 for several values of the damping ratio ζ. Most structural and mechanical vibration problems involve systems with damping values even smaller than the smallest value of $\zeta = 0.05$ shown in Fig. 4.6. In fact, values of ζ of 0.01 or even smaller are not uncommon. The larger damping values in Fig. 4.6 are included to illustrate the effect of damping, rather than because they are typical. If damping is increased still further, the peak of $|H_x(\omega)|$ eventually completely disappears. In fact, for $\zeta \geq 2^{-1/2}$, $S_{XX}(\omega)$ simply decays monotonically from its value at $\omega = 0$.

From Eq. 4.38 and Fig. 4.6, one can identify three key frequency ranges in the response of the SDF system. For $\omega \approx 0$, the response autospectral density $S_{XX}(\omega)$ is closely approximated by $S_{FF}(\omega)$ at that frequency divided by $m^2\omega_0^4 \equiv k^2$. Thus, the low-frequency response of the system is dependent only on the stiffness of the oscillator, behaving in the same way as if the response were pseudo-static. These low-frequency components are almost the same as they would be for a system without mass or damping, being governed by $kX(t) = F(t)$. At the other extreme, one finds that $S_{XX}(\omega)$ for $|\omega| >> \omega_0$ is approximated by $S_{FF}(\omega)/(m^2\omega^4)$. This implies that, at these frequencies, the autospectral density of the response acceleration has $S_{\ddot{X}\ddot{X}}(\omega) \approx S_{FF}(\omega)/m^2$. Thus, the high-frequency response components are almost the same as they would be for a system without spring or damping, being governed by $m\ddot{X}(t) = F(t)$. More important in most applications is the intermediate situation. If ζ is small, then the "resonant" response components for $\omega \approx \pm\omega_0$ are greatly amplified, so that $S_{XX}(\omega)$ is much greater than $S_{FF}(\omega)$. In this situation, the contributions to Eq. 4.38 depending on m and k nearly cancel out so that the response has $S_{XX}(\omega) \approx S_{FF}(\omega)/(2m\zeta\omega_0\omega)^2 = S_{FF}(\omega)/(c^2\omega^2)$ or $S_{\ddot{X}\ddot{X}}(\omega) \approx S_{FF}(\omega)/c^2$, which corresponds to a

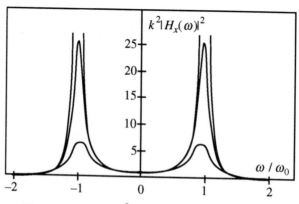

Figure 4.6 $|H_x(\omega)|^2$ for $\zeta = 0.05$, 0.10, and 0.20

governing equation of $c\dot{X}(t) = F(t)$. Thus, each of the terms in the governing differential equation plays a key role. The $kX(t)$ term dominates the low-frequency response, the $m\ddot{X}(t)$ term dominates the high-frequency response, and the $c\dot{X}(t)$ term dominates the resonant-frequency response, giving the height of the resonant peak of $S_{XX}(\omega)$. It is important to remember that the stochastic response process $\{X(t)\}$ generally consists of a superposition of all frequency components from zero to infinity, all occurring simultaneously. The comments in this paragraph are merely interpretations of the magnitudes of those various components, not discussions of different dynamic problems.

The plot in Fig. 4.6 shows that for small values of ζ the SDF system acts as a band-pass filter, giving substantial amplification only to components of $\{F(t)\}$ which are near $\pm\omega_0$. In fact, unless $S_{FF}(\omega)$ is much larger for some other frequencies than it is for $\omega = \omega_0$ it is clear that the response autospectral density $S_{XX}(\omega)$ will be dominated by the frequencies near $\pm\omega_0$. This leads to the common situation in which the stochastic response of the SDF system can be considered to be a narrowband process, which sometimes results in significant analytical simplifications. The excitation $\{F(t)\}$, on the other hand, can often be considered a broadband process or even approximated by an "equivalent" white noise, as explained in the following paragraph.

The justification for the approximation of $S_{FF}(\omega)$ by an "equivalent" white noise can be seen in Fig. 4.7. Since $S_{XX}(\omega)$ is the product of $S_{FF}(\omega)$ and $|H_x(\omega)|^2$ and the most significant portion of $S_{XX}(\omega)$ comes from the near-resonant frequencies with $\omega \approx \pm\omega_0$, we can see that $S_{XX}(\omega)$ would not be changed very significantly if the actual $S_{FF}(\omega)$ were replaced by some other broadband function which agreed with $S_{FF}(\omega)$ for $\omega \approx \pm\omega_0$. The simplest such approximation is the white noise with constant autospectral density $S_0 = S_{FF}(\omega_0)$. The most common usage for this approximation is in the computation of the response variance, for which the approximation gives

$$\sigma_X^2 = \int_{-\infty}^{\infty} S_{XX}(\omega)\,d\omega = \int_{-\infty}^{\infty} S_{FF}(\omega)|H_x(\omega)|^2\,d\omega \approx S_{FF}(\omega_0)\int_{-\infty}^{\infty}|H_x(\omega)|^2\,d\omega \qquad (4.39)$$

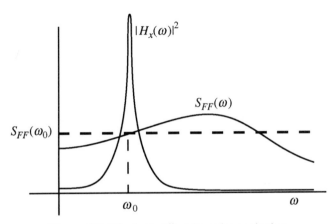

Figure 4.7 "Equivalent" white noise excitation

One can compute the value of this response variance by performing the frequency domain integration in the last term of Eq. 4.39, but the result, of course, must be the same as was found from time domain integration in Chapter 3 for the response to a delta-correlated excitation.[5] Taking the G_0

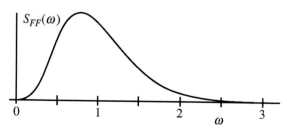

(a) Autospectral density of the excitation

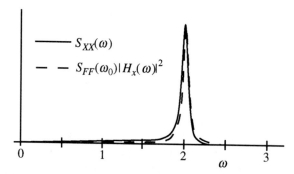

(b) Autospectral density of the response, $\omega_0 = 2$, $\zeta = 0.02$
Approximation is adequate.

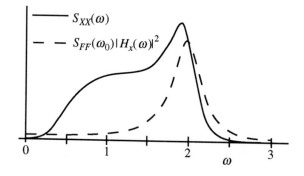

(c) Autospectral density of the response, $\omega_0 = 2$, $\zeta = 0.10$
Approximation is inadequate.

Figure 4.8 Satisfactory and unsatisfactory approximation by "equivalent" white noise

for the delta-correlated process to be $2\pi S_{FF}(\omega_0)$ gives the approximation of the SDF response variance as

$$\sigma_X^2 \approx \frac{\pi S_{FF}(\omega_0)}{2m^2 \zeta \omega_0^3} = \frac{\pi S_{FF}(\omega_0)}{ck} \qquad (4.40)$$

This approximation will generally be quite adequate so long as $S_{XX}(\omega)$ is sharply peaked so that its magnitude is much smaller for other frequencies than it is for $\omega \approx \pm\omega_0$. That is, the approximation can generally be used if $\{X(t)\}$ is narrowband. Although this is a common situation, it should be kept in mind that this condition is not always met. Figure 4.8 illustrates one such situation in which the approximation is not very good for $\zeta = 0.10$, although it does appear to be acceptable for $\zeta = 0.02$. Note that the excitation in this example is not very broadband, having a clear peak near $\omega = 1.0$.

4.8 Measures of Bandwidth

In Sections 4.4 and 4.5, we considered the basic characteristics of narrowband and broadband stochastic processes. Now, we wish to consider some ways of assigning a numerical value to this concept. That is, for any given autospectral density, one should be able to calculate a parameter which tells to what extent that particular process can be considered to be narrowband. One reason that we need such a concept is that some approximate methods of analysis of stochastic vibration and stochastic fatigue are based on the assumption that some process is narrowband, and a bandwidth parameter can help to define what that means.

Probably the most direct method of defining a bandwidth parameter was introduced by Vanmarcke (1972), who noted that the autospectral density $S_{XX}(\omega)$ from $\omega = 0$ to $\omega = \infty$ can be likened to a probability density function. In particular, it is a nonnegative function and it has a bounded integral if the $\{X(t)\}$ process has bounded variance. A probability density function might be considered to be narrow if the associated random variable has a small variance. More precisely, the dimensionless coefficient of variation, giving the ratio of the standard deviation to the mean of the random variable, gives a measure of how much the random variable is likely to differ from its mean. Similarly one can define moments of the autospectral density function, and from these calculate a parameter which gives the relative width of the autospectral density in the same way as the coefficient of variation of a random variable gives the relative width of its probability density function. We will define these spectral moments as

$$\lambda_m = \int_{-\infty}^{\infty} |\omega|^m S_{XX}(\omega)\,d\omega = 2\int_0^\infty \omega^m S_{XX}(\omega)\,d\omega \qquad (4.41)$$

For a very narrowband process concentrated at frequencies $\pm\omega_c$, we see that $\lambda_m \approx \omega_c^m \lambda_0$.

---

[5] Alternatively one can perform the frequency domain integration by noting that $|H_x(\omega)|^2$ has poles at $\omega = \pm\omega_d \pm i\zeta\omega_0$ and using the calculus of residues.

Note that in order to have a good analogy with a probability density function, we need to consider a function which has a unit integral; $S_{XX}(\omega)/\lambda_0$ has this property. Furthermore, the mth moment of $S_{XX}(\omega)/\lambda_0$ is simply λ_m/λ_0. The bandwidth parameter directly corresponding to coefficient of variation is, thus,

$$ s = \frac{\sqrt{(\lambda_2/\lambda_0)-(\lambda_1/\lambda_0)^2}}{\lambda_1/\lambda_0} = \sqrt{\frac{\lambda_0\lambda_2}{\lambda_1^2}-1} $$

The parameter s tends to zero for the limiting narrowband process, and it is always greater than zero for any other situation. Rather than using this parameter directly, though, it is common convention to convert to a parameter which is always in the range of zero to unity, and which tends to unity for a narrowband process. The commonly used parameter with this normalization is

$$ \alpha_1 = \frac{\lambda_1}{\sqrt{\lambda_0\lambda_2}} \tag{4.42} $$

which is easily shown to be $\alpha_1 = (s^2+1)^{-1/2}$. The fact that s^2 is non-negative demonstrates that $0 \le \alpha_1 \le 1$.

A slight variation on Eq. 4.42 can be used to define a more general family of bandwidth parameters, each of which has properties similar to α_1, but with some differences:

$$ \alpha_m = \frac{\lambda_m}{\sqrt{\lambda_0\lambda_{2m}}} \tag{4.43} $$

The normalization of α_m is the same as that of α_1 inasmuch as α_m is always in the unit interval. That is is nonnegative follows directly from its definition in terms of spectral moments, and one can show that $\alpha_m \le 1$ by demonstrating that $\lambda_m^2 \le \lambda_0\lambda_{2m}$. This latter bound follows from the fact that

$$ \int_{-\infty}^{\infty}\left(\sqrt{\frac{\lambda_{2m}}{\lambda_0}}-\omega^m\right)^2 S_{XX}(\omega)\,d\omega = \lambda_{2m}-2\lambda_m\sqrt{\frac{\lambda_{2m}}{\lambda_0}}+\lambda_{2m} $$

and this quantity must be greater than or equal to zero since the integrand is non-negative. Furthermore, consideration of $\lambda_j \approx \omega_c^j\lambda_0$ shows that α_m will also tend to unity for a narrowband process.

Note that there is little restriction on the quantity m in the definition of the α_m bandwidth parameter. In fact, it need not even be an integer. One consequence of different choices for m regards the types of autospectral density functions for which $\alpha_m = 0$. In particular, $\alpha_m = 0$ if the high-frequency autospectral density decays sufficiently slowly that $\lambda_{2m} = \infty$. For example, if

$S_{XX}(\omega)$ decays like $|\omega|^{-6}$ then $\lambda_{2m} = \infty$ for $m \geq 2.5$. One advantage of the α_1 parameter is the fact that it goes to zero only if the autospectral density does not decay more rapidly than $|\omega|^{-3}$, which covers quite a wide variety of situations. Out of this group, the other parameter that is widely used in practice corresponds to choosing $m = 2$. One disadvantage of α_2, compared to α_1, is its increased sensitivity to high-frequency components of the autospectral density, resulting in its being zero if the spectral density does not decay more rapidly than $|\omega|^{-5}$. In Example 4.8 we will see that some problems of practical interest fail to meet this criterion.

Probably the primary reason that investigators often choose to use the α_2 parameter has to do with certain special properties of the even-integer spectral moments. In particular, it is easy to verify that the three spectral moments λ_0, λ_2, and λ_4 used in defining α_2 each correspond to a variance term:

$$\lambda_0 = \int_{-\infty}^{\infty} S_{XX}(\omega)\,d\omega = \sigma_X^2 \tag{4.44}$$

$$\lambda_2 = \int_{-\infty}^{\infty} \omega^2 S_{XX}(\omega)\,d\omega = \sigma_{\dot{X}}^2 \tag{4.45}$$

and

$$\lambda_4 = \int_{-\infty}^{\infty} \omega^4 S_{XX}(\omega)\,d\omega = \sigma_{\ddot{X}}^2 \tag{4.46}$$

so that one can write

$$\alpha_2 = \frac{\lambda_2}{\sqrt{\lambda_0 \lambda_4}} = \frac{\sigma_{\dot{X}}^2}{\sigma_X \sigma_{\ddot{X}}} \tag{4.47}$$

Of course the sequence of relationships in Eqs. 4.44 - 4.46 can also be extended, showing that λ_{2m} corresponds to the variance of the mth derivative of $\{X(t)\}$, whenever m is a non-negative integer. When m is not an even integer, though, there is not such an easy physical interpretation of λ_m. In particular, the λ_1 spectral moment entering into the definition of α_1 is not easily identified with the variance of any physical quantity. If one knows the autospectral density function, then there is generally no difficulty in evaluating any of the bandwidth measures, but one can also find the value of α_2 without knowledge of the complete autospectral density function if the appropriate variances are known.

The α_2 bandwidth parameter also has another interpretation which may help to illuminate its significance as a meaningful parameter of the stochastic process. In particular, note that for a covariant stationary process the cross-covariance $K_{X\dot{X}} \equiv G_{X\dot{X}}(0) \equiv E\big([X(t) - \mu_X(t)][\dot{X}(t) - \mu_{\dot{X}}(t)]\big)$ must be a constant. In fact it is zero, but we will only use the fact that it is a constant, since that is enough to demonstrate that

$$\frac{d}{dt} E\big([X(t) - \mu_X(t)][\dot{X}(t) - \mu_{\dot{X}}(t)]\big) = G_{X\dot{X}}(0) + \sigma_{\dot{X}}^2 = 0$$

so that the cross-covariance of $X(t)$ and $\ddot{X}(t)$ is $K_{X\ddot{X}} \equiv G_{X\ddot{X}}(0) = -\sigma_{\dot{X}}^2$. This fact shows that the correlation coefficient of $X(t)$ and $\ddot{X}(t)$ is exactly

$$\rho_{X\ddot{X}} \equiv \frac{K_{X\ddot{X}}}{\sigma_X \sigma_{\ddot{X}}} = -\frac{\sigma_{\dot{X}}^2}{\sigma_X \sigma_{\ddot{X}}} = -\alpha_2 \tag{4.48}$$

The α_2 bandwidth parameter is precisely the negative of the correlation coefficient between the process and its second derivative. Since the time histories of a narrowband process are nearly harmonic we should certainly expect $X(t)$ and $\ddot{X}(t)$ to have almost perfect negative correlation in that situation, since the second derivative of any harmonic function is proportional to the negative of the original function. Nonetheless, it seems somewhat fortuitous that the magnitude of the correlation coefficient is exactly α_2.

****

Example 4.6: Find the values of the α_1 and α_2 bandwidth parameters for the covariant stationary process $\{X(t)\}$ of Section 4.4, with autospectral density function given by

$$S_{XX}(\omega) = S_0 \big(U[|\omega| - (\omega_c - b)] - U[|\omega| - (\omega_c + b)] \big)$$

The mth spectral density for this $S_{XX}(\omega)$ is easily evaluated as

$$\lambda_m = 2S_0 \int_{\omega_c - b}^{\omega_c + b} \omega^m \, d\omega = 2S_0 \frac{(\omega_c + b)^{m+1} - (\omega_c - b)^{m+1}}{m+1}$$

so that $\lambda_0 = 4S_0 b$, $\lambda_1 = 4S_0 \omega_c b$, $\lambda_2 = 4S_0 b(\omega_c^2 + b^2/3)$ and $\lambda_4 = 4S_0 b(\omega_c^4 + 2\omega_c^2 b^2 + b^4/5)$. Thus,

$$\alpha_1 = \frac{\lambda_1}{\sqrt{\lambda_0 \lambda_2}} = \frac{\omega_c}{\sqrt{\omega_c^2 + b^2/3}} = \left(1 + \frac{b^2}{3\omega_c^2}\right)^{-1/2}$$

and

$$\alpha_2 = \frac{\lambda_2}{\sqrt{\lambda_0 \lambda_4}} = \frac{\omega_c^2 + b^2/3}{\sqrt{\omega_c^4 + 2\omega_c^2 b^2 + b^4/5}} = \left(1 + \frac{b^2}{3\omega_c^2}\right)\left(1 + 2\frac{b^2}{\omega_c^2} + \frac{b^4}{5\omega_c^4}\right)^{-1/2}$$

For a narrowband situation with $b \ll \omega_c$, one can use power series expansions to simplify these relationships to

$$\alpha_1 \approx 1 - \frac{b^2}{6\omega_c^2} \qquad \alpha_2 \approx 1 - \frac{2b^2}{3\omega_c^2}$$

Thus, we see that both α_1 and α_2 do tend to unity as b tends to zero, but we also see that they approach unity at quite different rates, with $(1 - \alpha_2)$ being four times as large as $(1 - \alpha_1)$.

The most broadband situation we can investigate with this particular autospectral density function is the one with $b = \omega_c$, so that $S_{XX}(\omega)$ is flat from $-2\omega_c$ to $+2\omega_c$. In this situation, the bandwidth parameters are $\alpha_1 = (3/4)^{1/2} \approx 0.866$ and $\alpha_2 = 5^{1/2}/3 \approx 0.745$. One notes that these parameters are not very near zero, even though this would usually be considered to be quite a broadband autospectral density. In general, the bandwidth parameters approach zero only for autospectral densities with slowly decaying tails, and the values obtained here are representative of apparently broadband spectral densities on a bounded set of frequencies.

****

Example 4.7: Find the values of the α_1 and α_2 bandwidth parameters for a covariant stationary process $\{X(t)\}$ with autospectral density function given by $S_{XX}(\omega) = S_0 e^{-\gamma|\omega|}$, as in Example 4.1.

The mth spectral moment can be evaluated in terms of the gamma function (see Example A.31 of Appendix A for definition), or the factorial if m is an integer:

$$\lambda_m = 2S_0 \int_0^\infty \omega^m e^{-\gamma\omega}\, d\omega = \frac{2S_0}{\gamma^{m+1}} \int_0^\infty u^m e^{-u}\, du = \frac{2S_0}{\gamma^{m+1}} \Gamma(m+1) = \frac{2S_0 m!}{\gamma^{m+1}}$$

Thus, $\lambda_0 = 2S_0 \gamma^{-1}$, $\lambda_1 = 2S_0 \gamma^{-2}$, $\lambda_2 = 4S_0 \gamma^{-3}$, and $\lambda_4 = 48S_0 \gamma^{-5}$. Substituting these expressions into the definitions for α_1 and α_2 gives

$$\alpha_1 = \frac{\lambda_1}{\sqrt{\lambda_0 \lambda_2}} = 0.707 \quad \text{and} \quad \alpha_2 = \frac{\lambda_2}{\sqrt{\lambda_0 \lambda_4}} = 0.408$$

Note that the values of α_1 and α_2 are unaffected by the value of the parameter γ, which seems somewhat surprising. For γ tending to zero we see that the stochastic process tends to white noise, while for γ quite large it has significant autospectral density only near frequency zero. This latter situation might be considered narrowband in a sense, but it is a different sense than we have been discussing. In particular, this process does not have nearly harmonic time histories since it does not have spectral density concentrated at some non-zero ω_c, as in the other narrowband processes we have considered.

Since neither the α_1 nor the α_2 bandwidth parameter can give us information about the effect of the γ parameter, this example illustrates an important point. Namely, that it is very difficult to make definite statements about the behavior of time histories based only on a bandwidth parameter. Knowledge of the complete autospectral density function will always give much more comprehensive information than will a bandwidth parameter.

****

Example 4.8: Find the values of the α_1 and α_2 bandwidth parameters for the covariant stationary response $\{X(t)\}$ of a SDF oscillator excited by white noise.

Two of the four spectral moments of interest can be obtained almost trivially. In particular, λ_0 is the response variance, so Eq. 4.40 gives

$$\lambda_0 = \sigma_X^2 = \frac{\pi S_0}{2m^2 \zeta \omega_0^3} = \frac{\pi S_0}{ck}$$

in which S_0 is the autospectral density of the excitation. Similarly, λ_2 is the variance of the derivative of the response, and this can be found from Eq. 3.57 as

$$\lambda_2 = \sigma_{\dot{X}}^2 = \frac{\pi S_0}{2m^2 \zeta \omega_0} = \frac{\pi S_0}{mc}$$

The λ_1 spectral moment is not a variance quantity, but it can be evaluated as

$$\lambda_1 = 2\int_0^\infty |\omega| S_{XX}(\omega)\, d\omega = 2S_0 \int_0^\infty |\omega| |H_x(\omega)|^2\, d\omega = \frac{2S_0}{m^2}\int_0^\infty \frac{\omega\, d\omega}{(\omega^2 - \omega_0^2)^2 + (2\zeta\omega_0\omega)^2}$$

This integral over ω from zero to infinity is not appropriate for evaluation by the calculus of residues, which is useful for most of our frequency domain integrals, but it can be shown that the integrand is the differential of an arctangent function. The result can be written as

$$\lambda_1 = \frac{S_0}{2m^2 \zeta \omega_0 \omega_d}\left[\frac{\pi}{2} + \tan^{-1}\left(\frac{1 - 2\zeta^2}{2\zeta\sqrt{1-\zeta^2}}\right)\right] = \frac{S_0}{2m^2 \zeta \omega_0 \omega_d}\left[\pi - 2\tan^{-1}\left(\frac{\zeta}{\sqrt{1-\zeta^2}}\right)\right]$$

in which the slightly simpler final form has been derived by use of trigonometric identities. The λ_4 spectral moment is somewhat more of a problem. In particular, $|H_x(\omega)|^2$ behaves like ω^{-4} for $|\omega| \to \infty$ tending to infinity, and this shows that $\omega^4 |H_x(\omega)|^2$ is not integrable. Actually, this is not surprising since $\lambda_4 = \sigma_{\ddot{X}}^2$, and we noted in Chapter 3 that the variance of $\ddot{X}(t)$ is infinite for the SDF system excited by a delta-correlated process. Thus, we have $\lambda_4 = \infty$. From these spectral moments, we can write the bandwidth parameters as

$$\alpha_1 = \frac{1}{\sqrt{1-\zeta^2}}\left[1 - \frac{2}{\pi}\tan^{-1}\left(\frac{\zeta}{\sqrt{1-\zeta^2}}\right)\right]$$

and $\alpha_2 = 0$. Clearly, this is a problem for which our two bandwidth parameters give very different results. In fact the α_2 bandwidth parameter fails to give any useful information, since it gives $\alpha_2 = 0$, suggesting a broadband $\{X(t)\}$ process regardless of the value of the damping parameter ζ. The α_1 parameter, on the other hand, behaves as we might expect. In particular, it tends to unity as ζ tends to zero, indicating that the response process is narrowband in this situation, which agrees with the shape of the $S_{XX}(\omega) = S_0 |H_x(\omega)|^2$ function. For $\zeta \ll 1$, the limiting behavior is linear in ζ with $\alpha_1 \approx 1 - 2\zeta/\pi$.

According to Eq. 4.48, α_2 is also the negative of the correlation coefficient $\rho_{X\ddot{X}}$. The fact that $\alpha_2 = 0$ for the SDF response to white noise then means that $X(t)$ and $\ddot{X}(t)$, at the same instant of time t, are uncorrelated for this system. This follows directly, though, from the fact that the cross-covariance $K_{X\ddot{X}}$ is finite while the variance of $\ddot{X}(t)$ is infinite.

****

4.9 Amplitude and Phase of a Stochastic Process

In our discussion of narrowband processes, we used the concepts of amplitude and phase of a stochastic process without giving any precise definition of the terms. If a process does not have time histories which closely resemble harmonic functions, then such an intuitive idea of amplitude and phase is of little use, and even for narrowband situations it is helpful to have a precise definition of the terms. This can be done in more than one way, and we will begin with what we consider the most obvious approach.

The first requirement for the two new stochastic processes $\{A(t)\}$ and $\{\theta(t)\}$ is that they give

$$Y(t) = A(t)\cos[\omega_a t + \theta(t)] \qquad (4.49)$$

in which $\{Y(t)\}$ is a mean zero process defined as

$$Y(t) = X(t) - \mu_X(t)$$

for any differentiable, covariant stationary stochastic process $\{X(t)\}$, and ω_a is an average frequency for the process according to some appropriate definition. Taking the derivative of Eq. 4.49, then, gives

$$\dot{Y}(t) = -\omega_a A(t)\sin[\omega_a t + \theta(t)] + \dot{A}(t)\cos[\omega_a t + \theta(t)] - A(t)\dot{\theta}(t)\sin[\omega_a t + \theta(t)]$$

However, we now choose to impose the restriction that

$$\dot{A}(t)\cos[\omega_a t + \theta(t)] = A(t)\dot{\theta}(t)\sin[\omega_a t + \theta(t)] \qquad (4.50)$$

so that

$$\dot{Y}(t) = -\omega_a A(t)\sin[\omega_a t + \theta(t)] \qquad (4.51)$$

It is important to note that Eqs. 4.49 and 4.51 may be viewed as two simultaneous equations which define the new stochastic processes $\{A(t)\}$ and $\{\theta(t)\}$ in terms of the original stochastic processes $\{Y(t)\}$ and $\{\dot{Y}(t)\}$. Specifically, Eqs. 4.49 and 4.51 give the explicit expressions for the amplitude and phase processes as

$$A(t) = \left[Y^2(t) + Z^2(t)\right]^{1/2} \qquad (4.52)$$

and

$$\theta(t) = -\tan^{-1}\left[\frac{Z(t)}{Y(t)}\right] - \omega_a t \qquad (4.53)$$

in which $Z(t) = \dot{Y}(t)/\omega_a = -A(t)\sin[\omega_a t + \theta(t)]$.

Note that Eqs. 4.52 and 4.53 give a definition of amplitude and phase for any differentiable stochastic process $\{X(t)\}$, whether or not it is narrowband. This is in sharp contrast to the development in Section 4.4 in which we argued that Eq. 4.28, which is equivalent to Eq. 4.49, should hold for a narrowband process as sketched in Fig. 4.4. In the particular case with $\{X(t)\}$ being narrowband, we do anticipate that $\{A(t)\}$ and $\{\theta(t)\}$ in Eqs. 4.52 and 4.53 will be slowly varying so that the time histories of $\{Y(t)\}$ will have the almost harmonic behavior which we identified in Section 4.4 as being appropriate for a narrowband process. Variations in $A(t)$, of course, will be evident in a time history plot such as Fig. 4.4 as changes in the magnitude of the excursions of $X(t)$. Variations in $\theta(t)$ are not quite so readily apparent in a time history, but they can be related to changes in the time interval between zero crossings. That is, since $X(t) = 0$ whenever $\omega_a t + \theta(t)$ is an odd multiple of $\pi/2$, the zero crossings will be exactly uniformly spaced at intervals of $2\pi/\omega_a$ if $\theta(t)$ has no variation. Any observed variations in the spacing will provide evidence that $\dot{\theta}(t) \neq 0$ so that $\theta(t)$ is not constant. Of course, $\{A(t)\}$ and $\{\theta(t)\}$ can be considered to be slowly varying processes if $\dot{A}(t)$ and $\dot{\theta}(t)$ are small, so we will expect narrowband processes to have small values of these derivatives.

Before we explicitly investigate the derivatives of $\{A(t)\}$ and $\{\theta(t)\}$, we want to generalize their definitions. Even though the amplitude, or envelope, defined in Eqs. 4.52 and 4.53 has the desired characteristics, it turns out that it is only one of the possible choices. That is, there are other possible definitions of the amplitude of a process which will also closely agree with the intuitive concept of amplitude in the narrowband situation, and at least one of these has certain mathematical advantages.[6] We will begin investigating this general idea of amplitude of a stochastic process, by looking at the crucial features of $\{A(t)\}$ for a narrowband process. Switching to truncated time histories allows us to use Fourier transforms, as we did in Section 4.2. In particular, the inverse Fourier transform relationship for $Z_T(t)$ is

$$Z_T(t) = \int_{-\infty}^{\infty} \tilde{Z}_T(\omega)e^{i\omega t}\,d\omega = \int_{-\infty}^{\infty}[\tilde{Y}_T(\omega)/\omega_a]e^{i\omega t}\,d\omega = \int_{-\infty}^{\infty} i\frac{\omega}{\omega_a}\tilde{Y}_T(\omega)e^{i\omega t}\,d\omega$$

For a $\{Y(t)\}$ process which is narrowband with average frequency ω_a, the $\tilde{Y}_T(\omega)$ function will be nearly zero except in the neighborhood of the two points $\omega = \pm\omega_a$. At $\omega \approx +\omega_a$ it will give $\tilde{Z}_T(\omega) \approx i\tilde{Y}_T(\omega)$, and at $\omega \approx -\omega_a$ the relationship will be $\tilde{Z}_T(\omega) \approx -i\tilde{Y}_T(\omega)$. These two relationships, then, are the key features of the $\{Z(t)\}$ process since other frequencies contribute very little to its behavior. Thus, if we choose some different $\{Z(t)\}$ process for which the Fourier transform has this same behavior in the neighborhood of $\omega = \pm\omega_a$ then we can be assured that $A(t) = [Y^2(t) + Z^2(t)]^{1/2}$ will still behave like the amplitude of $Y(t)$ for a narrowband process. There are many $\{Z(t)\}$ processes which meet this general condition. In particular, using

---

[6]   This idea of alternate definitions of amplitude (or envelope) and their relationship to bandwidth was investigated by Winterstein and Cornell in 1985.

$$\tilde{Z}_T(\omega) = i\, g(\omega)\, \tilde{Y}_T(\omega) \tag{4.54}$$

for any odd $g(\cdot)$ function with $g(\omega_a) = 1$ makes

$$Z_T(t) = \int_{-\infty}^{\infty} i\, g(\omega)\, \tilde{Y}_T(\omega)\, e^{i\omega t}\, d\omega$$

be real and $\tilde{Z}_T(\omega)$ have the desired behavior in the neighborhood of $\omega = \pm\omega_a$. Using $g(\omega) = \omega / \omega_a$, as in Eqs. 4.49 and 4.51, is only one of infinitely many choices.

Based on Eq. 4.54 we can now investigate the spectral density functions related to the new process $\{Z(t)\}$. In particular, the autospectral density definition in Eq. 4.14 gives

$$S_{ZZ}(\omega) = g^2(\omega) S_{YY}(\omega) \equiv g^2(\omega) S_{XX}(\omega) \tag{4.55}$$

and Eq. 4.19 gives the cross-spectral density between $\{X(t)\}$ and $\{Z(t)\}$ as

$$S_{XZ}(\omega) = -i\, g(\omega) S_{XX}(\omega) \tag{4.56}$$

From these spectral densities, one can write various covariance relationships which will be useful in analyzing the behavior of the amplitude. Specifically,

$$G_{ZZ}(\tau) = \int_{-\infty}^{\infty} g^2(\omega) S_{XX}(\omega) e^{i\omega\tau}\, d\omega \tag{4.57}$$

and

$$G_{XZ}(\tau) = -i \int_{-\infty}^{\infty} g(\omega) S_{XX}(\omega) e^{i\omega\tau}\, d\omega \tag{4.58}$$

as well as related relationships for the derivative processes $\{\dot{X}(t)\}$ and $\{\dot{Z}(t)\}$:

$$G_{Z\dot{Z}}(\tau) = -i \int_{-\infty}^{\infty} \omega\, g^2(\omega) S_{XX}(\omega) e^{i\omega\tau}\, d\omega \tag{4.59}$$

$$G_{\dot{X}Z}(\tau) = -G_{X\dot{Z}}(\tau) = \int_{-\infty}^{\infty} \omega\, g(\omega) S_{XX}(\omega) e^{i\omega\tau}\, d\omega \tag{4.60}$$

and

$$G_{\dot{X}\dot{Z}}(\tau) = -i \int_{-\infty}^{\infty} \omega^2\, g(\omega) S_{XX}(\omega) e^{i\omega\tau}\, d\omega \tag{4.61}$$

Setting $\tau = 0$ in any of these relationships, of course, gives a variance or covariance term for a single instant of time t, and the fact that $g(\omega)$ is an odd function makes several of these terms be zero. In particular, we find that $X(t)$ and $Z(t)$ are uncorrelated at any instant of time since $K_{XZ} \equiv G_{XZ}(0) = 0$, and also $\dot{X}(t)$ and $\dot{Z}(t)$ are uncorrelated since $K_{\dot{X}\dot{Z}} \equiv G_{\dot{X}\dot{Z}}(0) = 0$. In addition we know that $X(t)$ and $\dot{X}(t)$ are uncorrelated $[K_{X\dot{X}} = 0]$ and $Z(t)$ and $\dot{Z}(t)$ are uncorrelated

$[K_{Z\dot{Z}} = 0]$, as is required for any covariant stationary process. The cross-covariances which are not zero are $K_{X\dot{Z}} \equiv G_{X\dot{Z}}(0)$ and $K_{Z\dot{X}} \equiv G_{Z\dot{X}}(0)$, and they differ from each other only in sign.

Considering now the derivatives of the general $\{A(t)\}$ and $\{\theta(t)\}$ processes, we find from Eqs. 4.52 and 4.53 that

$$\dot{A}(t) = \frac{Y(t)\dot{Y}(t) + Z(t)\dot{Z}(t)}{A(t)} \tag{4.62}$$

and

$$\dot{\theta}(t) = \frac{Z(t)\dot{Y}(t) - Y(t)\dot{Z}(t)}{A^2(t)} - \omega_a \tag{4.63}$$

We will look at the magnitude of these quantities in a mean squared sense, but we will use a simplified approach. For example, the expression for $[A(t)\dot{A}(t)]$ is simpler than the one for $\dot{A}(t)$, and since $A(t)$ is nonnegative we can see that $\dot{A}(t)$ will have a small mean squared value only if $[A(t)\dot{A}(t)]$ does as well. Thus, we can say that $A(t)$ will vary slowly in situations which give a small value for

$$E\left(A^2(t)\dot{A}^2(t)\right) = E[Y^2(t)\dot{Y}^2(t)] + 2E[Y(t)\dot{Y}(t)Z(t)\dot{Z}(t)] + E[Z^2(t)\dot{Z}^2(t)] \tag{4.64}$$

Note also that the $E[A^2(t)]$ must be a constant, since $\{X(t)\}$ and $\{Z(t)\}$ are covariant stationary processes, and this assures that $E[A(t)\dot{A}(t)] = 0$.

In order to achieve a similar simplification for the condition of small $\dot{\theta}(t)$, it is useful to investigate the quantity $[A^2(t)\dot{\theta}(t)]$. The mean value of this quantity, though, is not automatically zero, but rather depends on our choice of ω_a. Since the frequency of a harmonic function is only the rate of change of its argument, it is strictly accurate to call ω_a the average frequency only if we choose it such that $E[\dot{\theta}(t)] = 0$. That is, any nonzero mean value for $\dot{\theta}(t)$ should be counted as part of the average frequency. We will use the related, but simplified, condition that ω_a is defined such that $E[A^2(t)\dot{\theta}(t)] = 0$, which gives

$$\omega_a = \frac{E[Z(t)\dot{Y}(t)] - E[Y(t)\dot{Z}(t)]}{E[A^2(t)]} = \frac{2K_{Z\dot{X}}}{\sigma_X^2 + \sigma_Z^2} \tag{4.65}$$

The mean squared value of $[A^2(t)\dot{\theta}(t)]$, which should be small for a narrowband process, is given by

$$E\left(A^4(t)\dot{\theta}^2(t)\right) = E[Z^2(t)\dot{Y}^2(t)] - 2E[Y(t)\dot{Y}(t)Z(t)\dot{Z}(t)] + E[Y^2(t)\dot{Z}^2(t)]$$

$$- 2\omega_a E\left(A^2(t)[Z(t)\dot{Y}(t) - Y(t)\dot{Z}(t)]\right) + \omega_a^2 E[A^4(t)]$$

or

$$E\left(A^4(t)\dot{\theta}^2(t)\right) = E[Z^2(t)\dot{Y}^2(t)] - 2E[Y(t)\dot{Y}(t)Z(t)\dot{Z}(t)] + E[Y^2(t)\dot{Z}^2(t)]$$
$$-2\omega_a\left(E[Y^2(t)Z(t)\dot{Y}(t)] + E[Z^3(t)\dot{Y}(t)] - E[Y^3(t)\dot{Z}(t)] - E[Y(t)Z^2(t)\dot{Z}(t)]\right)$$
$$+\omega_a^2\left(E[Y^4(t)] + 2E[Y^2(t)Z^2(t)] + E[Z^4(t)]\right)$$

$$(4.66)$$

Even though the expectations in Eqs. 4.64 and 4.66 are simpler than those for $\dot{A}^2(t)$ and $\dot{\theta}^2(t)$, they still present certain difficulties. In particular, they show that one must investigate fourth moment properties of $Y(t)$, $Z(t)$ and their derivatives in order to find second-moment properties of $\dot{A}(t)$ and $\dot{\theta}(t)$. Thus, ordinary spectral density functions for $Y(t)$ and $Z(t)$, which are frequency decompositions of covariance, are not sufficient to give us these second moments of $\dot{A}(t)$ and $\dot{\theta}(t)$. One can rewrite the expressions to take advantage of what we do know by expressing the fourth moment quantities in terms of covariances and fourth cumulants. This amounts to writing the fourth moment as a sum of contributions from variance plus a part, called the fourth cumulant, which cannot be explained by variance. (See Appendix Section A.12 for more information on cumulants.) For any four mean zero random variables (R_1, R_2, R_3, R_4), this fourth moment relationship can be written as

$$E(R_1 R_2 R_3 R_4) = E(R_1 R_2)E(R_3 R_4) + E(R_1 R_3)E(R_2 R_4) + E(R_1 R_4)E(R_2 R_3) + \kappa_4(R_1, R_2, R_3, R_4)$$

in which the first three terms are products of covariances and the last term is the joint fourth cumulant. In most practical problems, it is found that the fourth cumulant contribution to the fourth moment is much less than than the contribution of the covariance terms. The primary reason for this is supposedly related to the so-called central limit theorem, which will be discussed briefly in Section 5.1.

We will now simplify Eqs. 4.64 and 4.66 by presuming that $\{X(t)\}$ and $\{Z(t)\}$ are jointly fourth cumulant stationary so that we can use the expansion of fourth moments in terms of covariances. We will also assume that the most significant contributions to the fourth moments in the equations come from the covariance terms, and will simply write $f_1(\kappa_4)$ and $f_2(\kappa_4)$ for the fourth cumulant contributions to Eqs. 4.64 and 4.66, respectively. We will also use the covariance relationships derived for the $\{Y(t)\}$ and $\{Z(t)\}$ processes from Eqs. 4.57 - 4.61 to simplify the results, giving

$$E\left(A^2(t)\dot{A}^2(t)\right) = \sigma_X^2\sigma_{\dot{X}}^2 - 2K_{Z\dot{X}}^2 + \sigma_Z^2\sigma_{\dot{Z}}^2 + f_1(\kappa_4)$$

$$(4.67)$$

and

$$E\left(A^4(t)\dot{\theta}^2(t)\right) = \sigma_Z^2\sigma_{\dot{X}}^2 + 6K_{Z\dot{X}}^2 + \sigma_X^2\sigma_{\dot{Z}}^2 - 8\omega_a K_{Z\dot{X}}(\sigma_{\dot{X}}^2 + \sigma_Z^2)$$
$$+\omega_a^2(3\sigma_X^4 + 3\sigma_Z^4 + 2\sigma_X^2\sigma_Z^2) + f_2(\kappa_4)$$

Substituting for ω_a based on Eq. 4.65 and simplifying converts the latter equation to

$$E\left(A^4(t)\dot{\theta}^2(t)\right) = \sigma_{\dot{Z}}^2\sigma_X^2 + \sigma_X^2\sigma_Z^2 + 2K_{Z\dot{X}}^2\frac{(\sigma_X^4 + \sigma_Z^4 - 6\sigma_X^2\sigma_Z^2)}{(\sigma_X^4 + \sigma_Z^4 + 2\sigma_X^2\sigma_Z^2)} + f_2(\kappa_4) \tag{4.68}$$

Having set up these general formulas related to the mean squared values of $\dot{A}(t)$ and $\dot{\theta}(t)$, we will now apply them to the special cases of the commonly used definitions of amplitude of a process. First we will take $g(\omega) = \omega / \omega_a$ so that $Z(t) = \dot{X}(t) / \omega_a$, as in Eq. 4.51. This particular amplitude, or envelope, is sometimes called the energy-based amplitude, since in many mechanical problems the potential energy is proportional to $Y^2(t)$ and the kinetic energy is proportional to $\dot{Y}^2(t)$, so that $A^2(t)$ defined according to Eq. 4.52 gives a quantity at least intuitively related to the total energy in the system. Using $Z(t) = \dot{X}(t) / \omega_a$ for this amplitude definition gives $K_{Z\dot{X}} = \sigma_{\dot{X}}^2 / \omega_a$ and $\sigma_Z^2 = \sigma_{\dot{X}}^2 / \omega_a^2$, and substituting these relationships into Eq. 4.65, then solving for the average frequency, gives

$$\omega_a = \sqrt{\frac{\lambda_2}{\lambda_0}} = \frac{\sigma_{\dot{X}}}{\sigma_X} \tag{4.69}$$

Furthermore, this then shows that $\sigma_Z = \sigma_X$ for this particular definition of $Z(t)$.

The expressions in Eqs. 4.67 and 4.68 related to the rates of change of $\{A(t)\}$ and $\{\theta(t)\}$ for the energy-based amplitude are

$$E\left(A^2(t)\dot{A}^2(t)\right) = \sigma_X^2\sigma_{\dot{X}}^2 - 2\frac{\sigma_{\dot{X}}^4}{\omega_a^2} + \frac{\sigma_X^2\sigma_{\dot{X}}^2}{\omega_a^4} + f_1(\kappa_4) = \frac{\sigma_X^4\sigma_{\dot{X}}^2}{\sigma_{\dot{X}}^2} - \sigma_X^2\sigma_{\dot{X}}^2 + f_1(\kappa_4)$$

which can be rewritten as

$$E\left(A^2(t)\dot{A}^2(t)\right) = \frac{\sigma_X^4\sigma_{\dot{X}}^2}{\sigma_{\dot{X}}^2}\left(1 - \frac{\sigma_{\dot{X}}^4}{\sigma_X^2\sigma_{\ddot{X}}^2}\right) + f_1(\kappa_4) = \frac{\sigma_X^4\sigma_{\dot{X}}^2}{\sigma_{\dot{X}}^2}\left(1 - \frac{\lambda_2^2}{\lambda_0\lambda_4}\right) + f_1(\kappa_4)$$

or

$$E\left(A^2(t)\dot{A}^2(t)\right) = \frac{\sigma_X^4\sigma_{\dot{X}}^2}{\sigma_{\dot{X}}^2}(1 - \alpha_2^2) + f_1(\kappa_4) \tag{4.70}$$

and

$$E\left(A^4(t)\dot{\theta}^2(t)\right) = \frac{\sigma_X^4}{\omega_a^2} + \frac{\sigma_X^2\sigma_{\dot{X}}^2}{\omega_a^2} - 2\frac{\sigma_X^4}{\omega_a^2} + f_2(\kappa_4) = \frac{\sigma_X^2\sigma_{\dot{X}}^2}{\omega_a^2}\left(1 - \frac{\sigma_X^4}{\sigma_X^2\sigma_{\ddot{X}}^2}\right) + f_2(\kappa_4)$$

or

$$E\left(A^4(t)\dot{\theta}^2(t)\right) = \frac{\sigma_X^4\sigma_{\dot{X}}^2}{\sigma_{\dot{X}}^2}(1 - \alpha_2^2) + f_2(\kappa_4) \tag{4.71}$$

in which α_2 is the bandwidth parameter defined in Eq. 4.47. The results in Eqs. 4.70 and 4.71 seem quite remarkable. Presuming that the $f_j(\kappa_4)$ terms are small, the expressions not only confirm that $\dot{A}(t)$ and $\dot{\theta}(t)$ are small in the mean squared sense for a narrowband process, but they also show that their magnitude is directly related to the α_2 parameter, which was originally defined in a quite different way, involving an inequality on an integral of the autospectral density function. Furthermore, the expressions show that the contribution of covariances to the mean squared value of $A^4(t)\dot{\theta}^2(t)$ is identical to that for $A^2(t)\dot{A}^2(t)$. This gives another meaning for the α_2 bandwidth parameter — it is the parameter which governs the rate of change of $\{A(t)\}$ and $\{\theta(t)\}$ for the energy-based amplitude definition of Eqs. 4.49 and 4.51, provided only that the fourth cumulant contribution is small.

The other particular choice which has been found to be very useful for the $g(\cdot)$ function in Eq. 4.54 is the signum function: $g(\omega) = \mathrm{sgn}(\omega) \equiv [U(\omega) - U(-\omega)]$. In fact, the $Z(t)$ defined in this way has a special name, being called the Hilbert transform of $Y(t)$.[7] The amplitude, or envelope, defined using the Hilbert transform of $Y(t)$ was apparently first introduced by Cramer and Leadbetter in 1967 and we will call it the Cramer and Leadbetter amplitude and denote it by the symbol $A_{CL}(t)$. One of the special features of the $\{Z(t)\}$ Hilbert transform process is that it has the same autospectral density as $\{Y(t)\}$, since it has $g^2(\omega) = 1$ everywhere. Among the implications of this are the fact that not only is $\sigma_Z^2 = \sigma_X^2 = \lambda_0$ (as was true for the $\{Z(t)\}$ process in the energy-based amplitude), but also $\sigma_{\dot{Z}}^2 = \sigma_{\dot{X}}^2 = \lambda_2$. In addition, Eq. 4.60 gives the cross-covariance of $Z(t)$ and $\dot{X}(t)$ as

$$K_{Z\dot{X}} = G_{Z\dot{X}}(0) = \int_{-\infty}^{\infty} \omega\,\mathrm{sgn}(\omega)\,S_{XX}(\omega)\,d\omega = \int_{-\infty}^{\infty} |\omega|\,S_{XX}(\omega)\,d\omega \equiv \lambda_1 \qquad (4.72)$$

Substituting these results into Eqs. 4.67 and 4.68, gives the terms related to the rates of change of $\{A(t)\}$ and $\{\theta(t)\}$ as

$$E\!\left(A_{CL}^2(t)\dot{A}_{CL}^2(t)\right) = 2\lambda_0\lambda_2 - 2\lambda_1^2 + f_1(\kappa_4) = 2\lambda_0\lambda_2(1 - \alpha_1^2) + f_1(\kappa_4) \qquad (4.73)$$

and

$$E\!\left(A_{CL}^4(t)\dot{\theta}_{CL}^2(t)\right) = 2\lambda_0\lambda_2 - 2\lambda_1^2 + f_2(\kappa_4) = 2\lambda_0\lambda_2(1 - \alpha_1^2) + f_2(\kappa_4) \qquad (4.74)$$

The average frequency, from Eq. 4.65, for this amplitude is

$$\omega_{CL} = \frac{\lambda_1}{\lambda_0} \qquad (4.75)$$

The similarities between the results in Eqs. 4.73 and 4.74 and those in Eqs. 4.70 and 4.71 are obvious. Just as the α_2 bandwidth parameter governs the rate of change of $\{A(t)\}$ and $\{\theta(t)\}$

---

[7] A time domain representation of the Hilbert transform is given by the rather ill behaved integral

$$Z(t) = \frac{1}{\pi}\int_{-\infty}^{\infty} \frac{Y(s)}{s - t}\,ds$$

for the energy-based amplitude, the α_1 parameter governs the rate of change of amplitude and phase for the Cramer and Leadbetter amplitude. Furthermore, we again find that the covariance contributions to mean squared values of $A^2(t)\dot{A}^2(t)$ and $A^4(t)\dot{\theta}^2(t)$ are identical.

Note that the average frequencies ω_a and ω_{CL} according to the energy-based and the Cramer and Leadbetter definitions, respectively, are not identical. This is not surprising, since they are defined to give a zero mean value for the rate of change of the phase according to two different definitions of the the phase of the process. For the limiting narrowband process, for which $\lambda_j \approx \omega_c^j \lambda_0$, both definitions give the average frequency as ω_c. Note also that Eq. 4.29 gave $\omega_c \approx \sigma_{\dot{X}} / \sigma_X$ for a narrowband process, while Eq. 4.69 gives $\omega_a \equiv \sigma_{\dot{X}} / \sigma_X$ for the energy-based definition of average frequency of any process. This demonstrates the consistency of the various frequency terms, as well as their differences.

****

Example 4.9: Investigate the average frequency and the rates of change of the amplitude and phase of the covariant stationary response $\{X(t)\}$ of a SDF oscillator excited by white noise, using both the energy-based and the Cramer and Leadbetter definition of the terms.

Since the quantities of interest are directly related to spectral moments and bandwidth parameters, we can simply use the results obtained in Example 4.8. For the Cramer and Leadbetter definition, Eq. 4.75 gives the average frequency as

$$\omega_{CL} = \frac{\omega_0}{\sqrt{1-\zeta^2}}\left[1 - \frac{2}{\pi}\tan^{-1}\left(\frac{\zeta}{\sqrt{1-\zeta^2}}\right)\right]$$

and Eq. 4.69 gives the energy-based average frequency as $\omega_a = \omega_0$. One may note that, in general, the Cramer and Leadbetter average frequency is less than the resonant frequency ω_0 for the oscillator, and for $\zeta \ll 1$ it tends to ω_0 like $\omega_{CL} \approx \omega_0(1 - 2\zeta/\pi)$. This limit certainly seems reasonable, based on the shape of the autospectral density. Rather surprisingly, the energy-based average frequency, though, is exactly equal to ω_0 regardless of the damping value.

Since the rates of change of the energy-based amplitude and phase are directly related to the value $(1-\alpha_2^2)$, the fact that $\alpha_2 = 0$ for the SDF response to white noise indicates that even when damping is very small this amplitude and phase are not slowly varying. This seems surprising, since sample time histories of $\{X(t)\}$ for this small damping situation surely do resemble slowly varying harmonic functions of time. Nonetheless, the energy-based amplitude and phase are defined in such a way that they are affected too much by the high-frequency components of $\{X(t)\}$, and they do not accurately reflect the narrowbanded nature of the autospectral density in the vicinity of the resonant peak for this particular system. The Cramer and Leadbetter amplitude and phase, on the other hand, do vary slowly for the small ζ situations for which sample time histories resemble slowly varying harmonic functions, since $(1-\alpha_1^2)$ goes to zero in this situation.

****

4.10   Calculating Autospectral Density from a Sample

In Example 4.5, we presented one method of calculating the autospectral density of process by filtering a time history, but noted that it is not the approach that is now commonly used. We will now present the usual method, which is based directly on the definition of spectral density in terms of a Fourier transform.

The concept of ergodicity discussed in Section 2.7 is very important when one needs to derive properties of a process by using measured data. In particular, if the process has the appropriate type of ergodicity, then one can use a time average over a single long time history to approximate an expected value, rather than needing an ensemble average over a large number of time histories. This property is usually assumed to hold unless there is evidence to the contrary. For autospectral density, on the other hand, we can show that ergodicity never holds in as strong a form as we might wish. Note that the definition of autospectral density in Eq. 4.14 involves both an expectation and a limit as the time interval T tends to infinity. What would be very desirable would be if ergodicity held in such a form that the expectation was unnecessary, and the calculation using a single time history of length T converged to the expected value as $T \to \infty$. Unfortunately, this is not true. We will now demonstrate that fact, and indicate how the matter can be remedied so that a useful estimate of the autospectral density can be obtained from a single time history.

The relationships appearing in Eq. 4.14 can be rewritten in a compact form by defining a new frequency domain stochastic process $\{Q_T(\omega)\}$ by

$$Q_T(\omega) = \frac{2\pi}{T} \tilde{Y}_T(\omega) \tilde{Y}_T(-\omega) \equiv \frac{1}{2\pi T} \int_{-T/2}^{T/2} \int_{-T/2}^{T/2} Y(t_1) Y(t_2) e^{-i\omega(t_1-t_2)} dt_1 dt_2 \qquad (4.76)$$

in which $\tilde{Y}_T(\omega)$ is the truncated Fourier transform of $Y(t) \equiv X(t) - \mu_X(t)$. The definition of autospectral density is then simply $S_{XX}(\omega) = E[Q_T(\omega)]$. If the variance of this $Q_T(\omega)$ random variable went to zero as $T \to \infty$, then we could be confident that $Q_T(\omega)$ calculated from a long time history would have a high probability of being a good approximation of $S_{XX}(\omega)$. Thus, we wish to investigate the variance of $Q_T(\omega)$, which can be found from its first and second moments. The second moment is given by

$$E[Q_T^2(\omega)] = \frac{1}{(2\pi T)^2} \int_{-T/2}^{T/2} \int_{-T/2}^{T/2} \int_{-T/2}^{T/2} \int_{-T/2}^{T/2} E[Y(t_1) Y(t_2) Y(t_3) Y(t_4)] e^{-i\omega(t_1-t_2+t_3-t_4)} dt_1 dt_2 dt_3 dt_4$$

$$(4.77)$$

We can now expand the fourth moment function in the integrand in terms of variance contributions and a fourth cumulant term (in the same way as we did in deriving Eqs. 4.67 and 4.68):

$$E[Y(t_1)Y(t_2)Y(t_3)Y(t_4)] = G_{XX}(t_1-t_2)G_{XX}(t_3-t_4) + G_{XX}(t_1-t_4)G_{XX}(t_3-t_2)$$
$$+ G_{XX}(t_1-t_3)G_{XX}(t_2-t_4) + \kappa_4[X(t_1),X(t_2),X(t_3),X(t_4)] \qquad (4.78)$$

One can show that the contribution to $E[Q_T^2(\omega)]$ of the third term in Eq. 4.78 tends to zero as $T \rightarrow \infty$. Also, the same is true for the fourth term provided that the fourth cumulant function of $\{X(t)\}$ meets the sort of restrictions required for second-moment ergodicity. Thus, we only need be concerned with the behavior of the first two terms in Eq. 4.78. The quadruple integral of each of these two terms can be separated into the product of two double integrals: grouped as (t_1,t_2) and (t_3,t_4) for the first term and as (t_1,t_4) and (t_3,t_2) for the second term. Furthermore, each of these double integrals is exactly of the form of $E[Q_T(\omega)]$. Substituting this result into the expression for $E[Q_T^2(\omega)]$ and taking the limit as T tends to infinity gives

$$\lim_{T\to\infty} E[Q_T^2(\omega)] = 2S_{XX}^2(\omega)$$

from which we find that the variance of $Q_T(\omega)$ tends to $S_{XX}^2(\omega)$. Thus, the standard deviation of $Q_T(\omega)$ tends to the same limit as the mean value, namely $S_{XX}(\omega)$. Clearly, this gives no basis for expecting a sample value of $Q_T(\omega)$ to have a high probability of being near the desired, but unknown, $S_{XX}(\omega)$.

The approach that is typically used to avoid the sampling difficulty is to replace $Q_T(\omega)$ by a smoothed version. We will illustrate this idea by using the simplest sort of smoothing, which is a simple average over an increment of frequency:

$$\bar{Q}_T(\omega) = \frac{2\pi}{T}\frac{1}{2\varepsilon}\int_{\omega-\varepsilon}^{\omega+\varepsilon} \tilde{Y}_T(\omega)\,\tilde{Y}_T(-\omega)\,d\omega_1 \qquad (4.79)$$

or

$$\bar{Q}_T(\omega) = \frac{1}{4\pi T\varepsilon}\int_{-T/2}^{T/2}\int_{-T/2}^{T/2} Y(t_1)Y(t_2)\int_{\omega-\varepsilon}^{\omega+\varepsilon} e^{-i\omega_1(t_1-t_2)}\,d\omega_1\,dt_1\,dt_2 \qquad (4.80)$$

The expected value of $\bar{Q}_T(\omega)$, of course, is not exactly $S_{XX}(\omega)$, but rather is an average of this autospectral density over the range $(\omega - \varepsilon,\ \omega + \varepsilon)$. This is generally not a significant problem if ε is chosen to be sufficiently small. The inclusion of this averaging, though, substantially changes the behavior of the mean squared value, as can be verified by writing out $E[\bar{Q}_T^2(\omega)]$ in the same way as Eq. 4.77:

$$E[\bar{Q}_T^2(\omega)] = \frac{1}{(4\pi T\varepsilon)^2}\int_{-T/2}^{T/2}\int_{-T/2}^{T/2}\int_{-T/2}^{T/2}\int_{-T/2}^{T/2} \left(E[Y(t_1)Y(t_2)Y(t_3)Y(t_4)] \right.$$

$$\left. \int_{\omega-\varepsilon}^{\omega+\varepsilon} e^{-i\omega_1(t_1-t_2)}\,d\omega_1\int_{\omega-\varepsilon}^{\omega+\varepsilon} e^{-i\omega_2(t_3-t_4)}\,d\omega_2 \right)dt_1\,dt_2\,dt_3\,dt_4$$

The change comes from the fact that the second term of the expansion in Eq. 4.78 behaves very differently for $E[\overline{Q}_T^2(\omega)]$ than for $E[Q_T^2(\omega)]$. After performing the integration with respect to frequency on this second term one obtains an integrand for the four-fold time integration of $G_{XX}(t_1-t_4)G_{XX}(t_3-t_2)/[(t_1-t_2)(t_3-t_4)]$ multiplied by a sum of four harmonic exponential terms. One can easily verify that this integrand goes to zero as any of the time arguments goes to infinity except in the special case when they all tend to infinity with $t_1 \approx t_2 \approx t_3 \approx t_4$. This behavior causes the four-fold time integral to give a term which grows like T as $T \to \infty$, but the division by T^2 then assures that this term contributes nothing to $E[\overline{Q}_T^2(\omega)]$ in the limit.

The first term of the expansion of Eq. 4.78, on the other hand, behaves basically as it did for $E[Q_T^2(\omega)]$. In particular, it gives the integrand of the time integration as $G_{XX}(t_1-t_2)G_{XX}(t_3-t_4)/[(t_1-t_2)(t_3-t_4)]$, and this term does not decay as time arguments go to infinity on the two-dimensional set $t_1 \approx t_2$, $t_3 \approx t_4$, with no restriction that the first pair of time arguments be close to the second pair. Thus, the integral of this term grows like T^2. In fact, it can be shown that its contribution to $E[\overline{Q}_T^2(\omega)]$ is exactly $(E[\overline{Q}_T^2(\omega)])^2$, so that the variance of $\overline{Q}_T(\omega)$ does tend to zero as $T \to \infty$. This guarantees that the $\overline{Q}_T(\omega)$ value obtained from a single sample time history has a high probability of being close to $E[\overline{Q}_T(\omega)]$ if T is sufficiently large. Interestingly, this limiting result holds no matter how small ε is chosen in the frequency averaging. Of course, the rate at which the variance of $\overline{Q}_T(\omega)$ tends to zero does depend on the value of ε, though.

The usual approach to estimation of $S_{XX}(\omega)$ from recorded data is, then, represented by Eq. 4.79, although a more sophisticated form of frequency averaging may be used in place of that shown in Eqs. 4.79 and 4.80. By taking the Fourier transform of a single time history of long duration and performing this averaging over a band of frequencies one can obtain an estimate of $S_{XX}(\omega)$. The bandwidth for the frequency averaging (2ε in Eq. 4.79) is often adjusted empirically so as to get an appropriately smooth estimate of $S_{XX}(\omega)$. This presumes, though, that one has some idea of the form of $S_{XX}(\omega)$ before the estimate is obtained. It must always be kept in mind that if the true $S_{XX}(\omega)$ function has a narrow peak which has a width not significantly exceeding 2ε, then the averaging in Eq. 4.79 will introduce significant error by greatly reducing that peak. Thus, considerable judgement should be exercised in estimating spectral densities.

It should also be noted that records of measured data are always of finite length. Furthermore, they represent observations only at discrete values of time rather than giving continuous time histories. This mandates that the integral Fourier transform considered here be replaced by the discrete Fourier transform (DFT), which is essentially a Fourier series in exponential form (see Eqs. D.6 and D.7 in Appendix D). Efficient numerical schemes, including use of the fast Fourier transform (FFT) algorithm, have been developed to expedite implementation of the basic ideas presented. More detail on the problems of spectral estimation, including extensions to more complicated situations, are given in books on signal processing, such as Bendat and Piersol (1966), Marple (1987), and Priestly (1988).

Exercises

****

Spectral Density and Variance

****

4.1 Let $\{X(t)\}$ be a covariant stationary stochastic process with autospectral density function

$$S_{XX}(\omega) = |\omega| e^{-\omega^2}$$

(a) Find the variance of the process, σ_X^2.

(b) Find the variance of the derivative process, $\sigma_{\dot{X}}^2$.

****

4.2 Consider a covariant stationary stochastic process $\{X(t)\}$ with autospectral density function

$$S_{XX}(\omega) = e^{-\omega^2/2}$$

(a) Find the variance of the process, σ_X^2.

(b) Find the variance of the derivative process, $\sigma_{\dot{X}}^2$.

****

4.3 Let $\{X(t)\}$ be a covariant stationary stochastic process with autospectral density function

$$S_{XX}(\omega) = S_0 \left|\frac{\omega}{\omega_0}\right|^c U(\omega_0 - |\omega|) + S_0 \left|\frac{\omega_0}{\omega}\right|^c [1 - U(\omega_0 - |\omega|)]$$

in which S_0, ω_0, and c are positive constants.

(a) Find the variance of the process, σ_X^2.

(b) Find the variance of the derivative process, $\sigma_{\dot{X}}^2$.

(c) Note any restrictions on c required for your answers to parts (a) and (b).

****

4.4 Let $\{X(t)\}$ be a mean zero covariant stationary stochastic process with autospectral density function $S_{XX}(\omega) = S_0 e^{-|\omega|}$.

(a) Find $E(X^2)$.

(b) Find $E(\dot{X}^2)$

(c) Find $E[X(t)\dot{X}(s)]$.

****

4.5 Consider a covariant stationary process $\{X(t)\}$ with autospectral density

$$S_{XX}(\omega) = S_0 U(\omega_0 - |\omega|) + S_0 \left|\frac{\omega_0}{\omega}\right|^4 [1 - U(\omega_0 - |\omega|)]$$

(a) Find the variance of the process, σ_X^2.

(b) Find the variance of the derivative process, $\sigma_{\dot{X}}^2$.

****

4.6 As a first step in modeling an earthquake ground motion I wish to find a stationary stochastic process $\{Z(t)\}$ simultaneously satisfying the following three conditions:

I: $E(\ddot{Z}^2) = 1.0$ (m/sec^2)^2,   II: $E(\dot{Z}^2) = 1.0$ (m/sec)^2, and   III: $E(Z^2) < \infty$

Consider each the three following autospectral density curves as candidates for this modeling.

For each autospectral density, state whether it is possible to choose real, positive constants a and b such as to satisfy conditions I, II, and III. If it is not possible to satisfy any particular condition, then explain the difficulty encountered. [Note: Evaluation of a or b is not required, only determination of whether solutions exist.]

(a) $S_{\ddot{Z}\ddot{Z}}(\omega) = aU(b-|\omega|)$

(b) $S_{\ddot{Z}\ddot{Z}}(\omega) = \dfrac{\omega^4}{a+b\omega^8}$

(c) $S_{ZZ}(\omega) = \dfrac{1}{a+b\omega^4}$

****

Autospectral Density and Probability Density
****

4.7 Assume that for a mean zero stationary process $\{X(t)\}$ you know a conditional probability density function of the form $p_{X(t+\tau)}(v|X(t)=u)$. Further, assume that

$$\lim_{\tau\to\infty} p_{X(t+\tau)}(v|X(t)=u) = p_X(v) \quad \text{for any } u \text{ value.}$$

Give an integral expression for the autospectral density function $S_{XX}(\omega)$ in terms of the conditional probability density function.

[Hint: As an intermediate step, use the result of Exercise 2.10 giving the autocovariance function $G_{XX}(\tau)$ in terms of the conditional probability density function.]
****

Narrowband and Broadband
****

4.8 Let $\{X(t)\}$ be a covariant stationary stochastic process with autospectral density function

$$S_{XX}(\omega) = S_0\left[\exp(-\gamma|\omega + \omega_0|) + \exp(-\gamma|\omega - \omega_0|)\right]$$

(a) Find the autocovariance function: $G_{XX}(\tau)$.

(b) Show that $\gamma \gg 1$ gives a narrowband process with $\rho_{XX}(\tau) \equiv G_{XX}(\tau)/\sigma_X^2 \approx \cos(\omega_0\tau)$.

(c) Show that $\gamma \to 0$ gives $G_{XX}(\tau) \to 0$ for $\tau \neq 0$, and $G_{XX}(0) \to \infty$, so that the autocovariance of $\{X(t)\}$ tends to that for a delta-correlated process.
****

Dynamic Response
****

4.9 Consider a linear system whose response $\{X(t)\}$ to an excitation $\{F(t)\}$ is governed by the differential equation $\ddot{X}(t) + c\dot{X}(t) = F(t)$, in which $c > 0$ is a constant.

Let $\{F(t)\}$ be covariant stationary with autospectral density $S_{FF}(\omega) = S_0 U(10c-|\omega|)$.

(a) Find the autospectral densities of $\{X(t)\}$ and $\{\dot{X}(t)\}$: $S_{XX}(\omega)$ and $S_{\dot{X}\dot{X}}(\omega)$.

(b) Are the variances σ_X^2 and $\sigma_{\dot{X}}^2$ finite? Briefly explain your answer.
****

4.10 Consider a linear system governed by the first-order differential equation

$$\dot{X}(t) + aX(t) = F(t) \quad \text{with} \quad 0 \leq a \leq 1$$

The excitation $\{F(t)\}$ is covariant stationary with non-zero mean $\mu_F = b$ and autocovariance function $K_{FF}(t+\tau,t) = e^{-|\tau|}$ for all t and τ.

(a) Find the autospectral density of $\{F(t)\}$: $S_{FF}(\omega)$.

(b) Find the autospectral density of $\{X(t)\}$: $S_{XX}(\omega)$.
****

4.11 Consider a building subjected to a wind force $\{F(t)\}$. The building is modeled as a linear

SDF system:
$$m\ddot{X}(t) + c\dot{X}(t) + kX(t) = F(t)$$
with $m = 200{,}000$ kg, $c = 8.0$ kN·s/m, and $k = 3{,}200$ kN/m. The force $\{F(t)\}$ has a mean of $\mu_F = 20$ kN and an autospectral density of $S_{FF}(\omega) = 500|\omega|e^{-\omega^2/2}$ (kN)^2 / (rad / s) for all ω.
(a) Find the mean squared force on the building: $E(F^2)$.
(b) Find $E(X) = \mu_X$.
(c) Estimate the response standard deviation σ_X by replacing $\{F(t)\}$ by a constant force of 20 kN plus an "equivalent" white noise excitation.
(d) A delicate instrument is to be mounted in the building. Find the autospectral density of its base acceleration: $S_{\ddot{X}\ddot{X}}(\omega)$.

****

4.12 Consider a linear system whose response $\{X(t)\}$ is governed by the differential equation
$$\ddot{X}(t) + 5\dot{X}(t) + 6X(t) = F(t)$$
The excitation $\{F(t)\}$ is a mean zero stationary white noise with autospectral density S_0.
(a) Find the harmonic transfer function $H_x(\omega)$.
(b) Find the autospectral density of the response: $S_{XX}(\omega)$. Sketch your answer.

****

4.13 Consider the response of two structures to an anticipated earthquake. The strong motion portion of the ground acceleration will be modeled as a mean zero stationary process with autospectral density
$$S_{FF}(\omega) = \frac{6}{400 + \omega^2} \quad \frac{(m/s^2)^2}{rad / s}$$
Each structure will be modeled as a single-degree-of-freedom system:
$$\ddot{X}_j + 2\zeta_j\omega_j\dot{X}_j + \omega_j^2 X_j = F(t)$$
with X_j representing the displacement at the top of structure j.
Structure number 1 has $\omega_1 = 10$ rad/s and $\zeta_1 = 0.01$. Structure number 2 is to be built near the first structure and is expected to have $\omega_2 = 15$ rad/s and $\zeta_2 = 0.005$.
(a) Use the concept of an equivalent white noise to estimate the stationary standard deviation of response of each structure: σ_{X_1} and σ_{X_2}.
(b) Let b = static clearance between the two structures. This gives $Y = b - X_1 + X_2$ as the clearance between the two structures during the earthquake (assuming that both are of the same height). Find stationary values of μ_Y and σ_Y, by assuming that X_1 and X_2 are independent.

****

Average Frequency and Bandwidth Parameters
****

4.14 Let $\{X(t)\}$ be the covariant stationary stochastic process of Exercise 4.8 with
$$S_{XX}(\omega) = S_0\left[\exp\left(-\gamma|\omega + \omega_0|\right) + \exp\left(-\gamma|\omega - \omega_0|\right)\right]$$
in which S_0, ω_0, and γ are positive constants.
(a) Find the spectral moments λ_0, λ_1, λ_2, and λ_4.
(b) Find the average frequencies $\omega_a = \sigma_{\dot{X}}/\sigma_X$ and $\omega_{CL} = \lambda_1/\lambda_0$.
(c) Find the bandwidth parameters α_1 and α_2.
(d) Discuss the behavior of the autospectral density, the bandwidth parameters, and the average

frequencies both for $\gamma \to \infty$ and for $\gamma \to 0$.
****

4.15 Let $\{X(t)\}$ be the covariant stationary stochastic process of Exercise 4.3 with

$$S_{XX}(\omega) = S_0 \left|\frac{\omega}{\omega_0}\right|^c U(\omega_0 - |\omega|) + S_0 \left|\frac{\omega_0}{\omega}\right|^c [1 - U(\omega_0 - |\omega|)]$$

in which S_0, ω_0, and c are positive constants.
(a) Find the spectral moments λ_0, λ_1, λ_2, and λ_4.
(b) Find the average frequencies $\omega_a = \sigma_{\dot{X}} / \sigma_X$ and $\omega_{CL} = \lambda_1 / \lambda_0$.
(c) Find the bandwidth parameters α_1 and α_2.
(d) Discuss the behavior of the autospectral density, the bandwidth parameters, and the average
    frequencies both for $c \to \infty$ and for $c \to 0$.
****

4.16 Let $\{X(t)\}$ be a covariant stationary stochastic process with autospectral density function
$$S_{XX}(\omega) = A|\omega|^b e^{-c|\omega|} \quad \text{in which } A, b, \text{ and } c \text{ are positive constants.}$$
(a) Find the spectral moments λ_0, λ_1, λ_2, and λ_4.
(b) Find the average frequencies $\omega_a = \sigma_{\dot{X}} / \sigma_X$ and $\omega_{CL} = \lambda_1 / \lambda_0$.
(c) Find the bandwidth parameters α_1 and α_2.
(d) For what values of b and/or c is this a narrowband process? What is the dominant frequency
    of this narrowband process?
****

4.17 Consider the covariant stationary process $\{X(t)\}$ of Exercise 4.5 with

$$S_{XX}(\omega) = S_0 U(\omega_0 - |\omega|) + S_0 \left|\frac{\omega_0}{\omega}\right|^4 [1 - U(\omega_0 - |\omega|)]$$

(a) Find the spectral moments λ_0, λ_1, λ_2, and λ_4.
(b) Find the average frequencies $\omega_a = \sigma_{\dot{X}} / \sigma_X$ and $\omega_{CL} = \lambda_1 / \lambda_0$.
(c) Find the bandwidth parameters α_1 and α_2.
****

4.18 Let $\{X(t)\}$ be a covariant stationary process with autocovariance function
$$G_{XX}(\tau) = \exp(-c\tau^2) \cos(a\tau)$$
(a) Find the average frequency $\omega_a = \sigma_{\dot{X}} / \sigma_X$
(b) Find the bandwidth parameter α_2.
(c) Under what conditions on the parameters a and c might $\{X(t)\}$ be considered a narrowband
    process?
****

4.19 Let $\{X(t)\}$ be a covariant stationary process with autocovariance function
$$G_{XX}(\tau) = \frac{\cos(a\tau)}{1 + c\tau^2}$$
(a) Find the average frequency $\omega_a = \sigma_{\dot{X}} / \sigma_X$
(b) Find the bandwidth parameter α_2.
(c) Under what conditions on the parameters a and c might $\{X(t)\}$ be considered a narrowband
    process?
****

Chapter 5

Gaussian and Non-Gaussian Stochastic Processes

5.1 Importance of the Gaussian Distribution

Gaussian stochastic processes play an extremely important role in practical engineering applications. In fact, the majority of applications of stochastic methods to the study of dynamical systems consider only Gaussian processes, even though Gaussian is only one special case out of infinitely many that could be studied. Similarly, the Gaussian special case occupies a very important, though not quite so dominant, position in the engineering application of random variable methods.

There are at least two important reasons why the Gaussian distribution is used so extensively in applied probability. One is convenience. The probability density functions and characteristic functions for Gaussian random variables have simple forms which often allow us to analytically evaluate quantities of interest. For example, all the moments of a Gaussian random variable X can be described by relatively simple relationships. Also important is the fact that the Gaussian probability density function $p_X(u)$ converges to zero sufficiently rapidly for $|u|$ tending to infinity that one is assured that the expectation of many functions of X will exist. For example, $E(e^{\alpha X})$ exists for all values of the parameter α, and that is not true for many other unbounded random variables X.

The second reason why the Gaussian distribution is often chosen in modeling real problems is that it often provides quite a good approximation to measured statistical data. Presumably, this is because of the central limit theorem. A very loose statement of this important theorem is that if a random variable X is the sum of a large number of separate components, then X is approximately Gaussian under weak restrictions on the joint distribution of the components. More precisely, if X_j is the sum of the terms $\{Z_1, Z_2, \cdots, Z_j\}$ then the probability density function for the standardized random variable $(X_j - \mu_{X_j})/\sigma_{X_j}$ tends to a Gaussian form as j tends to infinity under weak restrictions on the distribution of the Z_k components[1]. In practical modeling situations, one generally cannot prove that the central limit theorem is satisfied since one rarely has the information which would be needed about the joint probability distribution of the components contributing to the quantity of interest. On the other hand, the theorem does provide some basis for expecting a random variable X to be approximately Gaussian if many factors contribute to its randomness. Experience seems to confirm that this is a quite good assumption in a great variety of situations in virtually all areas of applied probability, including stochastic structural dynamics.

The definition of a Gaussian stochastic process is extremely simple. In particular, $\{X(t)\}$ is

---

[1] One adequate set of restrictions on the Z_k components is that they all be independent and identically distributed, but neither the condition of identical distribution or that of independence of all the Z_k terms is essential for the proof of the central limit theorem.

said to be a Gaussian stochastic process if any finite set of random variables $\{X(t_1),\ X(t_2),\cdots,$ $X(t_n)\}$ from that process is jointly Gaussian. Thus, understanding Gaussian processes requires knowledge of the properties of jointly Gaussian random variables. These properties are covered in some detail in Appendix B, for the benefit of readers who need a review. The following sections discuss how these random variable results translate into very special properties for Gaussian stochastic processes.

5.2   Gaussian Processes and Linear Dynamics

Since we are interested in linear systems governed by differential equations, we will now investigate the behavior of derivatives and integrals of a Gaussian process. The fact that the derivative $\dot{X}(t)$ is defined as a limit of $Y(t,h) = [X(t + h) - X(t)]/h$ proves that $\dot{X}(t)$ is a Gaussian random variable for any value of t, since a linear combination of jointly Gaussian random variables is always Gaussian. Furthermore, any set of random variables $\{\dot{X}(t_1),\dot{X}(t_2),\cdots,\quad \dot{X}(t_n)\}$ from the derivative process is defined to be a limit of the sequence $\{Y(t_1,h_1),Y(t_2,h_2),\cdots,\ Y(t_n,h_n)\}$, and the Gaussian nature of $\{X(t)\}$ assures that this latter sequence is jointly Gaussian. This proves that the $\{\dot{X}(t_1),\dot{X}(t_2),\cdots,\dot{X}(t_n)\}$ random variables are jointly Gaussian, and this is a statement that $\{\dot{X}(t)\}$ is a Gaussian process. Thus, the relationships for Gaussian random variables are sufficient to prove the very important result that the derivative of any Gaussian process is also a Gaussian process.

In a similar way one can show the integral of a Gaussian process is generally a Gaussian process, since it is the limit of a Riemann summation which is a linear combination of the Gaussian integrand values. This result applies to all three types of stochastic integrals defined in Section 2.9: the definite integral of $\{X(t)\}$ between deterministic limits is a Gaussian random variable; the integral of $\{X(s)\}$ from a deterministic lower limit to an upper limit of t gives a Gaussian stochastic process; and the convolution integral of $\{X(t)\}$ with a deterministic kernel also defines a new Gaussian process. In some situations, another random variable is added to a stochastic integral giving expressions such as

$$X(t) = X(t_0) + \int_{t_0}^{t} \dot{X}(s)\,ds \tag{5.1}$$

or

$$X(t) = X(t_0) + \int_{t_0}^{t} F(s)h(t-s)\,ds \tag{5.2}$$

In these situations, $X(t)$ is not simply a stochastic integral of $\{\dot{X}(t)\}$ and $\{F(t)\}$, respectively, so it might not be Gaussian. The condition for $\{X(t)\}$ in Eq. 5.1 to be a Gaussian process is that $X(t_0)$ and $\{\dot{X}(t_1),\cdots,\dot{X}(t_n)\}$ be jointly Gaussian, and in Eq. 5.2 it is that $X(t_0)$ and $\{F(t_1),\cdots,\ F(t_n)\}$ be jointly Gaussian. If the initial condition is deterministic rather than random, then Eq. 5.2 will always give a Gaussian process since a deterministic variable can always be considered as a Gaussian random variable with zero variance. Furthermore, this zero variance random variable can be considered to be independent of all other variables, which makes it jointly Gaussian with any other Gaussian random variable.

These results are obviously very pertinent to the study of the response process $\{X(t)\}$ for a linear system with an excitation process $\{F(t)\}$ which is Gaussian. In particular, the Duhamel convolution integral of Eq. 3.11 guarantees that $\{X(t)\}$ is almost sure to be a Gaussian process. The only exception is if $\{X(t)\}$ has some random initial condition which is not jointly Gaussian with $\{F(t)\}$. This exception can occur for either a non-Gaussian or Gaussian initial condition, but it is unusual in practice.

One of the major advantages of Gaussian random variables or stochastic processes is that they are completely described by their first and second moment properties. In particular, a Gaussian process $\{X(t)\}$ is completely described by its mean-value function $\mu_X(t)$ and either its autocorrelation function $\phi_{XX}(t_1,t_2)$ or its autocovariance function $K_{XX}(t_1,t_2)$. For a stationary Gaussian process, the information needed for a complete description becomes simply a constant μ_X and a function of a single variable, $R_{XX}(\tau)$ or $G_{XX}(\tau)$. From this very limited information, one can use the Gaussian probability density function (see Eq. B.14 of Appendix B) to write expressions for the probability of any event which is determined by the $\{X(t)\}$ process. Of course, the results that we have derived in Chapters 3 and 4 for the response to a general stochastic excitation still hold for the Gaussian situation, so they can be used to find the necessary first and second moment terms.

The analysis of Gaussian dynamics problems is simplified by the fact that uncorrelated jointly Gaussian random variables are also independent. For example, we have found in Section 2.8 that the random variables $X(t)$ and $\dot{X}(t)$ are uncorrelated at any time t for a covariant stationary response process $\{X(t)\}$. If this process is also Gaussian, though, this implies that $X(t)$ and $\dot{X}(t)$ are also independent, so that knowledge of the value of either $X(t)$ or $\dot{X}(t)$ gives absolutely no information about the probability distribution of the other quantity at the same instant of time. Thus, for example, it is trivial to find the conditional probability distribution of $\dot{X}(t)$ given $X(t) = u$ or the conditional probability distribution of $X(t)$ given $\dot{X}(t) = v$ if $\{X(t)\}$ is a covariant stationary Gaussian process. The reader is cautioned that the results discussed here do not imply that $\{X(t)\}$ and $\{\dot{X}(t)\}$ are independent processes. In particular, it is generally true that the random variables $X(t)$ and $\dot{X}(s)$ are correlated, and therefore dependent, for $t \neq s$. It is only at the same instant of time that one can be assured that $X(t)$ and $\dot{X}(t)$ are uncorrelated for a stationary process.

****

Example 5.1: Let the process $\{X(t)\}$ represent the response of the SDF oscillator shown, measured from the static equilibrium position. The excitation $\{F(t)\}$ is a stationary Gaussian white noise with a mean value of $\mu_F = 5$ kN, and an autospectral density of $S_0 = 4,000$ N^2 / (rad / s). Find the probability that the spring will be in tension at any instant of time t.

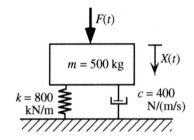

In static equilibrium we will have $X = 0$ (by definition of X) and a compression force of mg in the spring, where $g = 9.81$ m / s^2 is the acceleration of gravity. Thus, the spring will be in tension only at times when the dynamic $X(t)$ is less than $-mg/k = -6.13$ mm. We also know that $\{X(t)\}$ is Gaussian, since the excitation is a Gaussian process. Thus, we only need find the mean and variance of the stationary $X(t)$ in order to find the probability of the event of interest. We find that the mean displacement is $\mu_X = \mu_F/k = (5/800)$ m $= 6.25$ mm. The parameters of the dynamic response are

$$\omega_0 = \sqrt{\frac{k}{m}} = \sqrt{\frac{800 \text{ kN/m}}{500 \text{ kg}}} = 40 \text{ rad/s}$$

and

$$\zeta = \frac{c}{2\sqrt{km}} = \frac{400 \text{ N} \cdot \text{s/m}}{2\sqrt{(800 \text{ kN/m})(500 \text{ kg})}} = 0.01$$

so that the standard deviation of the response is

$$\sigma_X = \sqrt{\frac{\pi S_0}{ck}} = \sqrt{\frac{\pi(4,000 N^2 \cdot s)}{(400 \text{ N} \cdot \text{s/m})(800 \text{ kN/m})}} = 6.27 \text{ mm}$$

Now we can use the Gaussian cumulative distribution (as given in Eq. B.7 and Table B.1 of Appendix B) to write

$$P[X(t) < -6.13 \text{ mm}] = \Phi\left(\frac{-6.13 - 6.25 \text{ mm}}{6.27 \text{ mm}}\right) = \Phi(-1.975) = 0.0241$$

Thus, there is a 2.4% probability that this displacement, which has a mean value of $+6.25$ mm, will actually be more negative than -6.13 mm at any instant of time.

Note that if the response is also first-order ergodic (see Sec. 2.7), then we can also say that the spring will be in tension for 2.4% of the time in a long sample of response.
****

Example 5.2: Let $\{X(t)\}$ be a stationary Gaussian process. Find the joint probability density function for the three random variables $X(t)$, $\dot{X}(t)$, and $\ddot{X}(t)$ for any single value of t.

Since $\{X(t)\}$ is a stationary process, we know that $\dot{X}(t)$ and $\ddot{X}(t)$ are mean zero, so that μ_X is the only mean value which must be considered. Furthermore, stationarity requires that $X(t)$ and $\dot{X}(t)$ are uncorrelated, and that $\dot{X}(t)$ and $\ddot{X}(t)$ are uncorrelated. The only nonzero correlation is between $X(t)$ and $\ddot{X}(t)$, and it was shown in Section 4.8 that this correlation coefficient is $\rho_{X\ddot{X}} = -\alpha_2$. Thus, the covariance matrix is

$$\mathbf{G}_{YY} \equiv E\left([\vec{Y} - \vec{\mu}_Y][\vec{Y} - \vec{\mu}_Y]^T\right) = \begin{bmatrix} \sigma_X^2 & 0 & -\alpha_2\sigma_X\sigma_{\ddot{X}} \\ 0 & \sigma_{\dot{X}}^2 & 0 \\ -\alpha_2\sigma_X\sigma_{\ddot{X}} & 0 & \sigma_{\ddot{X}}^2 \end{bmatrix}$$

in which $\vec{Y} = [X(t), \dot{X}(t), \ddot{X}(t)]^T$. The general form of the Gaussian probability density function (see Eq. B.14 of Appendix B), then, gives

$$p_{\vec{Y}}(\vec{u}) = \frac{1}{(2\pi)^{3/2}\sqrt{|\mathbf{G}_{YY}|}}\exp\left[-\frac{1}{2}(\vec{u}-\vec{\mu}_Y)^T\mathbf{G}_{YY}^{-1}(\vec{u}-\vec{\mu}_Y)\right]$$

in which the determinant of the covariance matrix is $|\mathbf{G}_{YY}| = \sigma_X^2\sigma_{\dot{X}}^2\sigma_{\ddot{X}}^2(1-\alpha_2^2)$, and the inverse is

$$\mathbf{G}_{YY}^{-1} = \frac{1}{1-\alpha_2^2}\begin{bmatrix} \sigma_X^{-2} & 0 & \alpha_2\sigma_X^{-1}\sigma_{\ddot{X}}^{-1} \\ 0 & \sigma_{\dot{X}}^{-2}(1-\alpha_2^2) & 0 \\ \alpha_2\sigma_X^{-1}\sigma_{\ddot{X}}^{-1} & 0 & \sigma_{\ddot{X}}^{-2} \end{bmatrix}$$

Substituting these results gives $p_{X(t)\dot{X}(t)\ddot{X}(t)}(u_1,u_2,u_3) \equiv p_{\vec{Y}}(\vec{u})$ as

$$p_{\vec{Y}}(\vec{u}) = \frac{\exp\left(\dfrac{-1}{2}\left(\dfrac{u_2}{\sigma_{\dot{X}}}\right)^2 - \dfrac{1}{2(1-\alpha_2^2)}\left[\left(\dfrac{u_1-\mu_X}{\sigma_X}\right)^2 + 2\alpha_2\left(\dfrac{u_1-\mu_X}{\sigma_X}\right)\left(\dfrac{u_3}{\sigma_{\ddot{X}}}\right) + \left(\dfrac{u_3}{\sigma_{\ddot{X}}}\right)^2\right]\right)}{(2\pi)^{3/2}\sigma_X\sigma_{\dot{X}}\sigma_{\ddot{X}}\sqrt{1-\alpha_2^2}}$$

Note, though, that this function can be factored into the form

$$p_{X(t)\dot{X}(t)\ddot{X}(t)}(u_1,u_2,u_3) = p_{\dot{X}(t)}(u_2)p_{X(t)\ddot{X}(t)}(u_1,u_3)$$

with

$$p_{\dot{X}(t)}(u_2) = \frac{1}{\sqrt{2\pi}\sigma_{\dot{X}}}\exp\left[\frac{-1}{2}\left(\frac{u_2}{\sigma_{\dot{X}}}\right)^2\right]$$

and

$$p_{X(t)\ddot{X}(t)}(u_1,u_3) = \frac{\exp\left(\dfrac{-1}{2(1-\alpha_2^2)}\left[\left(\dfrac{u_1-\mu_X}{\sigma_X}\right)^2 + 2\alpha_2\left(\dfrac{u_1-\mu_X}{\sigma_X}\right)\left(\dfrac{u_3}{\sigma_{\ddot{X}}}\right) + \left(\dfrac{u_3}{\sigma_{\ddot{X}}}\right)^2\right]\right)}{2\pi\sigma_X\sigma_X\sqrt{1-\alpha_2^2}}$$

This demonstrates that the random variable $\dot{X}(t)$ is independent of the pair $\{X(t), \ddot{X}(t)\}$. Since any two Gaussian random variables are independent if they are uncorrelated, it was already known that $\dot{X}(t)$ was independent of $X(t)$ and independent of $\ddot{X}(t)$. For the Gaussian distribution, this is sufficient information to assert that $\dot{X}(t)$ is independent of the pair $\{X(t), \ddot{X}(t)\}$, although this is not generally true for other distributions.

****

5.3 Amplitude and Phase of a Gaussian Process

For a Gaussian process, one can both simplify and extend the results derived in Section 4.9

regarding the amplitude and phase of a general stochastic process. The most obvious simplification is in the expressions related to the rate of change of the amplitude and phase. In particular, all cumulants beyond second-order are zero for a Gaussian process, so that the $f_j(\kappa_4)$ terms in Eqs. 4.66 - 4.70 are exactly zero and those equations give

$$E\left(A^2(t)\dot{A}^2(t)\right) = E\left(A^4(t)\dot{\theta}^2(t)\right) = \frac{\lambda_0^2\lambda_4}{\lambda_2}(1-\alpha_2^2) = \frac{\sigma_X^4\sigma_{\dot{X}}^2}{\sigma_X^2}(1-\alpha_2^2) \tag{5.3}$$

for the energy-based definition using $X(t)$ and $\dot{X}(t)$, and

$$E\left(A_{CL}^2(t)\dot{A}_{CL}^2(t)\right) = E\left(A_{CL}^4(t)\dot{\theta}_{CL}^2(t)\right) = 2\lambda_0\lambda_2(1-\alpha_1^2) = 2\sigma_X^2\sigma_{\dot{X}}^2(1-\alpha_1^2) \tag{5.4}$$

for the Cramer and Leadbetter definition using $X(t)$ and its Hilbert transform. Thus, for a Gaussian process there is no approximation involved in saying that $A(t)$ and $\theta(t)$ are slowly varying if α_2 is near unity, or that $A_{CL}(t)$ and $\theta_{CL}(t)$ are slowly varying if α_1 is near unity.

For a Gaussian $\{X(t)\}$ process, we can also learn much more than we did in Section 4.9 about the $\{A(t)\}$ and $\{\theta(t)\}$ processes. In particular, we can derive relevant probability density functions, according to either definition of amplitude and phase, by using the relationships given by Eqs. 4.49, 4.51, 4.62, and 4.63 between the four random variables $A(t)$, $\theta(t)$, $\dot{A}(t)$, and $\dot{\theta}(t)$ and the four "original" variables of $Y(t)$, $Z(t)$, $\dot{Y}(t)$, and $\dot{Z}(t)$, in which $Y(t) = X(t) - \mu_X(t)$ and $\{Z(t)\}$ is either $\{\dot{Y}(t)/\omega_a\}$ or the Hilbert transform of $\{Y(t)\}$. Furthermore, these latter four variables are all mean zero, so that when $\{X(t)\}$ is a Gaussian process their joint probability density function depends only on their variance and covariance values, which were related to spectral moments in Section 4.9.

Before looking at the probability distributions of $\dot{A}(t)$ and $\dot{\theta}(t)$ we will consider the simpler problem of finding the probability density functions for $A(t)$ and $\theta(t)$. Note that we can consider the two equations

$$Y = A\cos(\omega_a t + \theta) \tag{5.5}$$

and

$$Z = -A\sin(\omega_a t + \theta) \tag{5.6}$$

to be the components of a vector nonlinear relationship between the original random variables and the new random variables: $(Y,Z) = \vec{g}(A,\theta)$, in which the t argument has been omitted since it is the same on each quantity. The joint probability density function for A and θ can then be written as (see Eq. A.35 in Appendix A)

$$p_{A,\theta}(u,\xi) = |J|\big|_{A=u,\theta=\xi}\, p_{Y,Z}[\vec{g}(u,\xi)]$$

in which $|J|$, which must be evaluated at the point of interest, is the absolute value of the Jacobian of the transformation:

$$|J| = \begin{Vmatrix} \dfrac{\partial Y}{\partial A} & \dfrac{\partial Y}{\partial \theta} \\ \dfrac{\partial Z}{\partial A} & \dfrac{\partial Z}{\partial \theta} \end{Vmatrix} = \begin{Vmatrix} \cos(\omega_a t + \theta) & -A\sin(\omega_a t + \theta) \\ -\sin(\omega_a t + \theta) & -A\cos(\omega_a t + \theta) \end{Vmatrix} = |-A| = A$$

Thus

$$p_{A,\theta}(u,\xi) = u\, p_{Y,Z}[u\cos(\omega_a t + \xi), -u\sin(\omega_a t + \xi)] \tag{5.7}$$

The joint probability density function on the right-hand side of Eq. 5.7 is simplified by the fact that Y and Z are independent random variables. This follows from the fact that X and Z are uncorrelated ($K_{XZ} = 0$) for any definition of amplitude and phase according to Eqs. 4.49 - 4.52, and the property of Gaussian random variables being uncorrelated only if they are independent. Thus, by using only the marginal Gaussian probability density functions of Y and Z we can write

$$p_{A,\theta}(u,\xi) = u\, p_Y[u\cos(\omega_a t + \xi)]\, p_Z[-u\sin(\omega_a t + \xi)] = \frac{u}{2\pi\lambda_0}\exp\left(-\frac{u^2}{2\lambda_0}\right)$$

in which the substitution $\sigma_X^2 = \sigma_Z^2 = \lambda_0$ has been used, since it is true for the Cramer and Leadbetter definition of amplitude and phase and also for the the energy-based definition with the average frequency taken as $\omega_a = (\lambda_2 / \lambda_0)^{1/2}$. Now we note that $p_{A,\theta}(u,\xi)$ can be factored into the product of two marginal probability density functions as

$$p_{A,\theta}(u,\xi) = p_A(u)\, p_\theta(\xi) \tag{5.8}$$

with

$$p_A(u) = \frac{u}{\lambda_0}\exp\left(-\frac{u^2}{2\lambda_0}\right) \quad \text{for} \quad u \geq 0 \tag{5.9}$$

and

$$p_\theta(\xi) = (2\pi)^{-1} \quad \text{for} \quad 0 \leq \xi \leq 2\pi \tag{5.10}$$

The fact that $p_{A,\theta}(u,\xi)$ can be factored as in Eq. 5.8 demonstrates that $A(t)$ and $\theta(t)$ are independent random variables at any time t. In addition, we see that the phase $\theta(t)$ is uniformly distributed over its possible values, while the amplitude $A(t)$ has what is called the Rayleigh distribution. It should be kept in mind that all these results apply equally well to the amplitude and phase defined according to the energy-based approach or the Cramer and Leadbetter approach.

The fact that $A(t)$ has the Rayleigh distribution is actually an example of a much more general result. For any two independent, mean zero, Gaussian random variables, R_1 and R_2, having the same variance σ_R^2, we can show that $(R_1^2 + R_2^2)^{1/2}$ has the Rayleigh distribution of Eq. 5.9 with $\lambda_0 = \sigma_R^2$. Thus, for example, the amplitude definition in Eq. 4.51 will give a Rayleigh distributed amplitude for a Gaussian $\{X(t)\}$ provided that $Z(t)$ is defined in such a way that it is: Gaussian, mean zero, independent of $Y(t)$, and has the same variance as $Y(t)$. The energy-based and

Cramer and Leadbetter amplitudes are two such examples. The Rayleigh distribution also arises in many other situations.

When we perform a similar analysis of the rates of change of amplitude and phase, we find that the results are somewhat different for the energy-based and Cramer and Leadbetter definitions. In either approach, the basic idea is to supplement Eqs. 5.5 and 5.6 with the additional relationships

$$\dot{Y} = -A(\omega_a + \dot{\theta})\sin(\omega_a t + \theta) + \dot{A}\cos(\omega_a t + \theta) \tag{5.11}$$

and

$$\dot{Z} = -A(\omega_a + \dot{\theta})\cos(\omega_a t + \theta) - \dot{A}\sin(\omega_a t + \theta) \tag{5.12}$$

in order to have the four original variables as a vector nonlinear function of the new variables: $(Y, Z, \dot{Y}, \dot{Z}) = \vec{g}(A, \theta, \dot{A}\dot{\theta})$, then use this relationship to find properties of the probability distribution of the A, θ, $\dot{A}$, and $\dot{\theta}$ variables.

For the Cramer and Leadbetter approach, we can use exactly the same procedure as before to derive the joint probability density function of A_{CL}, θ_{CL}, $\dot{A}_{CL}$, and $\dot{\theta}_{CL}$ from that for Y, Z, $\dot{Y}$, and $\dot{Z}$. In particular, Eqs. 5.5, 5.6, 5.11, and 5.12 allow the joint probability density function to be written in the form

$$p_{A_{CL},\theta_{CL},\dot{A}_{CL},\dot{\theta}_{CL}}(u,\xi,v,\psi) = |J|_{u,\xi,v,\psi} \, p_{Y,Z,\dot{Y},\dot{Z}}[\vec{g}(u,\xi,v,\psi)]$$

Taking the determinant of the 4 x 4 matrix of derivatives of $(Y, Z, \dot{Y}, \dot{Z})$ with respect to the components of $(A_{CL}, \theta_{CL}, \dot{A}_{CL}, \dot{\theta}_{CL})$ gives the Jacobian as $J = A_{CL}^2$. Thus,

$$p_{A_{CL},\theta_{CL},\dot{A}_{CL},\dot{\theta}_{CL}}(u,\xi,v,\psi) = u^2 \, p_{Y,Z,\dot{Y},\dot{Z}}[u\cos(\omega_a t + \xi), -u\sin(\omega_a t + \xi),$$
$$-u(\omega_a + \psi)\sin(\omega_a t + \xi) + v\cos(\omega_a t + \xi),$$
$$-u(\omega_a + \psi)\cos(\omega_a t + \xi) - v\sin(\omega_a t + \xi)] \tag{5.13}$$

The joint Gaussian probability density function for the $(Y,Z,\dot{Y},\dot{Z})$ random variables is fairly simple since, as shown in Section 4.9, $K_{XZ} = 0$, $K_{X\dot{Z}} = 0$, $K_{X\dot{X}} = 0$, $K_{Z\dot{Z}} = 0$, and $K_{X\dot{Z}} = -K_{Z\dot{X}}$ for any definition of amplitude and phase according to Eqs. 4.49 - 4.52. Furthermore, for the Cramer and Leadbetter definition we know that $\sigma_X^2 = \sigma_Z^2 = \lambda_0$, $\sigma_{\dot{Z}}^2 = \sigma_{\dot{X}}^2 = \lambda_2$, $K_{Z\dot{X}} = \lambda_1$, and ω_a should be replaced with $\omega_{CL} = \lambda_1 / \lambda_0$. Rather than directly using the general jointly Gaussian formula (given in Eq. B.14 of Appendix B) to write the joint probability density function for the four random variables $(Y,Z,\dot{Y},\dot{Z})$, we can note that $(Y,\dot{Z})$ and $(Z,\dot{Y})$ are independent pairs, since they are jointly Gaussian and neither of the components of the first pair is correlated with either of the components of the second pair. Thus, one can write the joint probability density function as the product of two probability density functions:

$$p_{Y,Z,\dot{Y},\dot{Z}}(w_1,w_2,w_3,w_4) = p_{Y,\dot{Z}}(w_1,w_4)p_{Z,\dot{Y}}(w_2,w_3)$$

and each of these functions has the simplified form which applies for two jointly Gaussian random variables (see Example B.2 in Appendix B). The correlation coefficient for $(Z,\dot{Y})$ is $\alpha_1 = \lambda_1/(\lambda_0\lambda_2)^{1/2}$, and it is $-\alpha_1$ for $(Y,\dot{Z})$, so that

$$p_{Z,\dot{Y}}[-u\sin(\eta),-u(\omega_a+\psi)\sin(\eta)+v\cos(\eta)] =$$

$$\frac{1}{2\pi\sqrt{\lambda_0\lambda_2-\lambda_1^2}}\exp\left\{\frac{-\lambda_0\lambda_2}{2(\lambda_0\lambda_2-\lambda_1^2)}\left[\frac{u^2\sin^2(\eta)}{\lambda_0}-2\frac{\lambda_1}{\lambda_0\lambda_2}\left(\begin{array}{c}u^2(\omega_a+\psi)\sin^2(\eta)\\-uv\sin(\eta)\cos(\eta)\end{array}\right)\right.\right.$$
$$\left.\left.+\frac{1}{\lambda_2}\left(\begin{array}{c}u^2(\omega_a^2+\psi^2)\sin^2(\eta)+v^2\cos^2(\eta)-\\2uv(\omega_a+\psi)\sin(\eta)\cos(\eta)+2u^2\omega_a\psi\sin^2(\eta)\end{array}\right)\right]\right\}$$

and

$$p_{Y,\dot{Z}}[u\cos(\eta),-u(\omega_a+\psi)\cos(\eta)-v\sin(\eta)] =$$

$$\frac{1}{2\pi\sqrt{\lambda_0\lambda_2-\lambda_1^2}}\exp\left\{\frac{-\lambda_0\lambda_2}{2(\lambda_0\lambda_2-\lambda_1^2)}\left[\frac{u^2\cos^2(\eta)}{\lambda_0}+2\frac{\lambda_1}{\lambda_0\lambda_2}\left(\begin{array}{c}-u^2(\omega_a+\psi)\cos^2(\eta)\\-uv\sin(\eta)\cos(\eta)\end{array}\right)\right.\right.$$
$$\left.\left.+\frac{1}{\lambda_2}\left(\begin{array}{c}u^2(\omega_a^2+\psi^2)\cos^2(\eta)+v^2\sin^2(\eta)+\\2uv(\omega_a+\psi)\sin(\eta)\cos(\eta)+2u^2\omega_a\psi\cos^2(\eta)\end{array}\right)\right]\right\}$$

in which the notation has been simplified by using $\eta \equiv \omega_a t+\xi$. After substitution and simplification, Eq. 5.13 gives

$$p_{A_{CL},\theta_{CL},\dot{A}_{CL},\dot{\theta}_{CL}}(u,\xi,v,\psi) = \frac{u^2}{(2\pi)^2(\lambda_0\lambda_2-\lambda_1^2)}\exp\left[-\frac{u^2}{2}\left(\frac{1}{\lambda_0}+\frac{\lambda_0\psi^2}{\lambda_0\lambda_2-\lambda_1^2}\right)-\frac{v^2}{2(1-\alpha_1^2)\lambda_2}\right]$$

A surprising fact about this function is that it can be factored into the product of three probability density functions as

$$p_{A_{CL},\theta_{CL},\dot{A}_{CL},\dot{\theta}_{CL}}(u,\xi,v,\psi) = p_{A_{CL},\dot{\theta}_{CL}}(u,\psi)p_{\theta_{CL}}(\xi)p_{\dot{A}_{CL}}(v) \qquad (5.14)$$

with

$$p_{\theta_{CL}}(\xi) = \frac{1}{2\pi} \quad \text{for} \quad 0 \le \xi \le 2\pi \qquad (5.15)$$

$$p_{\dot{A}_{CL}}(v) = \frac{1}{\sqrt{2\pi(1-\alpha_1^2)\lambda_2}}\exp\left\{\frac{-v^2}{2(1-\alpha_1^2)\lambda_2}\right\} \qquad (5.16)$$

and

$$p_{A_{CL},\dot{\theta}_{CL}}(u, \psi) = \frac{u^2}{\sqrt{2\pi\lambda_0(\lambda_0\lambda_2 - \lambda_1^2)}}\exp\left[-\frac{u^2}{2}\left(\frac{1}{\lambda_0} + \frac{\lambda_0\psi^2}{\lambda_0\lambda_2 - \lambda_1^2}\right)\right] \qquad (5.17)$$

We had already established in Eq. 5.8 that the phase angle $\theta_{CL}(t)$ is independent of the amplitude $A_{CL}(t)$, and is uniformly distributed over its possible values. Equation 5.14 gives us the additional information that it is also independent of $\dot{A}_{CL}(t)$ and $\dot{\theta}_{CL}(t)$. Also we see that the rate of change of the amplitude, $\dot{A}_{CL}(t)$, is independent of the other variables, and it is Gaussian. Its variance is $(1 - \alpha_1^2)\lambda_2$, which confirms our argument in Section 4.9 that $A_{CL}(t)$ is slowly varying if α_1^2 is near unity. Finally, the amplitude $A_{CL}(t)$ and the rate of change of phase, $\dot{\theta}_{CL}(t)$, are dependent. We already know, from Eq. 5.9, that $A_{CL}(t)$ has the Rayleigh distribution, and we could also confirm that marginal probability density function by integrating Eq. 5.17 with respect to ψ from $-\infty$ to ∞. Similarly, integrating Eq. 5.17 with respect to u from zero to infinity gives the marginal probability density function for $\dot{\theta}_{CL}(t)$ as

$$p_{\dot{\theta}}(\psi) = \frac{\lambda_0(\lambda_0\lambda_2 - \lambda_1^2)}{2[\lambda_0\lambda_2 - \lambda_1^2 + \lambda_0^2\psi^2]^{3/2}} = \frac{\lambda_0^{1/2}\lambda_2(1 - \alpha_1^2)}{2[\lambda_2(1 - \alpha_1^2) + \lambda_0\psi^2]^{3/2}} \qquad (5.18)$$

Note that this symmetric probability density function $p_{\dot{\theta}}(\psi)$ only decays like ψ^{-3} for $|\psi| \to \infty$. We previously argued that $\dot{\theta}_{CL}(t)$ is small if α_1^2 is near unity, and Eq. 5.18 does confirm that its probability density function approximates a Dirac delta function at the origin in that situation. On the other hand, we see that the mean squared value of $\dot{\theta}_{CL}(t)$ is infinite so long as $\alpha_1^2 \neq 1$, because of the relatively slow decay of the tails of the probability density function.

For the energy-based definition of amplitude and phase, these operations must be performed in a slightly different manner. The reason is that the choice of $Z(t)$ as $\dot{X}(t)/\omega_a$ makes $Z(t)$ and $\dot{X}(t)$ be perfectly correlated, so that their joint probability density is a degenerate Gaussian form. Another way of viewing the difference in this situation is to note that Eqs. 5.6 and 5.11, along with $Z(t) = \dot{X}(t)/\omega_a$, give $A\dot{\theta}\sin(\omega_a t + \theta) = \dot{A}\cos(\omega_a t + \theta)$.[2] This equation can be solved for one of the new variables in terms of the others, so that the set $(A, \theta, \dot{A}, \dot{\theta})$ can be completely described by the probability distribution for three of the variables. For example, if we solve for $\dot{\theta}$ as

$$\dot{\theta} = \frac{\dot{A}\cos(\omega_a t + \theta)}{A\sin(\omega_a t + \theta)}$$

then it will suffice to find the joint probability distribution of $(A, \theta, \dot{A})$. This can be done by using Eqs. 5.5, 5.6, and 5.12. After substituting for $\dot{\theta}$ in Eq. 5.12, this gives three nonlinear equations relating $(Y, \dot{Y}, \dot{Z})$ to $(A, \theta, \dot{A})$, and one can proceed in basically the same way as we have already done for situations with two and four nonlinear functions. The result can be written as

---

[2] It may be noted that this formula is identical to Eq. 4.50, which was invoked in defining the energy-based amplitude and phase.

$$p_{A,\theta,\dot{A}}(u,\xi,v) = \frac{\lambda_2^{1/2}u}{(2\pi)^{3/2}\lambda_0\sqrt{\lambda_0\lambda_4 - \lambda_2^2}\ |\sin(\eta)|}\exp\left[-\frac{u^2}{2\lambda_0} - \frac{\lambda_2 v^2}{2(\lambda_0\lambda_4 - \lambda_2^2)\sin^2(\eta)}\right] \quad (5.19)$$

in which $\eta = \omega_a t + \xi$ and we have already substituted $\omega_a = (\lambda_2/\lambda_0)^{1/2}$. This equation can be factored into

$$p_{A,\theta,\dot{A}}(u,\xi,v) = p_A(u)\,p_{\theta,\dot{A}}(\xi,v) \quad (5.20)$$

showing that in this situation $A(t)$ is independent of $(\theta,\dot{A})$, but this latter pair is dependent, with a joint probability density function of

$$p_{\theta,\dot{A}}(\xi,v) = \frac{\lambda_2^{1/2}}{(2\pi)^{3/2}\sqrt{\lambda_0\lambda_4 - \lambda_2^2}\ |\sin(\eta)|}\exp\left[-\frac{\lambda_2 v^2}{2(\lambda_0\lambda_4 - \lambda_2^2)\sin^2(\eta)}\right] \quad (5.21)$$

The distribution of $A(t)$, of course, is the same Rayleigh distribution given in Eq. 5.9, and one can also integrate Eq. 5.21 with respect to v to verify that it gives the marginal probability density function of $\theta(t)$ as $(2\pi)^{-1}$. There seems to be no simple form for the marginal probability density function for $\dot{A}$, corresponding to the integral of Eq. 5.21 with respect to η from zero to 2π. Figure 5.1 shows a plot of values obtained from numerical integration, with a Gaussian probability density function shown for comparison. It can easily be observed that $\dot{A}$ is not Gaussian, and also it can be shown that $p_{\dot{A}}(v)$ goes to infinity at $v = 0$.

One can also use the independence results derived here to extend some of the results presented in Section 4.9 regarding the rate of change of the amplitude and phase, according to the two definitions. For example, the fact that both definitions give $A(t)$ and $\dot{A}(t)$ as independent allows one to say that $E[A^2(t)\dot{A}^2(t)] = E[A^2(t)]E[\dot{A}^2(t)]$ and the Rayleigh distribution of Eq. 5.9 gives

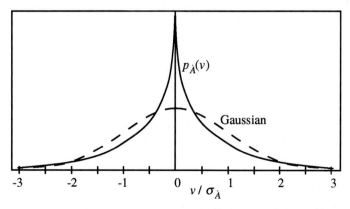

Figure 5.1 Probability density function for energy-based $\dot{A}(t)$

$E[A^2(t)] = 2\lambda_0$. Using these results in Eq. 5.3 gives

$$E[\dot{A}^2(t)] = \frac{\lambda_0 \lambda_4}{2\lambda_2}(1 - \alpha_2^2) = \frac{\sigma_X^2 \sigma_{\dot{X}}^2}{2\sigma_X^2}(1 - \alpha_2^2) = \frac{\sigma_{\dot{X}}^2}{2}\left(\frac{1}{\alpha_2^2} - 1\right) \tag{5.22}$$

for the energy-based amplitude, while Eq. 5.4 gives

$$E[\dot{A}_{CL}^2(t)] = \lambda_2(1 - \alpha_1^2) = \sigma_{\dot{X}}^2(1 - \alpha_1^2) \tag{5.23}$$

for the Cramer and Leadbetter definition. Note that Eq. 5.23 is identical to the value we had determined from the probability distribution in Eq. 5.16. Since $A(t)$ and $\dot{\theta}(t)$ are not independent, we cannot follow the same procedure to find $E[\dot{\theta}^2(t)]$ directly from $E[A^4(t)\dot{\theta}^2(t)]$ in Eqs. 5.3 and 5.4, in addition to the fact that we have already established that $E[\dot{\theta}_{CL}^2(t)]$ is infinite.

From Eqs. 5.22 and 5.23 we can see a significant difference between the two definitions of amplitude for a broadband process. In particular, Eq. 5.23 shows that the mean squared value of $\dot{A}_{CL}$ is always less than the variance of $\dot{X}$, since $0 < \alpha_1^2 < 1$. Thus, we can say that $A_{CL}(t)$ varies less rapidly than $X(t)$, even for a broadband process. On the other hand, Eq. 5.22 shows that the energy-based $A(t)$ varies less rapidly than $X(t)$ only if $\alpha_2^2 > 1/3$, which is not satisfied for some broadband processes.

The Cramer and Leadbetter amplitude also provides an example of an important distinction between Gaussian processes and more general processes also consisting of Gaussian random variables. In particular, recall that Eq. 5.16 shows that $\dot{A}_{CL}(t)$ is a Gaussian random variable at every time t. On the other hand, we know that $\{\dot{A}_{CL}(t)\}$ is not a Gaussian process, since the integral of a Gaussian process is also a Gaussian process, but we know that $A(t)$ does not have the Gaussian distribution. The explanation of this apparent dilemma is that a set of random variables $\{\dot{A}_{CL}(t_1), \dot{A}_{CL}(t_2), \cdots, \dot{A}_{CL}(t_n)\}$ is not generally jointly Gaussian, even though each of the random variables has a Gaussian marginal distribution. This rather confusing type of stochastic process consisting of a family of marginally Gaussian, but not jointly Gaussian, random variables arises in various areas of application. We shall encounter it again in Chapter 9 when we study the dynamics of nonlinear systems.

****

Example 5.3: Let $\{X(t)\}$ be a stationary, mean zero Gaussian process with a known standard deviation σ_X. Compare $P[A(t) > b]$ with $P[X(t) > b]$ for b values of σ_X, $3\sigma_X$, and $5\sigma_X$, in which $\{A(t)\}$ is either the energy-based or Cramer and Leadbetter amplitude of $\{X(t)\}$.

In order to evaluate $P[A(t) > b]$, we will need to integrate the probability density function for the Rayleigh distribution, which describes both $\{A(t)\}$ processes. This integration is simple, and gives the general result that $F_A(u) = 1 - e^{-u^2/(2\lambda_0)}$, or $P[A(t) > b] = e^{-b^2/(2\lambda_0)}$. The corresponding result for $\{X(t)\}$ is

$P[X(t) > b] = 1 - \Phi(b/\sigma_X) = \Phi(-b/\sigma_X)$. Substituting the three given values of b gives the probabilities shown in the table. The values show that it is much more likely that $A(t)$ will exceed any level b than it is that $X(t)$ will, and this difference becomes larger as b is made larger.

b	$P[X(t) > b]$	$P[A(t) > b]$	$P[A(t) > b]/P[X(t) > b]$
σ_X	0.159	0.607	3.82
$3\sigma_X$	1.35×10^{-3}	1.11×10^{-2}	8.23
$5\sigma_X$	2.87×10^{-7}	3.73×10^{-6}	13.0

****

Example 5.4: Let $\{X(t)\}$ and $\{Y(t)\}$ denote two orthogonal horizontal components of ground acceleration during an earthquake. Consider these two components to be independent, mean zero, Gaussian process with the same nonstationary variance $\sigma^2(t)$. Find the probability distribution of the absolute value of the horizontal ground acceleration, $Z(t) \equiv [X^2(t) + Y^2(t)]^{1/2}$.

Even though $\{X(t)\}$ and $\{Y(t)\}$ are nonstationary processes, we see that the random variables $X(t)$ and $Y(t)$ at any time t meet the conditions for $Z(t)$ to have the Rayleigh distribution of Eq. 5.9, with $\lambda_0 = \sigma^2(t)$. Based on the results in Example 5.3 we can also conclude that $P[Z(t) > b]$ is much greater than $P[X(t) > b]$ or $P[Y(t) > b]$, for any given b value and any instant of time t.

****

5.4 Approximations for Non-Gaussian Processes

In many practical problems, we use Gaussian stochastic process models even though we have no assurance that the physical quantity of interest is Gaussian. In Section 5.1 we mentioned that the central limit theorem provides one justification for this approach, inasmuch as it gives a basis for believing that many physical quantities are nearly Gaussian. Another justification for this approach is given by consideration of the cumulants of the processes (see Section A.12 of Appendix A and Section B.2 of Appendix B). In particular, one can say that the simplest approximation of any stochastic process is as a quantity with only one cumulant. However, the second cumulant is the covariance function, and only a deterministic quantity has a zero covariance function. Thus, using an approximation which has only a first cumulant results in approximating the stochastic process by a deterministic function. In fact, one can consider this to be a classical engineering approach, in which all quantities are considered to be deterministic. If one chooses the single-cumulant approximation on the basis of matching the first cumulant of the original stochastic process, then this deterministic approximation will replace any stochastic process by its mean-value function.

Going from a single-cumulant approximation of a stochastic process to a two-cumulant approximation gives a Gaussian process, since the Gaussian process has only two cumulant functions. Thus, using a Gaussian model for any problem can be viewed as a consistent approximation of the problem based on consideration of only the first and second cumulant functions. Since knowledge of first and second cumulants is equivalent to knowledge of first and second moments, one can also consider the Gaussian model to be the consistent model based only on the approximation of first and second moment functions. Of course, if one does have information about the higher moments or cumulants of the processes in a particular problem, then it may be possible to find a model which is better than the Gaussian one. In this section, we will briefly introduce the idea of using higher moment information to improve the modeling of some non-Gaussian stochastic processes.

The Hermite polynomials have proved to be very useful in approximating probability distributions, primarily because of their property of being orthogonal with respect to the normalized Gaussian probability density function. That is,

$$M_{jk} \equiv \int_{-\infty}^{\infty} H_{ej}(u) \, H_{ek}(u) \frac{e^{-u^2/2}}{\sqrt{2\pi}} \, du = 0 \quad \text{for} \quad j \neq k \tag{5.24}$$

in which $H_{ej}(u)$ denotes the jth Hermite polynomial. The normalization of $H_{ej}(u)$ is such that the coefficient on the highest-order term (i.e., the u^j term) is unity, and each polynomial contains only odd or even powers of u. That is, if j is odd, then $H_{ej}(u)$ contains only odd powers of u and if j is even then it contains only even powers. It can also be shown that setting $j = k$ in Eq. 5.24 gives $M_{jj} = j!$. One general relationship which can be used to obtain these polynomials is

$$H_{ej}(u) = (-1)^j \, e^{u^2/2} \frac{d^j e^{-u^2/2}}{du^j} \tag{5.25}$$

and the first few terms in the sequence are: $H_{e1}(u) = u$, $H_{e2}(u) = u^2 - 1$, $H_{e3}(u) = u^3 - 3u$, and $H_{e4}(u) = u^4 - 6u^2 + 3$. One can also introduce an order-zero polynomial of $H_{e0}(u) = 1$, which is consistent with the orthogonality and normalization of the other terms in the sequence.[3]

The best known approach for approximating a non-Gaussian probability density function involves writing the probability density function as a product of a Gaussian density function and a polynomial. This is a truncated version of an infinite expansion which we will briefly explore for the simple case of a non-Gaussian random variable X, which might correspond to $X(t)$ for some random process. The procedure can also be used, however, for joint probability density functions. For the non-Gaussian X, we will write

$$p_X(u) = \frac{1}{\sqrt{2\pi}\,\sigma_X} \exp\left(-\frac{(u-\mu_X)^2}{2\sigma_X^2}\right) \sum_{j=0}^{\infty} a_j H_{ej}\left(\frac{u-\mu_X}{\sigma_X}\right) \tag{5.26}$$

---

[3]   For more detail on the Hermite polynomials refer to a mathematical handbook such as Abramowitz and Stegun (1965).

The polynomial is written as a sum of Hermite polynomials because their orthogonality makes it relatively easy to obtain expressions for the a_j coefficients. In particular, multiplying both sides of Eq. 5.26 by the kth Hermite polynomial and integrating with respect to u gives

$$E\left[H_{ek}\left(\frac{X - \mu_X}{\sigma_X} \right) \right] = a_k M_{kk} = a_k k!$$

Substituting each of the first three Hermite polynomials into this expression shows that $a_0 = 1$ and $a_1 = a_2 = 0$, so that the equation can also be written as

$$p_X(u) = \frac{1}{\sqrt{2\pi}\,\sigma_X} \exp\left(-\frac{(u - \mu_X)^2}{2\sigma_X^2} \right)\left[1 + \sum_{j=3}^{\infty} a_j H_{ej}\left(\frac{u - \mu_X}{\sigma_X} \right) \right] \tag{5.27}$$

The next three coefficients in the series are simply related to the cumulants of X, as $a_j = \kappa_j(X)/(j!\sigma_X^j)$ for $j = 3$, 4, and 5. For $j \geq 6$ the relationships are more complicated. For example, $a_6 = [\kappa_6(X) + 10\kappa_3^2(X)]/(6!\sigma_X^6)$ and $a_7 = [\kappa_7(X) + 35\kappa_3(X)\kappa_4(X)]/(7!\sigma_X^7)$. In practice, though, one may not wish to consider terms with $j \geq 5$ anyway, because of difficulties related to the explosive nature of the tails of higher-order polynomials. In fact, attention is commonly limited to the $j = 3$ and $j = 4$ terms, which provide one antisymmetric and one symmetric correction to the Gaussian probability density function.

The names Gram-Charlier and Edgeworth are historically associated with series of the form of Eq. 5.27. Edgeworth and Charlier derived equivalent forms of the series by different approaches, and later Longuet-Higgins (1963) provided still another derivation. More information on these series approximations is given by Ochi (1986), Kendall and Stuart (1977), and Stratonovich (1963). In practice, a distinction is sometimes made between the Gram-Charlier series and the Edgeworth series. The difference has to do with the way that the series is truncated. For example, the Gram-Charlier series with fourth-order truncation is considered to be Eq. 5.27 with only $j = 3$ and $j = 4$ included in the summation, while the corresponding truncated Edgeworth series is at least sometimes considered to include a term of $10\kappa_3^2(X)/(6!\sigma_X^6)H_{e6}[(u - \mu_X)/\sigma_X]$, which depends only on the third cumulant, but comes from $j = 6$ in Eq. 5.27.

The idea of truncation of the Gram-Charlier or Edgeworth series can also be considered to give a new interpretation of the common practice of using the Gaussian distribution to approximate a quantity X which is known to be non-Gaussian. This approximation is the result of consideration of only the first term in the series of Eq. 5.27. By including more terms in the series, it is possible to approximate $p_X(u)$ by a function which gives the correct values of moments or cumulants of order higher than 2. Similarly, the first term in the Gram-Charlier or Edgeworth series for a joint probability density function of some random variables $X(t_1)$, $\cdots$, $X(t_n)$ is the nth-order Gaussian distribution, and inclusion of more terms allows matching of higher-order moment functions.

Unfortunately, the truncated Gram-Charlier or Edgeworth series also has major shortcomings. For example, consider keeping only the $j = 3$ term in the summation of Eq. 5.27, so that we obtain an approximation of $p_X(u)$ which can be written as

$$\hat{p}_X(u) = \frac{1}{\sqrt{2\pi}\,\sigma_X} \exp\left(-\frac{(u-\mu_X)^2}{2\sigma_X^2}\right)\left[1 + \frac{\kappa_3(X)}{6\sigma_X^3}\left[\left[\frac{u-\mu_X}{\sigma_X}\right]^3 - 3\frac{u-\mu_X}{\sigma_X}\right]\right] \tag{5.28}$$

Integrating $u^k \hat{p}_X(u)$ does give the proper moment, $E(X^k)$, for $k = 1, 2$, or 3, as desired. On the other hand, $\hat{p}_X(u)$ cannot truly be a probability density function because it is sometimes negative. In particular, if $\kappa_3(X)$ is positive, then very negative values of $(u - \mu_X)$ will give $\hat{p}_X(u) < 0$, and if $\kappa_3(X)$ is negative then the same difficulty arises for large positive u values. It is clear that any truncation with the highest-order term being odd will have this problem for large values of $|u|$, with one sign or the other. Keeping one more term in the series improves matters slightly. In particular, for $\kappa_4(X) > 0$ it is possible to have a situation for which $\hat{p}_X(u) > 0$ for all u. This truncation is not good for approximating a situation with $\kappa_4(X) < 0$, though, since then $\hat{p}_X(u)$ becomes negative for large values of $|u|$ with either sign.

In addition to the difficulty related to large values of $|u|$, it is possible for the truncated series to give a negative approximation of probability density for relative low values of u. For example, if $\kappa_3(X) > 3\sigma_X^3$ (giving skewness > 3) then Eq. 5.28 gives $\hat{p}_X(u) < 0$ when $(u - \mu_X)$ is in the neighborhood of σ_X. This possibility of having $\hat{p}_X(u) < 0$ for moderate values of u arises whether the truncation is at an odd or even power. It depends on the magnitudes of the cumulants of the original X distribution. For example, a symmetric distribution matching the fourth cumulant by including only $j = 4$ from the summation of Eq. 5.27 will give $\hat{p}_X(u) < 0$ for $u \approx \mu_X \pm \sqrt{3}$ if the kurtosis exceeds 7, so that $a_4 > 1/6$. Even if the cumulants are such that the truncated Gram-Charlier or Edgeworth series gives a positive $\hat{p}_X(u)$, it may well be found that the approximation is bimodal (i.e., has multiple peaks) even though the original $p_X(u)$ was unimodal. Because of these problems, it is often preferable to use an alternate relationship between the Gaussian distribution and a non-Gaussian X. It should be noted that the Gram-Charlier and Edgeworth series were originally developed to describe random variables which were very nearly Gaussian, and the difficulties that have been described primarily relate to attempts to use the series beyond their range of usefulness.

An alternate approach for describing a relationship between the Gaussian distribution and a non-Gaussian stochastic process $\{X(t)\}$ is to define a function $g(\cdot)$ such that $X(t) \approx g[Y(t)]$, with $\{Y(t)\}$ being Gaussian. Many $g(\cdot)$ functions will exist such that this relationship will give an exact marginal probability probability distribution, but in practice we will generally not be able to find an exact relationship so will use approximations. Even if we did succeed in exactly matching the marginal probability density functions of $X(t)$ and $g[Y(t)]$, that would not assure perfect matching of joint probability density functions involving multiple values of t. An approximate relationship may be chosen, for example, on the basis of fitting a certain number of moments or cumulants of $g[Y(t)]$ to those of $X(t)$, just as was discussed for the Gram-Charlier or Edgeworth series. Note, though, that there will not be any possibility of negative approximations for the probability density function in this instance, since $g[Y(t)]$ will be a real random variable with a nonnegative probability density function. In order to slightly simplify matters, we will replace the basic idea of $X(t) \approx g[Y(t)]$ with a different $g(\cdot)$ function which gives $X(t) \approx \mu_X(t) + \sigma_X(t)g[Y(t)]$. Matching the mean and variance values of $X(t)$ then gives $\{Y(t)\}$ as a stationary Gaussian process with mean zero and unit variance, even if $\{X(t)\}$ is nonstationary.

In addition to having the proper probability distribution, one wants any approximation of an $\{X(t)\}$ process to have time histories which resemble those of the original process. One way to achieve a very basic level of similarity between $X(t)$ and its approximation

$$\hat{X}(t) \equiv \mu_X(t) + \sigma_X(t)g[Y(t)]$$

is as follows. Let $\mu_X(t) + \sigma_X(t)Y(t)$ be the Gaussian process with the same mean-value function and the same autocovariance function as $\{X(t)\}$, and choose $g(\cdot)$ to be a monotonically increasing function. First, the matching of the mean-value and autocovariance functions of $\{X(t)\}$ and $\mu_X(t) + \sigma_X(t)Y(t)$ assures that the time histories of $Y(t)$ will have the same frequency content as those of $X(t)$. Thus, for example, $\{X(t)\}$ and $\{Y(t)\}$ have the same average frequency and the same values of any bandwidth parameters. The choice of a monotonic $g(\cdot)$ function, then, assures that $\hat{X}(t)$ will have a peak every time that $Y(t)$ does, and a crossing of the level $\mu_X(t) + \sigma_X(t)g(0)$ every time that $Y(t)$ crosses zero. This assures that the time histories of $\hat{X}(t)$ will resemble those of $X(t)$. In general, it is not feasible also to assure that the $\{\hat{X}(t)\}$ has the same autospectral density function as $\{X(t)\}$, but this type of monotonic mapping does assure similarity in the frequency content.

Using a monotonically increasing $g(\cdot)$ allows one to write the approximation of the marginal probability density function for $\hat{X}(t)$ as (see Section A.4 of Appendix A)

$$p_{\hat{X}(t)}(u) = \frac{1}{g'[g^{-1}(u)]\sigma_X(t)} p_{Y(t)} \left[\frac{g^{-1}(u) - \mu_X(t)}{\sigma_X(t)} \right]$$

or

$$p_{\hat{X}(t)}(u) = \frac{1}{g'[g^{-1}(u)]\sigma_X(t)} \frac{1}{\sqrt{2\pi}} \exp\left[-\frac{1}{2}\left(\frac{g^{-1}(u) - \mu_X(t)}{\sigma_X(t)} \right)^2 \right] \qquad (5.29)$$

in which $g^{-1}(\cdot)$ denotes the inverse of the $g(\cdot)$ function, and the $g'[g^{-1}(u)]$ term in the denominator denotes the derivative of the $g(\cdot)$ function evaluated at the location $g^{-1}(u)$. Obviously one can write a closed form expression for $p_{\hat{X}(t)}(u)$ only if both the inverse and the derivative of $g(\cdot)$ are available, but one can use numerical results for other situations.

It is useful to note the implications of certain characteristics of the $g(\cdot)$ function. For example, if the derivative of $g(\cdot)$ is zero at some point, then there will be a value of u for which $p_{\hat{X}(t)}(u)$ is infinite, and if the derivative of $g(\cdot)$ is infinite then there will be a value of u for which $p_{\hat{X}(t)}(u)$ is zero. Thus, one will generally wish to avoid $g(\cdot)$ functions having these extreme values of the derivative. The behavior of the derivative for very large values of either u or $-u$ is also of particular interest. For example, if this derivative value for large u is greater than unity, then Eq. 5.29 will give the right-hand tail of $p_{\hat{X}(t)}(u)$ as being smaller than that of the Gaussian distribution, as is appropriate if $\{X(t)\}$ is less likely to take on very large values than is a Gaussian process with the same mean and variance. Correspondingly, if the derivative for very large argument values is smaller than unity, then the tail of the $p_{\hat{X}(t)}(u)$ function will be enhanced so that the probability of $\hat{X}(t)$ taking on very large values will be greater than for the Gaussian distribution. The behavior of the derivative of $g(\cdot)$ for large negative values of u, obviously, has the same effects on the left-hand tail of $p_{\hat{X}(t)}(u)$. Attention to these limiting characteristics of $p_{\hat{X}(t)}(u)$, as well as the moments or cumulants, can help in the selection of an appropriate $g(\cdot)$ function.

One of the simplest choices for the $g(\cdot)$ function is a polynomial. For example, we might choose

$$\hat{X}(t) \equiv \mu_X(t) + \sigma_X(t)g[Y(t)] = \mu_X(t) + \sigma_X(t)\sum_{j=0}^{J} a_j Y^j(t) \tag{5.30}$$

and evaluate the a_j coefficients to make certain moments of $\hat{X}(t)$ be the same as those of $X(t)$. Note, though, that such a polynomial $g(u)$ will be dominated by its highest-order term $b_J u^J$ when $|u|$ is large. Thus, one can have a monotonically increasing $g(u)$ function only if J is odd and b_J is positive, in addition to restrictions on the values of the lower-order coefficients. With these restrictions, the polynomial $g(\cdot)$ function will always give a greater than Gaussian probability that $\hat{X}(t)$ will have large values of either sign. Thus, there are significant restrictions on the usefulness of the polynomial $g(\cdot)$ function. Nonetheless, it is a convenient approach for appropriate situations.

In order to simplify evaluation of moments of $\hat{X}(t)$, we will rewrite the general polynomial of Eq. 5.30 as a linear combination of Hermite polynomials. In particular, we will use

$$\hat{X}(t) = \mu_X(t) + \sigma_X(t)\sum_{j=1}^{J} b_j H_{ej}[Y(t)] \tag{5.31}$$

in which the argument of the Hermite polynomials has the normalized Gaussian probability distribution. One advantage of using the Hermite form is that the expected value of any $H_{ej}[Y(t)]$ term is zero, which assures that the mean of $\hat{X}(t)$ is the same as that of $X(t)$. Also, second-order moment terms such as $E\big(H_{ej}[Y(t)]H_{ek}[Y(t)]\big)$ are also zero for $j \neq k$, because of the orthogonality property of the Hermite polynomials. The only nonzero second-order terms are ones of the form $E\big(H_{ej}^2[Y(t)]\big) = j!$.

Using the properties of the Hermite polynomials, it is easily shown that $\hat{X}(t)$ and $X(t)$ will have the same variance if

$$\sum_{j=1}^{J} b_j^2 j! = 1 \tag{5.32}$$

Note that one can stop with this level of matching of $\hat{X}(t)$ and $X(t)$ by choosing $J = 1$ and $b_1 = 1$, which gives $\hat{X}(t)$ the Gaussian distribution. Once again, we see that the Gaussian approximation of $X(t)$ is the first stage in a sequence of possible approximations.

Improved $\hat{X}(t)$ approximations are usually found by matching higher moment values for the marginal distributions. It is convenient to write these higher moments in standardized form, for which the third and fourth-order terms are the skewness and kurtosis. Under the conditions of Eq. 5.32, so that $\sigma_{\hat{X}}(t) = \sigma_X(t)$, these terms are simply expected values of powers of the summation in Eq. 5.31. In principle, one can match any number of the standardized moments of $\hat{X}(t)$ to those of $X(t)$, provided that J is chosen sufficiently large. In practice, though, attention is usually limited to matching of the third and fourth-order terms, which can be accomplished with $J = 3$. The b_1, b_2, and b_3 coefficients are then evaluated from Eq. 5.32 and the equations for skewness and kurtosis. Substitution of the Hermite polynomials into the summation for $J = 3$, and evaluation of the resulting moments of the standardized Gaussian $Y(t)$ random variable, gives these equations as

$$E\left[\left(\frac{\hat{X}(t) - \mu_X(t)}{\sigma_X(t)}\right)^3\right] = 6b_1^2 b_2 + 36b_1 b_2 b_3 + 8b_2^3 + 108b_2 b_3^2 \tag{5.33}$$

and

$$E\left[\left(\frac{\hat{X}(t) - \mu_X(t)}{\sigma_X(t)}\right)^4\right] = 3b_1^4 + 24b_1^3 b_3 + 60b_1^2 b_2^2 + 252b_1^2 b_3^2 + 576b_1 b_2^2 b_3 + 1296b_1 b_3^3$$
$$+ 60b_2^4 + 2232b_2^2 b_3^2 + 3348b_3^4 \tag{5.34}$$

One can, of course, use numerical techniques to solve Eqs. 5.32 - 5.34, based on given values of skewness and kurtosis. Some caution is required, though, since such simultaneous nonlinear equations generally have multiple solutions. The solution branch that is sought gives real values for the three coefficients, and reduces to $b_1 = 1$, $b_2 = b_3 = 0$ for the situation with skewness of zero and kurtosis of three. Winterstein (1988) has given an approximate solution of the equations which is simple to use and gives a quite good fit over a wide range of skewness and kurtosis values. This approximation can be written as: $b_1 \approx (1 + 2c_2^2 + 6c_3^2)^{-1/2}$, $b_2 \approx b_1 c_2$, and $b_3 \approx b_1 c_3$, in which

$$c_3 = \frac{1}{18}\left[\sqrt{1 + \frac{3}{2}(kurtosis - 3)} - 1\right] \quad \text{and} \quad c_2 = \frac{skewness}{6(1 + 6c_3)} \tag{5.35}$$

For very small non-Gaussianity, it can be seen that these formulas reduce to a first-order approximation of $b_1 = 1$, $b_2 = skewness/6$, and $b_3 = (kurtosis - 3)/24$. As noted previously, this cubic polynomial approximation is useful only for $b_3 > 0$, so it is limited to processes with *kurtosis* > 3.

For *kurtosis* < 3, there is no polynomial $g(\cdot)$ such that $\hat{X}(t) = \mu_X(t) + \sigma_X(t)g[Y(t)]$ matches the third and fourth moments of $X(t)$. It is possible, though, to take $g(\cdot)$ to be the inverse of a cubic polynomial, such that the Gaussian $Y(t)$ in Eq. 5.30 is given by

$$Y(t) = g^{-1}\left(\frac{\hat{X}(t) - \mu_X(t)}{\sigma_X(t)}\right) = \sum_{j=1}^{3} c_j He_j\left(\frac{\hat{X}(t) - \mu_X(t)}{\sigma_X(t)}\right) \tag{5.36}$$

Evaluation of the coefficients to match moments of $X(t)$ and $\hat{X}(t)$ is complicated by the fact that the argument of the Hermite polynomials is not Gaussian, but Winterstein (1988) has shown that useful results can be obtained by using the first-order approximate solution of: $c_1 = 1$, $c_2 = -skewness/6$, $c_3 = (3 - kurtosis)/24$.

Overall, it seems that the approach of writing a non-Gaussian process $\{X(t)\}$ as a function of a Gaussian process has great versatility, while avoiding the difficulties of the Gram-Charlier and Edgeworth series. The polynomial forms of Eqs. 5.31 and 5.36 seem to be particularly useful, but there is also the possibility of using the same approach with any other monotonic $g(\cdot)$ function. Note, in particular, that the classic log-normal $X(t)$ results when $g(\cdot)$ is taken to be the exponential function. It should be noted that the treatment of non-Gaussian processes given here is quite introductory in nature. The study by Grigoriu (1995) is much more extensive, including its analysis of non-Gaussian processes which are nonlinear transformations of Gaussian processes.

****

Example 5.5: Let X be a random variable which has a Rayleigh distribution:

$$p_X(u) = \frac{u}{r^2}\exp\left(-\frac{u^2}{2r^2}\right)U(u)$$

Compare this probability density with the approximations which result from the Gram-Charlier approximation of Eq. 5.27 truncated to include only the $j = 3$ and $j = 4$ terms and from the use of Eqs. 5.29 and 5.31 with $J = 3$ and the coefficients evaluated from Eq.5.35.

First we need to evaluate the first four central moments of X. One way to find these is to write the general expression for the ordinary moments of the Rayleigh random variable as

$$E(X^j) = \int_0^\infty \frac{u^{j+1}}{r^2}\exp\left(-\frac{u^2}{2r^2}\right)du = \int_0^\infty (2r^2v)^{j/2}e^{-v}dv = (2r^2)^{j/2}\Gamma\left(\frac{j}{2}+1\right)$$

then use the relationships between the ordinary moments and the central moments:

$$\mu_X = \sqrt{2}\, r \Gamma\!\left(\frac{3}{2}+1\right) = \sqrt{\frac{\pi}{2}}\, r \approx 1.253 r$$

$$\sigma_X^2 = E(X^2) - \mu_X^2 = 2r^2 - \frac{\pi r^2}{2} = \frac{4-\pi}{2} r^2 \approx 0.4292\, r^2$$

$$skewness = E\!\left[\left(\frac{X-\mu_X}{\sigma_X}\right)^3\right] = \frac{E(X^3) - 3\mu_X E(X^2) + 2\mu_X^3}{\sigma_X^3} \approx 0.6311$$

$$kurtosis = E\!\left[\left(\frac{X-\mu_X}{\sigma_X}\right)^4\right] = \frac{E(X^4) - 4\mu_X E(X^3) + 6\mu_X^2 E(X^2) - 3\mu_X^4}{\sigma_X^4} \approx 3.2451$$

The coefficients for Eq. 5.27 are then $\kappa_3(X) \approx 0.6311\sigma_X^3$ and $\kappa_4(X) \approx 0.2451\sigma_X^4$, giving $a_3 = \kappa_3(X)/(3!\sigma_X^3) = skewness/3! \approx 0.105$ and $a_4 = \kappa_4(X)/(4!\sigma_X^4) = (kurtosis - 3)/4! \approx 0.102$. Note that $kurtosis > 3$, so that Eq. 5.31 is appropriate, rather than Eq. 5.36, for use in Eq. 5.29. Using the $skewness$ and $kurtosis$ values in Eq. 5.35 gives $c_3 \approx 0.009414$ and $c_2 \approx 0.09956$, so that $b_1 \approx 0.9900$, $b_2 \approx 0.09856$, and $b_3 \approx 0.009320$ for use in Eq. 5.31.

Knowing the values of a_3 and a_4, it is simply a matter of substitution to evaluate the Gram-Charlier probability density approximation from the truncated Eq. 5.27. Using Eq. 5.29, on the other hand, requires knowledge of the inverse and the derivative of the $g(\cdot)$ function. Taking the derivative of the polynomial $g(\cdot)$ is almost trivial, and the inverse can be written by using the general relationship that if

$$u = g(v) \equiv a_0 + a_1 v + a_2 v^2 + a_3 v^3 \qquad \text{then} \qquad v = g^{-1}(u) = D - w_0 - \frac{w_1}{D}$$

with

$$D = \left(w_0 w_2 + \sqrt{w_1^3 + (w_0 w_2 + \hat{u})^2} + \hat{u}\right)^{1/3}, \qquad w_0 = \frac{a_2}{3a_3}, \; w_1 = \frac{a_1}{3a_3} - \frac{a_2^2}{9a_3^2}, \; w_2 = \frac{a_1}{2a_3} - \frac{a_2^2}{9a_3^2}$$

and $\hat{u} = (u - a_0)/2a_3$.

The sketch shows the approximations of Eqs. 5.27 and 5.29, along with the true Rayleigh probability density function and the Gaussian density function having the same mean and variance. It is seen that, for this example, the results of Eqs. 5.27 and 5.29 are very similar. As expected, both of these non-Gaussian approximations are better than the Gaussian approximation, which has a symmetric density function. In particular, the peak of the non-Gaussian density approximations is not shifted as far to the right as is that of the Gaussian distribution. For large positive values of u, it is also seen that the Gaussian approximation tends to zero more rapidly than the Rayleigh distribution or the non-Gaussian approximations. Finally, we note that all three approximations of X have some probability of being negative, since the probability density $p_{\hat{X}}(u)$ is positive for some negative u values. This discrepancy is greater for the Gaussian approximation than it is for the non-Gaussian approximations. The magnified sketch for negative u values shows that the Gram-Charlier approximation also exhibits its classic shortcoming of giving a negative probability density function for some arguments.

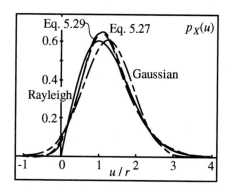 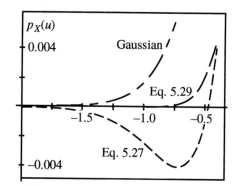

Note that this example is given strictly for the purpose of illustrating the applicability of Eqs. 5.27 and 5.29 to a particular non-Gaussian situation. If one knows the true probability distribution of a non-Gaussian quantity in a practical problem, then there is generally no need to use such approximations. The value of the approximations is that they can be used when the true distribution is unknown.

****

Example 5.6: Let X be a random variable which has a triangular probability distribution:

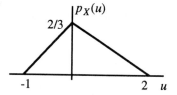

$$p_X(u) = \frac{2(u+1)[U(u+1)-U(u)]+(2-u)[U(u)-U(u-2)]}{3}$$

Compare this probability density with the approximations which result from the Gram-Charlier approximation of Eq. 5.27 truncated to include only the $j=3$ and $j=4$ terms and from the use of Eqs. 5.29 and 5.36 with the coefficients evaluated according to the first-order approximation.

First we need to evaluate the first four central moments of X. The general expression for the ordinary moments of the random variable can be written as

$$E(X^j) = \frac{2}{3}\int_{-1}^{0} u^j(u+1)\,du + \frac{1}{3}\int_{0}^{2} u^j(2-u)\,du = \frac{2[(2)^{j+1}-(-1)^{j+1}]}{3(j+1)(j+2)}$$

which gives

$$\mu_X = 1/3, \quad E(X^2)=1/2, \quad E(X^3)=1/2, \quad E(X^4)=11/15$$

Thus, $\mu_X = 1/3$, $\sigma_X^2 = 7/18$, $skewness = 2^{5/2}/7^{3/2} \approx 0.3054$, $kurtosis = 12/5 = 2.4$. This gives the coefficients for Eq. 5.27 as $a_3 = skewness/3! \approx 0.05091$, $a_4 = (kurtosis - 3)/4! = -0.025$. Note that $kurtosis < 3$, so that Eq. 5.36 is appropriate, rather than Eq. 5.31, for use in Eq. 5.29. The first-order approximation gives the values of the coefficients for use in Eq. 5.36 as the negative of the two coefficients already obtained: $c_2 = -a_3 \approx -0.05091$, $c_3 = -a_4 = 0.025$.

The approximation of Eq. 5.27 is easily evaluated by substitution of these parameter values. The results of Eq. 5.36 are also evaluated quite easily. The $g^{-1}(\cdot)$ function is directly given by Eq. 5.36:

$$g^{-1}(u) = u + c_2(u^2 - 1) + c_3(u^3 - 3u)$$

and the derivative can be evaluated as

$$\frac{1}{g'[g^{-1}(u)]} = \frac{d}{du}g^{-1}(u) = 1 + 2c_2u + c_3(3u^2 - 3)$$

The probability density function is then found by substituting these relationships into Eq. 5.29.

The sketch shows the approximations of Eqs. 5.27 and 5.29, along with the true original triangular probability density function and the Gaussian density function having the same mean and variance. It is seen that the results of Eqs. 5.27 and 5.29 are quite similar. As expected, both of these non-Gaussian approximations are somewhat better than the Gaussian approximation, which has a symmetric density function. Note that all three approximations of X give nonzero probabilities of $(X < -1)$ and $(X > 2)$, since the probability density $p_{\hat{X}}(u)$ is not identically zero in these regions. The magnified sketch for negative u values shows that the Gram-Charlier approximation also exhibits its classic shortcoming of giving a negative probability density function for some arguments. A similar, but less significant, discrepancy occurs for $(X > 2)$.

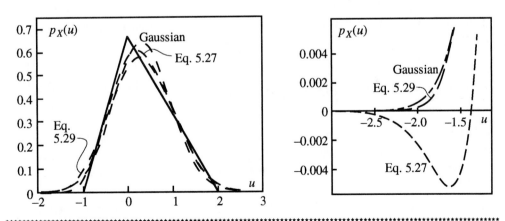

Example 5.7: Show that shot noise is not a Gaussian process, and evaluate its cumulant functions.

As defined in Example 3.6, shot noise $\{F(t)\}$ can be written as

$$F(t) = \sum_j F_j \delta(t - T_j)$$

in which $\{T_1, T_2, \cdots, T_j, \cdots\}$ are the arrival times for a Poisson process $\{N(t)\}$, and $\{F_1, F_2, \cdots, F_j, \cdots\}$ is a sequence of identically distributed random variables which are independent of each other and independent of the arrival times. We will consider the nonstationary situation with the arrival rate $\dot{\mu}_N(t)$ varying with time. As in Example 3.6, we will take the approach of first analyzing the process which is the integral of $\{F(t)\}$:

$$Q(t) = \int_0^t F(s)\,ds = \sum_j F_j U(t - T_j)$$

then we will use these results in describing $\{F(t)\} = \{\dot{Q}(t)\}$. The cumulant functions, of course, are an ideal tool for determining whether a stochastic process is Gaussian, since the cumulants of order higher than two are identically zero for a Gaussian process. Thus, we will proceed directly to seek the cumulant functions of $\{Q(t)\}$ and $\{F(t)\}$. It turns out that this is fairly easily done by consideration of the log-characteristic function of $\{Q(t)\}$. (See Section A.12 of Appendix A for information on cumulants and log-characteristic functions.)

The nth-order joint characteristic function of $\{Q(t)\}$ for times $\{t_1, \cdots, t_n\}$ can be written as

$$M_{Q(t_1), \cdots, Q(t_n)}(\theta_1, \cdots, \theta_n) \equiv E\left[\exp\left(i\sum_{j=1}^n \theta_j Q(t_j)\right)\right] = E\left[\exp\left(i\sum_{k=1}^n [Q(t_k) - Q(t_{k-1})]\sum_{l=k}^n \theta_l\right)\right]$$

in which $t_0 = 0$, so that $Q(t_0) = 0$ in the final form. The purpose of the rearranged form is to take advantage of the delta-correlated properties of shot noise. In particular, if we choose $t_0 < t_1 < t_2 < \cdots < t_n$, then the fact that $F(t)$ and $F(s)$ are independent random variables for $t \neq s$ allows us to say that $\{Q(t_1) - Q(t_0), Q(t_2) - Q(t_1), \cdots, Q(t_n) - Q(t_{n-1})\}$ is a sequence of independent random variables. This allows us to write the characteristic function as

$$M_{Q(t_1), \cdots, Q(t_n)}(\theta_1, \cdots, \theta_n) = \prod_{k=1}^n E\left[\exp\left(i[Q(t_k) - Q(t_{k-1})]\sum_{l=k}^n \theta_l\right)\right] = \prod_{k=1}^n M_{Q(t_k)-Q(t_{k-1})}\left(\sum_{l=k}^n \theta_l\right)$$

Thus, the joint characteristic function is exactly a product of marginal characteristic functions. By finding the characteristic function for an arbitrary increment $[Q(t_k) - Q(t_{k-1})]$ of the $\{Q(t)\}$ process, we can use that information to find the joint characteristic function describing the behavior of the process at various times.

For simplicity, we will derive the characteristic function for the increment from time zero to time t, that is, for the random variable $Q(t)$. This can most easily be evaluated in two steps. First, we evaluate the conditional expected value

$$E\big(\exp[i\theta\, Q(t)]|\,N(t) = r\big) = E\left[\exp\left(i\theta\sum_{j=1}^r F_j\right)\right] = E\big(e^{i\theta F_1}\cdots e^{i\theta F_r}\big) = M_F^r(\theta)$$

in which the final form follows from the fact that $\{F_1, F_2, \cdots, F_j, \cdots\}$ is a sequence of independent, identically distributed random variables, and $M_F(\theta)$ is used to designate the characteristic function of any member of the sequence. Using the probability values for the Poisson process then gives the characteristic function of interest as

$$M_{Q(t)}(\theta) = \sum_{r=1}^{\infty} E\big(\exp[i\theta\, Q(t)]\,|\, N(t) = r\big)\, P[N(t) = r] = \sum_{r=1}^{\infty} M_F^r(\theta)\, e^{-\mu_N(t)}\frac{\mu_N^r(t)}{r!}$$

which can be simplified as

$$M_{Q(t)}(\theta) = e^{-\mu_N(t)}\sum_{r=1}^{\infty}\frac{1}{r!}[M_F(\theta)\,\mu_N(t)]^r = \exp[-\mu_N(t) + M_F(\theta)\,\mu_N(t)]$$

since the summation is exactly the power series expansion for the exponential function. The log-characteristic function of $Q(t)$ is then

$$\log[M_{Q(t)}(\theta)] = \mu_N(t)[M_F(\theta) - 1]$$

The corresponding function for any arbitrary increment $[Q(t_k) - Q(t_{k-1})]$ is obtained by replacing $\mu_N(t)$ with $E[Q(t_k) - Q(t_{k-1})] = \mu_N(t_k) - \mu_N(t_{k-1})$.

The log-characteristic function for $Q(t)$ provides sufficient evidence to prove that $\{F(t)\}$ is not Gaussian. In particular, we can find the cumulants of $Q(t)$ by taking derivatives of $\log[M_{Q(t)}(\theta)]$ and evaluating them at $\theta = 0$. Since

$$\frac{d^j}{d\theta^j}\log[M_{Q(t)}(\theta)] = \mu_N(t)\frac{d^j}{d\theta^j}M_F(\theta)$$

though, this relates the cumulants of $Q(t)$ to the moments of F, since the derivatives of a characteristic function evaluated at $\theta = 0$ always give moment values. It is not possible for the higher-order even moment values of F, such as $E(F^4)$, to be zero so we can see that the higher-order even cumulants of $Q(t)$ are not zero. This proves that $Q(t)$ is not a Gaussian random variable, which then implies that $\{Q(t)\}$ cannot be a Gaussian stochastic process. If $\{Q(t)\}$ is not a Gaussian process, though, then its derivative, $\{F(t)\}$, is also not a Gaussian process.

Using these results, we can evaluate the joint log-characteristic function for $\{Q(t)\}$ as

$$\log[M_{Q(t_1),\cdots,Q(t_n)}(\theta_1,\cdots,\theta_n)] = \sum_{k=1}^{n}[\mu_N(t_k) - \mu_N(t_{k-1})]\left[M_F\left(\sum_{l=k}^{n}\theta_l\right) - 1\right]$$

The joint cumulant function for $\{Q(t_1),\cdots,Q(t_n)\}$ is then

$$\kappa_n[Q(t_1),\cdots,Q(t_n)] = \left[\frac{\partial^n}{\partial\theta_1\cdots\partial\theta_n}\log[M_{Q(t_1),\cdots,Q(t_n)}(\theta_1,\cdots,\theta_n)]\right]_{\theta_1=\cdots=\theta_n=0}$$

There is only one term in our expression for the log-characteristic function, though, which gives a non-zero value for the mixed partial derivative. In particular, only the first term contains all the arguments $(\theta_1,\cdots,\theta_n)$, so only it contributes to the cumulant function. Thus, we have

$$\kappa_n[Q(t_1),\cdots,Q(t_n)] = \mu_N(t_1)\left[\frac{\partial^n}{\partial\theta_1\cdots\partial\theta_n}M_F\left(\sum_{l=1}^{n}\theta_l\right)\right]_{\theta_1=\cdots=\theta_n=0} = \mu_N(t_1)E[F^n]$$

Recall that the only way in which t_1 was distinctive in the set $(t_1,\cdots,t_n)$ was that it was the minimum of the set. Thus, the general result for $\{Q(t)\}$ is

$$\kappa_n[Q(t_1),\cdots,Q(t_n)] = \mu_N[\min(t_1,\cdots,t_n)]E[F^n]$$

The linearity property of cumulants (see Eq. A.91 of Appendix A) then allows us to write the corresponding cumulant function for the $\{F(t)\}$ process as

$$\kappa_n[F(t_1),\cdots,F(t_n)] = E[F^n]\frac{\partial}{\partial t_1\cdots\partial t_n}\mu_N[\min(t_1,\cdots,t_n)] = E[F^n]\frac{\partial}{\partial t_1\cdots\partial t_n}\sum_{j=1}^{n}\mu_N(t_j)\prod_{\substack{k=1\\k\neq j}}^{n}U(t_k-t_j)$$

After performing the differentiation with respect to t_n and eliminating a number of terms which cancel, this expression can be simplified to

$$\kappa_n[F(t_1),\cdots,F(t_n)] = E[F^n]\dot\mu_N(t_n)\frac{\partial}{\partial t_1\cdots\partial t_{n-1}}\left[\prod_{k=1}^{n-1}U(t_k-t_n)\right] = E[F^n]\dot\mu_N(t_n)\prod_{k=1}^{n-1}\delta(t_k-t_n)$$

which confirms the fact that $\{F(t)\}$ is a delta-correlated process. One may also note that this expression includes the result in Example 3.6 giving the covariance function in the special case of stationary shot noise, for which $n=2$ and $\dot\mu_N(t_n)=\lambda$.

****

5.5 Higher-Order Spectral Density Functions

For a non-Gaussian stochastic process, the spectral density analysis presented in Chapter 4 is important but gives only a partial description of the process. For example, one can, in principle, take the inverse Fourier transform of the autospectral density of any covariant stationary process $\{X(t)\}$, and thereby find the autocovariance function of the process. If the process is Gaussian, then this autocovariance function along with the mean-value function gives complete information about the probability density functions of any random variables $\{X(t_1), \cdots X(t_n)\}$ from the process. For example, any higher-order moment functions of $\{X(t)\}$ have simple relationships to the mean and autocovariance functions. If $\{X(t)\}$ is non-Gaussian then the second moment information given by the autospectral density is still equally valid, but it gives much less information about the probability distributions of the process since the process is no longer completely determined by its mean and autocovariance functions. Higher-order spectral density functions provide a way to analyze this higher-moment (or cumulant) information in the frequency domain.

The higher-order spectral density functions are most conveniently defined in terms of Fourier transforms of higher-order cumulant functions, either for a single stochastic process or for several processes. We will write out some explicit results for the so-called bispectrum which relates to the third cumulant function. (See Sections A.11 and A.12 of Appendix A for more general information on cumulant functions.) We will write the result for three jointly stationary processes $\{X(t)\}$, $\{Y(t)\}$, and $\{Z(t)\}$, because that includes the simpler cases with two or all of the processes being the same. The definition of the bispectrum is related to the Fourier transforms of the processes at three different frequency values:

$$S_{XYZ}(\omega_1, \omega_2) = \lim_{T \to \infty} \frac{2\pi}{T} E\big([\tilde{X}_T(\omega_1) - \tilde{\mu}_{X_T}(\omega_1)][\tilde{Y}_T(\omega_2) - \tilde{\mu}_{Y_T}(-\omega_2)]$$

$$[\tilde{Z}_T(-\omega_1 - \omega_2) - \tilde{\mu}_{Z_T}(-\omega_1 - \omega_2)]\big) \qquad (5.37)$$

Note that the sum of the three frequency arguments in the right-hand side of Eq. 5.37 is zero. It can be shown that the corresponding expectation of the product of j Fourier transforms is generally bounded as T goes to infinity if the sum of the frequency arguments is not zero, but unbounded when the sum of the frequency arguments is zero. Thus, the useful information contained within the jth cumulant function normalized by T is given only by the special case with the sum of the frequency arguments equal zero. Equations 4.16 and 4.19 are also special cases of this general result. Another higher-order spectrum which is sometimes encountered in practical analysis of non-Gaussian processes is called the trispectrum, and is related to the Fourier transforms of processes at four frequency values which sum to zero.

From Eq. 5.37 one can show that

$$S_{XYZ}(\omega_1, \omega_2) = \frac{1}{4\pi^2} \int_{-\infty}^{\infty} \int_{-\infty}^{\infty} \kappa_3[X(t + \tau_1), Y(t + \tau_2), Z(t)] e^{-i\omega_1\tau_1 - i\omega_2\tau_2} d\tau_1 d\tau_2 \qquad (5.38)$$

The third cumulant, though, is identical to the third central moment, so can be written as

$$\kappa_3[X(t + \tau_1), Y(t + \tau_2), Z(t)] = E\big([X(t + \tau_1) - \mu_X(t + \tau_1)]$$

$$[Y(t + \tau_2) - \mu_Y(t + \tau_2)][Z(t) - \mu_Z(t)]\big)$$

From Eq. 5.38 we see that the bispectrum is exactly the second-order Fourier transform of the stationary third cumulant function, just as the ordinary autospectral density function is the Fourier transform of the autocovariance function. This same relationship also extends to higher-order spectral densities. In particular, the trispectrum is the third-order Fourier transform of the fourth-order cumulant function. It must be kept in mind, though, that the fourth-order and higher-order cumulants are not the same as the corresponding central moments (see Sec. A.11 of Appendix A).

****

Example 5.8: Find the higher-order autospectral density function for a general stationary delta-correlated process $\{F(t)\}$, and for the shot noise of Example 5.7 when the mean arrival rate is stationary.

Extending the notation of Section 3.5, we will say that the general stationary delta-correlated process $\{F(t)\}$ has higher-order cumulants given by

$$\kappa_n[X(t_1),\cdots,X(t_n)] = G_n \delta(t_1 - t_n)\delta(t_2 - t_n)\cdots\delta(t_{n-1} - t_n)$$

This includes Eq. 3.31 as the special case with $n = 2$, except that we used the notation G_0 rather than G_2 in that equation. We now write the nth-order autospectral density function as the $(n-1)$th-order Fourier transform of this cumulant. Thus, the general relationship that

$$S_n(\omega_1,\cdots,\omega_n) = \frac{1}{(2\pi)^{n-1}} \int_{-\infty}^{\infty}\cdots\int_{-\infty}^{\infty} \kappa_n[X(t+\tau_1),\cdots,X(t+\tau_{n-1}),X(t)]$$

$$\exp\left(-i\sum_{j=1}^{n-1}\omega_j\tau_j\right)d\tau_1\cdots d\tau_n$$

gives $S_n(\omega_1,\cdots,\omega_n) = G_n / (2\pi)^{n-1}$. Autospectral density functions of all orders are constants for a delta-correlated process.

For the shot noise of Example 5.7 we found that the nth-order cumulant was

$$\kappa_n[F(t_1),\cdots,F(t_n)] = E[F^n]\lambda\,\delta(t_1 - t_n)\delta(t_2 - t_n)\cdots\delta(t_{n-1} - t_n)$$

in which we have now replaced the nonstationary arrival rate $\dot{\mu}_N(t_n)$ with the constant λ. Thus, we see that $G_n = E[F^n]\lambda$ and $S_n(\omega_1,\cdots,\omega_n) = E[F^n]\lambda / (2\pi)^{n-1}$.
****

5.6   Average Frequency of a Non-Gaussian Process

We will first show that there is generally a discrepancy between the average frequency of a covariant stationary non-Gaussian $\{X(t)\}$ process and that of its approximation of $\{\hat{X}(t)\}$ defined by $\hat{X}(t)$

$= \mu_X(t) + \sigma_X g[Y(t)]$, as introduced in Section 5.4. We will then use the idea of the amplitudes of the two processes to interpret this result, and illustrate one shortcoming of the approximation. We will consider the energy-based average frequency, which is related to variances of a process, and its derivative according to Eq. 4.69. Thus, in order to find the average frequency of $\{\hat{X}(t)\}$ we differentiate the expression for $\hat{X}(t)$ to get

$$\dot{\hat{X}}(t) = \mu_{\dot{X}}(t) + \sigma_X g'[Y(t)]\dot{Y}(t)$$

The variance of $\dot{\hat{X}}(t)$ can then be written as

$$\sigma_{\dot{\hat{X}}}^2 = \sigma_X^2 E\left[\left(g'[Y(t)]\dot{Y}(t)\right)^2\right]$$

Since the $\{Y(t)\}$ process is Gaussian and stationary, though, we know that $Y(t)$ and $\dot{Y}(t)$ are independent random variables, so that this can be simplified to

$$\sigma_{\dot{\hat{X}}}^2 = \sigma_X^2 E\left[(g'[Y(t)])^2\right]\sigma_{\dot{Y}}^2 = E\left[(g'[Y(t)])^2\right]\sigma_{\dot{X}}^2 \tag{5.39}$$

in which the final form is based on the fact that $\sigma_{\dot{Y}} / \sigma_Y = \sigma_{\dot{X}} / \sigma_X$ since the two processes $\{\mu_X(t) + \sigma_X Y(t)\}$ and $\{X(t)\}$ have the same autocovariance function. In fact, these ratios of standard deviations are exactly the energy-based average frequencies of the two processes, which have been chosen to have the same autospectral density functions.

From Eq. 5.39 we see the result that the variance of $\{\dot{\hat{X}}(t)\}$ generally differs from that of $\{\dot{X}(t)\}$. Rewritten in terms of average frequency, the relationship is

$$\omega_{a,\hat{X}} = \sqrt{E\left[(g'[Y(t)])^2\right]}\, \omega_{a,X} \tag{5.40}$$

so that we have also illustrated the discrepancy between the average frequencies of the $\{X(t)\}$ and $\{\hat{X}(t)\}$ processes. The only situation in which there would be no discrepancy would be if the mean squared value of $g'[Y(t)]$ were unity, but we can readily verify that this condition is not generally true. For example, Fig. 5.2 shows sample values of $\omega_{a,\hat{X}} / \omega_{a,X}$ for the cubic polynomial of Eq. 5.31 with $J = 3$, with the coefficients chosen according to the approximations of Eqs. 5.35. It is seen that the discrepancy is not large if $\{X(t)\}$ is nearly Gaussian, but it does become substantial if skewness and kurtosis are large.

The discrepancies in derivative and average frequency values between $\{X(t)\}$ and $\{\hat{X}(t)\}$ can be interpreted by considering the time history behavior of the processes and their amplitudes in the narrowband situation. Recall the intuitive idea of the $\{A(t)\}$ amplitude process for a narrowband $\{X(t)\}$, as introduced in Section 4.9. In this case, $\{A(t)\}$ is slowly varying and $X(t)$ reaches the level $A(t)$ near the peak of each cycle and the level $-A(t)$ near each valley.

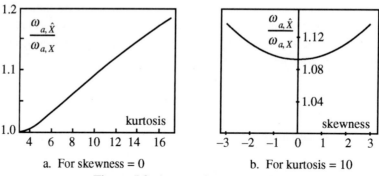

a. For skewness = 0   b. For kurtosis = 10

Figure 5.2 Average frequency ratio

Between these contacts with $\pm A(t)$, a time history of $\{X(t)\}$ resembles a harmonic function. In addition, let $\{B(t)\}$ be the amplitude of the Gaussian $\{Y(t)\}$ process with zero mean, unit variance, and the same autocovariance function as $\{[X(t) - \mu_X(t)]/\sigma_X(t)\}$. If we now define a new process $\{\hat{A}(t)\}$ by $\hat{A}(t) = \mu_X(t) + \sigma_X(t)g[B(t)]$, then it will have the nature of an amplitude of the non-Gaussian $\{X(t)\}$ process, even though it is not identical to $\{A(t)\}$. In particular, $\hat{A}(t)$ will be slowly varying and the $\hat{X}(t)$ approximation will be equal to $\pm\hat{A}(t)$ in the vicinity of each peak or valley of the narrowband $\hat{X}(t)$. The primary difference between the time histories of $X(t)$ and $\hat{X}(t)$ seems to be that the variation of $\hat{X}(t)$ within a cycle is not harmonic. In particular, consider the Fourier transform of a sample $\hat{X}(t) = g\big(B(t)\cos[\omega_{a,X}t + \theta(t)]\big)$ with the slowly varying $B(t)$ and $\theta(t)$ approximated as being constants. The nonlinear $g(\cdot)$ function will always cause this Fourier transform to contain higher harmonic terms, as well as the fundamental term with frequency $\omega_{a,X}$. This then results in the average frequency of $\hat{X}(t)$ being higher than $\omega_{a,X}$, just as we previously found.

****

Example 5.9: Let $\{X(t)\}$ be a stochastic process which is mean-value and covariant stationary and has the Rayleigh marginal probability distribution used in Example 5.5. Find the discrepancy between the average frequency of $\{X(t)\}$ and that of an approximation $\{\hat{X}(t)\}$ given by Eq. 5.31 with the coefficients determined according to Eq. 5.35.

In Example 5.5 we found that the nonlinear transformation for this problem is

$$\hat{X}(t) = \mu_X + \sigma_X g[Y(t)] = \mu_X + \sigma_X[0.990Y(t) + 0.0986\{Y^2(t) - 1\} + 0.00932\{Y^3(t) - 3Y(t)\}$$
$$= \mu_X + \sigma_X[0.962\,Y(t) + 0.0986Y^2(t) + 0.00932Y^3(t) - 0.0986].$$

We could, of course, substitute for μ_X and σ_X, but it would not be of any particular benefit. From the expression for $\hat{X}(t)$, we find that $\dot{\hat{X}}(t) = \sigma_X[0.962 + 0.1972Y(t) + 0.0280Y^2(t)]\dot{Y}(t)$ and squaring and taking the expected value gives $\sigma_{\dot{\hat{X}}}^2 = \sigma_{\dot{X}}^2(1.021)\sigma_{\dot{Y}}^2$, in which we have used the following facts: $E[Y(t)] = 0$, $E[Y^2(t)] = 1$, $E[Y^3(t)] = 0$, and $E[Y^4(t)] = 3$. Since $\sigma_{\dot{Y}}^2 = \omega_{a,\hat{X}}^2\sigma_{\dot{Y}}^2 = \omega_{a,\hat{X}}^2$ and $\sigma_{\hat{X}}^2 = \sigma_X^2$, we now have the result that $\omega_{a,\hat{X}}^2 \equiv \sigma_{\dot{\hat{X}}}^2/\sigma_{\hat{X}}^2 = 1.021\omega_{a,X}^2$. Taking the square root shows that in this example there is only about 1% difference between the average frequency of $\{\hat{X}(t)\}$ and that of $\{X(t)\}$.

****

Example 5.10: Let $\{X(t)\}$ be a stochastic process which is mean-value and covariant stationary and has the triangular marginal probability distribution used in Example 5.6. Find the discrepancy between the average frequency of $\{X(t)\}$ and that of an approximation $\{\hat{X}(t)\}$ given by Eq. 5.36 with the coefficients determined according to the first-order approximation.

In Example 5.6, we found the inverse of the nonlinear transformation $\hat{X}(t) = \mu_X + \sigma_X g[Y(t)]$ for this problem to be

$$Y(t) = \frac{\hat{X}(t) - \mu_X(t)}{\sigma_X(t)} - 0.05091\left[\left(\frac{\hat{X}(t) - \mu_X(t)}{\sigma_X(t)}\right)^2 - 1\right] + 0.025\left[\left(\frac{\hat{X}(t) - \mu_X(t)}{\sigma_X(t)}\right)^3 - 3\left(\frac{\hat{X}(t) - \mu_X(t)}{\sigma_X(t)}\right)\right]$$

for which the derivative is

$$\dot{Y}(t) = \frac{\dot{\hat{X}}(t)}{\sigma_X}\left[0.925 - 0.1018\left(\frac{\hat{X}(t) - \mu_X}{\sigma_X}\right) + 0.075\left(\frac{\hat{X}(t) - \mu_X}{\sigma_X}\right)^2\right]$$

The mean squared value can be evaluated by noting that $\hat{X}(t)$ and $\dot{\hat{X}}(t)$ are uncorrelated (because of stationarity) and the first four moments of $[\hat{X}(t) - \mu_X]/\sigma_X$ are, respectively, 0, 1, *skewness* ≈ 0.3054, and $(kurtosis - 3) = -0.6$. Thus,

$$E[\dot{Y}^2(t)] = \frac{E[\dot{\hat{X}}^2(t)]}{\sigma_X^2}\left[(0.925)^2 + (0.1018)^2 + 2(0.075) - 2(0.1018)(0.075)(0.3054) + (0.075)^2(-0.6)\right]$$

$$= 1.00795 E[\dot{\hat{X}}^2(t)]/\sigma_X^2$$

This then gives $\sigma_{\dot{\hat{X}}}^2 = \sigma_{\dot{Y}}^2\sigma_X^2/1.00795$. Since $\omega_{a,X}^2 = \omega_{a,Y}^2$, $\sigma_Y^2 = 1$, and $\sigma_{\hat{X}}^2 = \sigma_X^2$, we now have the result that

$$\omega_{a,X}^2 = \omega_{a,Y}^2 = \sigma_{\dot{Y}}^2/\sigma_Y^2 = \sigma_{\dot{Y}}^2 = 1.00795\sigma_{\dot{\hat{X}}}^2/\sigma_{\hat{X}}^2 = 1.00795\omega_{a,\hat{X}}^2$$

Taking the square root shows that in this example the average frequency of $\{\hat{X}(t)\}$ is about 0.4% lower than that of $\{X(t)\}$.

****

Exercises
****

Gaussian or Non-Gaussian Processes
****

5.1 Consider a covariant stationary stochastic process $\{X(t)\}$ for which the joint probability density function of the process and its derivative is given by

$$p_{X(t)\dot{X}(t)}(u,v) = \frac{1}{\pi\sqrt{2}\,|u|}\exp\left(-u^2 - \frac{v^2}{2u^2}\right)$$

(a) Find the marginal probability density function $p_{X(t)}(u)$.

(b) Find the conditional probability density function for the derivative, $p_{\dot{X}(t)}[v|X(t)=u]$.

(c) Find the conditional mean and variance of $\dot{X}(t)$, given the event $X(t)=u$.

(d) Is either $\{X(t)\}$ or $\{\dot{X}(t)\}$ a Gaussian process? Briefly explain your answer.
****

5.2 Consider a covariant stationary stochastic process $\{X(t)\}$ for which the joint probability density function of the process and its derivative is given by

$$p_{X(t)\dot{X}(t)}(u,v) = \frac{|u|}{2\pi}\exp\left(\frac{-u^2(1+v^2)}{2}\right)$$

(a) Find the marginal probability density function $p_{X(t)}(u)$.

(b) Find the conditional probability density function for the derivative, $p_{\dot{X}(t)}[v|X(t)=u]$.

(c) Find the conditional mean and variance of $\dot{X}(t)$, given the event $X(t)=u$.

(d) Is either $\{X(t)\}$ or $\{\dot{X}(t)\}$ a Gaussian process? Briefly explain your answer.
****

5.3 Consider a covariant stationary stochastic process $\{X(t)\}$ for which the joint probability density function of the process and its derivative is given by

$$p_{X(t)\dot{X}(t)}(u,v) = \frac{2\sqrt{|u|}}{\pi}e^{-2u^2-2|u|v^2}$$

(a) Find the marginal probability density function $p_{X(t)}(u)$.

(b) Find the conditional probability density function for the derivative, $p_{\dot{X}(t)}[v|X(t)=u]$.

(c) Find the conditional mean and variance of $\dot{X}(t)$, given the event $X(t)=u$.

(d) Is either $\{X(t)\}$ or $\{\dot{X}(t)\}$ a Gaussian process? Briefly explain your answer.
****

Gaussian Processes
****

5.4 Let the stationary, mean zero, Gaussian process $\{X(t)\}$ denote the stress in a particular member of a system responding to stochastic excitation. Let the standard deviation of X be $\sigma_X = 75$ MPa.

(a) Find $P[X(t) \geq 350$ MPa$]$ for any particular time t.

(b) Find $P[A(t) \geq 350$ MPa$]$ for both the energy-based definition and the Cramer and Leadbetter definition of the amplitude $\{A(t)\}$ of the stress.
****

5.5 Let the stationary Gaussian process $\{X(t)\}$ denote the deflection at the top of an offshore structure responding to a stochastic sea. Let the mean and standard deviation of X be $\mu_X = 3\,\text{m}$ and $\sigma_X = 2$ m.
(a) Find $P[X(t) \geq 10\text{ m}]$ for any particular time t.
(b) Find $P[\mu_X + A(t) \geq 10\text{ m}]$ for both the energy-based definition and the Cramer and Leadbetter definition of the amplitude $\{A(t)\}$ of the variable part of the deflection.
****

5.6 Let $\{X(t)\}$ denote the stationary response of a linear system governed by the first-order differential equation $\dot{X}(t) + bX(t) = F(t)$, with $b = 10\text{s}^{-1}$. The excitation $\{F(t)\}$ is a stationary, mean zero, Gaussian random process with an autospectral density function of

$$S_{FF}(\omega) = \frac{\omega_0^2}{50(\omega^2 + \omega_0^2)} \text{ (m/s)}^2/\text{(rad/s)} \quad \text{with} \quad \omega_0 = 20 \text{ rad/s}$$

(a) Find the standard deviation σ_X of the response.
(b) Find $P[|X(t)| \geq 0.1\text{ m}]$ for any particular time t.
(c) For both the energy-based definition and the Cramer and Leadbetter definition, find
    $P[A(t) \geq 0.1\text{ m}]$ for any particular time t.
[Example 4.2 relates this $S_{FF}(\omega)$ to autocovariance, so that time integrals can be used if desired.]
****

5.7 Let $\{X(t)\}$ be the mean zero covariant stationary Gaussian random process of Example 4.6 with autospectral density of $S_{XX}(\omega) = S_0 \left(U[|\omega| - (\omega_c - b)] - U[|\omega| - (\omega_c + b)] \right)$, in which $b < \omega_c$.
(a) Find $E(\dot{A}^2)$ for the energy-based definition of amplitude.
(b) Find $E(\dot{A}_{CL}^2)$ for the Cramer and Leadbetter definition of amplitude.
(c) Determine the ratio $E(\dot{A}^2)/E(\dot{A}_{CL}^2)$ for the limiting narrowband situation with $b << \omega_c$.
****

5.8 Let $\{X(t)\}$ be the mean zero covariant stationary Gaussian random process of Exercise 4.15 with autospectral density of $S_{XX}(\omega) = S_0|\omega/\omega_0|^c U(\omega_0 - |\omega|) + S_0|\omega_0/\omega|^c U(|\omega| - \omega_0)$.
(a) Find $E(\dot{A}^2)$ for the energy-based definition of amplitude.
(b) Find $E(\dot{A}_{CL}^2)$ for the Cramer and Leadbetter definition of amplitude.
(c) Determine the ratio $E(\dot{A}^2)/E(\dot{A}_{CL}^2)$ for the limiting narrowband situation with $c >> 1$.
****

Probability Density Function Approximation
****

5.9 Let X be a random variable which has an exponential distribution: $p_X(u) = \lambda e^{-\lambda u} U(u)$. Compare this probability density with the approximations which result from the Gram-Charlier approximation of Eq. 5.27 truncated to include only the $j = 3$ and $j = 4$ terms and from the use of Eqs. 5.29 and 5.31 with $J = 3$ and the coefficients evaluated from Eq.5.35.
****

5.10 Let X be a random variable which has a uniform probability distribution with $p_X(u) = 0.5[U(u+1) - U(u-1)]$. Compare this probability density with the approximations which result from the Gram-Charlier approximation of Eq. 5.27 truncated to include only the $j = 3$ and $j = 4$ terms and from the use of Eqs. 5.29 and 5.36 with the coefficients evaluated according to the first-order approximation.

****
5.11 Let X be a random variable which has a triangular probability distribution with $p_X(u) = 2u[U(u) - U(u-1)]$. Compare this probability density with the approximations which result from the Gram-Charlier approximation of Eq. 5.27 truncated to include only the $j = 3$ and $j = 4$ terms and from the use of Eqs. 5.29 and 5.36 with the coefficients evaluated according to the first-order approximation.
****

Average Frequency Discrepancy
****
5.12 Let $\{X(t)\}$ be a stochastic process which is mean-value and covariant stationary and has the exponential probability distribution used in Exercise 5.9. Find the discrepancy between the average frequency of $\{X(t)\}$ and that of an approximation $\{\hat{X}(t)\}$ given by Eqs. 5.29 and 5.31 with the coefficients determined according to Eq. 5.35.
****

5.13 Let $\{X(t)\}$ be a stochastic process which is mean-value and covariant stationary and has the uniform probability distribution used in Exercise 5.10. Find the discrepancy between the average frequency of $\{X(t)\}$ and that of an approximation $\{\hat{X}(t)\}$ given by Eq. 5.36 with the coefficients determined according to the first-order approximation.
****

5.14 Let $\{X(t)\}$ be a stochastic process which is mean-value and covariant stationary and has the triangular probability distribution used in Exercise 5.11. Find the discrepancy between the average frequency of $\{X(t)\}$ and that of an approximation $\{\hat{X}(t)\}$ given by Eq. 5.36 with the coefficients determined according to the first-order approximation.
****

Chapter 6

Occurrence Rates and Distributions of Extremes

6.1 Categories Of Problems

Unsatisfactory performance of an engineering system is often associated with the occurrence of an unusually large value of some quantity, such as displacement, force, strain, or stress. Thus, the analysis of extreme values of dynamic response is a crucial part of our investigation of stochastic vibration problems. Two different types of extreme problems will be investigated in this chapter, along with some related problems. The two types of extremes might be classified as "local" and "global," although we mean global in a somewhat limited sense in this instance. We will use the term "peak" distribution to refer to the probability distribution of the purely local extrema of a time history — the points where the first derivative of the time history is zero and the second derivative is negative. When needed we will use the term "valley" to refer to a local minimum of a time history, although we will do little specific analysis of valley distributions since a valley of a process is simply a peak of the negative of the process. The more global sort of extreme problem which we will consider involves the extreme value of some stochastic process $\{X(t)\}$ during a fixed time interval, such as $0 \leq t \leq T$. Henceforth, we will use the term "extreme distribution" only to refer to this more global problem, and not for the distribution of peaks.

The extreme probability distribution is one of the most important aspects of many stochastic analyses of engineering problems. Knowledge of this distribution for the appropriate process allows the analyst to calculate the probability that the displacement, stress, etc. has ever exceeded some critical level during the time interval of interest. Unfortunately, it is generally not possible to find this probability distribution other than in an approximate sense. The peak distribution provides less information, but is generally somewhat simpler to obtain. Intuitively one expects the peak distribution to be related to the distribution of the amplitude of the process, as analyzed in Chapters 4 and 5, although we will find that this relationship is not very precise unless the process is very narrowband. Since a peak or valley corresponds to an occurrence of $\dot{X}(t) = 0$, the study of peaks is also intimately related to the study of the occurrence of zero crossings. Moving from the simpler to the more complex problems, we will begin with the study of crossing rates, proceed to the study of peak distributions, and then investigate the extreme distribution problem.

6.2 Rate of Occurrence of Upcrossings

As noted, the study of the occurrence of crossings provides an introduction to some ideas which are helpful in studying peak distributions and extreme distributions. In addition, though, crossing rates are significant as a measure of the frequency of a stochastic process. In particular, if the only information available about a stochastic process is a sample time history, then the rates of occurrence of crossings and peaks are the most readily determined measures of the frequency of the process. In particular, it is much easier to count the frequency with which the process crosses its own mean value, for instance, or the frequency with which its peaks occur, than it is to estimate the spectral moments which are needed for evaluation of the average frequencies defined in Section 4.9.

We will let $v_X^+(u,t)$ denote the expected rate of occurrence of the event $X(t) = u$ with $\dot{X}(t) > 0$, and $v_X^-(u,t)$ denote the expected rate for the event $X(t) = u$ with $\dot{X}(t) < 0$. Commonly these are called the rate of upcrossings and the rate of downcrossings, respectively, of the level $X = u$, as shown in Fig. 6.1. The expected number of upcrossings during any time interval of finite length, for example, is then the integral of $v_X^+(u,t)$ over the interval. Particularly for a nonstationary process, the rate of upcrossings can probably be more clearly understood by relating it to the probability of occurrence of an upcrossing during a small time increment. The basic definition of the expected rate of occurrence of any event can be written as

$$\lim_{\Delta t \to 0} \frac{E(\text{number of occurrences in } [t,\ t + \Delta t])}{\Delta t}$$

During any infinitesimal time interval $[t, t + \Delta t]$, though, we expect there to be either one or zero occurrences so that the expected number of occurrences is the same as the probability of an occurrence. Thus, we can write the expected rate of upcrossings as

$$v_X^+(u,t) = \lim_{\Delta t \to 0} \frac{P(\text{an upcrossing of } u \text{ in } [t,\ t + \Delta t])}{\Delta t} \qquad (6.1)$$

The probability of an upcrossing during an infinitesimal interval can be investigated intuitively by considering the phase diagram shown in Fig. 6.2. In particular, we can argue that there can only be an upcrossing of the level u within the interval $[t, t + \Delta t]$ if $X(t)$ at the beginning of the interval is less than u, but close to u, and has a positive derivative. In particular, since Δt is infinitesimal, we might consider the derivative to be constant at the value $\dot{X}(t)$ throughout the time interval, and conclude that there will be an upcrossing within the interval only if $0 < u - X(t) < \dot{X}(t)\Delta t$, which translates into $u - \dot{X}(t)\Delta t < X(t) < u$. This event is shown shaded

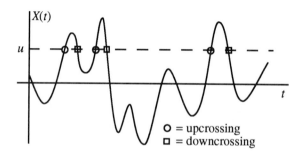

Figure 6.1 Crossing of the level $X(t) = u$

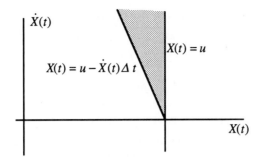

Figure 6.2 Phase diagram showing the event of an upcrossing

on the space of possible values of $X(t)$ and $\dot{X}(t)$ in Fig. 6.2. The probability of this event can now be found by integrating the joint probability density of $X(t)$ and $\dot{X}(t)$ over the shaded region:

$$P(\text{an upcrossing of } u \text{ in } [t,\ t + \Delta\, t]) \approx \int_0^\infty \int_{u-v\Delta\, t}^{u} p_{X(t),\dot{X}(t)}(w,v)\, dw\, dv$$

We now once again use the fact that $\Delta\, t$ is infinitesimal to argue that the w variable of integration is always almost the same as u. Replacing $p_{X(t),\dot{X}(t)}(w,v)$ with $p_{X(t),\dot{X}(t)}(u,v)$ in the integrand allows easy evaluation of the integral with respect to w, and gives

$$P(\text{an upcrossing of } u \text{ in } [t,\ t + \Delta\, t]) \approx \int_0^\infty (v\Delta\, t)\, p_{X(t),\dot{X}(t)}(u,v)\, dv$$

Substituting this expression into Eq. 6.1 now gives the expected rate of upcrossing as

$$v_X^+(u,t) = \int_0^\infty v\, p_{X(t),\dot{X}(t)}(u,v)\, dv \tag{6.2}$$

This derivation of Eq. 6.2 illustrates the ideas involved but is not very rigorous, particularly if the derivative $\dot{X}(t)$ is unbounded. That is, if the probability density function allows the possibility $\dot{X}(t)$ having arbitrarily large values, then one cannot precisely claim that the range of the w integration from $u - \dot{X}(t)\Delta\, t$ to u is small, as we did in replacing w with u in the integrand. In fact, for any given value of $\Delta\, t$ this distance of $\dot{X}(t)\Delta\, t$ tends to infinity as $\dot{X}(t)$ tends to infinity. It is possible to prove that the limiting process is legitimate in spite of this difficulty, but rather than pursue that approach we will give an alternate derivation of Eq. 6.2 which relies more on algebra and less on geometry.

Since $v_X^+(u,t)$ is the expected rate of upcrossings, we will take the point of view that it is the mean value of the derivative of a counting process $N_X^+(u,t)$ which gives the number of upcrossings since time zero.[1] Figure 6.3 illustrates this idea. First we define a process $Z(t) \equiv U[X(t) - u]$

---

[1] As for the Poisson process in Example 2.17, we can anticipate that this derivative of a counting

which steps back and forth between the level zero and unity, depending on whether $X(t)$ is less than or greater than u, respectively, as shown in part (b) of the figure. The derivative of this process is always zero or infinity, but it can be written formally as

$$\dot{Z}(t) \equiv \delta[X(t) - u]\dot{X}(t)$$

Each of the Dirac delta function pulses in $\dot{Z}(t)$ is shown in Fig. 6.3(c) as an arrow, pointing toward either plus infinity or minus infinity, depending on the sign of $\dot{X}(t)$. Of course, the integral across any one of these delta functions is a step of unit magnitude, since the integral of $\dot{Z}(t)$ is $Z(t)$. By eliminating the delta functions with negative multipliers and then integrating, we can obtain a process which counts the number of upcrossings. This elimination of negative pulses is easily done by multiplying by $U[\dot{X}(t)]$, as shown in Fig. 6.3(d). The resulting counting process, as shown in part (e) of the figure, is $N_X^+(u,t)$ which starts at zero and proceeds to increase by unit step values, since it contains all the positive steps and none of the negative steps of $Z(t)$.

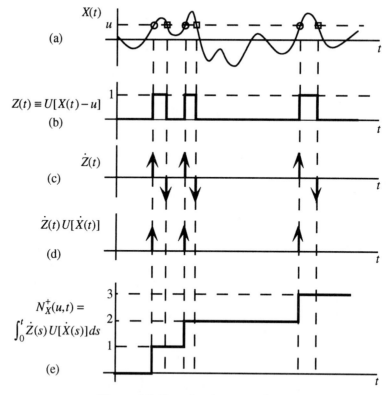

Figure 6.3 Counting the upcrossings

process may have a finite mean value, but an infinite mean squared value.

We can now say that $v_X^+(u,t)$ is the expected value of the derivative of $N_X^+(u,t)$. Thus, it is the expected value of the $\dot{Z}(t)U[\dot{X}(t)]$ process illustrated in part (d) of the figure. Thus, we merely need to substitute for $\dot{Z}(t)$ to obtain

$$v_X^+(u,t) = E\big(\dot{Z}(t)U[\dot{X}(t)]\big) = E\big(\delta[X(t)-u]\dot{X}(t)U[\dot{X}(t)]\big)$$

or

$$v_X^+(u,t) = \int_{-\infty}^{\infty}\int_{-\infty}^{\infty} p_{X(t),\dot{X}(t)}(w,v)\,\delta(w-u)\,v\,U(v)\,dw\,dv$$

Performing the integration with respect to w gives

$$v_X^+(u,t) = \int_{-\infty}^{\infty} p_{X(t),\dot{X}(t)}(u,v)\,v\,U(v)\,dv = \int_0^{\infty} v\,p_{X(t),\dot{X}(t)}(u,v)\,dv$$

which is identical to Eq. 6.2. Alternatively, we can factor the joint probability density function as a product of a marginal and conditional density function and write

$$v_X^+(u,t) = p_{X(t)}(u)\int_0^{\infty} v\,p_{\dot{X}(t)}[v\,|\,X(t)=u]\,dv \tag{6.3}$$

The relationships in Eqs. 6.2 and 6.3 are very general, applying to any stationary or nonstationary process with any probability distribution. It should be noted, though, that the value obtained may be infinite if the conditional probability density function of $\dot{X}(t)$ does not decay sufficiently rapidly. In particular, if $E[\dot{X}(t)|X(t)=u]$ does not exist then $v_X^+(u,t)$ may be infinite.

In order to obtain the downcrossing rate, we can essentially reverse the sign of $\dot{X}(t)$ to obtain

$$v_X^-(u,t) = -\int_{-\infty}^{0} v\,p_{X(t),\dot{X}(t)}(u,v)\,dv = p_{X(t)}(u)\int_{-\infty}^{0} |v|\,p_{\dot{X}(t)}[v\,|\,X(t)=u]\,dv \tag{6.4}$$

From Eqs. 6.2 and 6.4, we can also see that if the $\{X(t)\}$ process is second-order stationary then the crossing rates will be independent of time t. In this special case, we can drop the t argument and write $v_X^+(u)$ and $v_X^-(u)$ for the upcrossing and downcrossing rates.

****

Example 6.1: Find the expected rate of upcrossings for a covariant stationary Gaussian process $\{X(t)\}$.

For a covariant stationary process, we know that the random variables $X(t)$ and $\dot{X}(t)$ are uncorrelated. In addition, we know that Gaussian random variables are uncorrelated if and only if they are independent. Thus, we know that $X(t)$ and $\dot{X}(t)$ are independent for the present situation. This allows us to use the unconditional distribution of $\dot{X}(t)$ in place of the conditional distribution in Eq. 6.3, giving

$$v_X^+(u,t) = p_{X(t)}(u) \int_0^\infty \frac{v}{\sqrt{2\pi}\,\sigma_{\dot X}} \exp\left[-\frac{1}{2}\left(\frac{v - \mu_{\dot X}(t)}{\sigma_{\dot X}}\right)^2\right] dv$$

which can be integrated to give

$$v_X^+(u,t) = p_{X(t)}(u)\left[\mu_{\dot X}(t)\,\Phi\left(\frac{\mu_{\dot X}(t)}{\sigma_{\dot X}}\right) + \frac{\sigma_{\dot X}}{\sqrt{2\pi}}\right]$$

or

$$v_X^+(u,t) = \frac{1}{\sqrt{2\pi}\,\sigma_X} \exp\left[-\frac{1}{2}\left(\frac{u - \mu_X(t)}{\sigma_X}\right)^2\right]\left[\mu_{\dot X}(t)\,\Phi\left(\frac{\mu_{\dot X}(t)}{\sigma_{\dot X}}\right) + \frac{\sigma_{\dot X}}{\sqrt{2\pi}}\right]$$

in which $\Phi(\cdot)$ denotes the cumulative Gaussian distribution function.

Next consider the special case in which $\{X(t)\}$ is also mean-value stationary, and thereby strictly stationary. In this case we know that μ_X is a constant and $\mu_{\dot X} = 0$, so that our expression can be simplified to

$$v_X^+(u) = \frac{\sigma_{\dot X}}{2\pi\,\sigma_X} \exp\left[-\frac{1}{2}\left(\frac{u - \mu_X}{\sigma_X}\right)^2\right] \qquad \text{(for mean and covariant stationary)}$$

Recall that the average frequency of a process $\{X(t)\}$ according to the energy-based definition of phase in Section 4.9 was $\omega_a = \sqrt{\lambda_2 / \lambda_0} = \sigma_{\dot X} / \sigma_X$. Thus, we can say that

$$v_X^+(u) = \frac{\omega_a}{2\pi} \exp\left[-\frac{1}{2}\left(\frac{u - \mu_X}{\sigma_X}\right)^2\right]$$

The maximum value of this crossing rate occurs when $u = \mu_X$ and it is simply $v_X^+(\mu_X) = \omega_a/(2\pi)$. Thus, we see that the rate of upcrossings of $u = \mu_X$ by a stationary Gaussian process is simply the energy-based average frequency divided by 2π. The factor of 2π, of course, comes from the fact that ω_a represents a frequency in radians per second, while the rate of mean-upcrossings represents a frequency in cycles per second or Hz. The interesting thing is that these two quite distinct definitions of process frequency are exactly equivalent for a stationary Gaussian process.
****

Example 6.2: Find the expected rate of upcrossings for a Gaussian process $\{X(t)\}$ which is mean zero and has a nonstationary covariance.

The conditional Gaussian probability density function of $\dot X(t)$ can be written in the usual Gaussian form (see Example B.6 in Appendix B). In particular, we can write

$$p_{\dot{X}(t)}[v|X(t)=u] = \frac{1}{\sqrt{2\pi}\,\sigma_*(t)} \exp\left[-\frac{1}{2}\left(\frac{v-\mu_*(t)}{\sigma_*(t)}\right)^2\right]$$

in which $\mu_*(t)$ and $\sigma_*(t)$ are the conditional mean and standard deviation of $\dot{X}(t)$, and are given by

$$\mu_*(t) = \frac{K_{X\dot{X}}(t,t)}{K_{XX}(t,t)}u = \rho_{X\dot{X}}(t,t)\frac{\sigma_{\dot{X}}(t)}{\sigma_X(t)}u, \qquad \sigma_*(t) = \sigma_{\dot{X}}(t)\sqrt{1-\rho_{X\dot{X}}^2(t,t)}$$

Since the conditional distribution of $\dot{X}(t)$ is the usual Gaussian form, the integral in Eq. 6.3 is basically the same as in Example 6.1:

$$v_X^+(u,t) = \frac{1}{\sqrt{2\pi}\,\sigma_X(t)} \exp\left(\frac{-u^2}{2\sigma_X^2(t)}\right)\left[\mu_*(t)\,\Phi\left(\frac{\mu_*(t)}{\sigma_*(t)}\right) + \frac{\sigma_*(t)}{\sqrt{2\pi}}\right]$$

which can be rewritten as

$$v_X^+(u,t) = \frac{\sigma_{\dot{X}}(t)}{\sigma_X(t)}\exp\left(\frac{-u^2}{2\sigma_X^2(t)}\right)\left[\frac{\rho_{X\dot{X}}(t,t)u}{\sqrt{2\pi}\,\sigma_X(t)}\Phi\left(\frac{\rho_{X\dot{X}}(t,t)u}{\sqrt{1-\rho_{X\dot{X}}^2(t,t)}\,\sigma_X(t)}\right) + \frac{\sqrt{1-\rho_{X\dot{X}}^2(t,t)}}{2\pi}\right]$$

Note that setting the correlation coefficient equal to zero in this expression gives the same result as in Example 6.1. We may also note that a symmetric probability distribution for $X(t)$ and $\dot{X}(t)$, as in this example, gives the rate of downcrossings to be the same as the rate of upcrossings.

Setting $u=0$ in the previous expression gives the rate of zero-upcrossings or mean-upcrossings as

$$v_X^+(0,t) = v_X^+(\mu_X,t) = \frac{\sqrt{1-\rho_{X\dot{X}}^2(t,t)}}{2\pi}\frac{\sigma_{\dot{X}}(t)}{\sigma_X(t)}$$

showing that a correlation between $X(t)$ and $\dot{X}(t)$ results in a lowering of the process frequency, as measured by this upcrossing rate.

****

Example 6.3: Find the rate of upcrossings for the stationary dynamic response of a linear SDF oscillator excited by stationary Gaussian white noise.

This is a special case of Example 6.1, since the response process $\{X(t)\}$ is Gaussian and covariant stationary. Furthermore, the response is also mean-value stationary. Thus, the key parameters needed are the stationary standard deviations of $\{X(t)\}$ and $\{\dot{X}(t)\}$. From the results in Sections 3.7 and 4.7, we can write these as

$$\sigma_X = \sqrt{\frac{\pi S_0}{2\zeta\omega_0^3}}, \qquad \sigma_{\dot X} = \sqrt{\frac{\pi S_0}{2\zeta\omega_0}}$$

Thus, we obtain

$$v_X^+(u) = \frac{\omega_0}{2\pi}\exp\left[-\frac{1}{2}\left(\frac{u-\mu_X}{\sigma_X}\right)^2\right]$$

with a maximum value of $\omega_0/(2\pi)$. The expected frequency of mean-upcrossings for this process is exactly the nominal undamped frequency of the oscillator in Hz. If the excitation is not white noise or is not stationary, then this very convenient result will no longer be strictly true.

****

Example 6.4: Find the expected rate of upcrossings for the Cramer and Leadbetter amplitude of a Gaussian process which is mean zero.

From Eq. 6.3, the rate of upcrossings for $\{A_{CL}(t)\}$ is

$$v_{A_{CL}}^+(u,t) = p_{A_{CL}(t)}(u)\int_0^\infty v\, p_{\dot A_{CL}(t)}[v|A_{CL}(t)=u]dv$$

However, we know from the results in Section 5.6 that $A_{CL}(t)$ and $\dot A_{CL}(t)$ are independent, so that the conditioning in the integrand can be ignored. Furthermore, we know that $A_{CL}(t)$ has the Rayleigh distribution of Eq. 5.9, and that $\dot A_{CL}(t)$ is Gaussian with mean zero and variance $\sigma_{\dot A}^2 = (1-\alpha_1^2)\lambda_2 \equiv (1-\alpha_1^2)\sigma_{\dot X}^2$. Thus, we can write

$$v_{A_{CL}}^+(u,t) = \left(\frac{u}{\sigma_X^2}\exp\left[\frac{-u^2}{2\sigma_X^2}\right]\right)\int_0^\infty \frac{v\, e^{-v^2/(2\sigma_{\dot A}^2)}}{\sqrt{2\pi}\,\sigma_{\dot A}}dv = \left(\frac{u}{\sigma_X^2}\exp\left[\frac{-u^2}{2\sigma_X^2}\right]\right)\frac{\sigma_{\dot A}}{\sqrt{2\pi}}$$

or

$$v_{A_{CL}}^+(u,t) = \left(\frac{u}{\sigma_X^2}\exp\left[\frac{-u^2}{2\sigma_X^2}\right]\right)\left(\frac{\sigma_{\dot X}}{\sigma_X}\right)\sqrt{\frac{1-\alpha_1^2}{2\pi}} \qquad \text{for } u \geq 0$$

Note that the first expression on the right-hand side is a dimensionless quantity giving the form of the dependence on u. The upcrossing rate tends to zero as u approaches zero, which is consistent with the fact that $A(t)$ is almost always above such a small u value.

****

Example 6.5: Find the expected rate of upcrossings for the energy-based amplitude of a Gaussian process which is mean zero.

At first glance this calculation seems more difficult than the one in Example 6.4, since we do not have a closed form solution for the probability density function for the derivative of the energy-based amplitude. We do, however, know that $A(t)$ and $\dot A(t)$ are independent and that the probability density of the latter quantity is obtained from Eq. 5.21 as the integral

$$p_{\dot{A}}(v) = \int_0^{2\pi} \frac{\lambda_2^{1/2}}{(2\pi)^{3/2}\sqrt{\lambda_0\lambda_4 - \lambda_2^2}\,|\sin(\eta)|} \exp\left[-\frac{\lambda_2 v^2}{2(\lambda_0\lambda_4 - \lambda_2^2)\sin^2(\eta)}\right]d\eta$$

By reversing the order of integration we can integrate first with respect to v, then with respect to η, to obtain

$$\int_0^\infty v p_{\dot{A}}(v)\,dv = \frac{\sqrt{\lambda_0\lambda_4 - \lambda_2^2}}{(2\pi)^{3/2}\lambda_2^{1/2}}\int_0^{2\pi}|\sin(\eta)|\,d\eta = \frac{4\sqrt{\lambda_0\lambda_4 - \lambda_2^2}}{(2\pi)^{3/2}\lambda_2^{1/2}} = \frac{4\sigma_X\sigma_{\ddot{X}}\sqrt{1 - \alpha_2^2}}{(2\pi)^{3/2}\sigma_{\dot{X}}}$$

so that

$$v_A^+(u,t) = p_A(u)\frac{4\sigma_X\sigma_{\ddot{X}}\sqrt{1 - \alpha_2^2}}{(2\pi)^{3/2}\sigma_{\dot{X}}} = \left(\frac{u}{\sigma_X^2}\exp\left[\frac{-u^2}{2\sigma_X^2}\right]\right)\left(\frac{\sigma_{\ddot{X}}}{\sigma_{\dot{X}}}\right)\frac{4\sqrt{1 - \alpha_2^2}}{(2\pi)^{3/2}} \quad \text{for} \quad u \geq 0$$

Note that the u dependence is the same as in Example 6.4 for $A_{CL}(t)$, but the basic frequency of occurrence is now proportional to $(\sigma_{\ddot{X}}/\sigma_{\dot{X}})(1 - \alpha_2^2)^{1/2}$, as compared to $(\sigma_{\dot{X}}/\sigma_X)(1 - \alpha_1^2)^{1/2}$ for $A_{CL}(t)$.
****

Example 6.6 Find the approximate expected rate of crossings of the mean value level of a narrowband process $\{X(t)\}$ for which the joint probability density of $X(t)$ and $\dot{X}(t)$ is unknown.

For a narrowband process we can approximate $v_X^+[\mu_X(t),t]$ without knowledge of $p_{X(t),\dot{X}(t)}(u,v)$, even though our general formula for calculating the rate of upcrossings depends on the term. This follows directly from the nearly harmonic time history behavior of a narrowband process, as discussed in Section 4. 4. For $X(t) \approx \mu_X(t) + A(t)\cos[\omega_c t + \theta(t)]$ with slowly varying amplitude and phase, the crossings of the level $\mu_X(t)$ must occur with an approximate period of $2\pi/\omega_c$, as illustrated in Fig. 4.3. Thus, we must have an approximate expected rate of upcrossings of

$$v_X^+[\mu_X(t),t] \approx \frac{\omega_c}{2\pi}$$

****

6.3 Rate of Occurrence of Peaks and Valleys

Since a peak of $X(t)$ occurs whenever $\dot{X}(t) = 0$ and $\ddot{X}(t) < 0$, we can say that the rate of occurrence of peaks of $\{X(t)\}$ is exactly the rate of downcrossings of the level zero by $\{\dot{X}(t)\}$:

$$v_P(t) \equiv v_{\dot{X}}^-(0,t) = \int_{-\infty}^0 |w| p_{\dot{X}(t),\ddot{X}(t)}(0,w)\,dw = p_{\dot{X}(t)}(0)\int_{-\infty}^0 |w| p_{\ddot{X}(t)}(w\,|\,\dot{X}(t) = 0)\,dw \qquad (6.5)$$

Similarly, the rate of occurrence of valleys of $\{X(t)\}$ is $v_V(t) \equiv v_{\dot{X}}^+(0,t)$.

There is at least one important property of the occurrence rates which is not immediately obvious from Eqs. 6.2 and 6.5. In particular, one can easily verify that any sufficiently long continuous time history of a process must have at least as many peaks as it has upcrossings of any level. Specifically, at least one peak must occur between any two upcrossing of the same level u. From this fact, we can conclude that $\nu_P(t) \geq \nu_X^+(u,t)$ for any u for any process with continuous time histories. On the other hand, the rate of occurrence of peaks of a narrowband process is expected to be only slightly larger than the rate of upcrossings of the mean, based on the similarity of a narrowband time history to a harmonic function with slowly varying amplitude and phase (as discussed in Chapter 4). This property is commonly used to provide a measure of bandwidth which can readily be estimated from a time history. It is called the irregularity factor, and is defined as

$$IF = \nu_X^+(\mu_X,t) / \nu_P(t) \tag{6.6}$$

Like the α_1 and α_2 bandwidth parameters defined in Section 4.8, the range of possible values of IF is from zero to unity, and IF tends to unity for a narrowband process.

****

Example 6.7: Find the rate of peak occurrences and the irregularity factor for a Gaussian process $\{X(t)\}$ which is mean-value stationary and covariant stationary.

We can obtain the rate of occurrences of peaks by rewriting the results in Examples 6.1 so that they apply to the $\{\dot{X}(t)\}$ process. In particular, we know that $\mu_{\dot{X}} = 0$ since μ_X is stationary, and we know that the rate of upcrossings is the same as the rate of downcrossings. Thus, we deduce from Example 6.1 that the rate of peak occurrences is

$$\nu_P = \frac{\sigma_{\ddot{X}}}{2\pi\,\sigma_{\dot{X}}}$$

The irregularity factor is then easily written as

$$IF = \frac{N_X^+(\mu_X)}{\nu_P} = \frac{\sigma_{\dot{X}}^2}{\sigma_X \sigma_{\ddot{X}}}$$

Note, though , that this ratio is exactly the bandwidth parameter $\alpha_2 = \lambda_2 / \sqrt{\lambda_0 \lambda_4}$. Thus, we see that for a stationary Gaussian process, the irregularity factor is identical to α_2. This may be viewed as giving another interpretation of the meaning of α_2. For a non-Gaussian process we cannot expect IF and α_2 to be identical, but they will generally be quite similar.
****

Example 6.8: Find the rate of peak occurrences and the irregularity factor for the response $\{X(t)\}$ of a linear SDF oscillator excited by Gaussian white noise.

As in the previous example, we can rewrite the results in Examples 6.1 and 6.2 for the $\{\dot{X}(t)\}$ so that they apply to the $\{\dot{X}(t)\}$ process. When we do this, we find that $\nu_{\dot{X}}^-(u,t)$ for a Gaussian process depends on the standard deviation of the acceleration, $\sigma_{\ddot{X}}(t)$. We know, though, that $\sigma_{\ddot{X}}(t)$ is infinite for the response of the SDF oscillator to white noise excitation. Thus, we find that the rate of peak occurrences, $\nu_P(t)$, is infinite for the response of any linear SDF oscillator excited by Gaussian white noise. Correspondingly, the irregularity factor, IF, is zero for the response process. This latter result does agree with our finding in Example 6.7 that $IF = \alpha_2$ for a stationary Gaussian process, since we had previously found that α_2 is zero for the stationary response of the SDF oscillator excited by white noise.

****

The result in Example 6.7 illustrates an important feature of the occurrence rate for peaks. In particular, it was found that the peak occurrence rate is finite for a Gaussian process if and only if $\sigma_{\ddot{X}}$ is finite. We do not have a correspondingly simple rigorous result for non-Gaussian processes, but it is clear from Eq. 6.5 that the existence of a finite peak occurrence rate is dependent on the manner in which $p_{\dot{X}\ddot{X}}(0,w)$ converges to zero as w tends to infinity. In any problem in which $\sigma_{\ddot{X}}$ is infinite, we should anticipate the possibility that the peak occurrence rate may also be infinite. Similarly, if $\sigma_{\dot{X}}$ is infinite then there is the possibility that the crossing rates will be infinite, and this is easily shown to be a rigorous relationship for a Gaussian process.

Example 6.8 demonstrated that the response of an SDF oscillator to white noise excitation is one important example in which $\sigma_{\ddot{X}}$ is not finite, so that the peak occurrence rate is infinite if the excitation and response are Gaussian. This finding of infinite occurrence rates of peaks in a model of an important physical problem may seem somewhat surprising. That is, we certainly expect the number of peaks in a time history of a physical phenomenon to be finite, and we believe that many physical phenomena are approximated by the response of linear SDF oscillators excited by broadband stochastic excitations. This reveals a shortcoming in our modeling of the excitation as white noise. If we replaced the white noise excitation with an $\{F(t)\}$ which had a finite variance, then we would find that $\sigma_{\ddot{X}}$ would be finite and the peak occurrence rate would be finite. As explained in Sections 3.5 and 4.5, the use of a delta-correlated or white noise excitation always involves an approximation of the physical problem of interest. This approximation gives us many useful results about the response of an oscillator, but it fails to give us the rate of occurrence of peaks. The rate of occurrence of peaks depends quite heavily on the behavior of the high-frequency portion of the autospectral density, and for the SDF oscillator this is approximated by $S_{XX}(\omega) \approx S_{FF}(\omega) / (m^2 \omega^4)$ or $S_{\ddot{X}\ddot{X}}(\omega) \approx S_{FF}(\omega) / m^2$, as discussed in Section 4.6. In order to find the peak occurrence rate for the system, one must know how $S_{FF}(\omega)$ decays as ω becomes large. One can show that the same conclusion applies to any problem in which a broadband force is applied to a finite mass within a system made up of masses, springs, and dashpots, since such a system will always give a linear relationship between $S_{\ddot{X}\ddot{X}}(\omega)$ and $S_{FF}(\omega)$ in the high-frequency region.

6.4   Probability Distribution of Peaks

The probability distribution of the peaks of $\{X(t)\}$ can be found by a procedure which is basically the same as that used in deriving the rates of occurrence of crossings or peaks. That is, we will derive the probability distribution from consideration of occurrence rates. First we define $\nu_P[t;X(t) \le u]$ as the expected rate of occurrence of peaks not exceeding the level u. Next we recall that for an infinitesimal time interval Δt we can say that the expected number of occurrences in the interval is the same as the probability of one occurrence in the interval, since we can neglect the probability of two or more occurrences. Thus, we say that

$$v_P[t; X(t) \le u] \Delta t = P(\text{peak} \le u \text{ during } [t, t + \Delta t]) \qquad (6.7)$$

just as

$$v_P(t) \Delta t = P(\text{peak during } [t, t + \Delta t]) \qquad (6.8)$$

for $v_P(t)$ being the total expected rate of peak occurrences, which is the limit as u goes to infinity of $v_P[t; X(t) \le u]$. Furthermore, we can say that

$$P(\text{peak} \le u \text{ during } [t, t + \Delta t]) = P(\text{peak during } [t, t + \Delta t]) P(\text{peak} \le u | \text{peak during } [t, t + \Delta t])$$

The final, conditional probability, term in this expression is precisely what we will consider to be the cumulative distribution function for a peak at time t:

$$F_{P(t)}(u) \equiv P(\text{peak} \le u | \text{peak during } [t, t + \Delta t]).$$

From Eqs. 6.7 and 6.8 we then solve for this cumulative distribution function as

$$F_{P(t)}(u) = \frac{v_P[t; X(t) \le u]}{v_P(t)} \qquad (6.9)$$

Thus, we see that determining the probability distribution of the peaks depends on finding the rate of occurrence of peaks below any level u, and this is relatively straightforward. First, we note that $U[-\dot{X}(t)]$ is a process which has a positive unit step at each peak of $X(t)$ and has a negative unit step at each valley of $X(t)$. Thus, the derivative of this process, $-\ddot{X}(t)\delta[-\dot{X}(t)]$, has positive and negative unit Dirac delta functions at the peaks and valleys, respectively. By multiplying by $U[-\ddot{X}(t)]$, we can eliminate the negative Dirac delta functions in order to count only peaks. Similarly, we can multiply by $U[u - X(t)]$ in order to eliminate all peaks above the level u. In this way, we obtain the rate of occurrence of peaks not exceeding the level u as

$$v_P[t; X(t) \le u] = E\left(-\ddot{X}(t)\delta[-\dot{X}(t)]U[-\ddot{X}(t)]U[u - X(t)]\right) \qquad (6.10)$$

Substituting into Eq. 6.9 gives

$$F_{P(t)}(u) = \frac{E\left(-\ddot{X}(t)\delta[-\dot{X}(t)]U[-\ddot{X}(t)]U[u - X(t)]\right)}{E\left(-\ddot{X}(t)\delta[-\dot{X}(t)]U[-\ddot{X}(t)]\right)}$$

which can be rewritten in terms of joint probability density functions as

$$F_{P(t)}(u) = \frac{\int_{-\infty}^{\infty}\int_{-\infty}^{\infty}\int_{-\infty}^{\infty}(-z)\delta(-v)U(-z)U(u-w)\,p_{X(t),\dot{X}(t),\ddot{X}(t)}(w,v,z)\,dw\,dv\,dz}{\int_{-\infty}^{\infty}\int_{-\infty}^{\infty}(-z)\delta(-v)U(-z)\,p_{\dot{X}(t),\ddot{X}(t)}(v,z)\,dv\,dz}$$

or

$$F_{P(t)}(u) = \frac{\int_{-\infty}^{0}\int_{-\infty}^{u}|z|\,p_{X(t),\dot{X}(t),\ddot{X}(t)}(w,0,z)\,dw\,dz}{\int_{-\infty}^{0}|z|\,p_{\dot{X}(t),\ddot{X}(t)}(0,z)\,dz} \tag{6.11}$$

Taking a derivative with respect to u now gives the probability density function for the peak distribution as

$$p_{P(t)}(u) = \frac{\int_{-\infty}^{0}|z|\,p_{X(t),\dot{X}(t),\ddot{X}(t)}(u,0,z)\,dz}{\int_{-\infty}^{0}|z|\,p_{\dot{X}(t),\ddot{X}(t)}(0,z)\,dz} \tag{6.12}$$

Either Eq. 6.11 or 6.12 describes the probability distribution of any peak which occurs within the vicinity of time t. The probability that the peak is within any given interval can be found directly from Eq. 6.11 or from integration of Eq. 6.12, and Eq. 6.12 is also convenient for evaluating other quantities such as the mean value

$$\mu_P(t) \equiv E[P(t)] = \int_{-\infty}^{\infty} u\,p_{P(t)}(u)\,du$$

the mean squared value

$$E[P^2(t)] = \int_{-\infty}^{\infty} u^2\,p_{P(t)}(u)\,du$$

the variance, $\sigma_P^2(t) = E[P^2(t)] - \mu_P^2(t)$, etc. A word of caution about the notation may be in order at this point. We have written the various equations describing $P(t)$ in exactly the same way as we do for a continuously parametered process, but there is no continuously parametered $\{P(t)\}$ process. We presume that it is possible for a peak to occur at any t value, but there may or may not actually be a peak in the vicinity of a particular t. What we have derived is the conditional probability distribution and conditional moments of a peak $P(t)$ in the vicinity of t, given that such a peak exists.

From Eqs. 6.11 and 6.12, we note that in order to find the probability distribution of the peak $P(t)$, one must know the joint probability distribution of $X(t)$, $\dot{X}(t)$, and $\ddot{X}(t)$. This is as expected, since the occurrence of a peak $P(t)$ at level u requires the intersection of the events $X(t) = u$, $\dot{X}(t) = 0$, and $\ddot{X}(t) < 0$. The need for the joint probability distribution of three random variables, though, can make these expressions a little more difficult than most of the expressions we have considered so far. One special case in which the expressions are relatively simple is when the $\{X(t)\}$ process is Gaussian and stationary. In this case, we can note (just as we did in Section 5.6) that $\dot{X}(t)$ is independent of the pair $[X(t), \ddot{X}(t)]$. Thus, the only parameters in the joint distribution of the three Gaussian random variables are the three standard deviations and the correlation coefficient between $X(t)$ and $\ddot{X}(t)$. We found in Eq. 4.48, though, that this correlation coefficient is exactly the negative of the α_2 bandwidth parameter. Thus, we see that α_2, in addition to its other interpretations, is a parameter governing the distribution of the peaks of $\{X(t)\}$. The details of the peak distribution for a stationary Gaussian process are worked out in Example 6.9.

****

Example 6.9: Find the cumulative distribution function and the probability density function for the peaks of a stationary Gaussian process $\{X(t)\}$.

Since $\dot{X}(t)$ is independent of $[X(t), \ddot{X}(t)]$, we can factor $p_{\dot{X}(t)}(0)$ out of both the numerator and the denominator of Eq. 6.12, giving

$$p_P(u) = \frac{\int_{-\infty}^0 |z| p_{X,\ddot{X}}(u,z)\,dz}{\int_{-\infty}^0 |z| p_{\ddot{X}}(z)\,dz} = \frac{\sqrt{2\pi}}{\sigma_{\ddot{X}}} \int_{-\infty}^0 |z| p_{X,\ddot{X}}(u,z)\,dz$$

Using a conditional probability density function, this can be rewritten as

$$p_P(u) = \frac{\sqrt{2\pi}}{\sigma_{\ddot{X}}} p_X(u) \int_{-\infty}^0 |z| p_{\ddot{X}}(z|X=u)\,dz$$

We know that the conditional probability distribution is also Gaussian, so that we can write it as

$$p_{\ddot{X}}[z|X=u] = \frac{1}{\sqrt{2\pi}\,\sigma_*} \exp\left[-\frac{1}{2}\left(\frac{z-\mu_*}{\sigma_*}\right)^2\right]$$

in which the conditional mean and standard deviation of $\ddot{X}(t)$ are

$$\mu_* \equiv E[\ddot{X}(t)|X(t)=u] = \rho_{X(t),\ddot{X}(t)}\frac{\sigma_{\ddot{X}}}{\sigma_X}(u-\mu_X) = -\alpha_2\frac{\sigma_{\ddot{X}}}{\sigma_X}(u-\mu_X)$$

and

$$\sigma_* = \sigma_{\ddot{X}}\sqrt{1-\rho_{X(t),\ddot{X}(t)}^2} = \sigma_{\ddot{X}}\sqrt{1-\alpha_2^2}$$

Substitution of this Gaussian form gives

$$p_P(u) = \frac{-1}{\sigma_{\ddot{X}}\sigma_*} p_X(u) \int_{-\infty}^0 z\exp\left[-\frac{1}{2}\left(\frac{z-\mu_*}{\sigma_*}\right)^2\right]dz = p_X(u)\left[\frac{\sigma_*}{\sigma_{\ddot{X}}}\exp\left(-\frac{\mu_*^2}{2\sigma_*^2}\right) - \sqrt{2\pi}\frac{\mu_*}{\sigma_{\ddot{X}}}\Phi\left(-\frac{\mu_*}{\sigma_*}\right)\right]$$

or

$$p_P(u) = p_X(u) \left[\sqrt{1-\alpha_2^2} \exp\left(-\frac{\alpha_2^2 (u-\mu_X)^2}{2(1-\alpha_2^2)\sigma_X^2} \right) + \sqrt{2\pi} \frac{\alpha_2 (u-\mu_X)}{\sigma_X} \Phi\left(\frac{\alpha_2 (u-\mu_X)}{\sqrt{1-\alpha_2^2}\,\sigma_X} \right) \right]$$

Finally, substituting the Gaussian form for $p_X(u)$ gives the probability density function as

$$p_P(u) = \frac{\sqrt{1-\alpha_2^2}}{\sqrt{2\pi}\,\sigma_X} \exp\left(-\frac{(u-\mu_X)^2}{2(1-\alpha_2^2)\sigma_X^2} \right) + \frac{\alpha_2 (u-\mu_X)}{\sigma_X^2} \exp\left(-\frac{(u-\mu_X)^2}{2\sigma_X^2} \right) \Phi\left(\frac{\alpha_2 (u-\mu_X)}{\sqrt{1-\alpha_2^2}\,\sigma_X} \right)$$

The corresponding cumulative distribution function can be written as

$$F_P(u) = \Phi\left(\frac{u-\mu_X}{\sqrt{1-\alpha_2^2}\,\sigma_X} \right) - \alpha_2 \exp\left(-\frac{(u-\mu_X)^2}{2\sigma_X^2} \right) \Phi\left(\frac{\alpha_2 (u-\mu_X)}{\sqrt{1-\alpha_2^2}\,\sigma_X} \right)$$

These formulas are commonly referred to as the S. O. Rice distribution, in recognition of Rice's pioneering work on this problem in 1945.

The limiting forms of this distribution for $\alpha_2 = 1$ and $\alpha_2 = 0$ yield interesting results regarding the peak distribution. For the narrowband situation with α_2 approaching unity, we see that some of the arguments in $p_P(u)$ and $F_P(u)$ tend to infinity. For the $\Phi(\cdot)$ function we must take proper account of the sign of the infinite argument, since $\Phi(\infty) = 1$ and $\Phi(-\infty) = 0$. Thus, we obtain quite different results for $u > \mu_X$ than we do for $u < \mu_X$. For $\alpha_2 = 1$, we obtain

$$p_P(u) = \frac{(u-\mu_X)}{\sigma_X^2} \exp\left(-\frac{(u-\mu_X)^2}{2\sigma_X^2} \right) U(u-\mu_X)$$

and

$$F_P(u) = \left[1 - \exp\left(-\frac{(u-\mu_X)^2}{2\sigma_X^2} \right) \right] U(u-\mu_X)$$

In the special case when $\mu_X = 0$, this is exactly the Rayleigh distribution which in Section 5.3 we found to describe the amplitude of the Gaussian process. When $\mu_X \neq 0$, we see that the peak distribution has the same shape as the Rayleigh amplitude distribution, but it is shifted to make μ_X be the smallest possible peak value. The agreement of the peak distribution and the amplitude distribution of the limiting narrowband process agrees with our prior observations that a narrowband process can be considered a harmonic function with slowly varying amplitude and phase. Since the narrowband amplitude varies slowly we can say that each peak of the narrowband process is equal to the amplitude of the process at that instant of time, so it is not surprising that the two quantities have the same probability distribution.

For the opposite extreme situation with $\alpha_2 = 0$, the probability distribution of the peaks becomes

$$p_P(u) = \frac{1}{\sqrt{2\pi}\,\sigma_X} \exp\left(-\frac{(u-\mu_X)^2}{2\sigma_X^2}\right)$$

and

$$F_P(u) = \Phi\left(\frac{u-\mu_X}{\sigma_X}\right)$$

which is simply the Gaussian distribution of $X(t)$. In this broadband situation, we find that the distribution of peaks is the same as the distribution of the process itself. This may seem to be a surprising result, but it is consistent with the result we have previously obtained for the rate of occurrence of peaks. In particular, it is consistent with the fact found in Example 6.7, that α_2 is the same as the irregularity factor for a Gaussian process.

Thus, if $\alpha_2 = 0$ and the $\{X(t)\}$ process has finite crossing rates, then the rate of occurrence of peaks is infinite, as was shown in a particular case in Example 6.8. However, if the rate of occurrence of peaks is infinite, then it is reasonable to think that there may be peaks everywhere along the process, which should be expected to give the distribution of peaks to be the same as the distribution of $X(t)$. The sketch shows the probability density function for peaks for several values of α_2.

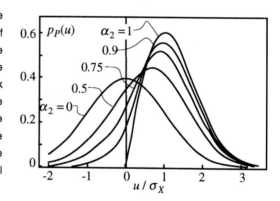

****

Example 6.10: Find the probability that any peak of a stationary process $\{X(t)\}$ is below the level μ_X, provided that the distribution of $\{X(t)\}$ is symmetric about the level μ_X.

Simply substituting $u = \mu_X$ in Eq. 6.11 gives an expression for this probability, but we can also present the information in a somewhat different way. In particular, based on consideration of any continuous time history, we can say that any valley of $X(t)$ below the level μ_X is followed either by a peak below μ_X or by an upcrossing of the level μ_X. Thus, the rates of occurrences of these three events must satisfy the following relationship

$$v_V(X < \mu_X) = v_P(X < \mu_X) + v_X^+(\mu_X)$$

Since $v_P(X < \mu_X) = v_P P[P(t) < \mu_X]$ and $v_V(X < \mu_X) = v_V P[V(t) < \mu_X]$, in which $V(t)$ denotes a valley of $X(t)$, and $v_V = v_P$ we can divide by v_P to obtain

$$P[V(t) < \mu_X] = P[P(t) < \mu_X] + IF$$

The hypothesized symmetry of the distribution of $\{X(t)\}$ now gives $P[V(t) < \mu_X] = P[P(t) > \mu_X] = 1 - P[P(t) < \mu_X]$, so that

$$P[P(t) < \mu_X] \equiv F_P(\mu_X) = \frac{1 - IF}{2}$$

Somewhat surprisingly, it is possible to obtain this one point on the cumulative distribution function for $P(t)$ from knowledge only of the occurrence rates for crossings and peaks.
****

The distribution of peaks of a narrowband process is sometimes approximated by a function which can be obtained from knowledge only of the crossing rates of $X(t)$. The rationale is that in the narrowband case we can ignore the possibility of peaks below μ_X or valleys above μ_X. With this simplification, we can say that a peak occurs within the interval $[u, u + \Delta u]$ if and only if there is an upcrossing of the level u which is not followed by an upcrossing of the level $u + \Delta u$. This approximation then implies that the expected number of peaks within an interval $[u, u + \Delta u]$ is the difference between the number of upcrossings of the level u and the number of upcrossing of the level $u + \Delta u$, and the expected rate of occurrence of peaks in the interval is the difference between the upcrossing rates:

$$v_P(u \le X \le u + \Delta u) \approx v_X^+(u) - v_X^+(u + \Delta u)$$

Saying that $p_{P(t)}(u) \Delta u$ is approximately $v_P(u \le X \le u + \Delta u) / v_P$, then taking the limit as Δu tends to zero gives

$$p_{P(t)}(u) \approx \frac{-1}{v_P} \frac{dv_X^+(u)}{du} U(u - \mu_X)$$

Integrating this equation over the set of possible values, though, does not generally give unity, so the function cannot truly be a probability density function. This problem is easily remedied, however, by noting that our choice to neglect peaks below μ_X and valleys above μ_X is consistent with saying that the rate of peak occurrences is the same as the rate of upcrossings of the level μ_X. This gives

$$p_{P(t)}(u) = \frac{-1}{v_X^+(\mu_X)} \frac{dv_X^+(u)}{du} U(u - \mu_X) \tag{6.13}$$

which is properly normalized for a probability density function.

For the special case of a stationary Gaussian process, one can use $v_X^+(u)$ from Example 6.1 to evaluate the $p_P(u)$ approximation in Eq. 6.13. The result is exactly the Rayleigh distribution which was shown in Example 6.9 to be the true answer in the limiting case with $\alpha_2 = 1$. Thus, the approximation does give the correct answer for the limiting narrowband process, but we can also see from the plot in Example 6.9 that the approximation can be in significant error for processes which still seem to be quite narrowband, such as when $\alpha_2 = 0.9$. The most notable error of the approximation may be the neglect of the peaks below the level μ_X, which was shown in Example 6.10 to be $(1 - IF)/2$ for a symmetric distribution. Thus, for a Gaussian process with $\alpha_2 = 0.9$, one should have 5% of the peaks occurring below the level μ_X, and these are neglected in the approximation. For a non-Gaussian process, as well, we can anticipate that Eq. 6.13 will be asymptotically correct for α_2 approaching unity, but may be inaccurate for other situations.

6.5 Extreme Value Distribution and Poisson Approximation

A simple way to formulate the extreme value problem is to define a new stochastic process $\{Y(t)\}$ which is the extreme value of $\{X(t)\}$ during the past. Specifically, we let

$$Y(t) = \max_{0 \le s \le t} X(s) \tag{6.14}$$

The extreme value distribution for $\{X(t)\}$ is then simply the distribution of the $Y(t)$ random variable. Note that even for a stationary $\{X(t)\}$ process, one must expect that $\{Y(t)\}$ will be nonstationary, since larger and larger values of $X(t)$ will generally occur if we extend the period of our observation. Letting $L_X(u,t)$ denote the cumulative distribution function of $Y(t)$ gives

$$L_X(u,t) = F_{Y(t)}(u) \equiv P[Y(t) \le u] \equiv P[X(s) \le u : 0 \le s \le t] \tag{6.15}$$

in which the notation on the final term means that the inequality $X(s) \le u$ holds for all the given s values. This $L_X(u,t)$ is sometimes called the probability of survival, which is certainly appropriate if u denotes a critical value for $\{X(t)\}$ corresponding to some failure mode of the system. The probability density function for the extreme value, of course, is simply the derivative

$$p_{Y(t)}(u) = \frac{\partial}{\partial u} L_X(u,t) \tag{6.16}$$

and from this information one can also calculate the mean, variance, etc. of the extreme value.

An alternative problem which is almost equivalent to extreme value analysis involves the random quantity called first-passage time. Let $T_X(u)$ denote the first time (after time zero) at which $X(t)$ has an upcrossing of the level u. That is, $X[T_X(u)] = u$, $\dot{X}[T_X(u)] > 0$, and there has been no upcrossing in the interval $0 \le t < T_X(u)$. For any given u value, this quantity $T_X(u)$ is a random variable, and one can consider the family of all such variables to constitute a form of stochastic process $\{T_X(u)\}$, although the index set u is not time or frequency in this instance, as it has been in our other stochastic processes.

The relationship between the first-passage time and the extreme value distribution becomes evident when we consider the event $\{X(s) \le u : 0 \le s \le t\}$ which appears in Eq. 6.15. We see that this event can also be written as $\{X(0) \le u, T_X(u) \ge t\}$, since $X(s)$ can be less than u throughout the time interval only if it starts below u and does not have an upcrossing during the time interval. Taking probabilities then gives

$$L_X(u,t) = P[X(0) \leq u] P[T_X(u) \geq t \mid X(0) \leq u] = L_X(u,0) P[T_X(u) \geq t \mid X(0) \leq u] \qquad (6.17)$$

and the final term is related to a conditional form of the cumulative distribution of $T_X(u)$. In many practical problems, we can further simplify the relationship by completely neglecting the conditioning on this final term. For example, in some problems of interest we may have $P[X(0) \leq u] = 1$, such as when the system is known to start at $X(0) = 0$, and conditioning by a sure event can always be neglected. In other situations we may not have any specific information that the distribution of $T_X(u)$ is independent of $X(0)$, but we anticipate that the effect of the conditioning will be significant only for small values of time.

Taking the derivative of Eq. 6.17 with respect to t gives

$$p_{T_X}[t \mid X(0) \leq u] = \frac{-1}{L_X(u,0)} \frac{\partial}{\partial t} L_X(u,t) \qquad (6.18)$$

which, along with Eq. 6.16, shows that the $L_X(u,t)$ function governs the conditional distribution of the first-passage time and the distribution of the extreme value in very similar ways. The primary difference is that one probability density function involves a partial derivative with respect to u while the other involves a partial derivative with respect to t.[2]

It is often convenient to write the probability of survival as some sort of exponential function of time. In particular, we will use the form

$$L_X(u,t) = L_X(u,0) \exp\left(-\int_0^t \eta_X(u,s) ds\right) \qquad (6.19)$$

Clearly it is always possible to find a function $\eta_X(u,s)$ such that one can write such an expression, but it is not immediately obvious why this is desirable. However, we will now derive an interpretation of $\eta_X(u,s)$ which will motivate some useful approximations of the problem. First, we note that the derivative of Eq. 6.19 gives

---

[2] The reader is cautioned that this close relationship between the extreme value problem and the first-passage problem is not always mentioned in the literature, with some authors using only one terminology and some using the other. It should also be mentioned that we are omitting one advanced method of analysis which specifically relates to the first-passage formulation of the problem. In particular, the moments of $T_X(u)$ for some simple systems can be found by recursively solving a series of differential equations, called the generalized Pontryagin equations (Lin and Cai 1995).

$$\frac{\partial}{\partial t} L_X(u,t) = -L_X(u,t)\, \eta_X(u,t)$$

which can be solved for $\eta_X(u,t)$ as

$$\eta_X(u,t) = \frac{-1}{L_X(u,t)}\frac{\partial}{\partial t} L_X(u,t) = \lim_{\Delta t \to 0}\frac{1}{\Delta t}\frac{L_X(u,t)-L_X(u,t+\Delta t)}{L_X(u,t)}$$

This can then be written in terms of probabilities of first passage as

$$\eta_X(u,t) = \lim_{\Delta t \to 0}\frac{1}{\Delta t}\frac{P[t \le T_X(u) < t+\Delta t \,|\, X(0) \le u]}{P[T_X(u) \ge t \,|\, X(0) \le u]} \tag{6.20}$$

Now we note that the numerator of Eq. 6.20 relates to the event of the first upcrossing being in the specified time interval. This event, though, is the intersection of the event of there being no upcrossing prior to t and the event of there being an upcrossing in the interval:

$$\{t \le T_X(u) < t+\Delta t\} = \{T_X(u) \ge t\} \cap \{\text{upcrossing in } [t, t+\Delta t]\}$$

Since the denominator of Eq. 6.20 is the probability of one of these two events, we can see that the ratio is the conditional probability

$$\eta_X(u,t) = \lim_{\Delta t \to 0}\frac{1}{\Delta t} P\big(\text{upcrossing in } [t,t+\Delta t]\,|\, X(0) \le u, \text{no upcrossing prior to } t\big)$$

or

$$\eta_X(u,t) = \lim_{\Delta t \to 0}\frac{E\big(\text{number of upcrossings in } [t,t+\Delta t]\,|\, X(0) \le u, \text{no upcrossing prior to } t\big)}{\Delta t}$$

$$\tag{6.21}$$

Thus, we see that $\eta_X(u,t)$ can be regarded as an occurrence rate. It is the conditional rate of upcrossings of the level u, given the initial condition and the fact that there has been no prior upcrossing. If exceedance of the level u is considered to correspond to a "failure" of the system, then $\eta_X(u,t)$ is what is called the hazard function in reliability theory, but our interest in the function is not limited to such an interpretation.

Unfortunately, it is not easy to calculate the conditional rate of upcrossings $\eta_X(u,t)$. In fact, we have no rigorous relationship between $\eta_X(u,t)$ and unconditional probability density functions for $X(t)$ and $\dot{X}(t)$ random variables.[3] One can use a conditional probability density function to write an expression for the conditional crossing rate $\eta_X(u,t)$ in the same way as we previously did for the unconditional crossing rate $v_X(u,t)$. In particular,

---

[3] The inclusion-exclusion series in Section 6.6 does rigorously describe the effect of the conditioning by the event of no prior upcrossings, but it ignores the initial condition at time zero.

$$\eta_X(u,t) = \int_0^\infty v\, p_{X(t),\dot{X}(t)}(u,v|\, X(0) \le u,\ \text{no upcrossings in } [0,t])\, dv$$

but this expression is not very useful for calculating values of $\eta_X(u,t)$, since the necessary conditional probability density function is generally unknown. Nonetheless, we will use it to gain some general information about the behavior of $\eta_X(u,t)$.

First we note that most physical processes have only a finite memory, in the sense that $X(t)$ and $X(t - \tau)$ can generally be considered to be independent if $\tau > T$ for some large T value. In this case, one can argue that

$$p_{X(t),\dot{X}(t)}(u,v|X(0) \le u,\ \text{no upcrossings in } [0,t]) \approx p_{X(t),\dot{X}(t)}(u,v|\text{no upcrossings in } [t - T,t])$$

for $t > T$. Some conditioning events are ignored in the second form, but they occurred prior to time $t - T$ and are out of the memory of $X(t)$. If $\{X(t)\}$ is a stationary process, though, this new form of conditional probability density is stationary, since it is independent of the choice of the origin of the time axis, except for the restriction that $t > T$. This means that $\eta_X(u,t)$ tends asymptotically to a stationary value $\eta_X(u)$ as $p_{X(t),\dot{X}(t)}(u,v|X(0) \le u,\ \text{no upcrossings in } [0,t])$ tends to $p_{X(t),\dot{X}(t)}(u,v|\text{no upcrossings in } [t - T,t])$. This asymptotic behavior of $\eta_X(u,t)$, then, implies that one can approximate $L_X(u,t)$ as

$$L_X(u,t) \approx L_0\, e^{-\eta_X(u)t} \qquad \text{for large } t \tag{6.22}$$

This same limiting behavior for large t will apply if $\{X(t)\}$ is a nonstationary process which has finite memory and becomes stationary with the passage of time.

The value L_0 in Eq. 6.22 is related to the behavior of $\eta_X(u,t)$ for small values of t. One extreme situation is when $\{X(t)\}$ is a nonstationary "zero-start" process with $X(0) = 0$ and $\dot{X}(0) = 0$. This is the case, for example, if $\{X(t)\}$ represents the response of an oscillator which starts from a condition of rest. In this case, we can see that $\eta_X(u,0) = 0$, and $\eta_X(u,t)$ increases from this initial value as $\sigma_X(t)$ and $\sigma_{\dot{X}}(t)$ grow, especially if $\mu_X(t)$ also grows. Another limiting condition is the "stationary-start" problem in which $\{X(t)\}$ is stationary for all time t. This stationary initial condition will usually give $\eta_X(u,t)$ for small time as being greater than the stationary value $\eta_X(u)$, because of the unconditional $p_{X(t),\dot{X}(t)}(u,v)$ being larger than the stationary conditional probability density $p_{X(t),\dot{X}(t)}(u,v|\text{no upcrossings in } [t - T,t])$. Thus, for the zero-start problem $\eta_X(u,t)$ will generally grow from zero toward its stationary value, and for the stationary-start problem it will generally decay toward the stationary value. This behavior gives the multiplier L_0 in Eq. 6.22 as being greater than unity for the zero-start problem, and as being less than $L_X(u,0) = P[X(0) \le u]$ for the stationary-start problem, as illustrated in Fig. 6.4.

Since it is not easy to calculate the conditional rate of upcrossings $\eta_X(u,t)$, we must use

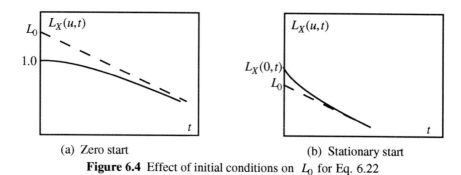

(a) Zero start   (b) Stationary start

Figure 6.4 Effect of initial conditions on L_0 for Eq. 6.22

approximations in solving practical problems.   We will present the most commonly used approximation here, and discuss methods giving somewhat better estimates in the following section.

Since $\eta_X(u,t)$ is a conditional upcrossing rate of the level u , we obviously expect it to be related to the unconditional upcrossing rate $v_X^+(u,t)$ which we studied in Section 6.2. In fact, the most widely used approximation of the extreme distribution problem results from simply neglecting the conditioning event in $\eta_X(u,t)$, and replacing it with $v_X^+(u,t)$:

$$\eta_X(u,t) \approx v_X^+(u,t) \tag{6.23}$$

giving

$$L_X(u,t) \approx L_X(u,0)\exp\left(-\int_0^t v_X^+(u,s)ds\right) \tag{6.24}$$

and, if $\{X(t)\}$ is a stationary process, this becomes

$$L_X(u,t) \approx L_X(u,0)\exp[-v_X^+(u)t] \quad \text{(for stationary process)} \tag{6.25}$$

which is exactly of the form of Eq. 6.22. The approximation of Eqs. 6.23, 6.24, and 6.25 is commonly called the Poisson approximation of the extreme value, or first-passage, problem. This name comes from the fact that if the crossing rate is independent of the past history of the process, then the lengths of time between upcrossings will be independent, and this makes the integer valued process which counts the number of upcrossings be a Poisson process.

From Eqs. 6.18 and 6.25, one can see that use of the Poisson approximation gives exactly an exponential distribution for the first-passage time of a stationary $\{X(t)\}$ process:

$$p_{T_X}(t) = v_X^+(u)\exp[-v_X^+(u)t] \quad \text{(for stationary process)} \tag{6.26}$$

The mean of this exponentially distributed random variable is

$$E[T_X(u)] = [v_X^+(u)]^{-1} \quad \text{(for stationary process)} \tag{6.27}$$

and the standard deviation also has this same value. One of the characteristics of the Poisson process is that the time between occurrences is exponentially distributed, and the Poisson approximation of the extreme value problem gives the same exponential distribution for the first-passage time as for the time between upcrossings for a stationary $\{X(t)\}$ process. (Example 2.13 gives more information on the Poisson process, and confirms this exponential value for the probability of no occurrences within an interval.)

The situation in which the Poisson approximation is most seriously in error is when the $\{X(t)\}$ process is very narrowband. In that situation, an upcrossing of level u at time t is very likely to be associated with another upcrossing approximately one period later, due to the slowly varying amplitude of $\{X(t)\}$. Such a relationship between the upcrossing times is inconsistent with the Poisson approximation that the times between upcrossings are independent. On the other hand, when u is very large it is found that the independence assumption seems to be better. It is difficult to find general results which apply to all $\{X(t)\}$ processes, but for Gaussian processes it has been demonstrated that $\eta_X(u,t)$ does tend asymptotically to $v_X^+(u,t)$ as u tends to infinity.[4] Thus, the Poisson approximation is best when the $\{X(t)\}$ process is very broadband and/or the level u is very large. In some other situations, it may be significantly in error.

From the form of Eq. 6.19 it is clear that an overestimation of $\eta_X(u,t)$ will result in an underestimation of $L_X(u,t)$. An error of this type is usually considered to be conservative, since it overestimates the probability of failure due to large excursions. Furthermore, it is usually assumed that $\eta_X(u,t) \le v_X^+(u,t)$, so that the Poisson approximation will underestimate $L_X(u,t)$. There are situations, though, in which this is not true, so some caution is required. In particular, we can see that if the level u is so small that $P[X(t) < u]$ is very small, but we are given the initial condition that $X(0) < u$, then it is very likely that $X(t)$ will quickly have an upcrossing of u. Mathematically this requires that $L_X(u,t)$ approach zero as $u \to -\infty$, for any finite t value. This is assured for any choice of $L_X(u,0)$, though, only if $\eta_X(u,t)$ tends to infinity for $u \to -\infty$, at least for $t \approx 0$. This unbounded $\eta_X(u,t)$ behavior surely violates the usual assumption that $\eta_X(u,t) \le v_X^+(u,t)$. On the other hand, the usual assumption does seem to be justified for the larger u values which are usually of primary importance.

In some problems one is concerned with the probability of large excursions of $X(t)$ in either the positive or negative direction, whereas all the development up until now has been concerned only with the probability that $X(t)$ remains below $+u$. On the other hand, the event of $X(t)$ remaining between $-u$ and $+u$ is exactly the same as the event of $|X(t)|$ remaining below the level u, and this allows us to apply the results to the new problem. Thus, we can write

---

[4] Nigam (1983) attributes this result to Cramer in 1966.

$$L_{|X|}(u,t) = L_{|X|}(u,0)\exp\left(-\int_0^t \eta_{|X|}(u,s)\,ds\right)$$ (6.28)

for the new probability. Particularly in the study of first-passage time, the terms "double-barrier problem" and "single-barrier problem" are often used to distinguish between the consideration of upcrossings by $|X(t)|$ and $X(t)$, respectively. Of course, one can also consider double-barrier problems in which the levels of interest are not symmetric, or even problems in which the constant level u is replaced by a given function $u(t)$. The Poisson approximation of the symmetric double-barrier problem of Eq. 6.28 is simply to replace $\eta_{|X|}(u,s)$ with $v_{|X|}^+(u,s) = v_X^+(u,s) + v_X^-(-u,s)$. If the distribution of $X(t)$ and $\dot{X}(t)$ is symmetric, this then gives $\eta_{|X|}(u,s) \approx 2\ v_X^+(u,s)$, so that the decay of L with increasing t is twice as fast as for the single-barrier problem.

6.6 Improved Estimates of the Extreme Value Distribution

First we will consider a modification of the usual Poisson estimate to account, in an approximate way, for the effect of the given information that $X(0) \le u$. This modification will improve our estimate of $\eta_X(u,t)$ for small values of u, but will have little effect for large u values. In particular, the modification will address the difficulty noted in Section 6.5 for situations in which u is so low that the expected time between upcrossings may be large, but this does not give a good estimate of the time until the first upcrossing. Recall that the usual Poisson assumption gives the probability distribution of the first-passage time $T_X(u)$ for a stationary process as being the same as that of a random variable T_{CR} representing the time interval between successive upcrossings of the level u. However, the time interval T_{CR} is composed of two segments: the time T_1 between the upcrossing and the following downcrossing, and the time T_2 between the downcrossing and the next upcrossing. Of these two segments, only T_2 is spent below the level u. Since $T_X(u)$ is also a time interval spent below u, it seems more appropriate to approximate it by T_2 instead of T_{CR}. Obviously, this will reduce our estimate of the time until first passage and give more conservative estimates of failure probabilities.

If we consider T_2 to be governed by a Poisson process, then the only parameter needed to describe its distribution is its arrival rate. Furthermore, we know that the arrival rate is the inverse of $E(T_2)$, as in Eq. 6.26, and since T_2 represents the portion of T_{CR} spent below the level u, we can say that $E(T_2) = E(T_{CR})P(X < u)$. Unfortunately, we do not have an exact result for $E(T_{CR})$ in this situation. If T_{CR} were governed by a Poisson process, then we would know that $E(T_{CR})$ was the same as the $[v_X^+(u)]^{-1}$ value, but the assumption that T_2 is governed by a Poisson process precludes the possibility that T_{CR} also has this Poisson property. Nonetheless, we will use this value as an approximation of $E(T_{CR})$ in order to obtain $E(T_2) \approx [v_X^+(u)]^{-1}P(X < u)$. Approximating $\eta_X(u)$ by the arrival rate for T_2 then gives $\eta_X(u) = v_X^+(u)/F_X(u)$ for a stationary process, and a consistent modification for a nonstationary $\{X(t)\}$ is

$$\eta_X(u,t) \approx \frac{v_X^+(u,t)}{F_{X(t)}(u)} \tag{6.29}$$

Clearly this approximation is almost identical to Eq. 6.23 for large values of u, since $F_X(u)$ is almost unity in that situation. For low values of u, on the other hand, it gives $\eta_X(u)$ as having a very large value and assures that $L_X(u,t)$ tends to zero as u tends to negative infinity, as desired.[5]

Next we will consider modifications to the Poisson approximation based directly on the narrowband limitation we previously noted. In particular, for a narrowband process with a slowly varying amplitude $\{A(t)\}$, a single upcrossing of the level u by $\mu_X(t) + A(t)$ is likely to be associated with several (almost uniformly spaced) upcrossings of u by $X(t)$. Since this is inconsistent with a Poisson assumption that the times between upcrossings are independent, a better approximation is desirable. The narrowband approximations that we will consider are based on consideration of the behavior of the $\{A(t)\}$ process.

Note that the developments up until now have placed little restriction on the probability distribution of $\{X(t)\}$. In particular, there has been no restriction that it be mean zero. Now, though, we are going to consider models based on the behavior of $\{A(t)\}$, and this amplitude is associated with the deviation of $X(t)$ from its mean value. Thus, it is more convenient to switch to a mean zero random process, and henceforth we will assume that $\mu_X(t) = 0$. Some information about a more general $\{Z(t)\}$ process can be obtained by writing it as $Z(t) = X(t) + \mu_Z(t)$. If $\{Z(t)\}$ is mean-value stationary, then it is simple to account for the constant μ_Z. In particular, we can say that the extreme of $Z(t)$ is simply μ_Z plus the extreme of $X(t)$. More care must be used if $\mu_Z(t)$ is not a constant, though, since in that situation we have $P[Z(s) \le u: 0 \le s \le t] = P[X(s) \le u - \mu_Z(s): 0 \le s \le t]$ which amounts to a problem with a variable barrier level. We will not explicitly address this more complicated problem.

The simplest extreme value approximation based on the amplitude process amounts to assuming that the extreme value of the mean zero $\{X(t)\}$ process is the same as that of $\{A(t)\}$. giving

$$L_X(u,t) \approx L_A(u,t) = L_A(0,t)\exp\left(-\int_0^t \eta_A(u,s)ds\right) \tag{6.30}$$

Since $X(t) \le A(t)$, we know that $L_X(u,t) \ge L_A(u,t)$ for all u and t. Thus, using Eq. 6.30 will always be conservative in the sense of overestimating the probability of large extreme values. Of course we still have the problem of determining $L_A(u,t)$, but it is generally assumed that the Poisson assumption is much better for $\{A(t)\}$ than for $\{X(t)\}$, and using it gives

$$\eta_A(u,t) \approx v_A^+(u,t) \tag{6.31}$$

---

[5] The approximation of Eq. 6.29 was obtained by Ditlevsen (1986) by a somewhat different method of reasoning.

The integration in Eq. 6.30, of course, is almost trivial if $v_A^+(u)$ is stationary.

It is also reasonable to apply an initial condition correction to $v_A^+(u,t)$, similar to what we did in obtaining Eq. 6.29. That is, if u is very small, then the initial condition of $A(0) < u$ implies that the time until first passage by $A(t)$ is likely to be short. By exactly the same reasoning as was used in obtaining Eq. 6.29, we say that

$$\eta_A(u,t) \approx \frac{v_A^+(u,t)}{F_{A(t)}(u)} \qquad\qquad (6.32)$$

This correction seems to be much more significant in this case than it was in Eq. 6.29, though. In particular, the $F_{A(t)}(u)$ term approaches zero as u approaches zero. Thus, the correction in Eq. 6.32 is evident at values of u which may be of practical importance.

It should be noted that use of the Poisson amplitude crossings model of Eqs. 6.30 - 6.32 is not exactly the same as introducing a new estimate of $\eta_X(u,t)$ in the general extreme value formula of Eq. 6.19. In particular, the multiplier of $L_X(u,0) \equiv P[X(0) \leq u]$ in Eq. 6.19 has been replaced by $P[A(0) \leq u]$. We will now investigate the implication of this difference. Consider an ensemble of possible time histories of $\{X(t)\}$. The term $P[X(0) \leq u]$ represents the fraction of the ensemble that is below the level u at time $t = 0$, excluding only samples with $X(0) > u$, while $P[A(0) \leq u]$ also excludes samples with $A(0) > u$, even if $X(0) \leq u$. For a narrowband process, though, we can see that it is very likely that the samples with $A(0) > u$ and $X(0) \leq u$ will have an upcrossing of u during the first cycle of the process. That is, if the narrowband process has average frequency ω_c, then it is very likely that an $X(t)$ sample with $A(0) > u$ will exceed u prior to time $2\pi/\omega_c$. Using an initial condition of $P[A(0) \leq u]$ on the probability distribution of the extreme values is equivalent to counting these time histories which are particularly likely to cross u during the first cycle as having crossed at time zero. This shift should be significant only if one is particularly interested in the details of the first cycle. The true conditional rate of upcrossings $\eta_X(u,t)$ is particularly large during the first cycle of a narrowband process, due to these samples which started with $A(0) > u$ having upcrossings, then it settles down to a lower value. By using an initial condition of $P[A(0) \leq u]$, one can obtain an approximate extreme value distribution which is good for $t \geq 2\pi/\omega_c$, without explicit consideration of the high early value of $\eta_X(u,t)$. Since the Poisson approximation ignores this high early value of the upcrossing rate, it seems appropriate that it should be used with an initial condition of $P[A(0) \leq u]$ if $\{X(t)\}$ is narrowband.

****

Example 6.11: Compare the $\eta_X(u)$ values from Eqs. 6.23, 6.29, 6.31, and 6.32 for a stationary, mean zero, Gaussian, narrowband $\{X(t)\}$ process with $\alpha_1 = 0.995$. (One particular process with this value of α_1 is the response to white noise of a SDF oscillator with approximately 0.8% of critical damping.)

The basic upcrossing rate $v_X^+(u)$ for this process has been found in Example 6.1 to be

$$v_X^+(u) = \frac{\omega_a}{2\pi} \exp\left(\frac{-u^2}{2\sigma_X^2}\right)$$

in which $\omega_a = \sigma_{\dot{X}} / \sigma_X$. This, then, is the approximation of $\eta_X(u)$ for Eq. 6.23. Since this is the most commonly used approximation, we will present our comparisons as the ratio of $\eta_X(u) / v_X^+(u)$ for each of the other three approximations.[6] The $\eta_X(u) / v_X^+(u)$ ratio for Eq. 6.29, of course, is simply $[F_X(u)]^{-1}$, which is $[\Phi_X(u / \sigma_X)]^{-1}$ for a mean zero Gaussian process.

The amplitude upcrossing rates for this process have been found in Examples 6.4 and 6.5, respectively, for the Cramer and Leadbetter and the energy-based definitions. They give

$$\frac{\eta_X(u)}{v_X^+(u)} = \frac{v_{A_{CL}}^+(u)}{v_X^+(u)} = \frac{u}{\sigma_X} \sqrt{2\pi} \sqrt{1 - \alpha_1^2} = 0.250 \frac{u}{\sigma_X} \qquad \text{for Cramer and Leadbetter definition}$$

and

$$\frac{\eta_X(u)}{v_X^+(u)} = \frac{v_A^+(u)}{v_X^+(u)} = \frac{u}{\sigma_X} \frac{4}{\sqrt{2\pi}} \sqrt{\frac{1}{\alpha_2^2} - 1} \qquad \text{for energy-based definition}$$

for Eq. 6.31. Since both of these amplitudes have the Rayleigh distribution, the additional term needed to obtain the $\eta_X(u)$ approximation of Eq. 6.32 is

$$F_A(u) = 1 - \exp\left(\frac{-u^2}{2\sigma_X^2}\right)$$

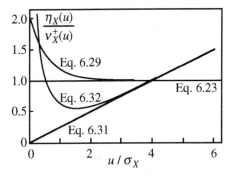

The sketch shows the values of $\eta_X(u) / v_X^+(u)$ versus u / σ_X for Eqs. 6.23 and 6.29, and for Eqs. 6.31 and 6.32 using the Cramer and Leadbetter amplitude.

From the plot we see that the different approximations all give quite different results for small values of u, which may not be of too much practical significance. We also see, though, that the X crossing and the amplitude crossing results behave very differently for the crucial region with large values of u. In particular, the $v_A^+(u) / v_X^+(u)$ ratio grows linearly with u so that the amplitude crossing result eventually becomes more conservative (i.e., predicts larger probabilities of upcrossings) than the original Poisson approximation.

No values are shown in the plot for the energy-based amplitude, since those values cannot be obtained without knowledge of α_2, and there is no unique relationship between α_2 and the information given in this problem. For example, we saw in Example 4.6 that an idealized rectangular autospectral density gave $(1 - \alpha_2) \approx 4(1 - \alpha_1)$, while Example 4.8 showed that $\alpha_2 = 0$ for the response of an SDF oscillator excited by white noise, regardless of the α_1 value. Thus, various results could be obtained for Eqs. 6.31 and 6.32 using the energy-based amplitude, depending on the form of the narrowband autospectral density. The

---

[6]   This normalization of the results has been commonly used since Crandall et al. in 1966.

shape of the plots for Eqs. 6.31 and 6.32 must be the same, though, for the energy-based and Cramer and Leadbetter amplitude definitions.

****

For very large u values, it is generally found that many of the upcrossings of u by $A(t)$ are not accompanied by upcrossings by $X(t)$. This leads to $v_A^+(u,t)$ being much larger than $v_X^+(u,t)$, as was found in Example 6.11 for the special case of a Gaussian process. On the other hand, we know that the conditional crossing rate $\eta_X(u,t)$ tends to $v_X^+(u,t)$ as u becomes very large, so there must be limits to the usefulness of Eqs. 6.31 and 6.32 for large u values. An improved approximation of the extreme value distribution can be obtained by estimating the fraction of the upcrossings by $A(t)$ which are accompanied by upcrossings by $X(t)$. The conditional crossing rate $\eta_X(u,t)$ can then be taken to represent the rate of occurrence of this subset of the amplitude upcrossings. Vanmarcke introduced such a scheme in 1972, and the following paragraph uses some of his approximations in deriving a very similar result.[7]

Using the same notation as we used before for the $\{X(t)\}$ process, let the random variable T_1 denote the time between an upcrossing of u by $A(t)$ and the subsequent downcrossing by $A(t)$. Then T_1 represents the duration of an interval with $A(t) > u$. If T_1 is large then it seems almost certain that $X(t)$ will have an upcrossing of u within the interval, but if T_1 is small then it seems quite likely that no upcrossing by $X(t)$ will occur. Following Vanmarcke, we will approximate this relationship by

$$P[\text{no upcrossing by } X(t)|T_1 = \tau] \approx [1 - v_X^+(0,t)\,\tau]U[1 - v_X^+(0,t)\,\tau] \qquad (6.33)$$

Considering $[v_X^+(0,t)]^{-1}$ to represent the period of an average cycle of the $\{X(t)\}$ process, this approximation amounts to saying that an upcrossing by $X(t)$ is sure if T_1 exceeds the period, and the probability of its occurrence grows linearly with T_1 for T_1 less than the period. Clearly this is a fairly crude approximation, but it should be substantially better than simply assuming that an upcrossing by $X(t)$ occurs in connection with every upcrossing by $A(t)$. In order to calculate the unconditional probability of an upcrossing in the T_1 interval, it is necessary to have a probability distribution for T_1. Consistent with the arrival times in the Poisson process, we will use Vanmarcke's approximation that this is the exponential distribution

$$p_{T_1}(\tau) = \frac{e^{-\tau/E(T_1)}}{E(T_1)}$$

and we will use the approximation that $E(T_1) = E(T_{CR})P[A(t) > u] \approx P[A(t) > u]/v_A^+(u,t)$, with T_{CR} being the time between successive upcrossing of u by $A(t)$. This gives

---

[7] The derivation, though, follows that of Madsen et al. (1986) much more closely than it does that of Vanmarcke. It should also be noted that extensions of the scheme were presented by Corotis et al. in 1972 and by Vanmarcke in 1975.

$$P[\text{upcrossing by } X(t) \text{ during } T_1] \approx \frac{P[A(t) > u] v_X^+(0,t)}{v_A^+(u,t)}\left[1 - \exp\left(\frac{-v_A^+(u,t)}{P[A(t) > u] v_X^+(0,t)}\right)\right]$$

and taking $\eta_X(u,t) \approx v_A^+(u,t) P[\text{upcrossing by } X(t) \text{ during } T_1]$ gives

$$\eta_X(u,t) \approx P[A(t) > u] v_X^+(0,t)\left[1 - \exp\left(\frac{-v_A^+(u,t)}{P[A(t) > u] v_X^+(0,t)}\right)\right] \tag{6.34}$$

Note that as u tends to zero the results of Eq. 6.34 approach those of Eq. 6.31 based on considering each upcrossing by $A(t)$ to correspond to a crossing by $X(t)$. The limiting behavior for large values of u may also be of interest. In this case we find that $P[A(t) > u]$ is very small so that $v_A^+(u,t) \gg P[A(t) > u] v_X^+(0,t)$ and Eq. 6.34 gives $\eta_X(u,t) \approx P[A(t) > u] v_X^+(0,t)$. For the special case of a Gaussian process it can be shown that this is identical to $v_X^+(u,t)$, so that the approximation agrees with the results from the assumption of Poisson crossings by $X(t)$. For a non-Gaussian process, these two results for large u values may not be identical, although they are expected to be quite similar.

As in our other approximations, we can expect to obtain better results for small u values by including the effect of the initial condition. Since this estimate of $\eta_X(u,t)$ is a modified version of the amplitude crossing rate, we will do this by dividing by $P[A(t) < u]$. This gives

$$\eta_X(u,t) \approx \frac{P[A(t) > u] v_X^+(0,t)}{P[A(t) < u]}\left[1 - \exp\left(\frac{-v_A^+(u,t)}{P[A(t) > u] v_X^+(0,t)}\right)\right] \tag{6.35}$$

and it is easily verified that Eq. 6.35 agrees with Eq. 6.32 in the limit for u near zero.

It should be noted that for a general $\{X(t)\}$ process, Eqs. 6.34 and 6.35 are not identical to Vanmarcke's results, and that his derivation uses somewhat more sophisticated assumptions about the behavior of $\{X(t)\}$. For the special case of the Gaussian process, though, it can be shown that $P[A_{CL}(t) > u] v_X^+(0,t) = v_X^+(u,t)$, so that Eq. 6.35 with the Cramer and Leadbetter definition of amplitude does become identical to Vanmarcke's form of

$$\eta_X(u,t) \approx v_X^+(u,t)\frac{1 - \exp\left(\dfrac{-v_{A_{CL}}^+(u,t)}{v_X^+(u,t)}\right)}{1 - \dfrac{v_X^+(u,t)}{v_X^+(0,t)}} \tag{6.36}$$

It can be expected that the two approximations will also give similar results for other processes which do not differ greatly from the Gaussian distribution.

* *

Example 6.12: Compare the $\eta_X(u)$ values from Eqs. 6.34, 6.35, and 6.36 with those obtained by other methods in Example 6.11 for a stationary, mean zero, Gaussian, narrowband $\{X(t)\}$ process with $\alpha_1 =$ 0.995.

Using the Gaussian relationships in Eq. 6.34 gives

$$\frac{\eta_X(u,t)}{v_X^+(u,t)} \approx 1 - \exp\left(\frac{-v_A^+(u,t)}{v_X^+(u,t)}\right) = 1 - \exp\left(-\sqrt{(2\pi)(1-\alpha_1^2)}\,\frac{u}{\sigma_X}\right)$$

and Eqs. 6.35 for the Cramer and Leadbetter definition of amplitude and 6.36 both give

$$\frac{\eta_X(u,t)}{v_X^+(u,t)} \approx \frac{1 - \exp\left(-\sqrt{(2\pi)(1-\alpha_1^2)}\,\dfrac{u}{\sigma_X}\right)}{1 - \exp\left(\dfrac{u^2}{2\sigma_X^2}\right)}$$

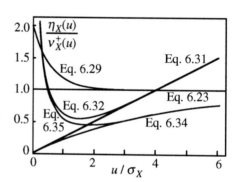

The plot confirms that for very large values of u Eqs. 6.34, 6.35 and 6.36 all tend to the original Poisson approximation of Eq. 6.23, while for small u values Eq. 6.34 tends to Eq. 6.31, and Eqs. 6.35 and 6.36 tend to Eq. 6.32.

* *

We indicated in Eq. 6.28 how the original Poisson crossings approximation should be applied to the double-barrier problem, in which one is concerned with the probability distribution of the extreme value of $|X(t)|$. Now we need to consider the same issue for the various modifications of the Poisson model which have been introduced in this section. Note that replacing X by $|X|$ has no effect at all on Eqs. 6.31 and 6.32, since these approximations are wholly based on the occurrence of amplitude crossings. That is, for the mean zero process, the extreme distribution of $A(t)$ can be considered to be an approximation of the extreme distribution of $|X(t)|$, just as we previously assumed that it approximated the extreme distribution of $X(t)$. In fact, it seems likely that the approximation will be better for $|X(t)|$ than for $X(t)$.

The approximations of Eqs. 6.29 and 6.34 - 6.36 are all affected by the change from single to double barrier, since these expressions all include probability or crossing rate information regarding $X(t)$. Looking first at the initial condition effect approximated by Eq. 6.29, we see that for the double-barrier problem we should use

$$\eta_{|X|}(u,t) \approx \frac{v_{|X|}^{+}(u,t)}{P[|X(t)| \le u]} \qquad (6.37)$$

The modification of the numerator is as in the original Poisson approximation of Eq. 6.28, but the change in the denominator is also significant. In particular, $P[|X(t)| \le u]$ will tend to zero for u approaching zero, so that this approximation will give $\eta_{|X|}(u,t)$ tending to infinity in this situation. Thus, consideration of the $|X(0)| \le u$ condition gives small u behavior which resembles that of Eq. 6.32 with the initial condition of $A(0) \le u$.

One must give a little more thought to the modification of Eqs. 6.34 - 6.36 to describe the double-barrier problem. Recall that these equations were designed to include the probability that an upcrossing by $A(t)$ is accompanied by an upcrossing by $X(t)$. For the double-barrier problem, then, we must approximate the probability that an upcrossing by $A(t)$ is accompanied by an upcrossing by $|X(t)|$. Just as we assumed that there would be an upcrossing by $X(t)$ during T_1 if $T_1 \ge [v_X^{+}(0,t)]^{-1}$ = one period of the process, we will now assume that there will be an upcrossing by $|X(t)|$ during T_1 if $T_1 \ge [v_X^{+}(0,t)]^{-1}/2$ = one-half period of the process. Thus, we replace $[v_X^{+}(0,t)]^{-1}$ in Eq. 6.33 with $[v_X^{+}(0,t)]^{-1}/2$, and this gives exactly the same replacement in Eqs. 6.34 and 6.35:

$$\eta_{|X|}(u,t) \approx 2\,P[A(t) > u]\,v_X^{+}(0,t)\left[1 - \exp\left(\frac{-v_A^{+}(u,t)}{2\,P[A(t) > u]\,v_X^{+}(0,t)}\right)\right] \qquad (6.38)$$

neglecting the initial condition at $t = 0$, and

$$\eta_{|X|}(u,t) \approx \frac{2\,P[A(t) > u]\,v_X^{+}(0,t)}{P[A(t) < u]}\left[1 - \exp\left(\frac{-v_A^{+}(u,t)}{2\,P[A(t) > u]\,v_X^{+}(0,t)}\right)\right] \qquad (6.39)$$

when the initial condition is included. The corresponding modification of Vanmarcke's formula in Eq. 6.36 is

$$\eta_{|X|}(u,t) \approx v_{|X|}^{+}(u,t)\frac{1 - \exp\left(\dfrac{-v_{A_{CL}}^{+}(u,t)}{v_{|X|}^{+}(u,t)}\right)}{1 - \dfrac{v_X^{+}(u,t)}{v_X^{+}(0,t)}} \qquad (6.40)$$

Similar to the single-barrier situation, Eqs. 6.39 using the Cramer and Leadbetter amplitude is identical to Eq. 6.40 for the case of a mean zero Gaussian $\{X(t)\}$ process.

For the special case of a stationary, mean zero, Gaussian $\{X(t)\}$ process, Vanmarcke (1975) has also offered an empirical correction which improves the approximation of the conditional crossing rate. For this situation, the $v_A^{+}(u) / v_X^{+}(u)$ ratio of Eq. 6.36 is given by

$$\frac{v_A^+(u)}{v_X^+(u)} = \sqrt{2\pi}\sqrt{1-\alpha_1^2}\,\frac{u}{\sigma_X}$$

in which the term $(1-\alpha_1^2)$ introduces the effect of the bandwidth of the process. Similarly, the $v_A^+(u)/v_{|X|}^+(u)$ ratio of Eq. 6.40 is one half this amount. As a correction for effects not included in the derivation of these equations, Vanmarcke has suggested replacing the $(1-\alpha_1^2)$ term with $(1-\alpha_1^2)^{1.2}$. This gives the modified Vanmarcke approximations as

$$\frac{\eta_X(u)}{v_X^+(u)} \approx \frac{1-\exp\left(-(1-\alpha_1^2)^{0.6}\sqrt{2\pi}\,\dfrac{u}{\sigma_X}\right)}{1-\exp\left(\dfrac{-u^2}{2\sigma_X^2}\right)}, \qquad \frac{\eta_{|X|}(u)}{v_{|X|}^+(u)} \approx \frac{1-\exp\left(-(1-\alpha_1^2)^{0.6}\sqrt{\dfrac{\pi}{2}}\,\dfrac{u}{\sigma_X}\right)}{1-\exp\left(\dfrac{-u^2}{2\sigma_X^2}\right)} \qquad (6.41)$$

for the single-barrier and double-barrier problems, respectively.

6.7   Inclusion-Exclusion Series for Extreme Value Distribution

Finally, we will derive the infinite series called the inclusion-exclusion relationship between $\eta_X(u,t)$ and probability density functions for $X(t)$ and $\dot{X}(t)$ random variables.[8] This cumbersome equation rigorously includes the fact that $\eta_X(u,t)$ is conditioned by the event of no prior upcrossings of the level u by $X(t)$. On the other hand, it ignores the additional conditioning in Eq. 6.21 by the event of $X(0) \leq u$. To emphasize this distinction, we will use the notation $\hat{\eta}_X(u,t)$ for the conditional rate of upcrossings given only the fact that there have been no prior upcrossings. The presentation will be simplified by using a notation for upcrossing events as follows: $B(t) = \{$event of an upcrossing of level u occurring in $[t, t + \Delta t]\}$ and $B^*(t) = \{$event of the first upcrossing of u occurring in $[t, t + \Delta t]\}$, giving $B^*(t)$ as a subset of $B(t)$. We will also introduce new terms $v_X^+(u,t,s_1,\cdots,s_j)$ and $\hat{\eta}_X(u,t,s_1,\cdots,s_j)$ which give the probabilities of intersections of these events as

$$P[B(t)B(s_1)\cdots B(s_j)] = v_X^+(u,t,s_1,\cdots,s_j)\Delta t\Delta s_1\cdots\Delta s_j \qquad (6.42)$$

and

$$P[B(t)B(s_1)\cdots B^*(s_j)] = \hat{\eta}_X(u,t,s_1,\cdots,s_j)\Delta t\Delta s_1\cdots\Delta s_j \quad \text{for} \quad s_j \leq s_{j-1} \leq \cdots \leq s_1 \leq t \qquad (6.43)$$

In the first relationship there is no requirement that there has not been an upcrossing (or many upcrossings) prior to time s_j, but the second relationship does include this restriction. If the first upcrossing is not at time s_j then it must be at some time s_{j+1} which is prior to s_j, so that we can say

---

The inclusion-exclusion relationship is attributed to S. O. Rice in 1944, but the derivation given here is due to Madsen et al. in 1986.

$$v_X^+(u,t,s_1,\cdots,s_j) = \hat{\eta}_X(u,t,s_1,\cdots,s_j) + \int_0^{s_j} \hat{\eta}_X(u,t,s_1,\cdots,s_j,s_{j+1})\,ds_{j+1}$$

or

$$\hat{\eta}_X(u,t,s_1,\cdots,s_j) = v_X^+(u,t,s_1,\cdots,s_j) - \int_0^{s_j} \hat{\eta}_X(u,t,s_1,\cdots,s_j,s_{j+1})\,ds_{j+1} \qquad (6.44)$$

The corresponding relationship for $j = 0$ is

$$\hat{\eta}_X(u,t) = v_X^+(u,t) - \int_0^t \hat{\eta}_X(u,t,s_1)\,ds_1 \qquad (6.45)$$

describing the fact that the occurrence of an upcrossing at time t implies that the first upcrossing is either at t or at some time $s_1 < t$. Now we can substitute for $\hat{\eta}_X(u,t,s_1)$ in Eq. 6.45 from Eq. 6.44 with $j = 1$ to give

$$\hat{\eta}_X(u,t) = v_X^+(u,t) - \int_0^t \left(v_X^+(u,t,s_1) - \int_0^{s_1} \hat{\eta}_X(u,t,s_1,s_2)\,ds_2 \right) ds_1$$

into which we can substitute for $\hat{\eta}_X(u,t,s_1,s_2)$ according to Eq. 6.44 with $j = 2$. Repetition of this procedure gives

$$\hat{\eta}_X(u,t) = v_X^+(u,t) + \sum_{j=1}^{\infty} (-1)^j \int_0^t \int_0^{s_1} \cdots \int_0^{s_{j-1}} v_X^+(u,t,s_1,\cdots,s_j)\,ds_j\cdots ds_2\,ds_1 \qquad (6.46)$$

which is the general inclusion-exclusion relationship.

In order to use the inclusion-exclusion relationship, of course, one must have knowledge of the rate of multiple upcrossings term, $v_X^+(u,t,s_1,\cdots,s_j)$, defined in Eq. 6.42. It is not difficult to write an integral expression for this term. In fact, a simple generalization of Eq. 6.2 gives

$$v_X^+(u,t,s_1,\cdots,s_j) = \int_0^{\infty}\cdots\int_0^{\infty}\int_0^{\infty}(v_j\cdots v_1 v)$$

$$p_{X(s_j),\dot{X}(s_j),\cdots,X(s_1),\dot{X}(s_1),X(t),\dot{X}(t)}(u,v_j,\cdots,u,v_1,u,v)\,dv_j\cdots dv_1\,dv \qquad (6.47)$$

Unfortunately it is usually quite difficult to carry out the integration in Eqs. 6.46 and 6.47 unless j is quite small.

It should be noted that using only the first term ($j = 0$) in Eq. 6.46 gives exactly the Poisson approximation. Furthermore, the alternating signs of the terms in Eq. 6.46 show that each truncation of the series gives either an upper or a lower bound on $\hat{\eta}_X(u,t)$. Thus, the Poisson approximation of $v_X^+(u,t)$ is an upper bound on $\hat{\eta}_X(u,t)$; including only the $j = 1$ term from the summation in Eq. 6.46 gives a lower bound; etc. It should be kept in mind, though, that we are not assured that the truncations of Eq. 6.46 will provide upper and lower bounds on the original $\eta_X(u,t)$, since $\hat{\eta}_X(u,t)$ omits conditioning by the event $X(0) \leq u$. In fact, we have already argued that $v_X^+(u,t)$ surely does not give an upper bound on $\eta_X(u,t)$ for very low u values.

6.8   Extreme Value of Gaussian Response of SDF Oscillator

Since we have no exact solution for the extreme value distribution, it is appropriate to use simulation data to verify which, if any, of the various approximate solutions may give accurate estimates. We will present this comparison only for the particular problem of stationary response of the SDF oscillator excited by mean zero, Gaussian, white noise. We will consider the double barrier problem, related to crossings of the level u by $|X(t)|$. One reason for the choice of this particular problem is the existence of multiple sets of simulation data which can be considered to provide quite reliable estimates of the true solution of the problem. The other reason is that this is a mathematical model often used in the estimation of first-passage failure in practical engineering problems.

Figure 6.5 shows analytical and simulation results for the situation of an oscillator with 1% of critical damping, which gives $\alpha_1 = 0.9937$. The plot includes the double barrier versions of most of the analytical approximations we have considered for $\eta_X(u) / v_{|X|}^+(u)$. It is seen that the simulated values of this ratio are smaller than the predictions from any of the approximations, except when u is less than about $1.2\sigma_X$. Also, the simulation data clearly show that the ratio has a minimum value when u is approximately $2\sigma_X$. Of all the approximations, it appears that the analytical form due to Vanmarcke comes closest to fitting the simulation data. In particular, the modified Vanmarcke form of Eq. 6.41 gives a minimum value of $\eta_X(u) / v_{|X|}^+(u)$ at approximately the right u value, and the values it gives for the ratio in this vicinity are better than those of any of the other approximations when u is approximately $2\sigma_X$. Of all the approximations, it appears that the analytical form due to

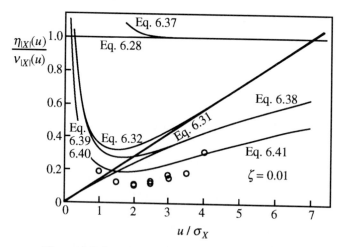

Figure 6.5 Simulation and various approximations

Vanmarcke comes closest to fitting the simulation data. In particular, the modified Vanmarcke form of Eq. 6.41 gives a minimum value of $\eta_X(u) / v^+_{|X|}(u)$ at approximately the right u value, and the values it gives for the ratio in this vicinity are better than those of any of the other approximations which have reasonable behavior for smaller u values. Nonetheless, it must be noted that there is sometimes a significant discrepancy between Vanmarcke's formula and the simulation data. For $u = 2\sigma_X$, for example, the approximation is about 70% above the value of 0.10 or 0.11 obtained from simulation, even though the modified approximation agrees almost perfectly with the data point from simulation for $u = 4\sigma_X$.[9] Any overestimation of $\eta_X(u)$ gives $L_X(u,t)$ values which decay more rapidly with increasing t than do the values from simulation. For $u = 2\sigma_X$ and large values of time, Vanmarcke's formula will significantly overpredict the probability that $X(t)$ has ever reached the level u. This discrepancy, though, is smaller when u is large, and this is often the region of primary interest.

Simulation data also exist for larger values of damping in the SDF system, and Fig. 6.6 presents some of these values. The only analytical result shown for comparison is Vanmarcke's modified formula given in Eq. 6.41, since this seems to be better than any of the other approximate methods that we have considered. It is noted that for $\zeta \geq 0.05$, the Vanmarcke approximation fits the simulation data very well, and the error consistently increases as the damping is decreased below this level. In all cases it appears that the Vanmarcke approximation is significantly better than the value of unity predicted by assuming that a Poisson process describes the crossings of the level u by $X(t)$.

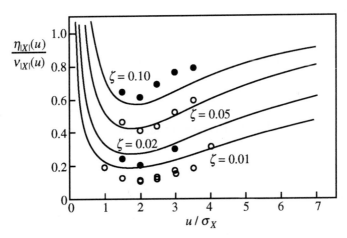

Figure 6.6 Simulation and Vanmarcke's modified approximation

---

[9]   Without the empirical correction added to Eq. 6.41, Vanmarcke's $\eta_X(u)$ values vary from 39% above the simulation result for $u = 4\sigma_X$ to 170% above for $u = 2\sigma_X$.

Exercises

****

General Upcrossing Rates

****

6.1 Each of the formulas gives the joint probability density of $X(t)$ and $\dot{X}(t)$ for a particular process $\{X(t)\}$ at a particular time t. For each process, find the expected rate of upcrossings $\nu_X^+(u,t)$ for all u values.

(a) $\quad p_{X(t)\dot{X}(t)}(u,v) = 9u^2 v^2 \, [U(u) - U(u-1)] \, [U(v) - U(v-1)]$

(b) $\quad p_{X(t)\dot{X}(t)}(u,v) = \dfrac{2u}{\sqrt{2\pi}\, b^2 \sigma_2} \exp\left(-\dfrac{v^2}{2\sigma_2^2}\right) [U(u) - U(u-b)]$

(c) $\quad p_{X(t)\dot{X}(t)}(u,v) = \dfrac{3(1-u^2)}{2\sqrt{2\pi}\,\sigma_2} \exp\left(-\dfrac{v^2}{2\sigma_2^2}\right) U(1-|u|)$

(d) $\quad p_{X(t)\dot{X}(t)}(u,v) = \dfrac{\lambda}{\sqrt{2\pi}\,\sigma_2} \exp\left(-\lambda\, u - \dfrac{v^2}{2\sigma_2^2}\right) U(u)$

****

6.2 Each of the formulas gives the joint probability density of $X(t)$ and $\dot{X}(t)$ for a particular covariant stationary stochastic process $\{X(t)\}$. For each process, find the expected rate of upcrossings $\nu_X^+(u)$ for all u values.

(a) $\quad p_{X(t)\dot{X}(t)}(u,v) = \dfrac{1}{\pi a^2}\, U(a^2 - u^2 - v^2)$

(b) $\quad p_{X(t)\dot{X}(t)}(u,v) = \dfrac{1}{2\sqrt{2\pi}\, b\sigma_2} \exp\left(-\dfrac{v^2}{2\sigma_2^2}\right) U(b-|u|)$

(c) $\quad p_{X(t)\dot{X}(t)}(u,v) = \dfrac{1}{2\pi\sigma_1\sigma_2} \exp\left(-\dfrac{u^2}{2\sigma_1^2} - \dfrac{v^2}{2\sigma_2^2}\right)$

(d) $\quad p_{X(t)\dot{X}(t)}(u,v) = \dfrac{3\alpha^{1/3}}{2\Gamma(1/3)} \sqrt{\dfrac{\gamma}{\pi}}\, e^{-\alpha|u|^3} e^{-\gamma v^2}$

****

6.3 The stationary response $\{X(t)\}$ for a certain nonlinear oscillator is characterized by

$$p_{X(t)\dot{X}(t)}(u,v) = A\exp\left(-\lambda\,|u| - \dfrac{v^2}{2\sigma_2^2}\right)$$

(a) Find the value of the constant A, in terms of λ and σ_2.
(b) Find $\nu_X^+(u)$ for all u values.
(c) Compare $\nu_X^+(u)$ with $\nu_Y^+(u)$ for a stationary Gaussian process $\{Y(t)\}$ with the same mean and autocovariance function as $\{X(t)\}$.

****

6.4 The stationary response $\{X(t)\}$ for a certain nonlinear oscillator is characterized by

$$p_{X(t)\dot{X}(t)}(u,v) = A\exp\left(-\alpha u^4 - \dfrac{v^2}{2\sigma_2^2}\right)$$

(a) Find the value of the constant A, in terms of α and σ_2.
(b) Find $\nu_X^+(u)$ for all u values.

(c) Compare $v_X^+(u)$ with $v_Y^+(u)$ for a stationary Gaussian process $\{Y(t)\}$ with the same mean and autocovariance function as $\{X(t)\}$.

HINT: $\Gamma(1/4) \approx 3.6256$ and $\Gamma(3/4) \approx 1.2254$.
****

Upcrossing Rates for Narrowband Processes
****

6.5 Let $\{X(t)\}$ be a mean zero covariant stationary stochastic process with the autospectral density function of Exercise 4.14:

$$S_{XX}(\omega) = S_0 \left[\exp(-\gamma|\omega + \omega_0|) + \exp(-\gamma|\omega - \omega_0|) \right]$$

in which S_0, ω_0, and γ are positive constants.

Under what limitations on S_0, ω_0, or γ can you use this information to approximate $v_X^+(0)$ without further information about the probability distribution of $\{X(t)\}$? Find this approximation, and explain your answer.
****

6.6 Let $\{X(t)\}$ be a mean zero covariant stationary stochastic process with the autospectral density function of Exercise 4.16: $S_{XX}(\omega) = A|\omega|^b e^{-c|\omega|}$, in which A, b, and c are positive constants.

Under what limitations on A, b, and c can you use this information to approximate $v_X^+(0)$ without further information about the probability distribution of $\{X(t)\}$? Find this approximation, and explain your answer.
****

6.7 Let $\{X(t)\}$ have a mean value function $\mu_X(t)$ and the covariant stationary autospectral density function introduced in Section 4.4:

$$S_{XX}(\omega) = S_0 \big(U[(|\omega|-(\omega_c - b)] - U[(|\omega|-(\omega_c + b)] \big)$$

in which S_0, ω_c, and b are positive constants.

Under what limitations on S_0, ω_c, and b can you use this information to approximate $v_X^+[\mu_X(t)]$ without further information about the probability distribution of $\{X(t)\}$? Find this approximation, and explain your answer.
****

Upcrossing Rates for Gaussian Processes
****

6.8 Let $\{X(t)\}$ be the mean zero, covariant stationary stochastic process of Exercise 6.5, but with the additional stipulation that it is Gaussian.

(a) Find $v_X^+(u)$ for all u values.

(b) Find the value of the irregularity factor *IF*.
****

6.9 Let $\{X(t)\}$ be the mean zero, covariant stationary stochastic process of Exercise 6.6, but with the additional stipulation that it is Gaussian.

(a) Find $v_X^+(u)$ for all u values.

(b) Find the irregularity factor *IF*.
****

6.10 Let $\{X(t)\}$ be the mean zero, covariant stationary stochastic process of Exercise 6.7, but with the additional stipulation that it is Gaussian.

(a) Find $v_X^+(u)$ for all u values.

(b) Find the irregularity factor IF.

****

6.11 Let $\{X(t)\}$ be a covariant stationary Gaussian stochastic process with a mean value function $\mu_X(t)$ and an autospectral density of

$$S_{XX}(\omega) = \frac{S_0}{\omega_0}|\omega|\,U(\omega_o - |\omega|)$$

Find $v_X^+(u)$ for all u values.

****

6.12 Let $\{X(t)\}$ be a covariant stationary Gaussian stochastic process with a constant mean value μ_X and an autospectral density of

$$S_{XX}(\omega) = |\omega|^5 e^{-|\omega|}$$

(a) Find $v_X^+(u)$ for all u values.

(b) Find the value of the irregularity factor IF.

****

6.13 Let $\{X(t)\}$ be a covariant stationary Gaussian stochastic process with a constant mean value μ_X and an autospectral density of

$$S_{XX}(\omega) = |\omega| e^{-\omega^2}$$

(a) Find $v_X^+(u)$ for all u values.

(b) Find the value of the irregularity factor IF.

****

6.14 Let $\{X(t)\}$ be a mean zero covariant stationary Gaussian stochastic process with the autocovariance function of Exercise 4.18:

$$G_{XX}(\tau) = e^{-c\tau^2} \cos(a\tau)$$

(a) Find $v_X^+(u)$ for all u values.

(b) Find the irregularity factor IF.

****

6.15 Let $\{X(t)\}$ be a mean zero covariant stationary Gaussian stochastic process with the autocovariance function of Exercise 4.19:

$$G_{XX}(\tau) = \frac{\cos(a\tau)}{1 + c\tau^2}$$

(a) Find $v_X^+(u)$ for all u values.

(b) Find the irregularity factor IF.

****

Peak Distribution

****

6.16 Let $\{X(t)\}$ be a covariant stationary mean zero Gaussian process with standard deviation σ_X, average frequency ω_a (energy-based definition), and irregularity factor $IF = 0.5$.

(a) Evaluate $P[P(t) > b\sigma_X]$ for $b = -1, 1, 3,$ and 5, in which $P(t)$ is any peak of $\{X(t)\}$.

(b) Compare the values in part (a) with those that would apply for the limiting cases of $IF = 0$ and $IF = 1$.

(c) Find the expected rate of occurrence of peaks with value greater than $3\sigma_X$.

(d) Find the expected rate of occurrence of peaks with value less than $-\sigma_X$.

****

Extreme Value Distribution — Poisson Approximation

****

6.17 (a) - (e) For the covariant stationary $\{X(t)\}$ processes having the probability density functions of Exercise 6.2, find estimates of $L_X(u,t)$ by using the Poisson approximation. Also give a qualitative sketch showing the shape of $L_X(u,t)$ versus u, both for $t = 0$ and for some $t > 0$.

****

6.18 Use the Poisson approximation to estimate the values of $L_X(u,t)$ and $L_Y(u,t)$ for the level $u = 3\sigma_X$ for the stationary $\{X(t)\}$ and $\{Y(t)\}$ processes of Exercise 6.3.

****

6.19 Use the Poisson approximation to estimate the values of $L_X(u,t)$ and $L_Y(u,t)$ for the level $u = 3\sigma_X$ for the stationary $\{X(t)\}$ and $\{Y(t)\}$ processes of Exercise 6.4.

****

6.20 Let $\{X(t)\}$ be a covariant stationary, Gaussian process with the autospectral density function of Exercises 6.6 and 6.9, with $b = 8$: $S_{XX}(\omega) = A|\omega|^8 e^{-c|\omega|}$.

(a) Use the Poisson approximation to obtain an estimate of the probability that the energy-based amplitude $A(t)$ will ever cross the level $u = 4\sigma_X$ during a time interval of length $50c/b$.

(b) Perform the same calculation for the Cramer and Leadbetter amplitude $A_{CL}(t)$.

****

6.21 Let $\{X(t)\}$ be a covariant stationary, Gaussian process with the autospectral density function of Exercise 4.17:

$$S_{XX}(\omega) = S_0 U(\omega_0 - |\omega|) + S_0 \left|\frac{\omega_0}{\omega}\right|^4 U(|\omega| - \omega_0).$$

(a) Use the Poisson approximation to estimate the probability that the Cramer and Leadbetter amplitude $A_{CL}(t)$ will ever cross the level $u = 4\sigma_X$ during a time interval of length $100/\omega_0$.

(b) Is it feasible to perform the same calculation for the energy-based amplitude $A(t)$?

****

6.22 Let $\{X(t)\}$ represent the dynamic response of an oscillator governed by
$$\ddot{X}(t) + 20\zeta\,\dot{X}(t) + 100\,X(t) = F(t)$$
with $\{F(t)\}$ being mean zero, Gaussian white noise with autospectral density $S_{FF}(\omega) = 400\,/\,\pi$ for all ω.

Use the Poisson approximation to estimate $L_X(u,t)$ for the following three situations:

(a) $\zeta = 0.20$, $u = 4$

(b) $\zeta = 0.20$, $u = 2$

(c) $\zeta = 0.05$, $u = 4$

(d) For which of the three situations $\{(a), (b), or (c)\}$ is the Poisson approximation the best?

(e) For which of the three situations $\{(a), (b), or (c)\}$ is the Poisson approximation the worst?

(f) Briefly explain your answers to parts (d) and (e).

****

6.23 Let $\{X(t)\}$ be a mean zero Gaussian process with autospectral density given by

$S_{XX}(\omega) = 4[U(|\omega|-10+b) - U(|\omega|-10-b)] \text{ mm}^2 / (\text{rad} / \text{s})$

Use the Poisson approximation to estimate $L_X(u,t)$ for the following three situations:

(a) $b = 4$ rad/s, $u = 32$ mm

(b) $b = 4$ rad/s, $u = 16$ mm

(c) $b = 1$ rad/s, $u = 16$ mm

(d) For which of the three situations {(a), (b), or (c)} is the Poisson approximation the best?

(e) For which of the three situations {(a), (b), or (c)} is the Poisson approximation the worst?

(f) Briefly explain your answers to parts (d) and (e).

****

6.24 Consider the problem of predicting the probability of structural yielding during a wind storm which has been modeled as a stationary segment of Gaussian stochastic force. It has been determined that the critical response is $X(t) \equiv$ distortion in the first story, and that yielding will occur if $|X(t)| \geq 60$ mm. The unyielded structural response $\{X(t)\}$ can be modeled as the response of a linear SDF system with a 1 second period ($\omega_0 = 2\pi$ rad/sec) and 2% of critical damping, and the stationary response has been found to have $\mu_X = 0$, $\sigma_X = 11$ mm. The duration of the wind storm is two hours. Estimate the probability that yielding will occur during the two-hour wind storm by using the Poisson approximation for the double-barrier problem.

****

6.25 You wish to estimate the probability that two buildings will collide during an earthquake. Let $X_1(t)$ be the response at the top of building no. 1 (the shorter building), $X_2(t)$ be the response of building no. 2 at the level of the top of building no. 1, and $Z(t) = X_1(t) - X_2(t)$. Collision will occur if $Z(t) \geq 20$ mm (i.e., there is 20 mm clearance in the static position). For your design earthquake you find that the standard deviations and average frequencies of the individual stationary responses are: $\sigma_{X_1} = 2.5$ mm, $v_{X_1}^+(0,t) = 1.0$ zero-upcrossings/s, $\sigma_{X_2} = 5.0$ mm, and $v_{X_2}^+(0,t) = 0.25$ zero-upcrossings/s. Consider $\{X_1(t)\}$ and $\{X_2(t)\}$ to be mean zero, independent, Gaussian processes.

(a) Find the standard deviation and the average frequency of zero-upcrossings by $\{Z(t)\}$.

(b) Find the rate of upcrossings by $\{Z(t)\}$ of the level $Z = 20$ mm.

(c) Assume that the probability of collision is the probability of at least one upcrossing of $Z = 20$ mm during 20 seconds of stationary response. Estimate this probability by using the Poisson approximation.

****

Extreme Value Distribution — Vanmarcke's Approximation

****

6.26 Let $\{X(t)\}$ be a covariant stationary Gaussian stochastic process with zero mean value and the autospectral density of Exercise 6.12: $S_{XX}(\omega) = |\omega|^5 e^{-|\omega|}$

(a) Estimate $L_{|X|}(u,t)$ for $u = 40$ and $t = 10$ by using the Poisson approximation.

(b) Estimate $L_{|X|}(u,t)$ for $u = 40$ and $t = 10$ by using Vanmarcke's modified approximation (Eq. 6.41).

****

6.27 Let $\{X(t)\}$ be a covariant stationary Gaussian stochastic process with zero mean value and the autospectral density of Exercise 6.13: $S_{XX}(\omega) = |\omega| e^{-\omega^2}$

(a) Estimate $L_{|X|}(u,t)$ for $u = 3$ and $t = 75$ by using the Poisson approximation.

(b) Estimate $L_{|X|}(u,t)$ for $u = 3$ and $t = 75$ by using Vanmarcke's modified approximation (Eq. 6.41).

****

Chapter 7

Linear Systems with Multiple Inputs and Outputs

7.1 Generalization of Scalar Formulation

Recall that Chapters 3 and 4 began with very general deterministic descriptions of the response of a linear system. In Chapter 3 this gave the time history of a single response as a Duhamel convolution integral of an impulse response function and the time history of a single excitation, and in Chapter 4 the Fourier transform of the response time history was found to be a simple product of a harmonic transfer function and the Fourier transform of the excitation. We now wish to extend those presentations to include consideration of multiple inputs and multiple outputs.

Let n_X be the number of response components in which we are interested, and let n_F be the number of different excitations of the system. We will denote the responses as $\{X_j(t)\}$ for $j = 1$, $\cdots$, n_X and the excitations as $\{F_l(t)\}$ for $l = 1, \cdots, n_F$, and begin by considering evaluation of the jth component of the response. In many situations the excitations may be components of force and the responses may be components of displacement, as illustrated in Fig. 7.1, but this is not necessary. The formulation is general enough to include any definitions of excitation and response, except that we will consider the system to be causal in the sense that the responses at time t are caused only by the excitations at prior times.

For the time domain formulation we use Eq. 3.2 to give the response due to the lth component of the excitation, and using superposition we can say that the total response is the sum over l of these components. Thus, we have

$$X_j(t) = \sum_{l=1}^{n_F} \int_{-\infty}^{+\infty} h_{j,l}(t-s)\, F_l(s)\, ds \qquad (7.1)$$

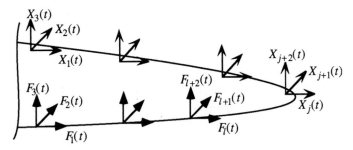

Figure 7.1 Selected components of force and displacement for an airplane wing

in which the impulse response function $h_{j,l}(t)$ is defined to be the response component $X_j(t)$ due to a Dirac delta function excitation with $F_l(t) = \delta(t)$ and $F_r(t) = 0$ for $r \neq l$. This equation, though, is simply the jth row of the matrix equation

$$\vec{X}(t) = \int_{-\infty}^{\infty} \mathbf{h}(t-s)\vec{F}(s)\,ds \tag{7.2}$$

Thus, Eq. 7.2 is a very convenient representation of the set of n_X equations which describe the n_X responses in terms of the n_F excitations. Each component of the rectangular matrix of dimension $n_X \times n_F$ is defined as stated. Alternatively, this information can be organized so that an entire column of the $\mathbf{h}(t)$ matrix is defined in one equation. Namely,

$$\vec{X}(t) = [h_{1,l}(t), \cdots, h_{n_X,l}(t)]^T$$

is the response to

$$\vec{F}(t) = [\underbrace{0, \cdots, 0}_{l-1}, \delta(t), \underbrace{0, \cdots, 0}_{n_F-l}]^T$$

and we define the entire $\mathbf{h}(t)$ matrix by considering different l values.

In a similar way we can use Eq. 4.31 to give the Fourier transform of the jth response due to the lth excitation component, and use superposition to obtain

$$\tilde{X}_j(\omega) = \sum_{l=1}^{n_F} H_{j,l}(\omega)\tilde{F}_l(\omega) \tag{7.3}$$

or

$$\vec{\tilde{X}}(\omega) = \mathbf{H}(\omega)\vec{\tilde{F}}(\omega) \tag{7.4}$$

in which the lth column of the harmonic transfer matrix is defined such that

$$\vec{X}(t) = [H_{1,l}(\omega), \cdots, H_{n_X,l}(\omega)]^T e^{i\omega t}$$

is the response to

$$\vec{F}(t) = [\underbrace{0, \cdots, 0}_{l-1}, e^{i\omega t}, \underbrace{0, \cdots, 0}_{n_F-l}]^T$$

Of course, we can show that the components of the impulse response function matrix and the harmonic transfer function matrix each satisfy a Fourier transform relationship like Eq. 4.32 for the scalar case. The collection of all these relationships can be written in matrix form as

$$\mathbf{H}(\omega) = \int_{-\infty}^{\infty} \mathbf{h}(t)e^{-i\omega t}\,dt$$

meaning that the scalar Fourier transform integral is applied to each component of the matrix in the integrand.

We can now use Eqs. 7.2 and 7.4 in writing matrix versions of various expressions in Chapters 3 and 4 which describe the response in terms of the excitation. The mean value relationships are obtained by simply taking the expectations of the two equations and take the form

$$\vec{\mu}_X(t) = \int_{-\infty}^{\infty} \mathbf{h}(t-s)\vec{\mu}_F(s)\,ds \tag{7.5}$$

and

$$\vec{\mu}_X(\omega) = \mathbf{H}(\omega)\vec{\mu}_F(\omega) \tag{7.6}$$

with the jth components of these equations being exactly equivalent to summations over l of Eqs. 3.12 and 4.31, respectively, written to give the jth component of response due to the lth component of excitation. Similarly, we can write the autocorrelation function for the response as

$$\boldsymbol{\phi}_{XX}(t,s) \equiv E[\vec{X}(t)\vec{X}^T(s)] = \int_{-\infty}^{\infty}\int_{-\infty}^{\infty} \mathbf{h}(t-u)\boldsymbol{\phi}_{FF}(u,v)\mathbf{h}^T(s-v)\,du\,dv \tag{7.7}$$

and the autocovariance as

$$\mathbf{K}_{XX}(t,s) = \boldsymbol{\phi}_{XX}(t,s) - \vec{\mu}_X(t)\vec{\mu}_X^T(s) = \int_{-\infty}^{\infty}\int_{-\infty}^{\infty} \mathbf{h}(t-u)\mathbf{K}_{FF}(u,v)\mathbf{h}^T(s-v)\,du\,dv \tag{7.8}$$

For the special case of stationarity, these relationships can be rewritten in any of several slightly simpler forms, including

$$\mathbf{R}_{XX}(\tau) = \int_{-\infty}^{\infty}\int_{-\infty}^{\infty} \mathbf{h}(u)\mathbf{R}_{FF}(\tau-u+v)\mathbf{h}^T(v)\,du\,dv \tag{7.9}$$

and

$$\mathbf{G}_{XX}(\tau) = \int_{-\infty}^{\infty}\int_{-\infty}^{\infty} \mathbf{h}(u)\mathbf{G}_{FF}(\tau-u+v)\mathbf{h}^T(v)\,du\,dv \tag{7.10}$$

Similarly, the autospectral density for the covariant stationary situation is described by

$$\mathbf{S}_{XX}(\omega) = \mathbf{H}(\omega)\mathbf{S}_{FF}(\omega)\mathbf{H}^{T*}(\omega) \tag{7.11}$$

The jth diagonal components of Eqs. 7.7 - 7.11 are exactly equivalent to summations over l of the expressions in Chapters 3 and 4 giving autocorrelation, autocovariance, and autospectral density of the jth response due to the lth component of excitation. Similarly, the off-diagonal components of these matrix equations give the cross-correlation, cross-covariance, and cross-spectral density functions for different components of response.

Note that if the value $\vec{X}(t_0)$ of the response at time t_0 is known, then one can write Eq. 7.2 in the alternative form of

$$\vec{X}(t) = \mathbf{g}(t - t_0)\vec{Y} + \int_{t_0}^{t} \mathbf{h}(t - s)\vec{F}(s)\,ds \tag{7.12}$$

in which the vector $\vec{Y}$ contains a complete set of initial conditions at time $t = t_0$, and $\mathbf{g}(t)$ is a matrix which gives the n_X time histories of free vibration (i.e., the homogeneous solution) to unit values of the possible initial conditions. Using this form of the time history solution makes it possible to write vector analogies of Eqs. 3.20 and 3.21, which described the input and output in the corresponding scalar situation. Vector versions of Eqs. 3.22 and 3.23 giving the conditional mean and conditional covariance of the response can also be obtained.

7.2 Multi-Degree of Freedom Systems

One of the most commonly encountered situations involving a linear system with multiple inputs and multiple outputs is the so-called multi-degree of freedom (MDF) system. As a simple generalization of Eq. 3.39, the MDF equation of motion can be written as

$$\mathbf{m}\ddot{\vec{X}}(t) + \mathbf{c}\dot{\vec{X}}(t) + \mathbf{k}\vec{X}(t) = \vec{F}(t) \tag{7.13}$$

in which $\mathbf{m}$, $\mathbf{c}$, and $\mathbf{k}$ are square matrices of dimension $n \times n$, and the location and orientation of the components of $\vec{F}(t)$ are identical to those of $\vec{X}(t)$.[1] Often we have problems of this type in which the elements of the $\mathbf{m}$, $\mathbf{c}$, and $\mathbf{k}$ matrices represent physical masses, dashpots, and springs, respectively, but this is not always the case. In our examples, it will always be true that $\mathbf{m}$, $\mathbf{c}$, and $\mathbf{k}$ will represent inertia terms, energy dissipation terms, and restoring force terms, but the exact meaning of any component will depend on the set of coordinates used in describing the problem. On the other hand, one can write certain energy expressions which are true for any choice of coordinates. In particular, the kinetic energy, the potential energy, and the rate of energy dissipation (power dissipation), respectively, are given by

$$KE(t) = \frac{1}{2}\dot{\vec{X}}^T(t)\,\mathbf{m}\,\dot{\vec{X}}(t) \tag{7.14}$$

$$PE(t) = \frac{1}{2}\vec{X}^T(t)\,\mathbf{k}\,\vec{X}(t) \tag{7.15}$$

and

$$PD(t) = \dot{\vec{X}}^T(t)\,\mathbf{c}\,\dot{\vec{X}}(t) \tag{7.16}$$

Also, it is always true that $\mathbf{m}$, $\mathbf{c}$, and $\mathbf{k}$ are symmetric matrices. Note that the nonnegativity of the kinetic and potential energy in Eq. 7.14 make it mandatory that $\mathbf{m}$ and $\mathbf{k}$, respectively, not have any

---

[1] For example, if $X_j(t)$ is the absolute displacement of story j in a multistory building, then $F_j(t)$ is the external force applied to that story, but if $X_j(t)$ is an interfloor displacement then $F_j(t)$ is the corresponding interfloor component of the applied force.

negative eigenvalues. For example, if $\mathbf{m}$ had a negative eigenvalue, then Eq. 7.14 would give a negative value of kinetic energy if $\dot{\vec{X}}(t)$ happened to be parallel to the corresponding eigenvector. Also, Eq. 7.16 requires that $\mathbf{c}$ have no negative eigenvalues if this matrix represents the effect of components like dashpots which can never, even for an instant, add energy to the system.

The meaning of the $\mathbf{m}$, $\mathbf{c}$, and $\mathbf{k}$ matrices is particularly simple when all the mass in the system is in n discrete masses, and the coordinates, denoted by the components of $\vec{X}(t)$, each represent one component of displacement of a mass relative to a fixed frame of reference. (If multidimensional motion of any mass is possible, then we will presume that orthogonal coordinates are used to describe the total motion.) In this case the $\mathbf{m}$ matrix is diagonal, since the total kinetic energy of Eq. 7.14 is simply a sum of quadratic terms, each depending on the magnitude of one mass and the square of a component of velocity of that mass. The terms of $\mathbf{k}$ and $\mathbf{c}$ can then be described as follows. The lth column of $\mathbf{k}$, which can be written as $[k_{1,l},\cdots,k_{n,l}]^T$, gives the vector of forces in the n coordinates which will result in a unit static displacement in coordinate $X_l(t)$, and zero displacement in every other coordinate. Similarly, a set of static forces $[c_{1,l},\cdots,c_{n,l}]^T$ would give a unit velocity in coordinate $X_l(t)$, and zero velocity in every other coordinate, if $\mathbf{k}$ were set equal to zero. In this situation with the coordinates representing displacement relative to a fixed frame of reference, $-k_{jl}$ usually represents a physical spring connecting the masses described by $X_j(t)$ and $X_l(t)$ for $j \neq l$, and k_{jj} represents the sum of all springs attached to the mass described by $X_j(t)$. In a corresponding way, the elements of $\mathbf{c}$ correspond to dashpots connected to the masses described by the coordinates. These statements are strictly true only if the springs and dashpots are aligned with the directions of the orthogonal coordinates.

****

Example 7.1:   Find the $\mathbf{m}$, $\mathbf{k}$, and $\mathbf{c}$ matrices for the two-degree-of-freedom (2DF) system shown.

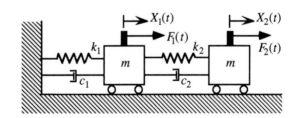

The equations of motion for the two masses are

$$m_1\ddot{X}_1(t) + c_1\dot{X}_1(t) + c_2[\dot{X}_1(t) - \dot{X}_2(t)] + k_1X_1(t) + k_2[X_1(t) - X_2(t)] = F_1(t)$$

and

$$m_2\ddot{X}_2(t) + c_2[\dot{X}_2(t) - \dot{X}_1(t)] + k_2[X_2(t) - X_1(t)] = F_2(t)$$

These equations exactly agree with Eq. 7.13 with

$$\mathbf{m} = \begin{bmatrix} m_1 & 0 \\ 0 & m_2 \end{bmatrix} \quad \mathbf{k} = \begin{bmatrix} k_1 + k_2 & -k_2 \\ -k_2 & k_2 \end{bmatrix} \quad \mathbf{c} = \begin{bmatrix} c_1 + c_2 & -c_2 \\ -c_2 & c_2 \end{bmatrix}$$

****

7.3 Uncoupled Modes of MDF Systems

The most common solution of Eq. 7.13 uses the eigenvectors and eigenvalues of the system. In particular, we will presume that $\mathbf{m}$ is positive definite (i.e., has no zero eigenvalues) so that its inverse exists. We will then let the matrix $\boldsymbol{\theta}$, of dimension $n \times n$, have columns which are the eigenvectors of $\mathbf{m}^{-1}\mathbf{k}$. This means that

$$\mathbf{m}^{-1}\mathbf{k}\boldsymbol{\theta} = \boldsymbol{\theta}\boldsymbol{\lambda} \tag{7.17}$$

in which $\boldsymbol{\lambda}$ is a diagonal matrix of dimension $n \times n$, with the (j,j) element being the eigenvalue corresponding to the eigenvector located in column j of $\boldsymbol{\theta}$. One important property of $\boldsymbol{\theta}$ is easily obtained by rewriting this equation as $\mathbf{k}\boldsymbol{\theta} = \mathbf{m}\boldsymbol{\theta}\boldsymbol{\lambda}$ and transposing that relationship to give $\boldsymbol{\theta}^T\mathbf{k} = \boldsymbol{\lambda}\boldsymbol{\theta}^T\mathbf{m}$, since $\mathbf{k}$, $\mathbf{m}$ and the diagonal matrix $\boldsymbol{\lambda}$ are all symmetric. Multiplying the first of these equations by $\boldsymbol{\theta}^T$ on the left, and the second equation by $\boldsymbol{\theta}$ on the right, gives $\boldsymbol{\theta}^T\mathbf{k}\boldsymbol{\theta} = \boldsymbol{\theta}^T\mathbf{m}\boldsymbol{\theta}\boldsymbol{\lambda}$ and $\boldsymbol{\theta}^T\mathbf{k}\boldsymbol{\theta} = \boldsymbol{\lambda}\boldsymbol{\theta}^T\mathbf{m}\boldsymbol{\theta}$. From these two equations we see that $(\boldsymbol{\theta}^T\mathbf{m}\boldsymbol{\theta})\boldsymbol{\lambda} = \boldsymbol{\lambda}(\boldsymbol{\theta}^T\mathbf{m}\boldsymbol{\theta})$, showing that the matrices $(\boldsymbol{\theta}^T\mathbf{m}\boldsymbol{\theta})$ and $\boldsymbol{\lambda}$ commute. Provided that the diagonal elements of $\boldsymbol{\lambda}$ are distinct, this condition can be satisfied only if $(\boldsymbol{\theta}^T\mathbf{m}\boldsymbol{\theta})$ is diagonal. If $\mathbf{m}^{-1}\mathbf{k}$ has one or more repeated eigenvalues, so that the elements of $\boldsymbol{\lambda}$ are not all distinct, then it is not required that $(\boldsymbol{\theta}^T\mathbf{m}\boldsymbol{\theta})$ be exactly diagonal, but it is possible to choose the eigenvectors such that it is diagonal. Thus, we will presume that

$$\boldsymbol{\gamma} \equiv \boldsymbol{\theta}^T\mathbf{m}\boldsymbol{\theta} \tag{7.18}$$

is diagonal. Furthermore, the fact that $\boldsymbol{\theta}^T\mathbf{k}\boldsymbol{\theta} = \boldsymbol{\theta}^T\mathbf{m}\boldsymbol{\theta}\boldsymbol{\lambda}$ shows that

$$\boldsymbol{\theta}^T\mathbf{k}\boldsymbol{\theta} = \boldsymbol{\gamma}\boldsymbol{\lambda} \tag{7.19}$$

which is also diagonal.

In order to simplify the MDF equation of motion, we now write $\vec{X}(t)$ as a linear expansion in terms of the eigenvectors of $\mathbf{m}^{-1}\mathbf{k}$. That is, we define a vector $\vec{Z}(t)$ such that

$$\vec{X}(t) = \boldsymbol{\theta}\vec{Z}(t) \tag{7.20}$$

giving the jth component of $\vec{Z}(t)$ as being the magnitude of the part of $\vec{X}(t)$ which is parallel to the jth eigenvector. Note that the component $\theta_{j,l}$ now gives the magnitude of the jth response component, $X_j(t)$, resulting from a unit magnitude of $Z_l(t)$. In the usual terminology, the lth eigenvector (i.e., column l of $\boldsymbol{\theta}$) is the lth mode shape and $Z_l(t)$ is the lth modal amplitude. Substituting Eq. 7.20 into Eq. 7.13 and multiplying the equation on the left by $\boldsymbol{\theta}^T$ gives

$$\boldsymbol{\theta}^T\mathbf{m}\boldsymbol{\theta}\ddot{\vec{Z}}(t) + \boldsymbol{\theta}^T\mathbf{c}\boldsymbol{\theta}\dot{\vec{Z}}(t) + \boldsymbol{\theta}^T\mathbf{k}\boldsymbol{\theta}\vec{Z}(t) = \boldsymbol{\theta}^T\vec{F}(t)$$

which can be rewritten according to Eqs. 7.18 and 7.19 as

$$\ddot{\vec{Z}}(t) + \boldsymbol{\beta}\dot{\vec{Z}}(t) + \boldsymbol{\lambda}\vec{Z}(t) = \boldsymbol{\gamma}^{-1}\boldsymbol{\theta}^T\vec{F}(t) \tag{7.21}$$

in which the new matrix β is defined such that

$$\theta^T c \theta = \gamma \beta \tag{7.22}$$

Although one can always rewrite the MDF equation of motion in the form of Eq. 7.21, this formulation is not particularly useful unless some limitations are placed on the c damping matrix. In particular, the formulation is very useful when β is diagonal, so that the jth row of the equation has the simple form

$$\ddot{Z}_j(t) + \beta_{j,j}\dot{Z}_j(t) + \lambda_{j,j}Z_j(t) = \left[\gamma^{-1}\theta^T \vec{F}(t)\right]_j = \frac{1}{\gamma_{j,j}}\sum_{k=1}^{n} \theta_{k,j} F_k(t) \tag{7.23}$$

Various terms are used to designate this special case in which β is diagonal. Caughey (1960a) called it the situation with classical normal modes, and other terms meaning the same thing include classical damping and uncoupled modes. The last term emphasizes the key fact shown in Eq. 7.23, that one can solve for $Z_j(t)$ from an equation which is completely uncoupled from the similar equations governing the behavior of $Z_l(t)$ for $l \neq j$.

For the situation with uncoupled modes, we see from Eq. 7.23 that the behavior of any modal amplitude $Z_j(t)$ is governed by a scalar, second-order, differential equation which is essentially the same as the SDF equation we have previously considered. In fact, if we define the modal frequency ω_j and the modal damping ζ_j such that $\omega_j^2 = \lambda_{j,j}$ and $2\zeta_j\omega_j = \beta_{j,j}$, then Eq. 7.23 takes exactly the form of Eq. 3.36

$$\ddot{Z}_j(t) + 2\zeta_j\omega_j\dot{Z}_j(t) + \omega_j^2 Z_j(t) = \frac{1}{\gamma_{j,j}}\sum_{l=1}^{n} \theta_{l,j} F_l(t) \tag{7.24}$$

Thus, modal analysis reduces the MDF problem to the solution of a set of SDF problems plus matrix algebra, provided that the system does have uncoupled modes.

We will now look at the conditions under which the modal equations do uncouple, giving Eq. 7.24 as the modal equations of motion. First we note that the matrix β commutes with the diagonal matrix λ so that $\lambda\beta = \beta\lambda$ if β is also diagonal. Furthermore, if all the elements of λ are distinct, then $\lambda\beta = \beta\lambda$ only if β is diagonal. Now we investigate the restrictions on m, c, and k which will result in $\lambda\beta = \beta\lambda$. Solving Eq. 7.19 for λ and Eq. 7.22 for β gives $\lambda\beta = \gamma^{-1}\theta^T k\theta\gamma^{-1}\theta^T c\theta$ and $\beta\lambda = \gamma^{-1}\theta^T c\theta\gamma^{-1}\theta^T k\theta$. From Eq. 7.18 we can note that $\gamma^{-1}\theta^T = \theta^{-1}m^{-1}$, and substitution of this relationship gives $\lambda\beta = \beta\lambda$ if and only if $km^{-1}c = cm^{-1}k$. Thus, we can be assured of having uncoupled modal equations if the coefficient matrices in the original equation of motion satisfy this condition. Alternatively, it can be stated that the modal equations uncouple if and only if $m^{-1}k$ and $m^{-1}c$ commute. If this condition is not met, then the β matrix will not be diagonal and different modal equations will be coupled by the damping terms, requiring simultaneous solution of the equations. It is possible to solve the problem when the uncoupling condition is not met, but it is necessary to use some different method. Two methods which do not require any restriction on the **m**, **c**, and **k** matrices are presented in Sections 7.5 and 7.6. Nonetheless, the most common method of solving the MDF problem is by uncoupling the equations, as presented here, or some equivalent variation of this method.

The general condition of $\mathbf{km}^{-1}\mathbf{c} = \mathbf{cm}^{-1}\mathbf{k}$ for uncoupled modal equations was presented by Caughey and O'Kelley in 1965. A much less general condition which is sufficient to assure the existence of uncoupled modes, but is not necessary for their existence, is called the Rayleigh condition, and is given by $\mathbf{c} = a_1\mathbf{m} + a_2\mathbf{k}$ for some scalar constants a_1 and a_2. It is easy to verify that the Rayleigh condition is a special case of $\mathbf{km}^{-1}\mathbf{c} = \mathbf{cm}^{-1}\mathbf{k}$. One can also show (Caughey, 1960a) that other sufficient conditions can be written as

$$\mathbf{c} = \mathbf{m}\sum_{j=0}^{J} a_j(\mathbf{m}^{-1}\mathbf{k})^j \tag{7.25}$$

which is sometimes called a generalized Rayleigh condition. It can be shown that if the upper limit J of the summation is chosen to be greater than or equal to $n-1$, then this condition is equivalent to the general condition of $\mathbf{km}^{-1}\mathbf{c} = \mathbf{cm}^{-1}\mathbf{k}$, but Eq. 7.25 is a more restrictive condition for any smaller value of J.

It seems somewhat surprising that it has been possible to model a great variety of physical systems by using the rather restrictive version of the MDF equations with uncoupled modes. One reason that this is true is that we often have little information about the precise form for the **c** damping matrix, so we are free to choose it in a way that simplifies the analysis. This is in contrast to the **m** and **k** matrices, whose elements typically are quite well approximated by calculations of physical mass and stiffness terms. In most cases, the energy dissipation in real structures occurs in very complicated, usually nonlinear, and poorly understood phenomena such as friction and local yielding. What is best known about the energy dissipation is its overall level, while the details of precisely where and how it occurs are ill defined. One of the most common ways of modeling the energy dissipation does not even begin with a choice of a **c** matrix, but rather with the choice of the ζ_j modal damping values. Knowing the λ_j values from analysis of **m** and **k**, one can then find the $\boldsymbol{\beta}$ matrix from $\beta_j = 2\zeta_j\omega_j$. If desired, one can also solve Eq. 7.22 for $\mathbf{c} = (\boldsymbol{\theta}^T)^{-1}\boldsymbol{\gamma}\boldsymbol{\beta}\boldsymbol{\theta}^{-1}$, which can be rewritten using Eq. 7.18 as $\mathbf{c} = \mathbf{m}\boldsymbol{\theta}\boldsymbol{\beta}\boldsymbol{\gamma}^{-1}\boldsymbol{\theta}^T\mathbf{m}$. Often, though, there is no need to know the values of the terms of **c**, so this calculation is not performed.

****

Example 7.2: Consider the system of Example 7.1 with: $m_1 = 2m$, $m_2 = m$, $k_1 = k_2 = k$, $c_1 = c_2 = c$, in which m, k, and c are scalar constants. Show that the system does have uncoupled modes, and find the $\boldsymbol{\theta}$, $\boldsymbol{\lambda}$, $\boldsymbol{\gamma}$, $\boldsymbol{\beta}$ matrices and the values of the modal frequencies and damping values.

Substituting the parameter values into **m**, **k**, and **c** from Example 7.1 gives

$$\mathbf{m} = m\begin{bmatrix} 2 & 0 \\ 0 & 1 \end{bmatrix} \qquad \mathbf{k} = k\begin{bmatrix} 2 & -1 \\ -1 & 1 \end{bmatrix} \qquad \mathbf{c} = c\begin{bmatrix} 2 & -1 \\ -1 & 1 \end{bmatrix}$$

Inverting $\mathbf{m}$ is almost trivial, and matrix multiplication then gives

$$\mathbf{km^{-1}c} = \mathbf{cm^{-1}k} = \frac{kc}{m}\begin{bmatrix} 3 & -2 \\ -2 & 1.5 \end{bmatrix}$$

The equality of $\mathbf{km^{-1}c}$ and $\mathbf{cm^{-1}k}$ assures us that uncoupled modes do exist. Eigenanalysis of $\mathbf{m^{-1}k}$ gives

$$\boldsymbol{\theta} = \begin{bmatrix} 0.6124 & 0.5773 \\ 0.8660 & -0.8165 \end{bmatrix} \quad \text{and} \quad \boldsymbol{\lambda} = \frac{k}{m}\begin{bmatrix} 0.2929 & 0 \\ 0 & 1.707 \end{bmatrix}$$

Note that the columns of $\boldsymbol{\theta}$ are the eigenvectors of $\mathbf{m^{-1}k}$, and their ordering is arbitrary. We have followed the common convention of putting the lower frequency mode in the first column. Note also that although the "directions" of the eigenvectors are unique, their "lengths" are arbitrary. A different scaling of these eigenvectors will affect the values of the components of $\boldsymbol{\gamma}$, but not of $\boldsymbol{\beta}$. Using $\boldsymbol{\theta}$ as written gives

$$\boldsymbol{\gamma} = m\begin{bmatrix} 1.5 & 0 \\ 0 & 1.333 \end{bmatrix} \quad \text{and} \quad \boldsymbol{\beta} = \frac{c}{m}\begin{bmatrix} 0.2929 & 0 \\ 0 & 1.707 \end{bmatrix}$$

Taking the square root of the elements of $\boldsymbol{\lambda}$ gives the modal frequencies as $\omega_1 = 0.5412(k/m)^{1/2}$ and $\omega_2 = 1.307(k/m)^{1/2}$. Dividing the elements of $\boldsymbol{\beta}$ by $2\omega_j$ gives the modal damping values as $\zeta_1 = 0.2706c/(km)^{1/2}$ and $\zeta_2 = 0.6533c/(km)^{1/2}$.

It may also be noted that one could have performed this analysis without checking to see if $\mathbf{km^{-1}c}$ and $\mathbf{cm^{-1}k}$ were equal. That is, one can perform the eigenanalysis to find $\boldsymbol{\theta}$, then evaluate $\boldsymbol{\beta}$ regardless of whether the system has uncoupled modes. The form of $\boldsymbol{\beta}$, in fact, then provides the information about the existence of uncoupled modes. A diagonal $\boldsymbol{\beta}$ matrix is sufficient evidence of uncoupled modes. Checking to see whether $\mathbf{km^{-1}c} = \mathbf{cm^{-1}k}$ allows one to avoid doing the eigenanalysis if the equations will not uncouple anyway.
****

Example 7.3: Show that the system of Example 7.1 with: $m_1 = 1000$ kg, $m_2 = 500$ kg, $k_1 = 500$ kN / m, $k_2 = 500$ kN / m, $c_1 = 0.5$ kN / (m / s), and $c_2 = 2.0$ kN / (m / s) does not have uncoupled modes.

Substituting the parameter values into $\mathbf{m}$, $\mathbf{k}$, and $\mathbf{c}$ from Example 7.1 gives

$$\mathbf{m} = \begin{bmatrix} 1000 & 0 \\ 0 & 500 \end{bmatrix} \text{kg}, \quad \mathbf{k} = \begin{bmatrix} 10^6 & -5 \times 10^5 \\ -5 \times 10^5 & 5 \times 10^5 \end{bmatrix}\frac{N}{m}, \quad \mathbf{c} = \begin{bmatrix} 2500 & -2000 \\ -2000 & 2000 \end{bmatrix}\frac{N \cdot s}{m}$$

From these matrices, we find that

$$\mathbf{cm}^{-1}\mathbf{k} = \begin{bmatrix} 4.5 & -3.25 \\ -4.0 & 3 \end{bmatrix} 10^6 \frac{N^2}{m \cdot s}$$

and

$$\mathbf{km}^{-1}\mathbf{c} = \begin{bmatrix} 4.5 & -4.0 \\ -3.25 & 3 \end{bmatrix} 10^6 \frac{N^2}{m \cdot s}$$

this shows that the system does not meet the condition necessary for uncoupled modes. If we had not performed this check and had proceeded to find $\boldsymbol{\theta}$, which is the same as in Example 7.2, then we would find that

$$\boldsymbol{\beta} = \begin{bmatrix} 0.4216 & -0.7071 \\ -0.7955 & 6.078 \end{bmatrix} \text{rad} / s$$

The fact that this matrix is not diagonal also shows that uncoupled modes do not exist.

****

Example 7.4: Consider the system of Example 7.1 with: $m_1 = m_2 = 100$ kg, $k_1 = k_2 = 500$ N / m. Find the values of dashpots in the system which will give uncoupled modes with 10% damping in the lower frequency mode and 1% damping in the higher frequency mode.

Using

$$\mathbf{m} = \begin{bmatrix} 100 & 0 \\ 0 & 100 \end{bmatrix} \text{kg}, \qquad \mathbf{k} = \begin{bmatrix} 200 & -100 \\ -100 & 100 \end{bmatrix} \frac{N}{m}$$

gives

$$\boldsymbol{\theta} = \begin{bmatrix} 0.5257 & 0.8507 \\ 0.8507 & -0.5257 \end{bmatrix}, \qquad \boldsymbol{\lambda} = \begin{bmatrix} 0.3820 & 0 \\ 0 & 2.618 \end{bmatrix} (\text{rad} / s)^2, \qquad \boldsymbol{\gamma} = \begin{bmatrix} 100 & 0 \\ 0 & 100 \end{bmatrix} \text{kg}$$

Setting $\omega_j = \lambda_j^{1/2}$ and $\beta_j = 2\zeta_j \omega_j$, with $\zeta_1 = 0.10$ and $\zeta_2 = 0.01$, gives

$$\boldsymbol{\beta} = \begin{bmatrix} 0.1236 & 0 \\ 0 & 0.0324 \end{bmatrix} \text{rad} / s$$

Solving Eq. 7.22 for $\mathbf{c}$ gives $\mathbf{c} = \boldsymbol{\theta}^{-1^T} \boldsymbol{\gamma} \boldsymbol{\beta} \boldsymbol{\theta}^{-1}$. Rather than invert a matrix which is not diagonal, we can now solve Eq. 7.18 for the inverse as $\boldsymbol{\theta}^{-1} = \boldsymbol{\gamma}^{-1} \boldsymbol{\theta}^T \mathbf{m}$ and use this to obtain

$$\mathbf{c} = \mathbf{m} \boldsymbol{\theta} \boldsymbol{\beta} \boldsymbol{\gamma}^{-1} \boldsymbol{\theta}^T \mathbf{m} = \begin{bmatrix} 5.758 & 4.081 \\ 4.081 & 9.839 \end{bmatrix}$$

(This technique for finding $\boldsymbol{\theta}^{-1}$ is not particularly important for a 2DF, but can save significant computational time for large matrices.) Using $\mathbf{m}$, $\mathbf{k}$, and $\mathbf{c}$ now gives the scalar equations of motion as

$$100\ddot{X}_1(t) + 5.758\dot{X}_1(t) + 4.081\dot{X}_2(t) + 200X_1(t) - 100X_2(t) = F_1(t)$$

and

$$100\ddot{X}_2(t) + 4.081\dot{X}_1(t) + 9.839\dot{X}_2(t) - 100X_1(t) + 100X_2(t) = F_2(t)$$

The sketch shows an arrangement of dashpots which will give these equations of motion. Note, in particular that the value of the dashpot connecting the two masses is negative.

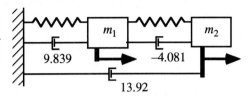

Obviously, a negative dashpot is not a physical element, but that is the result obtained any time that an off-diagonal element of **c** is found to be positive. Furthermore, it is not unusual for such a procedure of assigning values of modal damping to result in a model which does not correspond to physical dashpots. This anomaly does not cause any difficulty in the mathematical analysis, or any pathological behavior in the dynamic response, in spite of its apparent lack of logical explanation. One should always keep in mind, though, that even when it corresponds to positive dashpots, the linear damping matrix **c** usually represents no more than a crude approximation of the actual energy losses limiting the dynamic response of a real system.

****

7.4   Time Domain Stochastic Analysis of Uncoupled MDF Systems

There are two different approaches that one can now use in performing a time domain analysis of the MDF system described by Eq. 7.13, with uncoupled modal responses described by Eq. 7.24. The results, of course, are the same, but the manipulations are somewhat different. The first approach is to use the results of Section 7.3 strictly for the deterministic analysis which gives the **h**(t) matrix of impulse response functions needed in order to use the stochastic analysis of Section 7.1. The second approach is to consider the equations of Section 7.3 to be stochastic, and to use them directly in finding such things as the mean and covariance of the $\vec{X}(t)$ response process. We will look briefly at both approaches.

In order to find the **h**(t) matrix, we must consider the response of the MDF system of Eq. 7.13 to excitations which consist solely of a single Dirac delta function pulse. As noted in Section 7.1 the lth column of **h**(t), written as $[h_{1,l}(t),\cdots,h_{n,l}(t)]^T$, is the $\vec{X}(t)$ response vector when the only excitation is $\vec{F}_l(t) = \delta(t)$. We can now use Eq. 7.23 or 7.24 to obtain the excitation of the jth modal equation as $\theta_{l,j}\delta(t) / \gamma_{j,j}$. Thus, one finds that the modal responses to this Dirac delta function excitation at location l are $Z_j(t) = \theta_{l,j}\hat{h}_{j,j}(t)$ in which $\hat{h}_{j,j}(t)$ represents the impulse response function of the jth modal equation, and from our study of the SDF system is known to be given by

$$\hat{h}_{j,j}(t) = \frac{1}{\gamma_{j,j}\omega_{dj}} e^{-\zeta_j\omega_j t} \sin(\omega_{dj}t) U(t) \tag{7.26}$$

with the damped modal frequency ω_{dj} given by $\omega_{dj} \equiv \omega_j(1-\zeta_j^2)^{1/2}$. Note that $\gamma_{j,j}$ in this equation is analogous to the mass m in the SDF formula of Example 3.3. Thus, $\gamma_{j,j}$ can be considered to be the modal mass. Using these modal impulse responses in Eq. 7.20 gives any element of column l of $\mathbf{h}(t)$ as

$$h_{r,l}(t) = X_r(t) = \sum_{j=1}^{n} \theta_{r,j} Z_j(t) = \sum_{j=1}^{n} \theta_{r,j} \theta_{l,j} \hat{h}_{j,j}(t) \tag{7.27}$$

From this equation for a typical element of the $\mathbf{h}(t)$ matrix, we can write the entire relationship in matrix form as

$$\mathbf{h}(t) = \mathbf{\theta} \hat{\mathbf{h}}(t) \mathbf{\theta}^T \tag{7.28}$$

in which $\hat{\mathbf{h}}(t)$ denotes a diagonal matrix of the modal impulse response functions given in Eq. 7.26.

These equations show that it is quite straightforward to obtain the MDF impulse response function matrix from the analysis of uncoupled equations. The largest computational effort in the procedure is the eigenanalysis to find $\mathbf{\theta}$ and $\mathbf{\lambda}$. The other operations involve only arithmetic. Note that our original MDF equation allowed the possibility of an exciting force at each of the n coordinate points, and we have chosen to include all of the coordinates in our definition of $\mathbf{h}(t)$. Obviously this gives a square $\mathbf{h}(t)$ matrix. If we wish to perform our stochastic analysis only for a subset of the coordinate points, then we can consider k in Eq. 7.27 to vary only over that subset, and thereby obtain a rectangular $\mathbf{h}(t)$ matrix appropriate for those points. Once the $\mathbf{h}(t)$ matrix is determined, the mean value vector for the stochastic response can be found from Eq. 7.5, the autocorrelation matrix is given by Eq. 7.7 or 7.9, and the autocovariance matrix is given by Eq. 7.8 or 7.10.

Now we will consider the alternative of doing stochastic analysis directly on the equations of Section 7.3, rather than using those equations to obtain the $\mathbf{h}(t)$ matrix. From Eq. 7.20 we can write the mean value vector of the response as $\vec{\mu}_X(t) = \mathbf{\theta}\vec{\mu}_Z(t)$, the autocorrelation matrix as $\mathbf{\phi}_{XX}(t,s) \equiv E[\vec{X}(t)\vec{X}^T(s)] = \mathbf{\theta}\mathbf{\phi}_{ZZ}(t,s)\mathbf{\theta}^T$, and the autocovariance matrix as $\mathbf{K}_{XX}(t,s) = \mathbf{\theta}\mathbf{K}_{ZZ}(t,s)\mathbf{\theta}^T$. The time histories of the components of the stochastic $\vec{Z}(t)$ vector of modal responses are obtained from Eq. 7.23 as

$$Z_j(t) = \sum_{k=1}^{n} \theta_{k,j} \int_{-\infty}^{\infty} \hat{h}_{j,j}(t-s) F_k(s)\, ds \tag{7.29}$$

or in matrix form as

$$\vec{Z}(t) = \int_{-\infty}^{\infty} \hat{\mathbf{h}}(t-u)\mathbf{\theta}^T \vec{F}(u)\, du \tag{7.30}$$

Thus, the mean value vector for the response is

$$\vec{\mu}_X(t) = \mathbf{\theta}\vec{\mu}_Z(t) = \int_{-\infty}^{\infty} \mathbf{\theta}\hat{\mathbf{h}}(t-u)\mathbf{\theta}^T \vec{\mu}_F(u)\, du \tag{7.31}$$

and the autocorrelation matrix for $\{\vec{Z}(t)\}$ is

$$\boldsymbol{\phi}_{ZZ}(t,s) = \int_{-\infty}^{\infty} \int_{-\infty}^{\infty} \hat{\mathbf{h}}(t-u) \boldsymbol{\theta}^T \boldsymbol{\phi}_{FF}(u,v) \boldsymbol{\theta} \hat{\mathbf{h}}^T(s-v)\, du\, dv \qquad (7.32)$$

which gives the corresponding result for $\{ \vec{X}(t) \}$ as

$$\boldsymbol{\phi}_{XX}(t,s) = \boldsymbol{\theta} \int_{-\infty}^{\infty} \int_{-\infty}^{\infty} \hat{\mathbf{h}}(t-u) \boldsymbol{\theta}^T \boldsymbol{\phi}_{FF}(u,v) \boldsymbol{\theta} \hat{\mathbf{h}}^T(s-v)\, du\, dv\, \boldsymbol{\theta}^T \qquad (7.33)$$

Similarly the autocovariance matrices are

$$\mathbf{K}_{ZZ}(t,s) = \int_{-\infty}^{\infty} \int_{-\infty}^{\infty} \hat{\mathbf{h}}(t-u) \boldsymbol{\theta}^T \mathbf{K}_{FF}(u,v) \boldsymbol{\theta} \hat{\mathbf{h}}^T(s-v)\, du\, dv \qquad (7.34)$$

and

$$\mathbf{K}_{XX}(t,s) = \boldsymbol{\theta} \int_{-\infty}^{\infty} \int_{-\infty}^{\infty} \hat{\mathbf{h}}(t-u) \boldsymbol{\theta}^T \mathbf{K}_{FF}(u,v) \boldsymbol{\theta} \hat{\mathbf{h}}^T(s-v)\, du\, dv\, \boldsymbol{\theta}^T \qquad (7.35)$$

It must be remembered that the expressions in Eqs. 7.32 - 7.35 are called autocorrelation and autocovariance matrices because each refers to only one response vector process, either $\{ \vec{Z}(t) \}$ or $\{ \vec{X}(t) \}$. At the same time, most of the elements of these $\boldsymbol{\phi}$ and $\mathbf{K}$ matrices are cross-correlation and cross-covariance terms of the scalar components of the response vectors. Only the diagonal terms of the matrices are autocorrelation and autocovariance terms of the scalar components. For example, the typical term of Eq. 7.34 can be written as $[K_{ZZ}(t,s)]_{j,l} \equiv K_{Z_j(t),Z_l(s)}$.

One can learn much about the relative importance of the various modal contributions to the response by considering the special case of response at time t to a white noise excitation process $\{ \vec{F}(t) \}$. In particular, we will consider the case with $\vec{F}(t)$ and $\vec{F}(s)$ being uncorrelated vectors for $t \neq s$, so that $\mathbf{K}_{FF}(t,s) = 2\pi \mathbf{S}_0 \delta(t-s)$, in which the autospectral density matrix $\mathbf{S}_0$ is generally a full square matrix. This allows the various components of $\vec{F}(t)$ to be correlated at any one instant of time t, even though they are uncorrelated at distinct times. Substituting this relationship into Eq. 7.35 for the case of $t = s$ gives the matrix of response variance and covariance values as

$$\mathbf{K}_{XX}(t,t) = 2\pi \boldsymbol{\theta} \int_{-\infty}^{\infty} \hat{\mathbf{h}}(t-u) \boldsymbol{\theta}^T \mathbf{S}_0 \boldsymbol{\theta} \hat{\mathbf{h}}^T(t-u)\, du\, \boldsymbol{\theta}^T$$

so that the typical component can be written as

$$K_{X_j(t),X_l(t)} = 2\pi \sum_{r_1=1}^{n} \sum_{r_2=1}^{n} \sum_{r_3=1}^{n} \sum_{r_4=1}^{n} \theta_{j,r_1} \theta_{r_2,r_1} [S_0]_{r_2,r_3} \theta_{r_3,r_4} \theta_{l,r_4} q_{r_1,r_4} \qquad (7.36)$$

in which

$$q_{r_1,r_4} \equiv \int_{-\infty}^{\infty} \hat{h}_{r_1,r_1}(t-u) \hat{h}_{r_4,r_4}(t-u)\, du = \int_{0}^{\infty} \hat{h}_{r_1,r_1}(v) \hat{h}_{r_4,r_4}(v)\, dv$$

One can now substitute Eq. 7.26 into this integrand, and perform the integration to obtain

$$q_{r_1,r_4} = \frac{2(\zeta_{r_1}\omega_{r_1} + \zeta_{r_4}\omega_{r_4})}{\gamma_{r_1,r_1}\gamma_{r_4,r_4}[(\omega_{r_1}^2 - \omega_{r_4}^2)^2 + 4\omega_{r_1}\omega_{r_4}(\zeta_{r_1}\omega_{r_1} + \zeta_{r_4}\omega_{r_4})(\zeta_{r_1}\omega_{r_4} + \zeta_{r_4}\omega_{r_1})]} \tag{7.37}$$

Setting $r_4 = r_1$ in this expression gives the result for the special case as

$$q_{r_1,r_1} = \frac{1}{4\gamma_{r_1,r_1}^2 \zeta_{r_1}\omega_{r_1}^3} \tag{7.38}$$

which agrees exactly with the SDF result in Eq. 3.57.

Note that Eq. 7.38 is proportional to modal damping raised to the −1 power, so that it is quite large for the commonly considered systems with small damping values. Furthermore, Eq. 7.37 is of the order of modal damping to the +1 power, provided that ω_{r_1} and ω_{r_4} are well separated. Thus, if all modal frequencies are well separated, then the "off-diagonal" terms with $r_4 \neq r_1$ in Eq. 7.36 are much smaller than are the "on-diagonal" terms with $r_4 = r_1$. In particular, the ratio between the off-diagonal and on-diagonal terms is of order damping squared. For modal damping values of less than 10%, for instance, one can expect the off-diagonal terms to contribute relatively little to the covariance of the $\vec{X}(t)$ responses. Thus, in many situations it is possible to neglect the terms from Eq. 7.36 with $r_4 \neq r_1$ and approximate Eq. 7.35 by considering only the $r_4 = r_1$ terms, giving

$$K_{X_j(t),X_l(s)} \approx \sum_{r_1=1}^{n}\sum_{r_2=1}^{n}\sum_{r_3=1}^{n} \theta_{j,r_1}\theta_{r_2,r_1}[\mathbf{K}_{FF}(u,v)]_{r_2,r_3}\theta_{r_3,r_1}\theta_{l,r_1}q_{r_1,r_1} \tag{7.39}$$

Furthermore, one notes from Eq. 7.38 that the q_{r_1,r_1} modal contributions are proportional to the modal frequency raised to the −3 power, so that high-frequency modes will generally contribute much less to the response than will low-frequency modes. This leads to the possibility of further approximating Eq. 7.39 by limiting the range of r_1 to include only the lower frequency modes.

Although the approximations in the preceding paragraph have been justified by consideration of the response to white noise, they can also be considered to be appropriate for other broadband excitations. It is important to notice that if two modes have nearly equal frequency, then the approximation in Eq. 7.39 is not valid. For example, if $\omega_l - \omega_j$ is of the order of the damping, then the denominator in Eq. 7.37 is of the order of damping squared, so that $q_{j,l}$ may be equally as significant as $q_{j,j}$ and $q_{l,l}$. Thus, one must exercise some caution in deciding whether to use the simplification of Eq. 7.39. For situations with closely spaced modal frequencies, it is sometimes convenient to rewrite Eqs. 7.37 and 7.38 in the form of the correlation coefficient for the modal responses: $\rho_{r_1,r_4} = q_{r_1,r_4}(q_{r_1,r_1}q_{r_4,r_4})^{-1/2}$. Numerical investigations (Der Kiureghian, 1980) have shown that this correlation coefficient is approximately correct for non-white broadband excitations, and this has formed the basis for a complete-quadratic-combination (CQC) method for accurately computing the response of MDF systems, even if they have closely spaced modal frequencies

(Wilson et al., 1981). It should also be noted that some continuous stuctures are very likely to have closely spaced modal frequencies. Elishakoff (1995) has particularly noted this fact for shells. The cross-correlation of modal responses should generally be expected to be important in these situations.

∗∗

Example 7.5: Find the $\mathbf{h}(t)$ impulse response function matrix for the 2DF model of Example 7.2, with $m = 500\,\text{kg}$, $k = 500\,\text{kN/m}$, and $c = 500\,\text{N/(m/s)}$. This gives the matrices as

$$\mathbf{m} = \begin{bmatrix} 1000 & 0 \\ 0 & 500 \end{bmatrix}\text{kg}, \quad \mathbf{k} = \begin{bmatrix} 10^6 & -5\times10^5 \\ -5\times10^5 & 5\times10^5 \end{bmatrix}\frac{\text{N}}{\text{m}}, \quad \mathbf{c} = \begin{bmatrix} 1000 & -500 \\ -500 & 500 \end{bmatrix}\frac{\text{N}\cdot\text{s}}{\text{m}}$$

It was found in Example 7.2 that this system has uncoupled modes, and the results there give $\gamma_{1,1} = 750\,\text{kg}$, $\gamma_{2,2} = 666.7\,\text{kg}$, $\omega_1 = 17.11\,\text{rad/s}$, $\omega_2 = 41.32\,\text{rad/s}$, $\zeta_1 = 0.00856$ and $\zeta_2 = 0.02066$. Thus, Eq. 7.26 gives

$$\hat{\mathbf{h}}(t) = \begin{bmatrix} \dfrac{e^{-0.1464t}\sin(17.11t)}{(750)(17.11)} & 0 \\[2ex] 0 & \dfrac{e^{-0.8536t}\sin(41.31t)}{(666.7)(41.31)} \end{bmatrix} U(t)$$

Using Eq. 7.28 with

$$\boldsymbol{\theta} = \begin{bmatrix} 0.6124 & 0.5773 \\ 0.8660 & -0.8165 \end{bmatrix}$$

then gives the elements of $\mathbf{h}(t)$ as

$$h_{1,1}(t) = \theta_{1,1}^2 \hat{h}_{1,1}(t) + \theta_{1,2}^2 \hat{h}_{2,2}(t)$$
$$= [2.922\times10^{-5}e^{-0.1464t}\sin(17.11t) + 1.210\times10^{-5}e^{-0.8536t}\sin(41.31t)]U(t)$$

$$h_{1,2}(t) = h_{2,1}(t) = \theta_{1,1}\theta_{2,1}\hat{h}_{1,1}(t) + \theta_{1,2}\theta_{2,2}\hat{h}_{2,2}(t)$$
$$= [4.132\times10^{-5}e^{-0.1464t}\sin(17.11t) - 1.712\times10^{-5}e^{-0.8536t}\sin(41.31t)]U(t)$$

$$h_{2,2}(t) = h_{2,1}(t) = \theta_{2,1}^2 \hat{h}_{1,1}(t) + \theta_{2,2}^2 \hat{h}_{2,2}(t)$$
$$= [5.843\times10^{-5}e^{-0.1464t}\sin(17.11t) + 2.421\times10^{-5}e^{-0.8536t}\sin(41.31t)]U(t)$$

∗∗

Example 7.6: Find $E(X_1^2)$ for the stationary response of the oscillator shown when the base acceleration $\{\ddot{Y}(t)\}$ is mean zero, white noise with autospectral density $S_0 = 0.1\,(\text{m/s}^2)/(\text{rad/s})$.

The system is described by Eq. 7.13 with $\mathbf{m}$, $\mathbf{k}$, and, $\mathbf{c}$ as given in Example 7.5, $F_1(t) = -m_{1,1}\ddot{Y}(t)$, and $F_2(t) = -m_{2,2}\ddot{Y}(t)$.

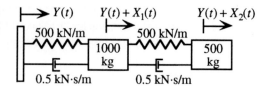

One approach is to write

$$X_1(t) = \int_0^\infty [m_{1,1}h_{1,1}(\tau) + m_{2,2}h_{1,2}(\tau)]\ddot{Y}(t - \tau)\,d\tau$$

Using $E[Y(u)Y(v)] = 2\pi S_0 \delta(u - v)$ and substituting the $h_{1,1}(s)$ and $h_{1,2}(s)$ impulse response functions found in Example 7.5 then gives

$$
\begin{aligned}
E(X_1^2) &= 2\pi S_0 \int_0^\infty [m_{1,1}h_{1,1}(s) + m_{2,2}h_{1,2}(s)]^2\,ds \\
&= 2\pi S_0 \bigg(2.488\times10^{-3}\int_0^\infty e^{-0.2929s}\sin^2(17.11s)\,ds \\
&\quad + 3.536\times10^{-4}\int_0^\infty e^{-1.0s}\sin(17.11s)\sin(41.31s)\,ds \\
&\quad + 1.257\times10^{-5}\int_0^\infty e^{-1.707s}\sin^2(41.31s)\,ds \bigg)
\end{aligned}
$$

Performing the integration then reduces this to

$$
\begin{aligned}
E(X_1^2) &= 2\pi S_0(4.246\times10^{-3} + 2.498\times10^{-7} + 3.680\times10^{-6}) = 2\pi S_0(4.250\times10^{-3}) \\
&= 2.671\times10^{-3}\,\mathrm{m}^2
\end{aligned}
$$

Note that the first term in this form is due to the first mode of the system, the second term is cross-modal, and the final term is due to the second mode. Clearly the cross-modal term and the second mode term are both much smaller than the first mode term, as discussed in conjunction with Eq. 7.39. The insignificance of the second mode is exaggerated in this example by the fact that the excitation has $F_1(t)$ in phase with $F_2(t)$, and this distribution of forces is not effective in exciting the second mode, which has $X_1(t)$ 180° out of phase with $X_2(t)$.

Alternatively, one can investigate the mean squared modal responses, then use these in finding $E(X_1^2)$. For the excitation given, the modal equations of Eq. 7.21 can be written as

$$\gamma\ddot{\vec{Z}}(t) + \gamma\beta\dot{\vec{Z}}(t) + \gamma\lambda\vec{Z}(t) = -\boldsymbol{\theta}^T\begin{bmatrix} m_{1,1} \\ m_{2,2}\end{bmatrix}\ddot{Y}(t) = -\begin{bmatrix} 1045 \\ 169.1\end{bmatrix}\ddot{Y}(t)$$

and this gives the modal responses as

$$E(Z_1^2) = \frac{\pi S_0}{2\gamma_{1,1}^2 \zeta_1 \omega_1^3}(1045)^2 = (2\pi S_0)(1.132 \times 10^{-2})$$

and

$$E(Z_2^2) = \frac{\pi S_0}{2\gamma_{2,2}^2 \zeta_2 \omega_2^3}(169.1)^2 = (2\pi S_0)(1.104 \times 10^{-5})$$

The cross-modal response from Eq. 7.37 is

$$E(Z_1 Z_2) = \frac{4\pi S_0(\zeta_1\omega_1 + \zeta_2\omega_2)(1045)(169.1)}{\gamma_{1,1}^2 \gamma_{2,2}^2[(\omega_1^2 - \omega_2^2)^2 + 4\omega_1\omega_2(\zeta_1\omega_1 + \zeta_2\omega_2)(\zeta_1\omega_2 + \zeta_2\omega_1)]} = (2\pi S_0)(3.532 \times 10^{-7})$$

Using the fact that $X_1(t) = \theta_{1,1} Z_1(t) + \theta_{1,2} Z_2(t)$ then gives

$$E(X_1^2) = \theta_{1,1}^2 E(Z_1^2) + 2\theta_{1,1}\theta_{1,2}E(Z_1 Z_2) + \theta_{1,2}^2 E(Z_2^2)$$
$$= 2\pi S_0(4.246 \times 10^{-3} + 2.498 \times 10^{-7} + 3.680 \times 10^{-6}) = 2.671 \times 10^{-3}\,\mathrm{m}^2$$

Note that exactly the same individual terms are summed in the two approaches to solving the problem. The difference is simply in the order of performance of the operations.
****

Example 7.7: Consider the system of Example 7.2 for the special case in which $F_2(t) \equiv 0$ and $F_1(t)$ is mean zero white noise with an autospectral density of S_0. Find the autocorrelation function of the $\{X_1(t)\}$ response component. That is, find $R_{X_1 X_1}(\tau) \equiv E[X_1(t + \tau)X_1(t)]$.

Using modal superposition, we note that $\mathbf{R}_{XX}(\tau) = \boldsymbol{\theta}\mathbf{R}_{ZZ}(\tau)\boldsymbol{\theta}^T$, so that

$$R_{X_1 X_1}(\tau) = \theta_{1,1}^2 R_{Z_1 Z_1}(\tau) + \theta_{1,1}\theta_{1,2}[R_{Z_1 Z_2}(\tau) + R_{Z_2 Z_1}(\tau)] + \theta_{1,2}^2 R_{Z_2 Z_2}(\tau)$$

The uncoupled modal equations of motion can be written as

$$\ddot{\vec{Z}}(t) + \boldsymbol{\beta}\dot{\vec{Z}}(t) + \boldsymbol{\lambda}\vec{Z}(t) = \boldsymbol{\gamma}^{-1}\boldsymbol{\theta}^T \vec{F}(t)$$

or

$$\ddot{Z}_1(t) + \beta_{1,1}\dot{Z}_1(t) + \lambda_{1,1}Z_1(t) = \theta_{1,1}F_1(t) / \gamma_{1,1}$$

$$\ddot{Z}_2(t) + \beta_{2,2}\dot{Z}_2(t) + \lambda_{2,2}Z_2(t) = \theta_{1,2}F_1(t) / \gamma_{2,2}$$

The responses of these equations can be written by Duhamel integrals as

$$Z_1(t) = \frac{\theta_{1,1}}{\gamma_{1,1}}\int_0^\infty F_1(t-s)\frac{e^{-\zeta_1\omega_1 s}}{\omega_{d1}}\sin(\omega_{d1}s)\,ds$$

$$Z_2(t) = \frac{\theta_{1,2}}{\gamma_{2,2}} \int_0^\infty F_1(t-s) \frac{e^{-\zeta_2\omega_2 s}}{\omega_{d2}} \sin(\omega_{d2} s)\, ds$$

The $R_{Z_1Z_1}(\tau)$ and $R_{Z_2Z_1}(\tau)$ terms are easily found, since they are essentially the same as the autocorrelation function of the response of a SDF system. The cross-modal terms, though, are new and must be evaluated. Multiplying appropriate versions of the two modal response integrals and taking the expected value gives

$$R_{Z_1Z_2}(\tau) \equiv E[Z_1(t+\tau)Z_2(t)] = \frac{\theta_{1,1}\theta_{1,2}}{\gamma_{1,1}\gamma_{2,2}} \int_0^\infty \int_0^\infty R_{F_1F_1}(t+\tau-s_1, t-s_2) \frac{e^{-\zeta_1\omega_1 s_1} e^{-\zeta_2\omega_2 s_2}}{\omega_{d1}\omega_{d2}}$$
$$\sin(\omega_{d1} s_1)\sin(\omega_{d2} s_2)\, ds_1\, ds_2$$

Substituting the white noise relationship of $R_{F_1F_1}(t+\tau-s_1, t-s_2) = 2\pi S_0 \delta(\tau-s_1+s_2)$ then reduces this to a single integral:

$$R_{Z_1Z_2}(\tau) = \frac{\theta_{1,1}\theta_{1,2}}{\gamma_{1,1}\gamma_{2,2}} \frac{2\pi S_0}{\omega_{d1}\omega_{d2}} e^{-\zeta_1\omega_1 \tau} \int_{\max(0,-\tau)}^\infty e^{-(\zeta_1\omega_1+\zeta_2\omega_2)s_2} \sin[\omega_{d1}(\tau+s_2)]\sin(\omega_{d2} s_2)\, ds_2$$

The integration can now be simplified by substituting the exponential form of the sine function terms. For $\tau > 0$, the result of integration and simplification can be written as

$$R_{Z_1Z_2}(\tau) = \frac{\theta_{1,1}\theta_{1,2}}{\gamma_{1,1}\gamma_{2,2}} \frac{2\pi S_0}{\omega_{d1}} e^{-\zeta_1\omega_1 \tau}$$
$$\frac{2\omega_{d1}(\zeta_1\omega_1+\zeta_2\omega_2)\cos(\omega_{d1}\tau) + [(\zeta_1\omega_1+\zeta_2\omega_2)^2 + \omega_{d2}^2 - \omega_{d1}^2]\sin(\omega_{d1}\tau)}{(\omega_1^2-\omega_2^2)^2 + 4\omega_1\omega_2(\zeta_1\omega_1+\zeta_2\omega_2)(\zeta_1\omega_2+\zeta_2\omega_1)]}$$

while $\tau < 0$ gives

$$R_{Z_1Z_2}(\tau) = \frac{\theta_{1,1}\theta_{1,2}}{\gamma_{1,1}\gamma_{2,2}} \frac{2\pi S_0}{\omega_{d2}} e^{\zeta_2\omega_2 \tau}$$
$$\frac{2\omega_{d2}(\zeta_1\omega_1+\zeta_2\omega_2)\cos(\omega_{d2}\tau) - [(\zeta_1\omega_1+\zeta_2\omega_2)^2 + \omega_{d1}^2 - \omega_{d2}^2]\sin(\omega_{d2}\tau)}{(\omega_1^2-\omega_2^2)^2 + 4\omega_1\omega_2(\zeta_1\omega_1+\zeta_2\omega_2)(\zeta_1\omega_2+\zeta_2\omega_1)]}$$

It should now be noted that stationarity allows us to write

$$R_{Z_2Z_1}(\tau) \equiv E[Z_2(t+\tau)Z_1(t)] = E[Z_2(t)Z_1(t-\tau)] = R_{Z_1Z_2}(-\tau)$$

so that $R_{Z_2Z_1}(\tau)$ can be found from the expressions already derived.

By modifying Eq. 3.54 to apply to our modal equations, we find the other two terms as

$$R_{Z_1 Z_1}(\tau) = \frac{\theta_{1,1}^2}{\gamma_{1,1}^2} \frac{\pi S_0}{2 \zeta_1 \omega_1^3} e^{-\zeta_1 \omega_1 |\tau|} \left[\cos(\omega_{d1}\tau) + \frac{\zeta_1 \omega_1}{\omega_{d1}} \sin(\omega_{d1}|\tau|) \right]$$

and

$$R_{Z_2 Z_2}(\tau) = \frac{\theta_{1,2}^2}{\gamma_{2,2}^2} \frac{\pi S_0}{2 \zeta_2 \omega_2^3} e^{-\zeta_2 \omega_2 |\tau|} \left[\cos(\omega_{d2}\tau) + \frac{\zeta_2 \omega_2}{\omega_{d2}} \sin(\omega_{d2}|\tau|) \right]$$

so that $R_{X_1 X_1}(\tau)$ can be found by superposition.

If damping values are small and the modal frequencies are well separated, then it is unlikely that $R_{Z_1 Z_2}(\tau)$ and $R_{Z_2 Z_1}(\tau)$ will contribute significantly to the value of $R_{X_1 X_1}(\tau)$. The argument is the same as was given in regard to Eqs. 7.37 and 7.38. Namely, the $R_{Z_1 Z_2}(\tau)$ and $R_{Z_2 Z_1}(\tau)$ terms are of the order of damping, while $R_{Z_1 Z_1}(\tau)$ and $R_{Z_2 Z_2}(\tau)$ are much larger since they are of the order of damping to the -1 power. Thus, one may often neglect the cross-modal terms, which is equivalent to assuming that the modal responses are independent of each other. Furthermore, if one modal frequency is much larger than the other then the presence of the frequency cubed term in the denominators of $R_{Z_1 Z_1}(\tau)$ and $R_{Z_2 Z_2}(\tau)$ may cause the higher frequency mode to contribute very little to the autocorrelation function, just as was previously noted for the covariance matrix for any single value of time.

****

7.5 Frequency Domain Analysis of MDF Systems

As with time domain analysis, there are two ways to proceed with the frequency domain analysis of MDF systems. In the first approach one uses deterministic analysis of the MDF equation of motion to obtain the harmonic transfer matrix $\mathbf{H}(\omega)$, then uses the equations of Section 7.1 for the stochastic analysis. The second approach consists of direct stochastic analysis of the MDF equations. Within the first approach, though, we will consider two different possible techniques for finding $\mathbf{H}(\omega)$.

If the MDF system has uncoupled modes, one can first use Eq. 7.24 to find a $\hat{\mathbf{H}}(\omega)$ harmonic transfer matrix describing the modal responses to a harmonic excitation, then use that in finding $\mathbf{H}(\omega)$ which describes the $\vec{X}(t)$ response vector. In particular, using an excitation which consists of only one harmonic component $F_l(t) = e^{i\omega t}$ in Eq. 7.24 gives

$$\ddot{Z}_j(t) + 2 \zeta_j \omega_j \dot{Z}_j(t) + \omega_j^2 Z_j(t) = \frac{\theta_{l,j}}{\gamma_{j,j}} e^{i\omega t}$$

and the fact that the response is defined to be $Z_j(t) = \hat{H}_{j,l}(\omega) e^{i\omega t}$ gives the (j,l) element of the harmonic transfer matrix as

$$\hat{H}_{j,l}(\omega) = \frac{\theta_{l,j}}{\gamma_{j,j}(\omega_j^2 - \omega^2 + 2i\zeta_j\omega_j\omega)}$$

Alternatively, this can be written in matrix form as

$$\hat{\mathbf{H}}(\omega) = \boldsymbol{\gamma}^{-1}[\boldsymbol{\lambda} - \omega^2\mathbf{I} + i\omega\boldsymbol{\beta}]^{-1}\boldsymbol{\theta}^T \qquad (7.40)$$

in which the inverse operations are essentially trivial, since they are for diagonal matrices. One can now use Eq. 7.20 to find the $\vec{X}(t)$ responses to the single harmonic excitation component, thereby obtaining the $\mathbf{H}(\omega)$ matrix as

$$\mathbf{H}(\omega) = \boldsymbol{\theta}\hat{\mathbf{H}}(\omega)\boldsymbol{\theta}^T = \boldsymbol{\theta}\boldsymbol{\gamma}^{-1}[\boldsymbol{\lambda} - \omega^2\mathbf{I} + i\omega\boldsymbol{\beta}]^{-1}\boldsymbol{\theta}^T \qquad (7.41)$$

One can also derive the $\mathbf{H}(\omega)$ matrix without using the modal equations. In particular, one can take the Fourier transform of Eq. 7.13, giving

$$[\mathbf{k} - \omega^2\mathbf{m} + i\omega\,\mathbf{c}]\vec{\tilde{X}}(\omega) = \vec{\tilde{F}}(\omega)$$

then solve this equation for $\vec{\tilde{X}}(\omega)$ as

$$\vec{\tilde{X}}(\omega) = [\mathbf{k} - \omega^2\mathbf{m} + i\omega\,\mathbf{c}]^{-1}\vec{\tilde{F}}(\omega)$$

Comparing this with Eq. 7.4 shows that

$$\mathbf{H}(\omega) = [\mathbf{k} - \omega^2\mathbf{m} + i\omega\,\mathbf{c}]^{-1} \qquad (7.42)$$

A little matrix manipulation shows that Eqs. 7.41 and 7.42 are exactly equivalent. In fact, Eq. 7.41 can be considered to be the version of Eq. 7.42 which results from using eigenanalysis to simplify the problem of inverting the matrix. In Eq. 7.41 the inversion is of a diagonal matrix, so is an almost trivial operation, whereas the inversion in Eq. 7.42 is of a general square matrix and will generally be done numerically unless the dimension n is quite small. On the other hand, use of Eq. 7.41 involves significant computation in the determination of $\boldsymbol{\theta}$ and $\boldsymbol{\lambda}$ from eigenanalysis. If one wants $\mathbf{H}(\omega)$ only for a single frequency, there is no particular advantage of one approach over the other. In general, though, one wants to know $\mathbf{H}(\omega)$ for many frequencies, and in this situation Eq. 7.41 is more efficient since the eigenanalysis need be performed only once, whereas the matrix to be inverted in Eq. 7.42 is different for each frequency value. If n is sufficiently small, then one can perform the inversion in Eq. 7.42 analytically as a function of ω, and this is a very practical approach to solving the problem.

There is one major advantage to Eq. 7.42, as compared to Eq. 7.41. Namely, Eq. 7.42 does not require the existence of uncoupled modes. Thus, one can use this approach for almost any $\mathbf{m}$, $\mathbf{c}$, and $\mathbf{k}$ matrices. Exceptions do exist, such as when ω is equal to a modal frequency of $\mathbf{m}^{-1}\mathbf{k}$ and $\mathbf{c}$ is such that it gives no damping in that particular mode, but these pathological cases are not of great practical interest.

One should note that Eq. 7.42 also gives us an alternative method for finding the matrix $\mathbf{h}(t)$ of impulse response functions. In particular, the inverse Fourier transform of Eq. 7.42 gives

$$\mathbf{h}(t) = \frac{1}{2\pi} \int_{-\infty}^{\infty} [\mathbf{k} - \omega^2 \mathbf{m} + i\omega \mathbf{c}]^{-1} e^{i\omega t} d\omega \tag{7.43}$$

If the system of interest has uncoupled modes then Eq. 7.43 will generally not be as efficient as the procedure in Section 7.4, but Eq. 7.43 does provide a possible method for finding $\mathbf{h}(t)$ for a system which does not have uncoupled modes. A more commonly used alternative method will be presented in Section 7.6.

In Section 7.4, we used the properties of the impulse response functions to show that terms involving the interaction of modes often contribute much less to response covariance than do terms involving only a single mode. Consideration of the harmonic transfer functions confirms this result and illustrates it in a rather graphic way. As in deriving Eqs. 7.36 - 7.38, we will limit our attention to the special case of a stationary white noise excitation with $\mathbf{K}_{FF}(t,s) = 2\pi \mathbf{S}_0 \delta(t-s)$. The autospectral density of the response from Eq. 7.11 is then given by

$$\mathbf{S}_{XX}(\omega) = \mathbf{H}(\omega)\mathbf{S}_0 \mathbf{H}^{T*}(\omega) = \boldsymbol{\theta}\boldsymbol{\gamma}^{-1}[\boldsymbol{\lambda} - \omega^2 \mathbf{I} + i\omega\,\boldsymbol{\beta}]^{-1}\boldsymbol{\theta}^T \mathbf{S}_0 \boldsymbol{\theta}[\boldsymbol{\lambda} - \omega^2 \mathbf{I} - 2i\omega\,\boldsymbol{\beta}]^{-1}\boldsymbol{\gamma}^{-1}\boldsymbol{\theta}^T$$

in which the final form has been obtained by use of Eq. 7.41. Using the notation $\lambda_{j,j} = \omega_j^2$ and $\beta_{j,j} = 2\zeta_j \omega_j$ then allows us to write the frequency domain equivalent of Eq. 7.36 as

$$S_{X_j,X_l}(\omega) = \sum_{r_1=1}^{n} \sum_{r_2=1}^{n} \sum_{r_3=1}^{n} \sum_{r_4=1}^{n} \theta_{j,r_1} \theta_{r_2,r_1} [S_0]_{r_2,r_3} \theta_{r_3,r_4} \theta_{l,r_4} \frac{w_{r_1}(\omega) w_{r_4}(\omega)}{\gamma_{r_1,r_1} \gamma_{r_4,r_4} \lambda_{r_1,r_1} \lambda_{r_4,r_4}}$$

in which

$$w_{r_1}(\omega) = \frac{\lambda_{r_1,r_1}}{\lambda_{r_1,r_1} - \omega^2 + i\omega \beta_{r_1,r_1}} = \frac{1}{1 - \dfrac{\omega^2}{\omega_{r_1}^2} + 2i\zeta_{r_1} \dfrac{\omega}{\omega_{r_1}}}$$

Note that the $w_r(\omega)$ functions have the form of the harmonic transfer function for a SDF system, since they result from the modal harmonic responses. When damping is small, the absolute values of these functions have very narrow peaks, as was shown in Figure 4.6. When ω_{r_1} and ω_{r_4} are not nearly equal, as illustrated in Fig. 7.2, the product of $w_{r_1}(\omega)$ and $w_{r_4}(\omega)$ must generally be very small in comparison with $w_{r_1}^2(\omega)$ and $w_{r_4}^2(\omega)$. Only if $|\omega_{r_1} - \omega_{r_4}|$ is of the order of the modal damping will the peaks overlap to such an extent that the contribution from $w_{r_1}(\omega) w_{r_4}(\omega)$ will be significant. This confirms our finding in Section 7.4 that if the damping in the system is small and

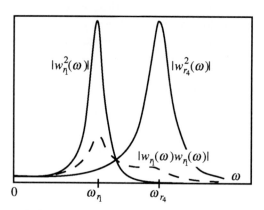

Figure 7.2 Cross-modal contributions to autospectral density

the modal frequencies are well separated, then the contributions of "cross-modal" terms are much smaller than contributions from single modes.

****

Example 7.8: Find the harmonic transfer matrix $\mathbf{H}(\omega)$ for the 2DF system of Example 7.5 with

$$\mathbf{m} = \begin{bmatrix} 1000 & 0 \\ 0 & 500 \end{bmatrix} \text{kg}, \quad \mathbf{k} = \begin{bmatrix} 10^6 & -5\times10^5 \\ -5\times10^5 & 5\times10^5 \end{bmatrix} \frac{\text{N}}{\text{m}}, \quad \mathbf{c} = \begin{bmatrix} 1000 & -500 \\ -500 & 500 \end{bmatrix} \frac{\text{N}\cdot\text{s}}{\text{m}}$$

and investigate the pole locations for $\mathbf{H}(\omega)$.

From the results in example 7.2, we know that

$$\gamma = \begin{bmatrix} 750. & 0 \\ 0 & 666.7 \end{bmatrix} \text{kg}, \quad \lambda = \begin{bmatrix} 292.9 & 0 \\ 0 & 1707. \end{bmatrix} (\text{rad}/\text{s})^2$$

$$\beta = \begin{bmatrix} 0.2929 & 0 \\ 0 & 1.707 \end{bmatrix} \text{rad}/\text{s}, \quad \theta = \begin{bmatrix} 0.6124 & 0.5773 \\ 0.8660 & -0.8165 \end{bmatrix}$$

Thus, we can obtain the harmonic transfer matrix from Eq. 7.41 as

$$\mathbf{H}(\omega) = \theta \gamma^{-1} \begin{bmatrix} 292.9 - \omega^2 + 0.2929i\omega & 0 \\ 0 & 1707. - \omega^2 + 1.707i\omega \end{bmatrix}^{-1} \theta^T$$

The components are

$$H_{1,1}(\omega) = \frac{H_{2,2}(\omega)}{2} = \frac{5 \times 10^{-4}}{292.9 - \omega^2 + 0.2929 i\omega} + \frac{5 \times 10^{-4}}{1707. - \omega^2 + 1.707 i\omega}$$

and

$$H_{1,2}(\omega) = H_{2,1}(\omega) = \frac{7.071 \times 10^{-4}}{292.9 - \omega^2 + 0.2929 i\omega} - \frac{7.071 \times 10^{-4}}{1707. - \omega^2 + 1.707 i\omega}$$

Alternatively, one can use Eq. 7.42 without eigenanalysis to find

$$\mathbf{H}(\omega) = \begin{bmatrix} 10^6 - 1000\omega^2 + 1000 i\omega & -0.5(10^6 + 1000 i\omega) \\ -0.5(10^6 + 1000 i\omega) & 0.5(10^6 - 1000\omega^2 + 1000 i\omega) \end{bmatrix}^{-1}$$

$$= \frac{1}{D} \begin{bmatrix} 1000 - \omega^2 + i\omega & 1000 + i\omega \\ 1000 + i\omega & 2(1000 - \omega^2 + i\omega) \end{bmatrix}$$

in which the denominator is $D = 1000(\omega^4 - 2i\omega^3 - 2000.5\omega^2 + 1000 i\omega + 5 \times 10^5)$, which can also be written as $D = 1000(292.9 - \omega^2 + 0.2929 i\omega)(1707 - \omega^2 + 1.707 i\omega)$. The reader can confirm that the two versions of $\mathbf{H}(\omega)$ are identical.

Recall that one could choose to evaluate the impulse response function matrix $\mathbf{h}(t)$ by using the inverse Fourier transform of $\mathbf{H}(\omega)$. In order to do this, one would probably use the calculus of residues, giving the integral as a sum of terms coming from the poles of $\mathbf{H}(\omega)$, that is, from the values of complex ω at which $\mathbf{H}(\omega)$ is infinite. Obviously these pole locations are the solutions of $D = 0$, and can be found from solving the two complex quadratic equations $\omega^2 - 0.2929 i\omega - 292.9 = 0$ and $\omega^2 - 1.707 i\omega - 1707 = 0$. The solutions are $\pm 17.11 + 0.2929 i$, and $\pm 41.31 + 1.707 i$. These values are a particular example of the general relationship that the poles of $\mathbf{H}(\omega)$ are located at $\pm \omega_j (1 - \zeta_j^2)^{1/2} + i\zeta_j \omega_j$. It is observed that the real values of the pole locations are the damped natural frequencies of the system, and the imaginary values are the modal damping values denoted by β_j. The fact that the imaginary parts are always positive is necessary in order that the system satisfy the causality condition of $\mathbf{h}(t) = 0$ for $t < 0$.

****

Example 7.9: Find the autospectral density matrix for the stationary response of the system of Example 7.6 with stationary, mean zero, white noise base acceleration with $S_0 = 0.1 \, (\mathrm{m}/\mathrm{s}^2)/(\mathrm{rad}/\mathrm{s})$.

The autospectral density matrix of the $\vec{F}(t)$ excitation process is

$$\mathbf{S}_{FF}(\omega) = S_0 \begin{bmatrix} m_{1,1}^2 & m_{1,1} m_{2,2} \\ m_{1,1} m_{2,2} & m_{2,2}^2 \end{bmatrix} = S_0 \begin{bmatrix} 10^6 & 5 \times 10^5 \\ 5 \times 10^5 & 2.5 \times 10^5 \end{bmatrix}$$

so using $\mathbf{H}(\omega)$ from Example 7.8 in Eq. 7.11 gives the response autospectral density as

$$\mathbf{S}_{XX}(\omega) = \frac{S_0}{D^2}(10^6) \begin{bmatrix} \omega^4 - 2998\omega^2 + 2.25 \times 10^6 & \omega^4 + 0.5 i\omega^3 - 3497\omega^2 + 3 \times 10^6 \\ \omega^4 + 0.5 i\omega^3 - 3497\omega^2 + 3 \times 10^6 & \omega^4 - 3996\omega^2 + 4 \times 10^6 \end{bmatrix}$$

in which D is as given in Example 7.8.

Note that the value of $E(X_1^2)$ found in Example 7.6 could also be evaluated from the integral of $[S_{XX}(\omega)]_{1,1}$ from $\omega = -\infty$ to $\omega = \infty$.

****

Example 7.10: Find the harmonic transfer matrix $\mathbf{H}(\omega)$ for the 2DF system of Example 7.3 with

$$\mathbf{m} = \begin{bmatrix} 1000 & 0 \\ 0 & 500 \end{bmatrix} \text{kg}, \quad \mathbf{k} = \begin{bmatrix} 1000 & -500 \\ -500 & 500 \end{bmatrix} \frac{\text{kN}}{\text{m}}, \quad \mathbf{c} = \begin{bmatrix} 2.5 & -2.0 \\ -2.0 & 2.0 \end{bmatrix} \frac{\text{kN} \cdot \text{s}}{\text{m}}$$

We know that this system does not have uncoupled modes, so we use Eq. 7.42 to obtain

$$\mathbf{H}(\omega) = [\mathbf{k} - \omega^2 \mathbf{m} + i\omega \mathbf{c}]^{-1} = 1000 \begin{bmatrix} 1000 + 2.5i\omega - \omega^2 & -500 - 2i\omega \\ -500 - 2i\omega & 500 + 2.0i\omega - 0.5\omega^2 \end{bmatrix}^{-1}$$

$$= \frac{1}{D} \begin{bmatrix} 1000 + 4i\omega - \omega^2 & 1000 + 4i\omega \\ 1000 + 4i\omega & 2000 + 5i\omega - 2\omega^2 \end{bmatrix}$$

in which $D = 1000(5 \times 10^5 + 2500i\omega - 2002\omega^2 - 6.5i\omega^3 + \omega^4)$.

Note that the impulse response function matrix $\mathbf{h}(t)$ can now be found as the inverse Fourier transform of $\mathbf{H}(\omega)$. Solving $D = 0$ gives the location of the four poles of $\mathbf{H}(\omega)$ as $\omega = \pm 17.12 + 0.2106i$ and $\omega = \pm 41.20 + 3.039i$, and the inverse Fourier transform integral can be evaluated by the calculus of residues. Rather than performing this operation, we will derive the impulse response functions for this system by an alternate method in Example 7.14.

****

7.6   State Space Formulation of Equations of Motion

We will now introduce a formulation which can be used for the MDF system, but which can also be used for more general problems. In particular, it is possible to write the equations of motion for any linear system of order n_Y as a set of n_Y first-order differential equations, or a single first-order differential equation for a vector with n_Y components:

$$\dot{\vec{Y}}(t) + \mathbf{A}\vec{Y}(t) = \vec{Q}(t) \tag{7.44}$$

in which $\vec{Y}(t)$ contains only expressions describing the response, $\mathbf{A}$ is a matrix determined from the coefficients in the original equations of motion, and the $\vec{Q}(t)$ vector involves only excitation terms in the original equations of motion. The vector $\vec{Y}(t)$ is called the state vector and its components are called state variables.

For a problem in which we begin with a coupled set of equations of motion which involve J variables $\{X_1(t), \cdots, X_J(t)\}$ with derivatives up to order n_j in the variable $X_j(t)$, the order of the system, and the dimension of the arrays in Eq. 7.46, will be

$$n_Y = \sum_{j=1}^{J} n_j \qquad (7.45)$$

In this case, the components of the state vector $\vec{Y}(t)$ will generally be taken as $X_j(t)$ and its first n_j-1 derivatives, for $j = 1, \cdots, J$. For example, the MDF system of Eq. 7.13 with matrices of dimension $n \times n$ is of order $n_Y = 2n$ and the state variables are usually taken as the components of $\vec{X}(t)$ and $\dot{\vec{X}}(t)$: $\vec{Y}^T(t) = [\vec{X}^T(t), \dot{\vec{X}}^T(t)]$. One can now solve Eq. 7.13 for $\ddot{\vec{X}}(t)$ as a function of the excitation terms and the state variables so that this equation of motion can be rewritten in the form of Eq. 7.44 with

$$A = \begin{bmatrix} 0 & -I \\ m^{-1}k & m^{-1}c \end{bmatrix}, \qquad \vec{Q}(t) = m^{-1} \begin{bmatrix} \vec{0} \\ \vec{F}(t) \end{bmatrix} \qquad (7.46)$$

in which each "element" shown in A is a submatrix of dimension $n \times n$, and each "element" of $\vec{Q}(t)$ is a vector of dimension n. Note that the upper rows of Eq. 7.44 are true with these definitions because the second n components of $\vec{Y}(t)$ are exactly the derivatives of the first n components. This formulation of the MDF problem is commonly called the Foss method.

****

Example 7.11: Find definitions of $\vec{Y}(t), A$, and $\vec{Q}(t)$ such that Eq. 7.44 describes the system with the equation of motion

$$\sum_{j=0}^{n} a_j \frac{d^j X(t)}{dt^j} = F(t)$$

We note that the order of the system is $n_Y = n$ and, as suggested, we take the state variables to be $X(t)$ and its first $n-1$ derivatives, so that

$$\vec{Y}(t) = \left[X(t), \dot{X}(t), \ddot{X}(t), \cdots, \frac{d^{n-1}X(t)}{dt^{n-1}} \right]^T$$

The original equation of motion then relates $\dot{Y}_n(t)$ to the components of $\vec{Y}(t)$ and the excitation $F(t)$. The other $n-1$ scalar equations come from the fact that the derivative of each of the other state variables is itself a state variable: $\dot{Y}_j(t) = Y_{j+1}(t)$ for $j = 1, \cdots, n-1$. Thus, we obtain Eq. 7.44 for the equation of motion if we choose

$$\mathbf{A} = \begin{bmatrix} 0 & -1 & 0 & \cdots & 0 \\ 0 & 0 & -1 & \ddots & \vdots \\ \vdots & \ddots & \ddots & \ddots & 0 \\ 0 & \cdots & 0 & 0 & -1 \\ a_0 & a_1 & a_2 & \cdots & a_{n-1} \\ a_n & a_n & a_n & & a_n \end{bmatrix}, \qquad \vec{Q}(t) = \frac{1}{a_n}\begin{bmatrix} 0 \\ 0 \\ \vdots \\ 0 \\ F(t) \end{bmatrix}$$

****

Example 7.12: Define terms such that the oscillator
shown is described by Eq. 7.44.

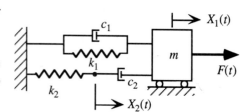

First we write the differential equations of motion in
terms of the variables $X_1(t)$ and $X_2(t)$ shown in the
sketch. One such form is

$$m\ddot{X}_1(t) + c_1\dot{X}_1(t) + k_1 X_1(t) + k_2 X_2(t) = F(t)$$

and

$$k_2 X_2(t) = c_2[\dot{X}_1(t) - \dot{X}_2(t)]$$

Since these equations involve up to the second derivative of $X_1(t)$, but only the first derivative of $X_2(t)$, we
define the state vector as $\vec{Y}(t) = [X_1(t), X_2(t), \dot{X}_1(t)]^T$. The first equation can then be written as
$m\dot{Y}_1(t) + c_1 Y_3(t) + k_1 Y_1(t) + k_2 Y_2(t) = F(t)$ and the second equation is $k_2 Y_2(t) = c_2[Y_3(t) - \dot{Y}_2(t)]$. The
third equation which will be needed is the general relationship that $\dot{Y}_1(t) = Y_3(t)$ for this state vector. Putting
these three equations into the standard form of Eq. 7.44 gives

$$\mathbf{A} = \begin{bmatrix} 0 & 0 & -1 \\ 0 & k_2/c_2 & -1 \\ k_1/m & k_2/m & c_1/m \end{bmatrix}, \qquad \vec{Q}(t) = \frac{1}{m}\begin{bmatrix} 0 \\ 0 \\ F(t) \end{bmatrix}$$

****

The fact that Eq. 7.44 involves only first-order derivatives allows its solution to be easily
written in terms of the matrix exponential function. In particular, the homogeneous solution of Eq.
7.44 can be written as $\vec{X}(t) = \exp[-t\mathbf{A}]$ in which the matrix exponential is defined as

$$\exp(\mathbf{B}) = \sum_{j=0}^{\infty}\frac{1}{j!}\mathbf{B}^j \qquad (7.47)$$

for any square matrix $\mathbf{B}$, and with $\mathbf{B}^0$ defined to be the identity matrix of the same dimension as $\mathbf{B}$,
and $\mathbf{B}^j = \mathbf{BB}^{j-1}$ for $j \geq 1$. This relationship gives the derivative with respect to t of $\exp[-t\mathbf{A}]$ as
$-\mathbf{A}\exp(-t\mathbf{A})$, so that the solution of Eq. 7.44 has the same form as for a scalar equation. Similarly,
the general inhomogeneous solution can be written as the convolution integral

$$\vec{Y}(t) = \int_{-\infty}^{t} \exp[-(t-s)\mathbf{A}]\vec{Q}(s)\,ds \tag{7.48}$$

Although the solution in Eq. 7.48 is mathematically correct, it is not very convenient for numerical computations. Simplified numerical procedures result from diagonalizing $\mathbf{A}$ by the use of eigenanalysis. We will use the notation of $\boldsymbol{\lambda}$ and $\boldsymbol{\theta}$ for the matrices of eigenvalues and eigenvectors of $\mathbf{A}$, just as we did for $\mathbf{m}^{-1}\mathbf{k}$ in Section 7.3. Thus, we have $\mathbf{A}\boldsymbol{\theta} = \boldsymbol{\theta}\boldsymbol{\lambda}$ and

$$\boldsymbol{\theta}^{-1}\mathbf{A}\boldsymbol{\theta} = \boldsymbol{\lambda} \tag{7.49}$$

as the diagonalized version of $\mathbf{A}$. Note that this expression uses the inverse of the eigenvector matrix. In Section 7.3 we were able to perform the corresponding diagonalization using only $\boldsymbol{\theta}^{T}$ instead of $\boldsymbol{\theta}^{-1}$, but that simplification depended on the symmetry of the matrices involved and does not generally apply to the state space formulation.[2] Solving Eq. 7.49 for $\mathbf{A}$ gives $\mathbf{A} = \boldsymbol{\theta}\boldsymbol{\lambda}\boldsymbol{\theta}^{-1}$, from which one can readily verify that $\mathbf{A}^{j} = \boldsymbol{\theta}\boldsymbol{\lambda}^{j}\boldsymbol{\theta}^{-1}$ for any jth power of the matrix, and Eq. 7.47 then shows that $\exp[-t\mathbf{A}] = \boldsymbol{\theta}\exp[-t\boldsymbol{\lambda}]\boldsymbol{\theta}^{-1}$. Thus, the general solution in Eq. 7.48 can be written as

$$\vec{Y}(t) = \int_{-\infty}^{t} \boldsymbol{\theta}\exp[-(t-s)\boldsymbol{\lambda}]\boldsymbol{\theta}^{-1}\vec{Q}(s)\,ds \tag{7.50}$$

in which $\exp[-(t-s)\boldsymbol{\lambda}]$ is a diagonal matrix involving only the scalar exponential function. In particular, the (j,j) element of $\exp[-(t-s)\boldsymbol{\lambda}]$ is given by $\exp[-(t-s)\lambda_{j,j}]$.

There are several possible approaches to performing the stochastic analysis of Eq. 7.44, just as there were in Sections 7.4 and 7.5 for the equations considered there. One approach is to use Eq. 7.50 only in finding the $\mathbf{h}(t)$ and $\mathbf{H}(\omega)$ matrices, for use in the equations of Section 7.1. For example we can say that $\vec{Y}(t) = [h_{1,l}(t), \cdots h_{n_Y,l}(t)]^{T}$ when the excitation is a single Dirac delta function pulse in the $Q_l(t)$ component of excitation, and Eqs. 7.48 and 7.50 then give

$$\mathbf{h}(t) = \exp[-t\mathbf{A}]U(t) = \boldsymbol{\theta}\exp[-t\boldsymbol{\lambda}]\boldsymbol{\theta}^{-1}U(t) \tag{7.51}$$

Similarly $\vec{Y}(t) = [H_{1,l}(\omega), \cdots H_{n_Y,l}(\omega)]^{T} e^{i\omega t}$ when the excitation is a single harmonic term $e^{i\omega t}$ in the $Q_l(t)$ component of excitation, and Eq. 7.44 directly gives $[\mathbf{A} + i\omega\,\mathbf{I}]\mathbf{H}(\omega) = \mathbf{I}$, so that

$$\mathbf{H}(\omega) = [\mathbf{A} + i\omega\,\mathbf{I}]^{-1} = \boldsymbol{\theta}[\boldsymbol{\lambda} + i\omega\,\mathbf{I}]^{-1}\boldsymbol{\theta}^{-1} \tag{7.52}$$

The alternative of direct stochastic analysis of Eq. 7.48 or 7.50 gives the mean value vector as

---

[2]   Rather than calculating $\boldsymbol{\theta}^{-1}$ directly, it is possible to first perform the eigenanalysis for $\mathbf{A}^{T}$, finding the matrix $\boldsymbol{\eta}$ such that $\mathbf{A}^{T}\boldsymbol{\eta} = \boldsymbol{\eta}\boldsymbol{\lambda}$, since $\mathbf{A}$ and $\mathbf{A}^{T}$ have the same eigenvalues. One then finds that $\boldsymbol{\eta}^{T}\boldsymbol{\theta}$ is diagonal so that it is easy to calculate $\boldsymbol{\theta}^{-1}$ from $\boldsymbol{\theta}^{-1} = (\boldsymbol{\eta}^{T}\boldsymbol{\theta})^{-1}\boldsymbol{\eta}^{T}$.

$$\vec{\mu}_Y(t) = \int_{-\infty}^{t} \exp[-(t-s)\mathbf{A}]\vec{\mu}_Q(s)\,ds = \int_{-\infty}^{t} \mathbf{\theta}\exp[-(t-s)\mathbf{\lambda}]\mathbf{\theta}^{-1}\vec{\mu}_Q(s)\,ds \qquad (7.53)$$

and the autocorrelation matrix as

$$\mathbf{\phi}_{YY}(t,s) = \int_{-\infty}^{s}\int_{-\infty}^{t} \exp[-(t-u)\mathbf{A}]\mathbf{\phi}_{QQ}(u,v)\exp[-(s-v)\mathbf{A}^T]\,du\,dv \qquad (7.54)$$

or

$$\mathbf{\phi}_{YY}(t,s) = \mathbf{\theta}\int_{-\infty}^{s}\int_{-\infty}^{t} \exp[-(t-u)\mathbf{\lambda}]\mathbf{\theta}^{-1}\mathbf{\phi}_{QQ}(u,v)\mathbf{\theta}\exp[-(s-v)\mathbf{\lambda}]\,du\,dv\,\mathbf{\theta}^{-1} \qquad (7.55)$$

Similarly, the autocovariance is

$$\mathbf{K}_{YY}(t,s) = \int_{-\infty}^{s}\int_{-\infty}^{t} \exp[-(t-u)\mathbf{A}]\mathbf{K}_{QQ}(u,v)\exp[-(s-v)\mathbf{A}^T]\,du\,dv \qquad (7.56)$$

or

$$\mathbf{K}_{YY}(t,s) = \mathbf{\theta}\int_{-\infty}^{s}\int_{-\infty}^{t} \exp[-(t-u)\mathbf{\lambda}]\mathbf{\theta}^{-1}\mathbf{K}_{QQ}(u,v)\mathbf{\theta}^{-1^T}\exp[-(s-v)\mathbf{\lambda}]\,du\,dv\,\mathbf{\theta}^T \qquad (7.57)$$

and these expressions all agree with what one would obtain from Eqs. 7.6 - 7.8 in Section 7.1. The autospectral density matrix for covariant stationary response is most easily obtained from Eq. 7.11 as

$$\mathbf{S}_{YY}(\omega) = [\mathbf{A} + i\omega\,\mathbf{I}]^{-1}\mathbf{S}_{QQ}(\omega)[\mathbf{A}^T - i\omega\,\mathbf{I}]^{-1} = \mathbf{\theta}[\mathbf{\lambda} + i\omega\,\mathbf{I}]^{-1}\mathbf{\theta}^{-1}\mathbf{S}_{QQ}(\omega)\mathbf{\theta}^{-1^T}[\mathbf{\lambda} - i\omega\,\mathbf{I}]^{-1}\mathbf{\theta}^T$$
$$(7.58)$$

Note that Eq. 7.44 gives $\vec{Y}(0)$ as the complete initial condition vector denoted by $\vec{Y}$ in Eq. 7.12. In addition, the initial value response matrix $\mathbf{g}(t)$ of Eq. 7.12 is $\mathbf{g}(t) \equiv \mathbf{h}(t)$, since a Dirac delta function in component j of $\vec{Q}(t)$ in Eq. 7.44 gives a unit initial value of component j of $\vec{Y}(t)$. Thus, the $\mathbf{h}(t)$ matrix described by Eq. 7.51 is the only matrix of system properties involved in using Eq. 7.12:

$$\vec{Y}(t) = \exp[-(t-t_0)\mathbf{A}]\vec{Y}(t_0) + \int_{t_0}^{t} \exp[-(t-s)\mathbf{A}]\vec{Q}(s)\,ds \qquad (7.59)$$

Direct stochastic analysis of this equation gives

$$\vec{\mu}_Y(t) = \exp[-(t-t_0)\mathbf{A}]\vec{\mu}_Y(t_0) + \int_{t_0}^{t} \exp[-(t-s)\mathbf{A}]\vec{\mu}_Q(s)\,ds \qquad (7.60)$$

and if the excitation $\vec{Q}(s)$ for $s > t_0$ is independent of the initial condtion $\vec{Y}(t_0)$, the covariance reduces to

$$\mathbf{K}_{YY}(t,s) = \exp[-(t-t_0)\mathbf{A}]\mathbf{K}_{YY}(t_0,t_0)\exp[-(s-t_0)\mathbf{A}^T]$$
$$+\int_{t_0}^{s}\int_{t_0}^{t} \exp[-(t-u)\mathbf{A}]\mathbf{K}_{QQ}(u,v)\exp[-(s-v)\mathbf{A}^T]\,du\,dv$$
$$(7.61)$$

The corresponding expressions for the conditional mean and covariance, analogous to Eqs. 3.52 and 3.53 for the scalar problem, are

$$E[\vec{Y}(t)|\vec{Y}(t_0) = \vec{w}] = \exp[-(t-t_0)\mathbf{A}]\vec{w} + \int_{t_0}^{t} \exp[-(t-s)\mathbf{A}]E[\vec{Q}(s)|\vec{Y}(t_0) = \vec{w}]ds \qquad (7.62)$$

and

$$\mathbf{K}[\vec{Y}(t),\vec{Y}(s)|\vec{Y}(t_0) = \vec{w}] = \int_{t_0}^{s}\int_{t_0}^{t} \exp[-(t-t_0)\mathbf{A}]\mathbf{K}[\vec{Q}(u),\vec{Q}(v)|\vec{Y}(t_0) = \vec{w}]\exp[-(s-t_0)\mathbf{A}^T]du\,dv \qquad (7.63)$$

For the special case of a delta-correlated excitation, of course, many of these relationships are somewhat simplified. For example, if we use the notation that $\mathbf{K}_{QQ}(u,v) = 2\pi\mathbf{S}_0(u)\delta(u-v)$, then the autocovariance of the response can be written as a single, rather than double, integral as

$$\mathbf{K}_{YY}(t,s) = 2\pi\int_{-\infty}^{t} \exp[-(t-u)\mathbf{A}]\mathbf{S}_0(u)\exp[-(s-u)\mathbf{A}^T]du \qquad (7.64)$$

or

$$\mathbf{K}_{YY}(t,s) = \exp[-(t-t_0)\mathbf{A}]\mathbf{K}_{YY}(t_0,t_0)\exp[-(s-t_0)\mathbf{A}^T]$$
$$+2\pi\int_{t_0}^{t} \exp[-(t-u)\mathbf{A}]\mathbf{S}_0(u)\exp[-(s-u)\mathbf{A}^T]du \qquad (7.65)$$

and the conditional mean and conditional variance expressions are

$$E[\vec{Y}(t)|\vec{Y}(t_0) = \vec{w}] = \exp[-(t-t_0)\mathbf{A}]\vec{w} + \int_{t_0}^{t} \exp[-(t-s)\mathbf{A}]\vec{\mu}_Q(s)ds \qquad (7.66)$$

and

$$\mathbf{K}[\vec{Y}(t),\vec{Y}(s)|\vec{Y}(t_0) = \vec{w}] = 2\pi\int_{t_0}^{t} \exp[-(t-u)\mathbf{A}]\mathbf{S}_0(u)\exp[-(s-u)\mathbf{A}^T]du \qquad (7.67)$$

All of these expressions can be diagonalized by use of $\mathbf{\theta}$ and $\mathbf{\theta}^{-1}$.

****

Example 7.13: Find the state space formulations of the impulse response matrix $\mathbf{h}(t)$ and the harmonic transfer matrix $\mathbf{H}(\omega)$ for the 2DF system of Examples 7.5 and 7.8, using the state space vector $\vec{Y}(t) = [X_1(t), X_2(t), \dot{X}_1(t), \dot{X}_2(t)]^T$.

Using the $\mathbf{m}$, $\mathbf{k}$, and $\mathbf{c}$ matrices previously given, we find that

$$\mathbf{m}^{-1}\mathbf{k} = \begin{bmatrix} 1000 & -500 \\ -1000 & 1000 \end{bmatrix}, \qquad \mathbf{m}^{-1}\mathbf{c} = \begin{bmatrix} 1 & -0.5 \\ -1 & 1 \end{bmatrix}$$

Thus, Eq. 7.46 gives $\mathbf{A}$ for use in Eq. 7.44 as

$$\mathbf{A} = \begin{bmatrix} 0 & 0 & -1 & 0 \\ 0 & 0 & 0 & -1 \\ 1000 & -500 & 1 & -0.5 \\ -1000 & 1000 & -1 & 1 \end{bmatrix}$$

Eigenanalysis of $\mathbf{A}$ yields

$$\boldsymbol{\theta} = \begin{bmatrix} -0.2213 - 0.0936i & -0.2213 + 0.0936i & 0.1094 - 0.02090i & 0.1094 + 0.02090i \\ -0.3129 - 0.1324i & -0.3129 + 0.1324i & -0.1547 + 0.02955i & -0.1547 - 0.02955i \\ -1.569 + 3.800i & -1.569 - 3.800i & -0.9566 - 4.500i & -0.9566 + 4.500i \\ -2.219 + 5.375i & -2.219 - 5.375i & 1.353 + 6.363i & 1.353 - 6.363i \end{bmatrix}$$

and the non-zero elements of $\boldsymbol{\lambda}$ are $\lambda_{1,1} = 0.1464 + 17.11i$, $\lambda_{2,2} = 0.1464 - 17.11i$, $\lambda_{3,3} = 0.8536 + 41.31i$, and $\lambda_{4,4} = 0.8536 - 41.31i$. Note that the eigenvalues of $\mathbf{A}$ are $\pm i\omega_j(1 - \zeta_j^2)^{1/2} + \zeta_j\omega_j$. That is, the imaginary parts of the eigenvalues are the same as the damped frequencies of the uncoupled modes, and the real parts are the same as the corresponding elements of the modal damping matrix $\boldsymbol{\beta}$. This, then, shows that the absolute values of the eigenvalues are like undamped natural frequencies. These general relationships will be found to be true for any MDF system with uncoupled modes.

The impulse response matrix can now be obtained from $\mathbf{h}(t) = \boldsymbol{\theta} e^{-t\boldsymbol{\lambda}} \boldsymbol{\theta}^{-1} U(t)$, in which the non-zero elements of $e^{-t\boldsymbol{\lambda}}$ are given by $[e^{-t\boldsymbol{\lambda}}]_{j,j} = e^{-t\lambda_{j,j}}$ and the harmonic transfer matrix can be obtained from $\mathbf{H}(\omega) = \boldsymbol{\theta}[\boldsymbol{\lambda} + i\omega\,\mathbf{I}]^{-1}\boldsymbol{\theta}^{-1}$, in which the non-zero elements of $[\boldsymbol{\lambda} + i\omega\,\mathbf{I}]^{-1}$ are given by $[\boldsymbol{\lambda} + i\omega\,\mathbf{I}]_{j,j}^{-1} = 1/(\lambda_{j,j} + i\omega)$. In both of these expressions, one needs the matrix

$$\boldsymbol{\theta}^{-1} = \begin{bmatrix} -0.9618 + 0.3972i & -0.6801 + 0.2809i & -0.02369 - 0.0560i & -0.01675 - 0.03960i \\ -0.9618 - 0.3972i & -0.6801 - 0.2809i & -0.02369 + 0.0560i & -0.01675 + 0.03960i \\ 2.197 + 0.4670i & -1.553 - 0.3302i & -0.01020 + 0.05339i & 0.007215 - 0.03775i \\ 2.197 - 0.4670i & -1.553 + 0.3302i & -0.01020 - 0.05339i & 0.007215 + 0.03775i \end{bmatrix}$$

For example, the $X_1(t)$ response to a Dirac delta function in $Q_3(t) = F_1(t) / m_1$ is

$$h_{1,3}(t) = [0.01461i\,e^{(-0.1464 - 17.11i)t} - 0.01461i\,e^{(-0.1464 + 17.11i)t}$$
$$-0.006052i\,e^{(-0.8536 + 41.31i)t} + 0.006052i\,e^{(-0.8536 - 41.31i)t}]U(t)$$

or

$$h_{1,3}(t) = [0.02922\,e^{-0.1464t}\sin(17.11t) + 0.01210\,e^{-0.8536t}\sin(41.31t)]U(t)$$

The corresponding harmonic transfer term giving the magnitude of the $X_1(t)$ response to a unit amplitude harmonic function in $Q_3(t)$ is

$$H_{1,3}(\omega) = \frac{0.01461i}{0.1464 + 17.11i + i\omega} - \frac{0.01461}{0.1464 - 17.11i + i\omega}$$
$$+ \frac{0.006052}{0.8536 + 41.31i + i\omega} - \frac{0.006052}{0.8536 - 41.31i + i\omega}$$

or

$$H_{1,3}(\omega) = \frac{0.5}{292.9 + 0.2929i\omega - \omega^2} + \frac{0.5}{1707 + 1.707i\omega - \omega^2}$$
$$= \frac{1000 + i\omega - \omega^2}{5 \times 10^5 + 1000i\omega - 2000.5\omega^2 - 2i\omega^3 + \omega^4}$$

These expressions are in perfect agreement with the results in Examples 7.5 and 7.8 for the $X_1(t)$ response to given Dirac delta and unit harmonic functions for $F_1(t)$.

For this $\mathbf{A}$ matrix of dimension 4×4, it is also feasible to obtain the expressions for $\mathbf{H}(\omega)$ by analytically inverting $[\mathbf{A} + i\omega\,\mathbf{I}]$ without using eigenanalysis. This approach is generally not feasible, though, if the dimension of $\mathbf{A}$ is not quite small.

****

Example 7.14: Find the state space formulations of the impulse response matrix $\mathbf{h}(t)$ and the harmonic transfer matrix $\mathbf{H}(\omega)$ for the 2DF system of Examples 7.3 and 7.10, using the state space vector $\vec{Y}(t) = [X_1(t), X_2(t), \dot{X}_1(t), \dot{X}_2(t)]^T$.

The approach is the same as in Example 7.13, but some of the results are different because of the fact that this system does not have uncoupled modes. From the $\mathbf{m}$, $\mathbf{k}$, and $\mathbf{c}$ matrices we find that

$$\mathbf{m}^{-1}\mathbf{k} = \begin{bmatrix} 1000 & -500 \\ -1000 & 1000 \end{bmatrix}, \quad \mathbf{m}^{-1}\mathbf{c} = \begin{bmatrix} 2.5 & -2.0 \\ -4.0 & 4.0 \end{bmatrix}$$

so that

$$\mathbf{A} = \begin{bmatrix} 0 & 0 & -1 & 0 \\ 0 & 0 & 0 & -1 \\ 1000 & -500 & 2.5 & -2.0 \\ -1000 & 1000 & -4.0 & 4.0 \end{bmatrix}$$

Eigenanalysis of $\mathbf{A}$ yields

$$\mathbf{\theta} = \begin{bmatrix} 0.1951 + 0.5574i & 0.1951 - 0.5574i & -0.1889 + 0.2604i & -0.1889 - 0.2604i \\ 0.2613 + 0.7923i & 0.2613 - 0.7923i & 0.2518 - 0.3810i & 0.2518 + 0.3810i \\ 9.499 - 3.457i & 9.499 + 3.457i & 11.30 + 6.990i & 11.30 - 6.990i \\ 13.51 - 4.640i & 13.51 + 4.640i & -16.46 - 9.216i & -16.46 + 9.216i \end{bmatrix}$$

and the non-zero elements of $\boldsymbol{\lambda}$ are $\lambda_{1,1} = 0.2106 + 17.12i$, $\lambda_{2,2} = 0.2106 - 17.12i$, $\lambda_{3,3} = 3.039 + 41.20i$, and $\lambda_{4,4} = 3.039 - 41.20i$. The $\boldsymbol{\theta}^{-1}$ matrix which is needed for the calculation of $\mathbf{h}(t)$ or $\mathbf{H}(\omega)$ is found to be

$$\boldsymbol{\theta}^{-1} = \begin{bmatrix} 0.1206 - 0.4080i & 0.1145 - 0.2763i & 0.02352 + 0.007761i & 0.01652 + 0.005782i \\ 0.1206 + 0.4080i & 0.1145 + 0.2763i & 0.02352 - 0.007761i & 0.01652 - 0.005782i \\ -0.3922 - 0.6720i & 0.2811 + 0.4734i & 0.01569 - 0.01036i & -0.01081 + 0.007836i \\ -0.3922 + 0.6720i & 0.2811 - 0.4734i & 0.01569 + 0.01036i & -0.01081 - 0.007836i \end{bmatrix}$$

For example, the $X_1(t)$ response to a Dirac delta function in $Q_3(t) = F_1(t) / m_1$ can be found as the (1,1) term of $\mathbf{h}(t) = \boldsymbol{\theta} \exp[-t\boldsymbol{\lambda}] \boldsymbol{\theta}^{-1} U(t)$. The result is

$$h_{1,3}(t) = [(2.641 \times 10^{-4} - 0.01463i)e^{(-0.2106+17.12i)t}$$
$$+ (2.641 \times 10^{-4} + 0.01463i)e^{(-0.2106-17.12i)t}$$
$$- (2.641 \times 10^{-4} + 0.006042i)e^{(-3.039+41.20i)t}$$
$$- (2.641 \times 10^{-4} - 0.006042i)e^{(-3.039-41.20i)t}$$

or

$$h_{1,3}(t) = 5.8282 \times 10^{-4}[e^{-0.2106t}\cos(17.12t) - e^{-3.039t}\cos(41.20t)]$$
$$+ 0.02925e^{-0.2106t}\sin(17.12t) + 0.01208e^{-3.039t}\sin(41.20t)]$$

Similarly, the $X_1(t)$ response to a Dirac delta function in $Q_4(t) = F_2(t) / m_2$ is

$$h_{1,4}(t) = 9.404 \times 10^{-7}[-e^{-0.2106t}\cos(17.12t) + e^{-3.039t}\cos(41.20t)]$$
$$+ 0.02067e^{-0.2106t}\sin(17.12t) - 0.008587e^{-3.039t}\sin(41.20t)]$$

Note the presence of cosine terms in these impulse response functions. Such terms never appear in the impulse response functions for a system with uncoupled modes, since they do not appear in the impulse response function for a SDF system. The primary effect of a damping matrix which does not allow the modes to be uncoupled is a change in phase of the motions. Note that the imaginary parts of the eigenvalues play the role in the impulse response function of the damped natural frequency ω_d in the SDF system and the real parts of the eigenvalues are like $\zeta\omega_0$ values. As when the system does have uncoupled modes, the absolute values of the eigenvalues of $\mathbf{A}$ are like undamped natural frequencies. In fact, one can show that they are precisely the square roots of the eigenvalues of $\mathbf{m}^{-1}\mathbf{k}$ for this problem.

In a similar fashion, the harmonic transfer matrix from $\mathbf{H}(\omega) = \boldsymbol{\theta}[\boldsymbol{\lambda} + i\omega\,\mathbf{I}]^{-1}\boldsymbol{\theta}^{-1}$ gives

$$H_{1,3}(\omega) = \frac{2.641 \times 10^{-4} + 0.01463i}{0.2106 + 17.12i + i\omega} + \frac{2.641 \times 10^{-4} - 0.01463i}{0.2106 - 17.12i + i\omega}$$

$$- \frac{2.641 \times 10^{-4} - 0.006042i}{3.0394 + 41.20i + i\omega} - \frac{2.641 \times 10^{-4} + 0.006042i}{3.0394 - 41.20i + i\omega}$$

or

$$H_{1,3}(\omega) = \frac{0.5008 + 5.282 \times 10^{-4} i\omega}{293.0 + 0.4213i\omega - \omega^2} + \frac{0.4962 - 5.282 \times 10^{-4} i\omega}{1706 + 6.079i\omega - \omega^2}$$

$$= \frac{1000 + 4i\omega - \omega^2}{5 \times 10^5 + 2500i\omega - 2002\omega^2 - 6.5i\omega^3 + \omega^4}$$

for the magnitude of the $X_1(t)$ response to a unit amplitude harmonic in $Q_3(t) = F_1(t) / m_1$, and

$$H_{1,4}(\omega) = \frac{500 + 2i\omega}{5 \times 10^5 + 2500i\omega - 2002\omega^2 - 6.5i\omega^3 + \omega^4}$$

for the $X_1(t)$ response to a unit amplitude harmonic in $Q_4(t) = F_2(t) / m_2$. Again, these results are in perfect agreement with the results in Example 7.10 for the $X_1(t)$ response to unit amplitude harmonic behavior of $F_1(t)$ and $F_2(t)$. One may also note that the poles of the harmonic response function which were found in Example 7.10 are the same as i times the eigenvalues of $\mathbf{A}$, emphasizing the similarity between the two approaches.

Note that there is considerable redundancy in the state space formulation of the eigenanalysis for a MDF system. In particular, the complex conjugates of the eigenvalues and eigenvectors are also eigenvalues and eigenvectors, respectively (unless the damping values are very large). This fact can be used in reducing the amount of computation required (Igusa et al., 1984; Veletsos and Ventura, 1986).

****

Example 7.15: Find the autospectral density function and the mean squared value of the stationary $\{X_1(t)\}$ response of the system of Examples 7.3, 7.10, and 7.14 when it is subjected to the mean zero, white noise base acceleration excitation of Examples 7.6 and 7.9 with $S_0 = 0.1 (\text{m}/\text{s}^2) / (\text{rad}/\text{s})$.

From Eq. 7.46 we see that the excitation vector in the state space formulation is $\vec{Q}(t) = -\ddot{Y}(t)[0, 0, 1, 1]^T$. The autospectral density matrix for this excitation is then

$$\mathbf{S}_{QQ} = S_0 \begin{bmatrix} 0 & 0 & 0 & 0 \\ 0 & 0 & 0 & 0 \\ 0 & 0 & 1 & 1 \\ 0 & 0 & 1 & 1 \end{bmatrix}$$

and the $(1,1)$ component of Eq. 7.58 gives

$$[S_{YY}(\omega)]_{1,1} = \sum_{j_1=1}^{4} \sum_{j_2=1}^{4} \sum_{j_3=1}^{4} \sum_{j_4=1}^{4} \frac{\theta_{1,j_1}\theta_{j_1,j_2}^{-1}[S_{QQ}]_{j_2,j_3}\theta_{j_4,j_3}^{-1}\theta_{1,j_4}}{(\lambda_{j_1,j_1}+i\omega)\,(\lambda_{j_4,j_4}-i\omega)}$$

$$= S_0 \sum_{j_1=1}^{4} \sum_{j_2=3}^{4} \sum_{j_3=3}^{4} \sum_{j_4=1}^{4} \frac{\theta_{1,j_1}\theta_{j_1,j_2}^{-1}\theta_{j_4,j_3}^{-1}\theta_{1,j_4}}{(\lambda_{j_1,j_1}+i\omega)\,(\lambda_{j_4,j_4}-i\omega)}$$

One can also separate this expression into a portion from the low-frequency mode, a portion from the high-frequency mode, and a cross-modal part. For example, the contribution of the low-frequency mode comes from summing j_1 and j_4 only from 1 to 2, since the first two eigenvalues of $\mathbf{A}$ correspond to the low-frequency mode. This first mode contribution is

$$S_0 \sum_{j_1=1}^{2} \sum_{j_2=3}^{4} \sum_{j_3=3}^{4} \sum_{j_4=1}^{2} \frac{\theta_{1,j_1}\theta_{j_1,j_2}^{-1}\theta_{j_4,j_3}^{-1}\theta_{1,j_4}}{(\lambda_{j_1,j_1}+i\omega)\,(\lambda_{j_4,j_4}-i\omega)} = \frac{2.780\times10^{-7}\omega^2+0.7303}{(\omega^4-585.4\omega^2+8.584\times10^4)}$$

The denominator of this expression can also be written as $[\omega^2-|\lambda_{1,1}|^2]^2+[2\operatorname{Re}(\lambda_{1,1})\omega]^2$, which is the form commonly used for the SDF autospectral density. Similarly, the high-frequency contribution comes from summing j_1 and j_4 only from 3 to 4, and is given by

$$S_0 \sum_{j_1=3}^{4} \sum_{j_2=3}^{4} \sum_{j_3=3}^{4} \sum_{j_4=3}^{4} \frac{\theta_{1,j_1}\theta_{j_1,j_2}^{-1}\theta_{j_4,j_3}^{-1}\theta_{1,j_4}}{(\lambda_{j_1,j_1}+i\omega)\,(\lambda_{j_4,j_4}-i\omega)} = \frac{2.780\times10^{-7}\omega^2+0.02029}{(\omega^4-3376\omega^2+2.912\times10^6)}$$

and this denominator can be rewritten as $[\omega^2-|\lambda_{3,3}|^2]^2+[2\operatorname{Re}(\lambda_{3,3})\omega]^2$. The cross-modal contribution is

$$S_0 \sum_{j_1=1}^{2} \sum_{j_2=3}^{4} \sum_{j_3=3}^{4} \sum_{j_4=3}^{4} \frac{\theta_{1,j_1}\theta_{j_1,j_2}^{-1}\theta_{j_4,j_3}^{-1}\theta_{1,j_4}}{(\lambda_{j_1,j_1}+i\omega)\,(\lambda_{j_4,j_4}-i\omega)} + S_0 \sum_{j_1=3}^{4} \sum_{j_2=3}^{4} \sum_{j_3=3}^{4} \sum_{j_4=1}^{2} \frac{\theta_{1,j_1}\theta_{j_1,j_2}^{-1}\theta_{j_4,j_3}^{-1}\theta_{1,j_4}}{(\lambda_{j_1,j_1}+i\omega)\,(\lambda_{j_4,j_4}-i\omega)}$$

which does not have a particularly simple form, and the total autospectral density, after simplification, is

$$[S_{YY}(\omega)]_{1,1} = \frac{\omega^4-2964\omega^2+2.25\times10^6}{(\omega^4-585.4\omega^2+8.584\times10^4)(\omega^4-3376\omega^2+2.912\times10^6)}$$

Although $[S_{YY}(\omega)]_{1,1}$ does have peaks in the vicinities of both natural frequencies of the undamped system, the high-frequency peak is so small that it is not very significant. As in Example 7.6, the insignificance of the higher mode is exaggerated by the phase of the excitation components. The sketches show the two peaks of the autospectral density. Note, though, that the vertical scale is greatly different in the two sketches.

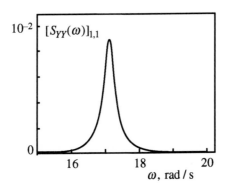 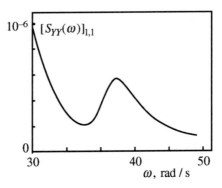

The mean squared response for this stationary, mean zero problem can be found from Eq. 7.64, which can be rewritten as

$$\mathbf{K}_{YY}(t,t) = 2\pi \int_0^\infty \exp[-u\mathbf{A}]\mathbf{S}_{QQ}\exp[-u\mathbf{A}^T]\,du$$

$$= 2\pi\mathbf{\theta}\int_0^\infty \exp[-u\mathbf{\lambda}]\mathbf{\theta}^{-1}\mathbf{S}_{QQ}\mathbf{\theta}^{-1^T}\exp[-u\mathbf{\lambda}^T]\,du\,\mathbf{\theta}^T$$

The $(1,1)$ component is then

$$E(X_1^2) = [\mathbf{K}_{YY}(t,t)]_{1,1} = 2\pi S_0 \sum_{j_1=1}^{4} \sum_{j_2=3}^{4} \sum_{j_3=3}^{4} \sum_{j_4=1}^{4} \theta_{1,j_1}\theta_{j_1,j_2}^{-1}\theta_{j_4,j_3}^{-1}\theta_{1,j_4} \int_0^\infty e^{-u\lambda_{j_1,j_1}}e^{-u\lambda_{j_4,j_4}}\,du$$

$$= 2\pi S_0 \sum_{j_1=1}^{4} \sum_{j_2=3}^{4} \sum_{j_3=3}^{4} \sum_{j_4=1}^{4} \frac{\theta_{1,j_1}\theta_{j_1,j_2}^{-1}\theta_{j_4,j_3}^{-1}\theta_{1,j_4}}{\lambda_{j_1,j_1}+\lambda_{j_4,j_4}} = 1.860\times10^{-3}\,\mathrm{m}^2$$

By separating the terms with j_1 and j_4 equal to 1 or 2 from those with j_1 and j_4 equal to 3 or 4, one can find that the low frequency modal contribution is $1.859\times10^{-3}\,\mathrm{m}^2$, the high frequency modal contribution is $6.289\times10^{-7}\,\mathrm{m}^2$, and the cross-modal contribution is $9.569\times10^{-7}\,\mathrm{m}^2$. In this case we see that the cross-modal term is slightly larger than the second modal term, but both are small compared to the contribution of the first mode.

****

Exercises

****

Response Modes and Impulse Response Functions

****

7.1 Let the vector process $\{\vec{X}(t)\}$ be the response of a system with the equation of motion

$$\ddot{\vec{X}}(t) + \begin{bmatrix} 0.10 & -0.02 \\ -0.02 & 0.10 \end{bmatrix}\dot{\vec{X}}(t) + \begin{bmatrix} 26 & -10 \\ -10 & 26 \end{bmatrix}\vec{X}(t) = \begin{bmatrix} F_1(t) \\ F_2(t) \end{bmatrix}$$

(a) Demonstrate that this system has uncoupled modes.

(b) Find the $\mathbf{h}(t)$ impulse response function matrix.

****

7.2   Consider the 2DF system shown in Example 7.1 with $m_1 = 400\,\text{kg}$, $m_2 = 300\,\text{kg}$, $k_1 = 1200\,\text{N}\,/\,\text{m}$, $k_2 = 600\,\text{N}\,/\,\text{m}$, $c_1 = 100\,\text{N}\cdot\text{s}\,/\,\text{m}$, and $c_2 = 50\,\text{N}\cdot\text{s}\,/\,\text{m}$.

(a) Demonstrate that this system has uncoupled modes.

(b) Find the $\mathbf{h}(t)$ impulse response function matrix.

****

7.3 Find the $\mathbf{h}(t)$ impulse response function matrix for the 2DF system of Example 7.4.

****

7.4 Consider the 2DF system shown in Example 7.1 with $m_1 = m_2 = m$, $k_1 = 2k$, $k_2 = k$, $c_1 = 3c$, and $c_2 = c$, in which m, k, and c are scalar constants.

(a) Find the $\mathbf{m}$, $\mathbf{k}$, and $\mathbf{c}$ matrices such that this system is described by Eq. 7.13.

(b) Show that the system does not have uncoupled modes.

****

7.5 Consider a three-degree-of-freedom system for which the mass matrix is $\mathbf{m} = m\mathbf{I}$, with m being a scalar constant and $\mathbf{I}$ being the 3×3 identity matrix. The modes are uncoupled and the mode shapes, natural frequencies, and damping values are given by

$$[3,5,6]^T \qquad\qquad [2,0,-1]^T \qquad\qquad [1,-3,2]^T$$

$$\omega_1 = \sqrt{\frac{k}{m}} \qquad\qquad \omega_2 = 2\sqrt{\frac{k}{m}} \qquad\qquad \omega_3 = 3\sqrt{\frac{k}{m}}$$

$$\zeta_1 = 0.05 \qquad\qquad \zeta_2 = 0.05 \qquad\qquad \zeta_3 = 0.05$$

in which k is another scalar constant.

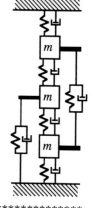

(a) Find the stiffness matrix $\mathbf{k}$.

(b) Find the damping matrix $\mathbf{c}$.

(c) On a sketch of the model, such as the one shown, indicate the values of the individual springs and dashpots.

****

7.6 Consider a three-degree-of-freedom system for which the mass matrix is $\mathbf{m} = m\mathbf{I}$, with m being a scalar constant and $\mathbf{I}$ being the 3×3 identity matrix. The modes are uncoupled and the mode shapes, natural frequencies, and damping values are given by

$$[2,5,9]^T \qquad\qquad [9,0,-2]^T \qquad\qquad [2,-17,9]^T$$

$$\omega_1 = \sqrt{\frac{k}{m}} \qquad\qquad \omega_2 = \sqrt{\frac{3k}{m}} \qquad\qquad \omega_3 = \sqrt{\frac{5k}{m}}$$

$$\zeta_1 = 0.01 \qquad\qquad \zeta_2 = 0.01 \qquad\qquad \zeta_3 = 0.02$$

in which k is another scalar constant.

(a) Find the stiffness matrix **k**.

(b) Find the damping matrix **c**.

(c) On a sketch of the model, such as the one shown in Exercise 7.5, indicate the values of the individual springs and dashpots.

****

7.7 Let the stationary processes $\{X_1(t)\}$ and $\{X_2(t)\}$ represent the motions at two different points in a complicated system. Let the correlation matrix for $\{X_1(t)\}$ and $\{X_2(t)\}$ be given by

$$\mathbf{R}_{XX}(\tau) \equiv E[\vec{X}(t+\tau)\vec{X}^T(t)] = \begin{bmatrix} g(\tau)+g(2\tau) & 2g(\tau)-g(2\tau) \\ 2g(\tau)-g(2\tau) & 4g(\tau)+g(2\tau) \end{bmatrix}$$

in which $g(\tau) = e^{-b|\tau|}[\cos(\omega_0\tau) + (b/\omega_0)\sin(\omega_0|\tau|)]$, for constants b and ω_0. Let $\{Y(t)\}$ denote the relative motion between the two points: $Y(t) = X_2(t) - X_1(t)$

(a) Find the mean squared value of $Y(t)$: $E(Y^2)$.

(b) Find the cross-correlation function $R_{\dot{X}_1 Y}(\tau) \equiv E[\dot{X}_1(t+\tau)Y(t)]$.

****

Harmonic Transfer Functions

****

7.8 Consider the 2DF system of Exercise 7.1.

(a) Find the $\mathbf{H}(\omega)$ harmonic transfer function matrix.

(b) Show that $\mathbf{H}(\omega)$ has poles at ω values of $\pm\omega_j(1-\zeta_j^2)^{1/2} + i\zeta_j\omega_j$.

****

7.9 Consider the 2DF system of Exercise 7.2.

(a) Find the $\mathbf{H}(\omega)$ harmonic transfer function matrix.

(b) Show that $\mathbf{H}(\omega)$ has poles at ω values of $\pm\omega_j(1-\zeta_j^2)^{1/2} + i\zeta_j\omega_j$.

****

7.10 Consider the 2DF system of Example 7.4 and Exercise 7.3.

(a) Find the $\mathbf{H}(\omega)$ harmonic transfer function matrix.

(b) Show that $\mathbf{H}(\omega)$ has poles at ω values of $\pm\omega_j(1-\zeta_j^2)^{1/2} + i\zeta_j\omega_j$.

****

7.11 Consider the 2DF system of Exercise 7.4 with $m = 1.0$ kg, $k = 50$ N/m, $c = 1.0 \text{N} \cdot \text{s}/\text{m}$.

(a) Find the $\mathbf{H}(\omega)$ harmonic transfer function matrix.

(b) Locate the poles of $\mathbf{H}(\omega)$, and use this information to identify appropriate values of the modal parameters ω_j and ζ_j.

****

MDF Response

****

7.12 For the 2DF system of Exercises 7.1 and 7.8, let $F_1(t) \equiv 0$ and $\{F_2(t)\}$ be a stationary, mean zero white noise with autospectral density of 10.0.

(a) Find the autospectral density function $S_{X_1 X_1}(\omega)$ for the stationary $\{X_1(t)\}$ response component.

(b) Approximate the autocorrelation function $R_{X_1 X_1}(\tau)$ for the stationary $\{X_1(t)\}$ response component by treating the modal responses as though they were independent.

(c) Approximate the mean squared value $E(X_1^2)$ for the stationary $\{X_1(t)\}$ response component by using the same simplification as in part (b).

****

7.13 For the 2DF system of Exercise 7.2, let $\{F_1(t)\}$ and $\{F_2(t)\}$ be independent stationary white noise processes with mean value vector and autospectral density matrix of

$$\vec{\mu}_F = \begin{bmatrix} 4 \\ 6 \end{bmatrix} N \quad \text{and} \quad S_{FF}(\omega) = \begin{bmatrix} 1.0 & 0 \\ 0 & 1.5 \end{bmatrix} \frac{N^2 \cdot s}{rad}$$

(a) Find the mean value vector $\vec{\mu}_X$ for the stationary response.
(b) Find the autospectral density function $S_{X_2 X_2}(\omega)$ for the stationary $\{X_2(t)\}$ response component.
(c) Approximate the autocovariance function $G_{X_2 X_2}(\tau)$ for the stationary $\{X_2(t)\}$ response component by treating the modal responses as though they were independent.
(d) Approximate the mean squared value $E(X_2^2)$ for the stationary $\{X_2(t)\}$ response component by using the same simplification as in part (b).

****

7.14 Consider the 2DF system of Examples 7.2, 7.5, and 7.8, with an excitation having $F_2(t) \equiv 0$, and $\{F_1(t)\}$ being a stationary, mean zero process with autospectral density of
$$S_{F_1 F_1}(\omega) = 10^{-18} \omega^{20} e^{-0.5|\omega|} N \cdot s / rad$$
Find the autospectral density of the stationary $\{X_1(t)\}$ response component and show a sketch of it.

****

7.15 Consider a 2DF linear system which has modes with natural circular frequencies of $\omega_1 = 3$ rad/s and $\omega_2 = 8$ rad/s and damping values of $\zeta_1 = 0.01$ and $\zeta_2 = 0.02$. The excitation is a broadband Gaussian process. Analysis has shown that the stationary modal responses are mean zero and have standard deviation values of $\sigma_{Z_1} = 20$ mm and $\sigma_{Z_2} = 10$ mm. The response of primary interest is $X(t) = Z_1(t) + Z_2(t)$.
(a) Estimate the standard deviation σ_X of stationary response.
(b) Estimate the standard deviation of velocity $\sigma_{\dot{X}}$ of stationary response.
(c) Approximate the rate of upcrossings of the level zero by the stationary $\{X(t)\}$ process.
(d) Approximate the probability that $X(t)$ will ever exceed the level 85 mm during 30 seconds of the stationary response, based on the Poisson approximation.
NOTE: If you find that insufficient information is given, then use reasonable approximations and explain your reasoning.

****

7.16 A linear structure has been studied by modal analysis. It has been determined that only four modes contribute significantly to the dynamic response. The undamped natural frequencies and damping ratios of these modes are:

$\omega_1 = 3$ rad/s $\qquad \omega_2 = 5$ rad/s $\qquad \omega_3 = 8$ rad/s $\qquad \omega_4 = 12$ rad/s
$\zeta_1 = 0.01$ $\qquad\qquad \zeta_2 = 0.01$ $\qquad\qquad \zeta_3 = 0.02$ $\qquad\qquad \zeta_4 = 0.03$

The process $\{X(t)\}$ represents a critical distortion in the structure, and it has been written as the sum of the four modal contributions: $X(t) = Y_1(t) + Y_2(t) + Y_3(t) + Y_4(t)$. The excitation has been modeled as a broadband, zero mean, Gaussian process and analysis of the dynamic response of the structure has given: $\sigma_{Y_1} = 20$ mm, $\sigma_{Y_2} = 10$ mm, $\sigma_{Y_3} = 2$ mm, and $\sigma_{Y_4} = 0.5$ mm.

(a) Approximate the rate of upcrossings of the level zero by the stationary $\{X(t)\}$ process.

(b) As an estimate of the risk of damage, approximate the probability that $|X(t)|$ will ever exceed the level of 135 mm during 10^6 seconds of stationary response. Use the Poisson approximation.

****

State-Space Formulation

****

7.17 Consider the 2DF system of Exercises 7.1 and 7.8.

(a) Find the matrix $\mathbf{A}$ and the vector $\vec{Q}(t)$ such that this system is described by Eq. 7.44 with a state vector of $\vec{Y}(t) = [X_1(t), X_2(t), \dot{X}_1(t), \dot{X}_2(t)]^T$.

(b) Verify that the eigenvalues of $\mathbf{A}$ are $\pm i\omega_j(1-\zeta_j^2)^{1/2} + \zeta_j\omega_j$. That is, show that the values $\lambda = \pm i\omega_j(1-\zeta_j^2)^{1/2} + \zeta_j\omega_j$ satisfy the eigenvalue relationship $|\mathbf{A} - \lambda\mathbf{I}| = 0$.

****

7.18 Consider the 2DF system of Exercises 7.2 and 7.9.

(a) Find the matrix $\mathbf{A}$ and the vector $\vec{Q}(t)$ such that this system is described by Eq. 7.44 with a state vector of $\vec{Y}(t) = [X_1(t), X_2(t), \dot{X}_1(t), \dot{X}_2(t)]^T$.

(b) Verify that the eigenvalues of $\mathbf{A}$ are $\pm i\omega_j(1-\zeta_j^2)^{1/2} + \zeta_j\omega_j$. That is, show that the values $\lambda = \pm i\omega_j(1-\zeta_j^2)^{1/2} + \zeta_j\omega_j$ satisfy the eigenvalue relationship $|\mathbf{A} - \lambda\mathbf{I}| = 0$.

****

7.19 Consider the 2DF system of Example 7.4 and Exercises 7.3 and 7.10.

(a) Find the matrix $\mathbf{A}$ and the vector $\vec{Q}(t)$ such that this system is described by Eq. 7.44 with a state vector of $\vec{Y}(t) = [X_1(t), X_2(t), \dot{X}_1(t), \dot{X}_2(t)]^T$.

(b) Verify that the eigenvalues of $\mathbf{A}$ are $\pm i\omega_j(1-\zeta_j^2)^{1/2} + \zeta_j\omega_j$. That is, show that the values $\lambda = \pm i\omega_j(1-\zeta_j^2)^{1/2} + \zeta_j\omega_j$ satisfy the eigenvalue relationship $|\mathbf{A} - \lambda\mathbf{I}| = 0$.

****

7.20 Consider the 2DF system of Exercise 7.11.

(a) Find the matrix $\mathbf{A}$ and the vector $\vec{Q}(t)$ such that this system is described by Eq. 7.44 with a state vector of $\vec{Y}(t) = [X_1(t), X_2(t), \dot{X}_1(t), \dot{X}_2(t)]^T$.

(b) Verify that the eigenvalues of $\mathbf{A}$ are ω_p / i for ω_p denoting the pole locations found in Exercise 7.11.

****

7.21 Consider the system of Example 7.12 with $k_1 = k_2 = k$, $c_1 = 0.01(km)^{1/2}$, and $c_2 = 50c_1$, in which m and k are scalar constants.

(a) Find the eigenvalues and eigenvectors of the state space matrix $\mathbf{A}$. (Two eigenvalues are complex and one is real.)

(b) Find the harmonic transfer matrix $\mathbf{H}(\omega)$ for the state space vector $\vec{Y}(t) = [X_1(t), X_2(t), \dot{X}_1(t)]^T$.

****

Chapter 8

State-Space Analysis

8.1 Basic Concept

In this chapter we will study linear stochastic vibration problems by using techniques which are fundamentally different than those we have used so far. These state-space techniques provide an alternative to the general time domain and frequency domain approaches which were introduced in Chapters 3 and 4. The state-space equations will involve derivatives with respect to time, but the evaluation of response moments will not involve writing the response as a time domain integral, as in the so-called "time domain methods" introduced in Chapter 3. Fourier domain integrals also will not be required in the state-space analysis.

The basic idea of state-space analysis is to find and solve one or more equations which govern the behavior of response measures of interest. The equations are generally differential in nature, but they are algebraic for some stationary response measures. The derivation of these equations is accomplished directly from the differential equation of motion governing the time histories of the process, without the use of either the Duhamel integral or Fourier transforms. The simplest situation is when the response measures studied are moment functions of the response process, and a closely related approach uses cumulant functions. A similar, but more complicated, approach involves study of a probability density function of the process using the so-called Fokker-Planck or Kolmogorov equation. We will consider all three of these situations, but we will give more emphasis to the simpler equations involving moments or cumulants.

It is usually easiest to apply the methods presented in this chapter by using the state-space formulation of the equations of motion, as introduced in Section 7.6. Hence the name state-space analysis for this category of methods. However, it is definitely not essential to use the state-space form and we will illustrate the idea by deriving moment equations for the second moments of a single-degree-of-freedom oscillator directly from the original form of the equation of motion. The important point to be remembered is that the state-space formulation of the equations of motion and the methods of state-space analysis are separate concepts which can be efficiently used together, but also can be used separately.

****

Example 8.1: Let $\{X(t)\}$ denote the response of the SDF oscillator governed by the equation of motion

$$m\ddot{X}(t) + c\dot{X}(t) + kX(t) = F(t)$$

For such a second-order differential equation it is customary to designate $X(t)$ and $\dot{X}(t)$ as "state variables" and to seek equations involving the moments of these variables. To do this we begin by multiplying each side of the equation by $X(t)$, then taking the expected value. This gives

$$mE[X(t)\ddot{X}(t)] + cE[X(t)\dot{X}(t)] + kE[X^2(t)] = E[X(t)F(t)]$$

Note that the second and third terms involve only moment functions of the state variables, but the first term involves $\ddot{X}(t)$. This first term can be rewritten, though, in terms of moment functions involving $X(t)$ and $\dot{X}(t)$ as

$$E[X(t)\ddot{X}(t)] = \frac{d}{dt}E[X(t)\dot{X}(t)] - E[\dot{X}^2(t)]$$

Thus, the equation becomes

$$m\frac{d}{dt}E[X(t)\dot{X}(t)] - mE[\dot{X}^2(t)] + cE[X(t)\dot{X}(t)] + kE[X^2(t)] = E[X(t)F(t)]$$

This involves only moment functions of the state variables and the term $E[X(t)F(t)]$ related to the excitation. Similarly one can multiply the original equation of motion by $\dot{X}(t)$ and take the expected value to obtain

$$mE[\dot{X}(t)\ddot{X}(t)] + cE[\dot{X}^2(t)] + kE[X(t)\dot{X}(t)] = E[\dot{X}(t)F(t)]$$

and this can be rewritten in terms of the three moment functions of the state variables as

$$\frac{m}{2}\frac{d}{dt}E[\dot{X}^2(t)] + cE[\dot{X}^2(t)] + kE[X(t)\dot{X}(t)] = E[\dot{X}(t)F(t)]$$

Note also that the general relationship

$$\frac{d}{dt}E[X^2(t)] = 2E[X(t)\dot{X}(t)]$$

gives a third equation involving only the moment functions of the state variables. If we can somehow find the values of $E[X(t)F(t)]$ and $E[\dot{X}(t)F(t)]$, then this will give us three simultaneous linear differential equations which can be solved to find the three moment functions $E[X^2(t)]$, $E[X(t)\dot{X}(t)]$, and $E[\dot{X}^2(t)]$. In vector form these equations can be written as

$$\frac{d}{dt}\begin{bmatrix} E[X^2(t)] \\ E[X(t)\dot{X}(t)] \\ E[\dot{X}^2(t)] \end{bmatrix} + \begin{bmatrix} 0 & -2 & 0 \\ k/m & c/m & -1 \\ 0 & 2k/m & 2c/m \end{bmatrix}\begin{bmatrix} E[X^2(t)] \\ E[X(t)\dot{X}(t)] \\ E[\dot{X}^2(t)] \end{bmatrix} = \frac{1}{m}\begin{bmatrix} 0 \\ E[X(t)F(t)] \\ 2E[\dot{X}(t)F(t)] \end{bmatrix}$$

A tremendous simplification in this equation results for the special case of second-moment stationary response of the oscillator. In particular, if $E[X^2(t)]$, $E[X(t)\dot{X}(t)]$, and $E[\dot{X}^2(t)]$ are all independent of time, then the derivative terms completely drop out of the equations. For this problem in the original scalar form, this gives $-mE[\dot{X}^2(t)] + kE[X^2(t)] = E[X(t)F(t)]$ and $cE[\dot{X}^2(t)] = E[\dot{X}(t)F(t)]$.

****

Several important general properties of state-space moment analysis are illustrated by Example 8.1. First, we note that no assumptions or approximations were made in deriving the moment equations, other than the assumed existence of the various expected values. Thus, the equations will be true for almost any $\{F(t)\}$ excitation of the oscillator. In addition, we find that we cannot derive one moment equation which governs the behavior of only one moment function. In particular, we found in the example that three simultaneous equations governed the behavior of the three possible second moment functions of the state variables. This coupling of all moments of the same order is typical of the method. For linear systems, though, it is also typical that the moment equations of one order are uncoupled from those of any other order. For instance, no mean value terms or third moment terms appear in the second-order moment equations of Example 8.1. Furthermore, the property that one need solve only algebraic equations, rather than differential equations, when evaluating stationary moments of the state variables is also a general feature of the method. Finally, we note that Example 8.1 illustrates the fact that in order to make the moment equations useful, we must find a way to evaluate certain cross-product values involving both the excitation and the response. We will begin to address this problem in Section 8.4, after introducing a systematic procedure for deriving state-space moment equations.

8.2 Derivation of State-Space Moment and Cumulant Equations

Let the $\{X(t)\}$ process be the response of a general linear system excited by a stochastic process $\{F(t)\}$. A state-space moment equation results any time that we multiply a governing stochastic differential equation of motion by some power of $X(t)$ and take the expected value of both sides of the resulting equation. For example, if the equation of motion can be written as

$$\sum_{j=0}^{n} a_j \frac{d^j X(t)}{dt^j} = F(t) \tag{8.1}$$

and we multiply both sides of this equation by $X^k(t)$ and then take the expected value, we obtain

$$\sum_{j=0}^{n} a_j E\left[X^k(t) \frac{d^j X(t)}{dt^j} \right] = E[X^k(t)F(t)] \tag{8.2}$$

We call this a moment equation, since each term is of the form of a cross-product of $\{X(t)\}$ and either $\{F(t)\}$ or a derivative of $\{X(t)\}$. Other moment equations can be obtained by multiplying Eq. 8.1 by different terms. For example, a multiplier which is the kth power of the same process at some other time s gives

$$\sum_{j=0}^{n} a_j E\left[X^k(s) \frac{d^j X(t)}{dt^j} \right] = E[X^k(s)F(t)]$$

Similarly, we could multiply by a power of some derivative of $\{X(t)\}$, such as

$$\sum_{j=0}^{n} a_j E\left[\dot{X}^k(t) \frac{d^j X(t)}{dt^j} \right] = E[\dot{X}^k(t)F(t)]$$

or we could multiply by a term involving some different process, including $\{F(t)\}$:

$$\sum_{j=0}^{n} a_j E\left[F^k(t) \frac{d^j X(t)}{dt^j} \right] = E[F^{k+1}(t)]$$

Each of these procedures gives a moment equation, as does any combination of the procedures.

We can obtain cumulant equations in the same way. Specifically, the cumulant has a linearity property which can be written as

$$\kappa_{n+1}\left(W_1, \cdots, W_n, \sum a_j Z_j\right) = \sum a_j \kappa_{n+1}(W_1, \cdots, W_n, Z_j) \qquad (8.3)$$

for any set of W_l and Z_j random variables (see Section A.12 of Appendix A). This shows that the joint cumulant having one argument which is a linear combination of random variables can be written as a linear combination of joint cumulants. In our state-space analysis we are generally dealing with equations of motion which have the form of linear combinations of random variables. The linearity property of cumulants allows us to deal with joint cumulants of such equations of motion and other random variables. For example, if we take the joint cumulant of Eq. 8.1 and $X^k(t)$ we obtain

$$\sum_{j=0}^{n} a_j \kappa_2\left(X^k(t), \frac{d^j X(t)}{dt^j} \right) = \kappa_2\left(X^k(t), F(t) \right) \qquad (8.4)$$

which is the cumulant version of the moment equation in Eq. 8.2. Each of our other moment equations can also be rewritten as a cumulant equation by simply replacing each moment term by the corresponding cumulant term.

Clearly it is not difficult to derive moment or cumulant equations. In fact, we can easily derive any number of such equations. The significant issues which must be addressed relate to the identification of sets of equations which can be solved, and to the simplification of the cross-product or cross-cumulant terms involving both the response and the excitation, such as the right-hand side of Eqs. 8.2 and 8.4.

As mentioned in Section 8.1, it is usually easier to implement the state-space analysis techniques by using the state-space formulation of the equations of motion. Thus, we will direct our attention to systems governed by the equation

$$\dot{\vec{Y}}(t) + \mathbf{A}\vec{Y}(t) = \vec{Q}(t) \qquad (8.5)$$

in which $\vec{Y}(t)$ is the state vector of dimension n_Y describing the response, $\mathbf{A}$ is a constant square matrix, and the $\vec{Q}(t)$ vector describes the excitation. As shown in Section 7.6, this form of the equations can be applied to all systems for which the original equations of motion are linear ordinary differential equations. Even though use of this form of the equations is somewhat arbitrary, it is standard for state-space analysis, and it emphasizes the similarity of the various problems, regardless of the original form of the equations.

8.3   Equations for First and Second Moments and Covariance

We now wish to identify the sets of moment equations governing the behavior of the response moments of a particular order. In order to do this, we must find moment equations including only those particular moments and their derivatives with respect to time. In this section we will look only at the mean value function, the second moment autocorrelation function, and the second cumulant covariance function, which is also the second-order central moment function. First and second moment information, of course, gives a complete description of a Gaussian process, so one reason for emphasizing first and second moment analysis is the fact that many problems can be modeled by stochastic processes which are Gaussian or nearly Gaussian. Furthermore, as emphasized in Chapter 5, first and second moment information can be considered to be the first two steps, and usually the most important steps, in a more accurate description of a non-Gaussian process. In this chapter, as in Chapters 3 and 4, it is important to remember that the analysis of first and second moments or cumulants is in no way limited to the Gaussian problem. That is, the analysis of mean and covariance, for example, is carried out in exactly the same way for Gaussian and non-Gaussian processes, the difference in the two situations is whether this first and second cumulant information gives a complete or a partial description of the process of interest.

The simplest, and almost trivial, case of state-space moment analysis is for the first moment, or mean value function. Simply taking the expected value of both sides of Eq. 8.5 gives the first-order moment equation as

$$\dot{\vec{\mu}}_Y(t) + \mathbf{A}\vec{\mu}_Y(t) = \vec{\mu}_Q(t) \qquad (8.6)$$

in which we have used the standard notation $\vec{\mu}_Y(t) = E[\vec{Y}(t)]$ and $\vec{\mu}_Q(t) = E[\vec{Q}(t)]$. Thus, the mean values of the state variables are governed by a first-order vector differential equation which has the same form as Eq. 8.5. There is a major difference, though, in that Eq. 8.6 is not a stochastic equation. That is, neither $\vec{\mu}_Y(t)$ or $\vec{\mu}_Q(t)$ is stochastic, so this is a set of ordinary deterministic differential equations. The solution of Eq. 8.6 is, of course, exactly the same as was given in Section 7.6 for time domain analysis of the state-space formulation of the equations of motion:

$$\vec{\mu}_Y(t) = \int_{-\infty}^{t} \exp[-(t-s)\mathbf{A}]\vec{\mu}_Q(s)\,ds = \int_{-\infty}^{t} \mathbf{\theta}\exp[-(t-s)\mathbf{\lambda}]\mathbf{\theta}^{-1}\vec{\mu}_Q(s)\,ds \qquad (8.7)$$

in which $\mathbf{\lambda}$ and $\mathbf{\theta}$ are the matrices of eigenvectors and eigenvalues such that $\mathbf{A\theta} = \mathbf{\theta\lambda}$.

Next we will consider the derivation of equations involving the second moments of the state variables. Since the state variables are the n_Y components of the state vector $\vec{Y}(t)$, it is convenient to write the second moments as the matrix $\mathbf{\phi}_{YY}(t,t) \equiv E[\vec{Y}(t)\vec{Y}^T(t)]$ (as in Section A.8 of Appendix A). This matrix contains all the second moment terms, with the (j,k) component being $E[Y_j(t)Y_k(t)]$. Note that the matrix is symmetric, so that there are exactly $n_Y(n_Y + 1)/2$ distinct scalar moment functions to be determined in the second-order moment analysis. The state-space analysis of these second moments will involve finding and solving a matrix differential equation governing the behavior of the $\mathbf{\phi}_{YY}(t,t)$ matrix. This differential equation must involve the derivative of the matrix with respect to time, and this can be written as

$$\frac{d}{dt}\mathbf{\phi}_{YY}(t,t) = E[\dot{\vec{Y}}(t)\vec{Y}^T(t)] + E[\vec{Y}(t)\dot{\vec{Y}}^T(t)] \qquad (8.8)$$

An equation involving the first term on the right-hand side of Eq. 8.8 is easily obtained by multiplying Eq. 8.5 on the right by $\vec{Y}^T(t)$ and then taking the expected value, giving

$$E[\dot{\vec{Y}}(t)\vec{Y}^T(t)] + \mathbf{A}\mathbf{\phi}_{YY}(t,t) = E[\vec{Q}(t)\vec{Y}^T(t)] \equiv \mathbf{\phi}_{QY}(t,t)$$

Similarly, the transpose of this relationship describes the behavior of the final term in Eq. 8.8. Using the fact that $\mathbf{\phi}_{YY}(t,t)$ is symmetric, this is

$$E[\vec{Y}(t)\dot{\vec{Y}}^T(t)] + \mathbf{\phi}_{YY}(t,t)\mathbf{A}^T = E[\vec{Y}(t)\vec{Q}^T(t)] \equiv \mathbf{\phi}_{YQ}(t,t)$$

Adding these two equations and using Eq. 8.8, then gives

$$\frac{d}{dt}\mathbf{\phi}_{YY}(t,t) + \mathbf{A}\mathbf{\phi}_{YY}(t,t) + \mathbf{\phi}_{YY}(t,t)\mathbf{A}^T = \mathbf{\phi}_{QY}(t,t) + \mathbf{\phi}_{YQ}(t,t) \qquad (8.9)$$

This is a compact form of the general set of equations for the second moments of the state variables. The left-hand side of the equation involves the matrix $\mathbf{\phi}_{YY}(t,t)$ of unknowns, its first derivative, and the system matrix $\mathbf{A}$. The right-hand side involves cross-products of the response variables and the excitation terms. Use of Eq. 8.9 for the study of moments is commonly called Lyapunov analysis.

The matrix differential equation derived for the second moments is, of course, equivalent to a set of simultaneous scalar differential equations. In particular, the (j,k) component of Eq. 8.9 is

$$\frac{d}{dt}E[Y_j(t)Y_k(t)] + \sum_{l=1}^{n_Y} A_{jl}E[Y_l(t)Y_k(t)] + \sum_{l=1}^{n_Y} E[Y_j(t)Y_l(t)]A_{kl} = E[Q_j(t)Y_k(t)] + E[Y_j(t)Q_k(t)]$$

The symmetry of the equation makes the (k,j) component identical to this (j,k) component, so that the total number of unique scalar differential equations is $n_Y(n_Y + 1)/2$. Thus, the number of equations is equal to the number of unknowns, and a solution is possible if the cross-products on the right-hand side are determined. Uniqueness of the solution will be assured by knowledge of $n_Y(n_Y + 1)/2$ initial conditions, such as the initial values of all components of $\phi_{YY}(t,t)$. Note that Example 8.1 was a special case of second moment analysis with $n_Y = 2$, giving the number of equations and unknowns as 3.

Equations 8.6 and 8.9 both confirm the fact that if the moments of interest are stationary, then one needs only to solve algebraic equations rather than differential equations. This is quite possibly the most important feature of state-space analysis. Neither the time domain analysis of Chapter 3 or the frequency domain analysis of Chapter 4 gives such a tremendous simplification for the stationary problem.

****

Example 8.2: Find the coupled scalar equations for the first-order and second-order moments for the third-order system of Example 7.12 governed by the equations

$$m\ddot{X}_1(t) + c_1\dot{X}_1(t) + k_1 X_1(t) + k_2 X_2(t) = F(t)$$

and

$$k_2 X_2(t) = c_2[\dot{X}_1(t) - \dot{X}_2(t)]$$

In Example 7.2 we showed that this system is governed by Eq. 8.5 with

$$\vec{Y}(t) = \begin{bmatrix} X_1(t) \\ X_2(t) \\ \dot{X}_1(t) \end{bmatrix}, \quad A = \begin{bmatrix} 0 & 0 & -1 \\ 0 & k_2/c_2 & -1 \\ k_1/m & k_2/m & c_1/m \end{bmatrix}, \quad \vec{Q}(t) = \frac{1}{m}\begin{bmatrix} 0 \\ 0 \\ F(t) \end{bmatrix}$$

Using Eq. 8.6 for the equations for the first-order moments gives three first-order differential equations as

$$\dot{\mu}_{Y_1}(t) - \mu_{Y_3}(t) = 0$$

$$c_2\dot{\mu}_{Y_2}(t) + k_2\mu_{Y_2}(t) - c_2\mu_{Y_3}(t) = 0$$

and

$$m\dot{\mu}_{Y_3}(t) + k_1\mu_{Y_1}(t) + k_2\mu_{Y_2}(t) + c_1\mu_{Y_3}(t) = \mu_F(t)$$

Alternatively, one can simply take the expected values of the original coupled equations of motion, giving two equations, with one of them involving a second derivative:

$$m\ddot{\mu}_{X_1}(t) + c_1\dot{\mu}_{X_1}(t) + k_1\mu_{X_1}(t) + k_2\mu_{X_2}(t) = \mu_F(t)$$

and

$$k_2\mu_{X_2}(t) = c_2[\dot{\mu}_{X_1} - \dot{\mu}_{X_2}].$$

Since $n_Y = 3$, we see that there will be six simultaneous scalar equations governing six second moment terms. One can use either the upper triangular portion or the lower triangular portion of Eq. 8.9 to give these six equations. Proceeding through the lower triangular components in the order (1,1), (2,1), (2,2), (3,1), (3,2), (3,3) gives the equations as

$$\frac{d}{dt}E[Y_1^2(t)] - 2E[Y_1(t)Y_3(t)] = 0$$

$$\frac{d}{dt}E[Y_1(t)Y_2(t)] + \frac{k_2}{c_2}E[Y_1(t)Y_2(t)] - E[Y_1(t)Y_3(t)] - E[Y_2(t)Y_3(t)] = 0$$

$$\frac{d}{dt}E[Y_2^2(t)] + 2\frac{k_2}{c_2}E[Y_2^2(t)] - 2E[Y_2(t)Y_3(t)] = 0$$

$$\frac{d}{dt}E[Y_1(t)Y_3(t)] + \frac{k_1}{m}E[Y_1^2(t)] + \frac{k_2}{m}E[Y_1(t)Y_2(t)] + \frac{c_1}{m}E[Y_1(t)Y_3(t)]$$

$$- E[Y_3^2(t)] = \frac{1}{m}E[F(t)Y_1(t)]$$

$$\frac{d}{dt}E[Y_2(t)Y_3(t)] + \frac{k_1}{m}E[Y_1(t)Y_2(t)] + \frac{k_2}{m}E[Y_2^2(t)] + \left(\frac{k_2}{c_2} + \frac{c_1}{m}\right)E[Y_2(t)Y_3(t)]$$

$$- E[Y_3^2(t)] = \frac{1}{m}E[F(t)Y_2(t)]$$

and

$$\frac{d}{dt}E[Y_3^2(t)] + 2\frac{k_1}{m}E[Y_1(t)Y_3(t)] + 2\frac{k_2}{m}E[Y_2(t)Y_3(t)] + 2\frac{c_1}{m}E[Y_3^2(t)] = \frac{2}{m}E[F(t)Y_3(t)]$$

Note that it is also possible to write these relationships as

$$\frac{d}{dt}\vec{V}(t) + \mathbf{B}\vec{V}(t) = \vec{\psi}(t)$$

with $\vec{V}(t) = \left[E[Y_1^2(t)], E[Y_1(t)Y_2(t)], E[Y_2^2(t)], E[Y_1(t)Y_3(t)], E[Y_2(t)Y_3(t)], E[Y_3^2(t)]\right]^T$ being a vector containing all the second moment functions, and

$$\mathbf{B} = \begin{bmatrix} 0 & 0 & 0 & -2 & 0 & 0 \\ 0 & \dfrac{k_2}{c_2} & 0 & -1 & -1 & 0 \\ 0 & 0 & 2\dfrac{k_2}{c_2} & 0 & -2 & 0 \\ \dfrac{k_1}{m} & \dfrac{k_2}{m} & 0 & \dfrac{c_1}{m} & 0 & -1 \\ 0 & \dfrac{k_1}{m} & \dfrac{k_2}{m} & 0 & \left(\dfrac{k_2}{c_2}+\dfrac{c_1}{m}\right) & -1 \\ 0 & 0 & 0 & 2\dfrac{k_1}{m} & 2\dfrac{k_2}{m} & 2\dfrac{c_1}{m} \end{bmatrix}, \quad \vec{\psi}(t) = \frac{1}{m}\begin{bmatrix} 0 \\ 0 \\ 0 \\ E[F(t)Y_1(t)] \\ E[F(t)Y_2(t)] \\ 2E[F(t)Y_3(t)] \end{bmatrix}$$

Clearly, this version of the second moment equations has the same form as the vector equation of motion in Eq. 8.5, or the vector first moment equation in Eq. 8.6.

****

In many cases, we will find it to be more convenient to use state-space covariance analysis rather than second moment analysis. This is primarily because of simplifications which result in the cross-product terms involving both the excitation and the response, as on the right-hand side of Eq. 8.9. In addition, if the process of interest is Gaussian then the mean value and covariance functions give us the elements needed for the standard form of the probability density function. It turns out to be very simple to derive a state-space covariance equation similar to Eq. 8.9. One approach is to substitute Eqs. 8.6 and 8.9 into the general relationship describing the covariance matrix as $\mathbf{K}_{YY}(t,t) = \boldsymbol{\phi}_{YY}(t,t) - \vec{\mu}_Y(t)\vec{\mu}_Y^T(t)$, which shows that $\mathbf{K}_{YY}(t,t)$ is governed by an equation which is almost identical in form to Eq. 8.9:

$$\frac{d}{dt}\mathbf{K}_{YY}(t,t) + \mathbf{A}\mathbf{K}_{YY}(t,t) + \mathbf{K}_{YY}(t,t)\mathbf{A}^T = \mathbf{K}_{QY}(t,t) + \mathbf{K}_{YQ}(t,t) \tag{8.10}$$

Alternatively, Eq. 8.10 can be derived by the use of the relationships for cumulants, without consideration of first and second moments. First we note that the cross-covariance matrix for any two vectors can be viewed as a two-dimensional array of the possible second-order cross-cumulants of the components of the vectors, so that we can write $\mathbf{K}_{QY}(t,s) = \kappa_2[\vec{Q}(t),\vec{Y}(s)]$, for example. The linearity property of cumulants, as given in Eq. 8.3, can then be used to show that the derivative of $\mathbf{K}_{YY}(t,t)$ can be written as

$$\frac{d}{dt}\mathbf{K}_{YY}(t,t) = \mathbf{K}_{\dot{Y}Y}(t,t) + \mathbf{K}_{Y\dot{Y}}(t,t)$$

which is the cumulant version of Eq. 8.8. Furthermore, Eq. 8.3 allows us to take the joint cumulant of $\vec{Y}(t)$ and Eq. 8.5 and write the result as

$$\mathbf{K}_{\dot{Y}Y}(t,t) + \mathbf{A}\mathbf{K}_{YY}(t,t) = \mathbf{K}_{QY}(t,t)$$

Adding this equation to its transpose gives exactly Eq. 8.10. The advantage of this second derivation is that it can also be directly generalized to obtain higher-order cumulant equations which will be considered in Section 8.6.

8.4 Simplifications for Delta-Correlated Excitation

We will now investigate the right-hand side of our state-space equations, beginning with the simplest situation. In particular, we will begin by looking at the state-space covariance equation in which the excitation vector stochastic process $\{\vec{Q}(t)\}$ is delta-correlated. We will not impose any stationarity condition, nor will we assume that different components of $\vec{Q}(t)$ are independent of each other at any time t. We do require, though, that $\vec{Q}(t)$ and $\vec{Q}(s)$ are independent vector random variables for $t \neq s$, meaning that knowledge of the values of $\vec{Q}(t)$ gives no information about the possible values of $\vec{Q}(s)$ or about the probability distribution on those possible values. For this situation, the covariance matrix of the excitation process is given by

$$\mathbf{K}_{QQ}(t,s) = 2\pi \mathbf{S}_0(t)\delta(t-s) \tag{8.11}$$

in which $\mathbf{S}_0(t)$ is the nonstationary autospectral density matrix for $\{\vec{Q}(t)\}$.

As the first step in simplifying the right-hand side of Eq. 8.10, we write $\vec{Y}(t)$ in the form

$$\vec{Y}(t) = \vec{Y}(t_0) + \int_{t_0}^{t} \dot{\vec{Y}}(u)\,du$$

in which t_0 is any time prior to t. Into this expression we now substitute $\dot{\vec{Y}}(u)$ as determined from Eq. 8.5, giving

$$\vec{Y}(t) = \vec{Y}(t_0) + \int_{t_0}^{t} \vec{Q}(u)\,du - \mathbf{A}\int_{t_0}^{t}\vec{Y}(u)\,du \tag{8.12}$$

Transposing this equation and substituting it into $\mathbf{K}_{QY}(t,t) \equiv E[\vec{Q}(t)\vec{Y}^T(t)]$ gives

$$\mathbf{K}_{QY}(t,t) = \mathbf{K}_{QY}(t,t_0) + \int_{t_0}^{t}\mathbf{K}_{QQ}(t,u)\,du - \int_{t_0}^{t}\mathbf{K}_{QY}(t,u)\,du\,\mathbf{A}^T \tag{8.13}$$

We will show, though, that only one of the terms on the right-hand side of this equation makes a significant contribution to $\mathbf{K}_{QY}(t,t)$ when $\{\vec{Q}(t)\}$ is a delta-correlated process.

First, we note that since $\vec{Q}(t)$ is independent of $\vec{Q}(u)$ for $u < t$, we can also argue from the principle of cause and effect that $\vec{Q}(t)$ is independent of $\vec{Y}(t_0)$ for $t \geq t_0$. That is, the response at time t_0 is due to the portion of the excitation which occurred at times up to time t_0, but $\vec{Q}(t)$ is independent of that portion of the excitation. Thus, $\vec{Q}(t)$ is independent of $\vec{Y}(t_0)$. This independence, of course, implies that $\mathbf{K}_{QY}(t,t_0) = 0$, eliminating the first term on the right-hand side of Eq. 8.13. Furthermore, this same property allows the last term in Eq. 8.13 to be written as

$$\int_{t_0}^{t} \mathbf{K}_{QY}(t,u)\,du = \int_{t-\Delta t}^{t} \mathbf{K}_{QY}(t,u)\,du$$

in which the time increment Δt can be taken to be arbitrarily small. We note, though, that if $\mathbf{K}_{QY}(t,t)$ is finite, then this term is an integral of a finite integrand over an infinitesimal interval, so it also is equal to zero. Thus, if the second term on the right-hand side of Eq. 8.13 is finite, then it is exactly equal to $\mathbf{K}_{QY}(t,t)$, since the other terms on the right-hand side of the equation are then zero.

Simply substituting the Dirac delta function autocovariance of Eq. 8.11 into the integral in the second term on the right-hand side of Eq. 8.13 does demonstrate that the term is finite, but it leaves some ambiguity about its value. In particular, we can write an integral similar to, but more general than, the one of interest as

$$\int_{t_0}^{s} \mathbf{K}_{QQ}(t,u)\,du = 2\pi \int_{t_0}^{s} \mathbf{S}_0(t)\,\delta(t-u)\,du = 2\pi\,\mathbf{S}_0(t)\,U(s-t)$$

For $s < t$ this integral is zero, and for $s > t$ it is $2\pi \mathbf{S}_0(t)$. We, though, are interested in the situation with $s = t$, for which the upper limit of the integral is precisely aligned with the Dirac delta pulse of the integrand. This is the one value for the upper limit for which there is uncertainty about the value of the integral. We will resolve this problem in the following paragraph, but for the moment we will simply write

$$\mathbf{K}_{QY}(t,t) = \int_{t_0}^{t} \mathbf{K}_{QQ}(t,u)\,du \tag{8.14}$$

Note that the transpose of this term also appears in Eq. 8.10, and we will write it in similar fashion as

$$\mathbf{K}_{YQ}(t,t) = \int_{t_0}^{t} \mathbf{K}_{QQ}(u,t)\,du$$

so that Eq. 8.10 can be written as

$$\frac{d}{dt}\mathbf{K}_{YY}(t,t) + \mathbf{A}\mathbf{K}_{YY}(t,t) + \mathbf{K}_{YY}(t,t)\mathbf{A}^{T} = \int_{t_0}^{t} \mathbf{K}_{QQ}(t,u)\,du + \int_{t_0}^{t} \mathbf{K}_{QQ}(u,t)\,du \tag{8.15}$$

In order to resolve the ambiguity about the value of the right-hand side of Eq. 8.15, we note that the Dirac delta form of the autocovariance in Eq. 8.11 gives

$$\int_{t_0}^{t}\int_{t_0}^{t} \mathbf{K}_{QQ}(u,v)\,du\,dv = 2\pi \int_{t_0}^{t} \mathbf{S}_0(v)\,U(t-v)\,dv = 2\pi \int_{t_0}^{t} \mathbf{S}_0(v)\,dv \tag{8.16}$$

for any $t_0 < t$. Taking the derivative with respect to t of each side of this expression gives

$$\int_{t_0}^{t} \mathbf{K}_{QQ}(t,v)\,dv + \int_{t_0}^{t} \mathbf{K}_{QQ}(u,t)\,du = 2\pi \mathbf{S}_0(t) \tag{8.17}$$

Note, though, that the left-hand side of this expression is exactly the same as the right-hand side of Eq. 8.15. Thus, we find that the state-space covariance equation can be written as

$$\frac{d}{dt}\mathbf{K}_{YY}(t,t) + \mathbf{A}\,\mathbf{K}_{YY}(t,t) + \mathbf{K}_{YY}(t,t)\,\mathbf{A}^T = 2\pi \mathbf{S}_0(t) \tag{8.18}$$

for the special case when the excitation $\{\vec{Q}(t)\}$ is delta-correlated with nonstationary autospectral density matrix $\mathbf{S}_0(t)$.

Similarly, we can simplify the right-hand side of Eq. 8.9 describing the second moments of the state variables by noting that

$$\boldsymbol{\phi}_{QY}(t,t) + \boldsymbol{\phi}_{YQ}(t,t) = \mathbf{K}_{QY}(t,t) + \mathbf{K}_{YQ}(t,t) + \vec{\mu}_Q(t)\vec{\mu}_Y^T(t) + \vec{\mu}_Y(t)\vec{\mu}_Q^T(t)$$

so that we have

$$\frac{d}{dt}\boldsymbol{\phi}_{YY}(t,t) + \mathbf{A}\,\boldsymbol{\phi}_{YY}(t,t) + \boldsymbol{\phi}_{YY}(t,t)\,\mathbf{A}^T = 2\pi \mathbf{S}_0(t) + \vec{\mu}_Q(t)\vec{\mu}_Y^T(t) + \vec{\mu}_Y(t)\vec{\mu}_Q^T(t) \tag{8.19}$$

for the response to the nonstationary delta-correlated excitation. This equation is somewhat less convenient than Eq. 8.18, since it involves the mean value vector as well as the autospectral density matrix of the excitation. This is the first demonstration of any significant difference between moment equations and cumulant equations. The difference in complexity between moment and cumulant equations becomes more substantial when the terms considered are of higher order.

In order to obtain a solution of either Eq. 8.18 or Eq. 8.19, it is useful to rewrite it in the standard form used in Example 8.2. That is, we arrange all our unknown covariance or cross-product terms into a vector $\vec{V}(t)$ and find a matrix $\mathbf{B}$ and a vector $\vec{\psi}(t)$ such that

$$\frac{d}{dt}\vec{V}(t) + \mathbf{B}\vec{V}(t) = \vec{\psi}(t) \tag{8.20}$$

for which the solution can be written as

$$\vec{V}(t) = \int_{-\infty}^{t} \exp[-(t-s)\mathbf{B}]\,\vec{\psi}(s)\,ds = \exp[-(t-t_0)\mathbf{B}]\vec{V}(t_0) + \int_{t_0}^{t} \exp[-(t-s)\mathbf{B}]\,\vec{\psi}(s)\,ds \tag{8.21}$$

One can also use the eigenvalues and eigenvectors of $\mathbf{B}$ in order to rewrite this expression with only scalar exponential functions, as was done in Eq. 7.50.

One way of defining the elements of our arrays so as to convert Eq. 8.18 into the form of Eq. 8.20 is as follows. Let

$$V_{r(j,l)}(t) = [\mathbf{K}_{YY}(t,t)]_{j,l} \equiv K_{Y_j(t),Y_l(t)}, \qquad \psi_{r(j,l)}(t) = 2\pi[\mathbf{S}_0(t)]_{j,l}$$

with the function $r(j,l)$ defined as

$$r(j,l) = \frac{(j-1)(j)}{2} + l \qquad \text{for} \qquad j \geq l$$

This arranges the lower triangular elements of the symmetric matrices $\mathbf{K}_{YY}(t,t)$ and $\mathbf{S}_0(t)$ into the vectors $\vec{V}(t)$ and $\vec{\psi}(t)$, respectively. The diagonal elements of the $\mathbf{B}$ matrix are then given by

$$B_{r(j,l),r(j,l)} = A_{j,j} + A_{l,l}$$

and the off-diagonal elements are

$$\begin{aligned}
B_{r(j,l),r(u,v)} &= A_{j,v} && \text{for} && u = l \neq j \\
&= A_{j,u} && \text{for} && v = l \neq j \\
&= A_{l,v} && \text{for} && u = j \neq l \\
&= A_{l,u} && \text{for} && v = j \neq l \\
&= 2A_{j,v} && \text{for} && u = j = l \\
&= 2A_{j,u} && \text{for} && v = j = l \\
&= 0 && \text{otherwise}
\end{aligned}$$

with the restrictions that $j \geq l$ and $u \geq v$.

By applying a different initial condition, one can also use Eq. 8.20 in obtaining the conditional variance of response consistent with given values of $\vec{Y}(t_0) = \vec{y}$ at some particular time t_0. That is, if the vector $\vec{V}(t)$ is composed of the components of conditional covariance of the response:

$$V_{r(j,l)}(t) = Cov[Y_j(t), Y_l(t) | \vec{Y}(t_0) = \vec{y}]$$

then the given condition on $\vec{Y}(t_0)$ will give $\vec{V}(t_0) = \vec{0}$ in Eq. 8.21. By combining this with the conditional mean of the response, as given in Eq. 8.7, one can find the conditional autocorrelation function as well.

****

Example 8.3: Using state space analysis find the variance of the response of
the system shown, in which $\{F(t)\}$ is a stationary white noise process with
autospectral density S_0.

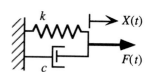

Putting the equation of motion into the form of Eq. 8.5, we have

$$\dot{X}(t) + \frac{k}{c}X(t) = \frac{1}{c}F(t)$$

so that $n_Y = 1$ and the state "vector" $\vec{Y}(t)$ is the scaler $X(t)$, the "matrix" $\mathbf{A}$ has dimension $(1,1)$ and is k/c,
and the "vector" $\vec{Q}(t)$ is the scalar $F(t)/c$. The covariance "matrix" in Eq. 8.18 is then the scalar variance
$K_{XX}(t) \equiv \sigma_X^2(t)$, and the relationship is

$$\frac{d}{dt}\sigma_X^2(t) + 2\frac{k}{c}\sigma_X^2(t) = \frac{2\pi S_0}{c^2}$$

Note the presence of the c^2 term in the denominator of the right-hand side of the expression. This comes
from the fact that the autospectral density of $Q(t) = F(t)/c$ is S_0/c^2.

For the special case of stationary response, the variance is found by setting the derivative term equal to zero
and solving the algebraic equation to get

$$\sigma_X^2 = \frac{\pi S_0}{kc}$$

If we are given an initial condition of $\sigma_X^2(t_0)$ for some particular time, then we can solve the differential
equation to give

$$\sigma_X^2(t) = \sigma_X^2(t_0)e^{-2kt/c} + \frac{2\pi S_0}{c^2}\int_{t_0}^{t} e^{-2k(t-s)/c}\,ds = \sigma_X^2(t_0)e^{-2kt/c} + \frac{\pi S_0}{kc}[1 - e^{-2k(t-t_0)/c}]$$

as in Eq. 8.21. For the common situation in which the system is at rest at time $t = 0$, we have

$$\sigma_X^2(t) = \frac{\pi S_0}{kc}[1 - e^{-2kt/c}]$$

All of these results, of course, are exactly the same as one would obtain by the time domain and frequency
domain analysis techniques of Chapters 3 and 4, respectively. Their determination by state space analysis,
though, is somewhat simpler, particularly for the stationary situation.
****
Example 8.4: Find the response variance and covariance values for a SDF oscillator with a mean zero,
stationary, delta-correlated excitation with autospectral density of S_0.

The state-space description of the system is Eq. 8.5 with $n_Y = 2$,

$$\vec{Y}(t) = \begin{bmatrix} X(t) \\ \dot{X}(t) \end{bmatrix}, \quad \mathbf{A} = \begin{bmatrix} 0 & -1 \\ k/m & c/m \end{bmatrix}, \quad \vec{Q}(t) = \frac{1}{m} \begin{bmatrix} 0 \\ F(t) \end{bmatrix}$$

Thus, the state-space equation for the covariance, from Eq. 8.18, is

$$\frac{d}{dt}\mathbf{K}_{YY}(t,t) + \begin{bmatrix} 0 & -1 \\ k/m & c/m \end{bmatrix}\mathbf{K}_{YY}(t,t) + \mathbf{K}_{YY}(t,t)\begin{bmatrix} 0 & k/m \\ -1 & c/m \end{bmatrix} = \frac{2\pi S_0}{m^2}\begin{bmatrix} 0 & 0 \\ 0 & 1 \end{bmatrix}$$

with

$$\mathbf{K}_{YY}(t,t) = \begin{bmatrix} K_{XX}(t,t) & K_{X\dot{X}}(t,t) \\ K_{\dot{X}X}(t,t) & K_{\dot{X}\dot{X}}(t,t) \end{bmatrix} = \begin{bmatrix} \sigma_X^2(t) & \rho_{X\dot{X}}(t,t)\sigma_X(t)\sigma_{\dot{X}}(t) \\ \rho_{X\dot{X}}(t,t)\sigma_X(t)\sigma_{\dot{X}}(t) & \sigma_{\dot{X}}^2(t) \end{bmatrix}$$

The three scalar equations of interest can be taken to be the (1,1), (2,1), and (2,2) elements of this matrix equation, giving

$$\frac{d}{dt}K_{XX}(t,t) - 2K_{X\dot{X}}(t,t) = 0$$

$$\frac{d}{dt}K_{X\dot{X}}(t,t) + \frac{k}{m}K_{XX}(t) + \frac{c}{m}K_{X\dot{X}}(t) - K_{\dot{X}\dot{X}}(t) = 0$$

and

$$\frac{d}{dt}K_{\dot{X}\dot{X}}(t,t) + 2\frac{k}{m}K_{X\dot{X}}(t) + 2\frac{c}{m}K_{\dot{X}\dot{X}}(t) = \frac{2\pi S_0}{m^2}$$

Writing $\vec{V}(t) = [K_{XX}(t,t), K_{X\dot{X}}(t,t), K_{\dot{X}\dot{X}}(t,t)]^T$ then gives $\dot{\vec{V}}(t) + \mathbf{B}\vec{V}(t) = \vec{\psi}(t)$, as in Eq. 8.20, with

$$\mathbf{B} = \frac{1}{m}\begin{bmatrix} 0 & -2m & 0 \\ k & c & -m \\ 0 & 2k & 2c \end{bmatrix}, \quad \vec{\psi}(t) = \frac{2\pi S_0}{m^2}\begin{bmatrix} 0 \\ 0 \\ 1 \end{bmatrix}$$

For the special case of stationary response we can find the covariance values, then, by solving the algebraic equation $\mathbf{B}\vec{V}(t) = \vec{\psi}(t)$. For this system of dimension three, one can do this by inverting the $\mathbf{B}$ matrix as

$$\mathbf{B}^{-1} = \frac{1}{2kc}\begin{bmatrix} km+c^2 & 2mc & m^2 \\ -kc & 0 & 0 \\ k^2 & 0 & km \end{bmatrix}$$

so that the stationary solution is $\vec{V}(t) = \mathbf{B}^{-1}\vec{\psi}(t) = \frac{\pi S_0}{mkc}[m,0,k]^T$. Thus, one has

$$\sigma_X^2 = \frac{\pi S_0}{kc}, \quad \sigma_{\dot{X}}^2 = \frac{\pi S_0}{mc}, \quad \rho_{X\dot{X}}(t,t) = 0$$

Note also that for this SDF it is very easy to solve the component equations of $\mathbf{B}\vec{V}(t) = \vec{\psi}(t)$ to find the stationary response relationships without any matrix operations. In particular, the first component equation gives $K_{\dot{X}\dot{X}}(t,t) = 0$, which reduces the third equation to $K_{\dot{X}\dot{X}}(t,t) = \pi S_0 / (mc)$, and the second equation gives $K_{XX}(t,t) = (m/k)K_{\dot{X}\dot{X}}(t,t) = \pi S_0 / (kc)$.

Let us now presume that we are given an initial condition of $\mathbf{K}_{XX}(t_0,t_0)$ at some specific time t_0. This will result in a nonstationary $\mathbf{K}_{XX}(t,t)$ matrix, and in order to find this nonstationary solution we will need to use the matrix exponential. This, however, requires that we diagonalize $\mathbf{B}$. Thus, we perform the eigenanalysis of $\mathbf{B}$, obtaining

$$\boldsymbol{\theta} = \begin{bmatrix} 1 & 1 & 1 \\ -\zeta\omega_0 & -\zeta\omega_0 - i\omega_d & -\zeta\omega_0 + i\omega_d \\ \omega_0^2 & -\omega_0^2(1-2\zeta^2) + 2i\zeta\omega_0\omega_d & -\omega_0^2(1-2\zeta^2) - 2i\zeta\omega_0\omega_d \end{bmatrix}$$

and

$$\boldsymbol{\theta}^{-1} = \frac{\omega_0}{\omega_d^2} \begin{bmatrix} \frac{\omega_0}{2} & \zeta & \frac{1}{2\omega_0} \\ \left(\frac{1-2\zeta^2}{4}\omega_0 + \frac{i\zeta\omega_d}{2}\right) & \left(\frac{-\zeta}{2} + \frac{i\omega_d}{2\omega_0}\right) & \left(\frac{-1}{4\omega_0}\right) \\ \left(\frac{1-2\zeta^2}{4}\omega_0 - \frac{i\zeta\omega_d}{2}\right) & \left(\frac{-\zeta}{2} - \frac{i\omega_d}{2\omega_0}\right) & \left(\frac{-1}{4\omega_0}\right) \end{bmatrix}$$

in which we have used the common notation of $k/m = \omega_0^2$, $c/m = 2\zeta\omega_0$ and $\omega_d^2 = \omega_0^2(1-\zeta^2)$. These equations give $\boldsymbol{\theta}^{-1}\mathbf{B}\boldsymbol{\theta} = \boldsymbol{\lambda}$, with $\boldsymbol{\lambda}$ being diagonal, and its non-zero elements being the eigenvalues of $\mathbf{B}$: $\lambda_{1,1} = 2\zeta\omega_0$, $\lambda_{2,2} = 2\zeta\omega_0 + 2i\omega_d$, and $\lambda_{3,3} = 2\zeta\omega_0 - 2i\omega_d$. This allows us to use $\exp[-t\mathbf{B}] = \boldsymbol{\theta}\exp[-t\boldsymbol{\lambda}]\boldsymbol{\theta}^{-1}$ in the general nonstationary solution of Eq. 8.21 to obtain

$$\vec{V}(t) = \exp[-(t-t_0)\mathbf{B}]\vec{V}(t_0) + \int_{t_0}^{t} \exp[-(t-s)\mathbf{B}]\vec{\psi}(s)\,ds$$

$$= \boldsymbol{\theta}\exp[-(t-t_0)\boldsymbol{\lambda}]\boldsymbol{\theta}^{-1}\vec{V}(t_0) + \boldsymbol{\theta}\int_{t_0}^{t} \exp[-(t-s)\boldsymbol{\lambda}]\boldsymbol{\theta}^{-1}\vec{\psi}(s)\,ds$$

Noting that $\vec{\psi}(s)$ has only one non-zero component, one can write out a component of $\vec{V}(t)$ as

$$V_j(t) = \sum_{r_1=1}^{3}\sum_{r_2=1}^{3}\theta_{j,r_1} e^{-(t-t_0)\lambda_{r_1}} \theta_{r_1,r_2}^{-1} V_{r_2}(t_0) + \frac{2\pi S_0}{m^2}\sum_{r=1}^{3}\theta_{j,r}\theta_{r,3}^{-1}\int_{t_0}^{t} e^{-(t-s)\lambda_r}\,dt$$

and perform the integration to obtain

$$V_j(t) = \sum_{r_1=1}^{3}\sum_{r_2=1}^{3}\theta_{j,r_1} e^{-(t-t_0)\lambda_{r_1}} \theta_{r_1,r_2}^{-1} V_{r_2}(t_0) + \frac{2\pi S_0}{m^2}\sum_{r=1}^{3}\frac{\theta_{j,r}\theta_{r,3}^{-1}}{\lambda_r}\left(1 - e^{-(t-t_0)\lambda_r}\right)$$

Substituting in the values for the complex θ and λ terms, of course, results in real expressions for components of $\vec{V}(t)$. The imaginary parts of λ in the exponentials lead to sine and cosine terms.

For the special case in which the initial condition is $\vec{V}(0) = \vec{0}$, the results are

$$\sigma_X^2(t) = V_1(t) = \frac{\pi S_0}{2m^2 \zeta \omega_0^3}\left(1 - e^{-2\zeta\omega_0 t}\left[\frac{\omega_0^2}{\omega_d^2} + \frac{\zeta\omega_0}{\omega_d}\sin(2\omega_d t) - \frac{\omega_0^2}{\omega_d^2}\cos(2\omega_d t)\right]\right)$$

$$\rho_{X\dot{X}}(t,t)\sigma_X(t)\sigma_{\dot{X}}(t) = V_2(t) = \frac{\pi S_0}{2m^2 \omega_d^2}\left(e^{-2\zeta\omega_0 t}\left[1 - \cos(2\omega_d t)\right]\right)$$

and

$$\sigma_{\dot{X}}^2(t) = V_3(t) = \frac{\pi S_0}{2m^2 \zeta \omega_0}\left(1 - e^{-2\zeta\omega_0 t}\left[\frac{\omega_0^2}{\omega_d^2} - \frac{\zeta\omega_0}{\omega_d}\sin(2\omega_d t) - \frac{\omega_0^2}{\omega_d^2}\cos(2\omega_d t)\right]\right)$$

These results are identical to those in Eqs. 3.45, 3.49, and 3.50 of Chapter 3, where they were obtained by using the Duhamel integral time domain solution of the original equation of motion.
****

Example 8.5: Find the mean and variance of the response $X(t)$ for a SDF oscillator with $m = 100$ kg, $k = 10,000$ N / m, and $c = 40$ N·s / m. The excitation is a nonstationary delta-correlated process having a mean value of $\mu_F(t) = \mu_0 t e^{-\alpha t} U(t)$ with $\mu_0 = 1000$ N / s and $\alpha = 0.8$ s^{-1} and a covariance function of $K_{FF}(t,s) = 2\pi S_0 t e^{-\alpha t}\delta(t-s)U(t)$ with $S_0 = 1000$ N^2 / rad. The system is at rest at time $t = 0$.

The state-space description of the system is identical to that in Example 8.4. Thus, the equation for the mean value of the response is

$$\dot{\vec{\mu}}_Y(t) + \mathbf{A}\vec{\mu}_Y(t) = \dot{\vec{\mu}}_Y(t) + \begin{bmatrix} 0 & -1 \\ 100 & 0.4 \end{bmatrix}\vec{\mu}_Y(t) = \frac{\mu_0}{m}te^{-\alpha t}U(t)\begin{bmatrix}0\\1\end{bmatrix} = 10te^{-0.8t}U(t)\begin{bmatrix}0\\1\end{bmatrix}$$

and its solution can be written as

$$\vec{\mu}_Y(t) = \frac{\mu_0}{m}\int_0^t \exp[-(t-s)\mathbf{A}]se^{-\alpha s}ds\begin{bmatrix}0\\1\end{bmatrix} = 10\int_0^t \mathbf{\theta}\exp[-(t-s)\mathbf{\lambda}]\mathbf{\theta}^{-1}se^{-0.8s}ds\begin{bmatrix}0\\1\end{bmatrix}$$

in which $\mathbf{\theta}$ is the matrix of dimension 2×2 with columns which are the eigenvectors of $\mathbf{A}$, and $\mathbf{\lambda} = \mathbf{\theta}^{-1}\mathbf{A}\mathbf{\theta}$. Eigenanalysis yields the necessary matrices as

$$\mathbf{\theta} = \begin{bmatrix} 1 & 1 \\ -\zeta\omega_0 - i\omega_d & -\zeta\omega_0 + i\omega_d \end{bmatrix} = \begin{bmatrix} 1 & 1 \\ -0.2 - 9.998i & -0.2 + 9.998i \end{bmatrix}$$

and

$$\mathbf{\theta}^{-1} = \frac{1}{2}\begin{bmatrix} 1 + i\zeta\omega_0/\omega_d & i/\omega_d \\ 1 - i\zeta\omega_0/\omega_d & -i/\omega_d \end{bmatrix} = \begin{bmatrix} 0.5 + 0.01000i & 0.05001i \\ 0.5 - 0.01000i & -0.05001i \end{bmatrix}$$

with the nonzero elements of λ being the eigenvalues of $\mathbf{A}$: $\lambda_{1,1} = \zeta\omega_0 + i\omega_d = 0.2 + 9.998i$ and $\lambda_{2,2} = 0.2 - 9.998i$. The first term of the vector $\vec{\mu}_Y(t)$ is the mean of $X(t)$, and it is given by

$$\mu_X(t) = 10 \sum_{r=1}^{2} \theta_{1,r} \theta_{r,2}^{-1} \int_0^t e^{-(t-s)\lambda_{r,r}} s e^{-\alpha s} ds$$

which reduces to

$$\mu_X(t) = (0.001192 + 0.09968t)e^{-0.8t}$$
$$-e^{-0.09899t}[0.001192\cos(9.9980t) + 0.009899\sin(9.9980t)]$$

This result is sketched, along with $\mu_F(t)/k$, which is the static response that one would obtain by neglecting the dynamics of the system.

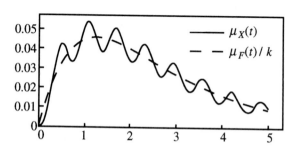

For the covariance analysis we can use the $\boldsymbol{\theta}$, $\boldsymbol{\theta}^{-1}$, and λ matrices derived in Example 8.4, for which our parameters give

$$\boldsymbol{\theta} = \begin{bmatrix} 1 & 1 & 1 \\ -0.2 & -0.2 - 9.998i & -0.2 + 9.998i \\ 100 & -99.92 + 3.999i & -99.92 - 3.999i \end{bmatrix}$$

and

$$\boldsymbol{\theta}^{-1} = \begin{bmatrix} 0.5002 & 0.002000 & 0.005002 \\ (0.2499 + 0.01000i) & (-0.001000 + 0.05001i) & (-0.002501) \\ (0.2499 - 0.01000i) & (-0.001000 - 0.05001i) & (-0.002501) \end{bmatrix}$$

and the elements of the diagonal λ matrix are $\lambda_{1,1} = 0.4$, $\lambda_{2,2} = 0.4 + 20.00i$, and $\lambda_{3,3} = 0.4 - 20.00i$. In the vector equation $\dot{\vec{V}}(t) + \mathbf{B}\vec{V}(t) = \vec{\psi}(t)$, the definitions of $\vec{V}(t)$ and $\mathbf{B}$ are the same as before and

$$\vec{\psi}(t) = \frac{2\pi S_0}{m^2} t e^{-\alpha t}[0,0,1]^T = 0.2\pi t e^{-0.8t}[0,0,1]^T$$

For our initial condition we have $\mathbf{K}_{YY}(0,0) = \mathbf{0}$, or $\vec{V}(0) = \vec{0}$, so the nonstationary solution is

$$\vec{V}(t) = \int_0^t \exp[-(t-s)\mathbf{B}]\vec{\psi}(s)\,ds$$

which has components described by

$$V_j(t) = 0.2\pi \sum_{r=1}^{3} \theta_{j,r} \theta_{r,3}^{-1} \int_0^t s e^{-0.8s} e^{-(t-s)\lambda_r} dt$$

The first component of this vector gives us

$$\sigma_{X}^2(t) = V_1(t) = (-1.965\times10^{-2} - 7.854\times10^{-3}t)e^{-0.8t}$$
$$+e^{-0.4t}[1.964\times10^{-2} + 7.851\times10^{-6}\cos(20.00t) - 3.142\times10^{-7}\sin(20.00t)]$$

The sketch compares this result with a "pseudo-stationary" approximation based only on stationary analysis. This approximation is simply $\pi S_0(t)/(kc)$, in which $K_{FF}(t+\tau,t) = 2\pi S_0(t)\delta(\tau)$ gives the definition of $S_0(t)$.

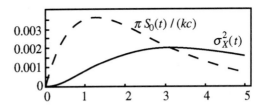

Thus, the approximation is obtained by substituting the nonstationary autospectral density of the excitation into the formula giving the response variance for a stationary excitation. This approximation is good for problems in which $S_0(t)$ varies slowly, but it is not surprising that it fails for the present problem.

****

Example 8.6: Consider the 2DF system shown with: $m_1 = 2m$, $m_2 = m$, $k_1 = k_2 = k$, $c_1 = c_2 = 0.01(km)^{1/2}$, in which m and k are scalar constants. Find the stationary covariance matrix for the response to independent white noise excitations $\{F_1(t)\}$ and $\{F_2(t)\}$ with autospectral density values of S_1 and S_2, respectively.

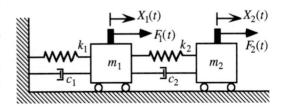

This system is governed by $\dot{\vec{Y}}(t) + \mathbf{A}\vec{Y}(t) = \vec{Q}(t)$ with $n_Y = 4$ and

$$\vec{Y}(t) = \begin{bmatrix} X_1(t) \\ X_2(t) \\ \dot{X}_1(t) \\ \dot{X}_2(t) \end{bmatrix}, \quad \mathbf{A} = \begin{bmatrix} 0 & 0 & -1 & 0 \\ 0 & 0 & 0 & -1 \\ k/m & -k/(2m) & 0.01\sqrt{k/m} & -0.005\sqrt{k/m} \\ -k/m & k/m & -0.01\sqrt{k/m} & 0.01\sqrt{k/m} \end{bmatrix}, \quad \vec{Q}(t) = \frac{1}{m}\begin{bmatrix} 0 \\ 0 \\ F_1(t)/2 \\ F_2(t) \end{bmatrix}$$

From Eq. 8.18 we see that the stationary covariance matrix is the solution of $\mathbf{A}\mathbf{K}_{YY}(t,t) + \mathbf{K}_{YY}(t,t)\mathbf{A}^T = 2\pi\mathbf{S}_0$ with

$$S_0 = \frac{2\pi}{m^2}\begin{bmatrix} 0 & 0 & 0 & 0 \\ 0 & 0 & 0 & 0 \\ 0 & 0 & S_1/4 & 0 \\ 0 & 0 & 0 & S_2 \end{bmatrix}$$

Writing out the 10 scalar equations representing the lower triangular part of this array, then arranging them into the form of Eq. 8.20 gives $\mathbf{B}\vec{V} = \vec{\psi}$ with

$$\vec{V} = [K_{X_1,X_1}, K_{X_2,X_1}, K_{X_2,X_2}, K_{\dot{X}_1,X_1}, K_{\dot{X}_1,X_2}, K_{\dot{X}_1,\dot{X}_1}, K_{\dot{X}_2,X_1}, K_{\dot{X}_2,X_2}, K_{\dot{X}_2,\dot{X}_1}, K_{\dot{X}_2,\dot{X}_2}]^T$$

$$\mathbf{B} = \begin{bmatrix}
0 & 0 & 0 & -2 & 0 & 0 & 0 & 0 & 0 & 0 \\
0 & 0 & 0 & 0 & -1 & 0 & -1 & 0 & 0 & 0 \\
0 & 0 & 0 & 0 & 0 & 0 & 0 & -2 & 0 & 0 \\
\frac{k}{m} & \frac{-k}{2m} & 0 & 0.01\sqrt{\frac{k}{m}} & 0 & -1 & -0.005\sqrt{\frac{k}{m}} & 0 & 0 & 0 \\
0 & \frac{k}{m} & \frac{-k}{2m} & 0 & 0.01\sqrt{\frac{k}{m}} & 0 & 0 & -0.005\sqrt{\frac{k}{m}} & -1 & 0 \\
0 & 0 & 0 & \frac{2k}{m} & \frac{-k}{m} & 0.02\sqrt{\frac{k}{m}} & 0 & 0 & -0.01\sqrt{\frac{k}{m}} & 0 \\
\frac{-k}{m} & \frac{k}{m} & 0 & -0.01\sqrt{\frac{k}{m}} & 0 & 0 & 0.01\sqrt{\frac{k}{m}} & 0 & -1 & 0 \\
0 & \frac{-k}{m} & \frac{k}{m} & 0 & -0.01\sqrt{\frac{k}{m}} & 0 & 0 & 0.01\sqrt{\frac{k}{m}} & 0 & -1 \\
0 & 0 & 0 & \frac{-k}{m} & \frac{k}{m} & -0.01\sqrt{\frac{k}{m}} & \frac{k}{m} & \frac{-k}{2m} & 0.02\sqrt{\frac{k}{m}} & -0.005\sqrt{\frac{k}{m}} \\
0 & 0 & 0 & 0 & 0 & 0 & \frac{-2k}{m} & \frac{2k}{m} & -0.02\sqrt{\frac{k}{m}} & 0.02\sqrt{\frac{k}{m}}
\end{bmatrix}$$

and

$$\vec{\psi} = (2\pi/m^2)[0,0,0,0,0,S_1/4,0,0,0,S_2]^T.$$

The solution can be written as

$$\vec{V} = \pi\left[\frac{75S_1 + 150S_2}{k\sqrt{km}}, \frac{100S_1 + 200S_2}{k\sqrt{km}}, \frac{150S_1 + 300S_2}{k\sqrt{km}}, 0, \frac{-0.25S_1 + 0.50S_2}{km}, \right.$$
$$\left. \frac{25S_1 + 50S_2}{m\sqrt{km}}, \frac{0.25S_1 - 0.50S_2}{km}, 0, \frac{25S_1 + 50S_2}{m\sqrt{km}}, \frac{50S_1 + 100S_2}{m\sqrt{km}}\right]^T$$

or

$$\vec{V} = \pi \left[\frac{0.75S_1 + 1.50S_2}{kc}, \frac{S_1 + 2S_2}{kc}, \frac{1.50S_1 + 3S_2}{kc}, 0, \frac{-0.25S_1 + 0.50S_2}{km}, \right.$$

$$\left. \frac{0.25S_1 + 0.50S_2}{mc}, \frac{0.25S_1 - 0.50S_2}{km}, 0, \frac{0.25S_1 + 0.50S_2}{mc}, \frac{0.50S_1 + S_2}{mc} \right]^T$$

Note that the fourth and eighth terms are exactly zero. These terms, though, are $K_{\dot{X}_1,X_1}$ and $K_{\dot{X}_2,X_2}$, and we already knew that these terms are zero since they are proportional to the derivatives of K_{X_1,X_1} and K_{X_2,X_2}.

Although the size of the $\mathbf{B}$ matrix becomes cumbersome to write out, a computer can easily handle the problem of solving the linear equation $\mathbf{B}\vec{V} = \vec{\psi}$ for quite large systems when numerical values are given for all the parameters. We have kept the symbolic representation of some parameters in the solution here only to emphasize the similarity between these results and the analytical solutions we have previously obtained for the SDF system.

****

Example 8.7: Consider the stationary response of a SDF oscillator with $m = 100$ kg, $k = 10,000$ N / m, and $c = 40$ N·s / m. The excitation is a stationary delta-correlated process having a mean value of $\mu_F(t) = \mu_0 = 1000$ N and a covariance function of $K_{FF}(t,s) = 2\pi S_0 \delta(t-s) = 2\pi (1000 \text{ N}^2 \cdot \text{s} / \text{rad})$ $\delta(t-s)$. Find the conditional mean and variance of the responses $X(t)$ and $\dot{X}(t)$ given that $X(5) = 0.05$ m and $\dot{X}(5) = 0$.

This oscillator has the same parameters as the one considered in Example 8.5, so we can use basically the same state-space equations as were developed there. Only the excitation and the initial conditions are different. Thus, the conditional mean value of the response is found from the solution of

$$\frac{d}{dt} E\left(\vec{Y}(t) | \vec{Y}(5) = [0.05,0]^T\right) + \mathbf{A}\, E\left(\vec{Y}(t) | \vec{Y}(5) = [0.05,0]^T\right) = \frac{\mu_0}{m} \begin{bmatrix} 0 \\ 1 \end{bmatrix} = \begin{bmatrix} 0 \\ 10 \end{bmatrix}$$

The result for $t \geq 5$ can be written as

$$E\left(\vec{Y}(t) | \vec{Y}(5) = [0.05,0]^T\right) = \exp[-(t-5)\mathbf{A}] \begin{bmatrix} 0.05 \\ 0 \end{bmatrix} + \int_5^t \exp[-(t-s)\mathbf{A}]ds \begin{bmatrix} 0 \\ 10 \end{bmatrix}$$

or

$$E\left(\vec{Y}(t) | \vec{Y}(5) = [0.05,0]^T\right) = \exp[-(t-5)\mathbf{A}] \begin{bmatrix} 0.05 \\ 0 \end{bmatrix} + \mathbf{A}^{-1}(\mathbf{I} - \exp[-(t-5)\mathbf{A}]) \begin{bmatrix} 0 \\ 10 \end{bmatrix}$$

$$= \theta \exp[-(t-5)\lambda]\theta^{-1} \begin{bmatrix} 0.05 \\ 0 \end{bmatrix} + \theta\lambda^{-1}\theta^{-1} \begin{bmatrix} 0 \\ 10 \end{bmatrix} - \theta\lambda^{-1}\exp[-(t-5)\lambda]\theta^{-1} \begin{bmatrix} 0 \\ 10 \end{bmatrix}$$

Using the θ and λ matrices of dimension 2×2 from Example 8.5 gives

$$E[X(t)| X(5) = 0.05, \dot{X}(5) = 0]$$

$$= 0.1 - e^{-0.2(t-5)}\left(0.05\cos[9.998(t-5)] + 10^{-3}\sin[9.998(t-5)]\right)$$

and

$$E[\dot{X}(t)|X(5) = 0.05, \dot{X}(5) = 0] = 0.5001\,e^{-0.2(t-5)}\sin[9.998(t-5)]$$

The values are shown in the sketches.

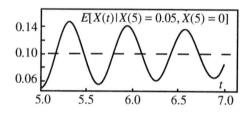

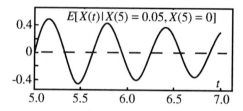

For the conditional analysis of covariance, we will let the elements of $\vec{V}(t)$ be the conditional covariances, with the initial condition that $\vec{V}(5) = \vec{0}$. The governing equation is the same as in Example 8.5: $\dot{\vec{V}}(t) + \mathbf{B}\vec{V}(t) = \vec{\psi}(t)$, with the excitation vector being

$$\vec{\psi}(t) = \frac{2\pi S_0}{m^2}[0,0,1]^T = 0.2\pi[0,0,1]^T$$

This gives the solution for $t \geq 5$ as

$$\vec{V}(t) = \frac{2\pi S_0}{m^2}\int_5^t \exp[-(t-s)\mathbf{B}]\vec{\psi}(s)\,ds$$

$$= 0.2\pi\int_5^t \exp[-(t-s)\mathbf{B}]ds[0,0,1]^T = 0.2\pi\,\mathbf{B}^{-1}(\mathbf{I} - \exp[-(t-5)\mathbf{B}])[0,0,1]^T$$

$$= 0.2\pi\,\mathbf{\theta}\lambda^{-1}\mathbf{\theta}^{-1}[0,0,1]^T - \mathbf{\theta}\lambda^{-1}\exp[-(t-5)\lambda]\mathbf{\theta}^{-1}[0,0,1]^T$$

Using the $\mathbf{\theta}$ and λ matrices describing $\mathbf{B}$ from Example 8.5 gives the first element of $\vec{V}(t)$ as

$$Var[X(t)|X(5) = 0.05, \dot{X}(5) = 0] = \frac{\pi}{1000}\left[2.5 - e^{-0.4(t-5)}(2.501 + 0.001\cos[20.00(t-5)]\right.$$

$$\left. -0.05001\sin[20.00(t-5)])\right]$$

and the third element of $\vec{V}(t)$ as

$$Var[\dot{X}(t)|X(5) = 0.05, \dot{X}(5) = 0] = \frac{\pi}{10}\left[2.5 - e^{-0.4(t-5)}(2.501 + 0.001\cos[20.00(t-5)]\right.$$

$$\left. +0.05001\sin[20.00(t-5)])\right]$$

These results are not sketched here since they have exactly the same form as the zero-start variances shown in Fig. 3.7 of Chapter 3.

****

8.5 Energy Balance and Covariance

Although we have derived the state-space moment equations in Section 8.4 without explicit consideration of energy, it turns out that in many situations the equations for second moments can also be viewed as energy balance relationships. This is not surprising since potential energy and kinetic energy are second moment properties of displacement and velocity, respectively, in spring-mass systems. Furthermore, a rate of energy addition or dissipation is a cross-product of a force and a velocity. We will illustrate this idea by consideration of the SDF oscillator.

First let us consider the equation of motion for the SDF oscillator rewritten as

$$m\ddot{X}(t) = F(t) - c\dot{X}(t) - kX(t)$$

The terms on the right-hand side of this equation are the forces on the mass, which is moving with velocity $\dot{X}(t)$. Thus, if we multiply the equation by $\dot{X}(t)$ we will obtain an equation involving the power into the mass from these three terms:

$$m\dot{X}(t)\ddot{X}(t) = F(t)\dot{X}(t) - [c\dot{X}(t)]\dot{X}(t) - k X(t)\dot{X}(t) = PA(t) - PD(t) - k X(t)\dot{X}(t) \qquad (8.22)$$

in which we have used the notation $PA(t)$ for the power added by the force $F(t)$ and $PD(t)$ for the power dissipation in the dashpot. The final term in the equation is the rate of energy transfer from the spring to the mass, and not surprisingly it is the same as the rate of decrease of the potential energy $PE(t) = k X^2(t) / 2$ in the spring:

$$k X(t)\dot{X}(t) = \frac{d}{dt} PE(t)$$

Similarly, the first term in Eq. 8.22 is the rate of change of the kinetic energy $KE(t) = m \dot{X}^2(t) / 2$ in the mass:

$$m X(t)\dot{X}(t) = \frac{d}{dt} KE(t)$$

Thus, one can rearrange the terms of Eq. 8.22 to give

$$\frac{d}{dt}[KE(t) + PE(t)] + PD(t) = PA(t)$$

which simply states that the power added is the sum of the power dissipated and the rate of change of the energy in the system. This confirms the well-known fact that Newtonian mechanics (force equals mass times acceleration) results in conservation of energy. The expectation of this energy balance relationship is one of our state space equations for the SDF system. Precisely this equation appears in our investigation of the SDF system in Example 8.1, and when Eq. 8.20 is used to give the second moments of the system, this energy balance relationship is the third scalar equation of $\dot{\vec{V}}(t) + \mathbf{B}\vec{V}(t) = \vec{\psi}(t)$.

It is particularly interesting to study the $PA(t)$ term for the special case in which the excitation of a SDF or MDF system is mean zero, white noise. That is, we want to find terms like $E[F_j(t)\dot{X}_j(t)]$ with $F_j(t)$ being a force applied to a mass m_j, and $\dot{X}_j(t)$ being the velocity of the mass. Let n be the number of degrees of freedom of the system, so that $n_Y = 2n$ is the dimension of the state vector. Using the usual state-space formulation with

$$\vec{Y}(t) = [X_1(t), \cdots X_n(t), \dot{X}_1(t), \cdots, \dot{X}_n(t)]^T, \qquad \vec{Q}(t) = [0, \cdots 0, F_1(t)/m_1, \cdots, F_n(t)/m_n]^T$$

the mean zero white noise condition can be written as

$$\boldsymbol{\phi}_{QQ}(t,s) = \mathbf{K}_{QQ}(t,s) = 2\pi \mathbf{S}_0(t)\delta(t-s)$$

with

$$\mathbf{S}_0(t) = \begin{bmatrix} 0 & \cdots & 0 & 0 & \cdots & 0 \\ \vdots & & \vdots & \vdots & & \vdots \\ 0 & \cdots & 0 & 0 & \cdots & 0 \\ 0 & \cdots & 0 & S_{1,1}(t) & \cdots & S_{1,n}(t) \\ \vdots & & \vdots & \vdots & & \vdots \\ 0 & \cdots & 0 & S_{n,1}(t) & \cdots & S_{n,n}(t) \end{bmatrix} \tag{8.23}$$

Note that $S_{j,l}(t)$ is the nonstationary cross-spectral density of the force components $F_j(t)$ and $F_l(t)$, and it is the $(n+j,n+l)$ component of $\mathbf{S}_0(t)$.

Our notation now gives the expected rate of energy addition by the jth component of force in the MDF system as

$$E[PA_j(t)] \equiv E[F_j(t)\dot{X}_j(t)] = m_j E[Q_{n+j}(t)Y_{n+j}(t)]$$

Thus, it is proportional to a diagonal term from the matrix $E[\vec{Q}^T(t)Y(t)] = \mathbf{K}_{QY}(t,t)$. We can find the value of such terms, though, from Eqs. 8.14 and 8.17. In particular, those equations give

$$\mathbf{K}_{QY}(t,t) + \mathbf{K}_{YQ}(t,t) = \int_{t_0}^t \mathbf{K}_{QQ}(u,t)\,dv + \int_{t_0}^t \mathbf{K}_{QQ}(t,v)\,du = 2\pi \mathbf{S}_0(t) \tag{8.24}$$

and for the special case of diagonal elements this becomes

$$[\mathbf{K}_{QY}(t,t)]_{r,r} \equiv [\mathbf{K}_{YQ}(t,t)]_{r,r} = [\pi \mathbf{S}_0(t)]_{r,r}$$

Since the $(n+j,n+j)$ term of $\mathbf{S}_0(t)$ is $S_{j,j}(t)/m_j^2$, we now have

$$E[PA_j(t)] = \frac{\pi S_{j,j}(t)}{m_j} \tag{8.25}$$

From Eq. 8.25 we note the rather remarkable property that the rate of energy addition to the MDF system by a mean zero, delta-correlated force depends only on the autospectral density of the force and the magnitude of the mass to which it is applied. It is unaffected by the magnitudes of any springs or dashpots in the system and of any other masses or forces in the system. Furthermore, it does not depend on the level of response of the system. For example, if a stationary excitation of this type is applied to a system which is initially at rest, then the rate of energy addition does not change as the response builds up from zero to its stationary level. This property of $E[PA_j(t)]$ also extends to the situation in which the excitation of the system is a base motion, rather than a force. When the base of an MDF system is moved with motion $\{Z(t)\}$, and $\{\vec{X}(t)\}$ represents motion relative to the base motion, then we know that the equations of motion have the same form as usual, but with excitation components of $F_j(t) = -m_j \ddot{Z}(t)$. For this situation, Eq. 8.25 gives the rate of energy addition by $F_j(t)$ as $E[PA_j(t)] = m_j \pi S_{\ddot{Z}\ddot{Z}}(t)$. Again, this rate depends only on the autospectral density of the excitation and the magnitude of the mass m_j, and is unaffected by response levels or by other parameters of the system.

It should be noted that Eq. 8.25 is generally not true if the processes involved are not mean zero. In particular, the rate of energy addition to a mass m_j by a delta-correlated force $\{F_j(t)\}$ with a non-zero mean is

$$E[PA_j(t)] = \frac{\pi S_{jj}(t)}{m_j} + \mu_{F_j}(t)\mu_{\dot{X}_j}(t)$$

In this situation we see that the rate of energy addition generally does depend on the response and may vary with time even if the excitation is stationary.

In addition to the energy balance, we can note that Eq. 8.24 gives important general relationships regarding the covariance of force and response components of an MDF system. For example, for j and $l \le n$, the $(n+j,n+l)$ component of Eq. 8.24 gives

$$m_j K_{F_j(t),\dot{X}_l(t)} + m_l K_{\dot{X}_j(t),F_l(t)} = 2\pi S_{j,l}(t) \tag{8.26}$$

which includes the special case of

$$K_{F_j(t),\dot{X}_j(t)} = \frac{\pi S_{j,j}(t)}{m_j} \tag{8.27}$$

for the covariance, at any instant of time, between a particular component of force and the velocity of the mass to which it is applied. A similar, and even simpler, result can be found regarding the covariance of $F_j(t)$ and any component of response $X_l(t)$ at the same instant of time, by considering the $(n+j,l)$ component of Eq. 8.24. Noting that $Q_j(t) \equiv 0$ and that Eq. 8.23 gives $[S_0(t)]_{n+j,l} = 0$ also, we obtain

$$K_{F_j(t),X_l(t)} \equiv K_{Q_{n+j}(t),Y_l(t)} = 0 \tag{8.28}$$

Thus, we see that $F_j(t)$ is uncorrelated with any component of response $X_l(t)$ at the same instant of time, including the $X_j(t)$ component giving the motion at the point of application of the force.

It should be emphasized that the covariance relationships in Eqs. 8.26, 8.27 and 8.28 apply only for a delta-correlated force applied directly to a mass. If a delta-correlated force is applied at some other point in a system or to a system with no mass, as in Example 8.3, then the relationships generally do not hold.

****

Example 8.8: Verify that the rate of energy addition of Eq. 8.25 is consistent with the expressions derived in Chapter 3 for the nonstationary variance values of displacement and velocity for the response of the SDF system with a stationary delta-correlated excitation.

We can write the response variance values of Eqs. 3.41 and 3.45 as

$$\sigma_X^2(t) = \frac{G_0}{2kc}\left\{1 - e^{-2\zeta\omega_0 t}\left[\frac{\omega_0^2}{\omega_d^2} + \frac{\zeta\omega_0}{\omega_d}\sin(2\omega_d t) - \frac{\zeta^2\omega_0^2}{\omega_d^2}\cos(2\omega_d t)\right]\right\}U(t)$$

and

$$\sigma_{\dot{X}}^2(t) = \frac{G_0}{2mc}\left\{1 - e^{-2\zeta\omega_0 t}\left[\frac{\omega_0^2}{\omega_d^2} - \frac{\zeta\omega_0}{\omega_d}\sin(2\omega_d t) - \frac{\zeta^2\omega_0^2}{\omega_d^2}\cos(2\omega_d t)\right]\right\}U(t)$$

for a system which is at rest at time $t = 0$, and has an excitation $\{F(t)\}$ with autocovariance of $K_{FF}(t,s) = G_0\delta(t-s)U(t)$. Letting the mean values be zero, we use these expressions in calculating the total energy in the oscillator as

$$E[PE(t) + KE(t)] = \frac{k\sigma_X^2(t)}{2} + \frac{m\sigma_{\dot{X}}^2(t)}{2} = \frac{G_0}{2c}\left\{1 - e^{-2\zeta\omega_0 t}\left[\frac{\omega_0^2}{\omega_d^2} - \frac{\zeta^2\omega_0^2}{\omega_d^2}\cos(2\omega_d t)\right]\right\}U(t)$$

The fact that the $(\zeta\omega_0 / \omega_d)\sin(2\omega_d t)$ term, which appears in both $\sigma_X^2(t)$ and $\sigma_{\dot{X}}^2(t)$, does not appear in the total energy expression shows that this term represents a transfer of energy back and forth between potential energy and kinetic energy, with no net energy addition. Taking the derivative of the expected total energy, and noting that $2\zeta\omega_0 = c/m$, gives the rate of energy growth in the system as

$$\frac{d}{dt}E[PE(t)+KE(t)]=\frac{G_0}{2m}e^{-2\zeta\omega_0 t}\left[\frac{\omega_0^2}{\omega_d^2}-\frac{\zeta\omega_0}{\omega_d}\sin(2\omega_d t)-\frac{\zeta^2\omega_0^2}{\omega_d^2}\cos(2\omega_d t)\right]U(t)$$

Adding this expression to $c\sigma_{\dot{X}}^2(t)$, which is the expected rate of energy dissipation in the system, gives the rate of energy addition as $E[PA(t)] = G_0/(2m)U(t)$. Recalling that the relationship between G_0 and the autospectral density is $G_0 = 2\pi S_0$ confirms Eq. 8.25.

Similar verification is given by Example 3.6, in which a delta-correlated force is applied to a mass which has no restoring force elements. In this case, we found that the variance of $\dot{X}(t)$ grows linearly with time as $\sigma_{\dot{X}}^2(t) = (G_0/m^2)tU(t)$. When the mean of velocity is zero, this indicates a linear growth of kinetic energy. Since this particular system has no potential energy or energy dissipation, the rate of growth of the kinetic energy is the rate of energy addition by the force. The value found is $G_0/(2m)$, which again confirms Eq. 8.25.

****

Example 8.9: Verify that Eq. 8.25 gives the rate of energy addition by a mean zero shot noise force applied to a mass m which is connected to some springs and dashpots.

As in Example 3.6, we will write the shot noise force as

$$F(t) = \sum_l F_l \delta(t - T_l)$$

in which $\{T_1, T_2, \cdots, T_l, \cdots\}$ is the sequence of arrival times for a Poisson process $\{N(t)\}$, and $\{F_1, F_2, \cdots, F_l, \cdots\}$ is a sequence of identically distributed random variables which are independent of each other and of the arrival times. We will first consider the energy added to the mass by the single pulse $F_l \delta(t - T_l)$. Let the mass have displacement and velocity $X(T_l^-)$ and $\dot{X}(T_l^-)$, respectively, immediately prior to the pulse. Since the forces from attached springs and dashpots will be finite, they will cause no instantaneous changes in $X(t)$ or $\dot{X}(t)$, but the pulse of applied force will cause an instantaneous change in $\dot{X}(t)$. Thus, the state of the mass m immediately after the impulse will be $X(T_l^+) = X(T_l^-)$, and $\dot{X}(T_l^+) = \dot{X}(T_l^-) + F_l/m$. Since displacement is continuous, there will be no change in the potential energy of the system during the instant of the excitation pulse, but the kinetic energy will instantaneously change by an amount

$$\Delta KE = \frac{m}{2}[\dot{X}^2(T_l^+) - \dot{X}^2(T_l^-)] = m\dot{X}(T_l^-)F_l + \frac{F_l^2}{2m}$$

Since F_l is independent of the past excitation, it is also independent of $\dot{X}(T_l^-)$. In conjunction with the fact that $E(F_l) = 0$ for a mean zero process, this gives $E[\dot{X}(T_l^-)F_l] = 0$. Thus, the expected value of the energy added by the pulse $F_l \delta(t - T_l)$ can be written as $E(F^2)/(2m)$. Even though the energy added by any one individual pulse does depend on the velocity of the mass at the time of the pulse, this term may be either positive or negative and its expected value is zero. The expected rate of energy addition per unit time by the force on mass m is then $E(F^2)/(2m)$ multiplied by the expected rate of arrival of the pulses. We can find the autospectral density of this shot noise process from the results in Example 3.6, and it again confirms that the expected rate of energy addition is as given in Eq. 8.25.

****

8.6 Higher Moments and Cumulants Using Kronecker Notation

One difficulty in investigating higher moments or cumulants of the state variables is the choice of an appropriate notation. We found that the first moments could be written as the vector $\vec{\mu}_Y(t)$ and the second moments could be written as the matrix $E[\vec{Y}(t)\vec{Y}^T(t)]$. Continuing with this approach gives the third moments as constituting a third-order tensor, the fourth moments as a fourth-order tensor, etc. There is no problem with writing expressions for components such as the (j,k,l,m) component of the fourth-order tensor as $E[Y_j(t)Y_k(t)Y_l(t)Y_m(t)]$, but there is some difficulty with the presentation of general relationships for the higher moments, comparable to Eqs. 8.6 and 8.9 for the first and second moments, as well as with the organization of the large amount of information involved. That is, $\vec{\mu}_Y(t)$, $\phi_{YY}(t,t)$, and $\mathbf{K}_{YY}(t,t)$ provide very convenient one- and two-dimensional arrays when that is needed, but we have not defined a notation or appropriate algebra for arrays with dimension higher than two. Since Kronecker notation provides one convenient way to handle this matter, we will now present its basic concepts.

The fundamental operation of Kronecker algebra is the product denoted by $\otimes$ and defined as follows: If $\mathbf{A}$ is a rectangular matrix of dimension (n_A, r_A) and $\mathbf{B}$ is a rectangular matrix of dimension (n_B, r_B), then $\mathbf{C} = \mathbf{A} \otimes \mathbf{B}$ is a rectangular matrix of dimension $(n_A n_B, r_A r_B)$ with components $C_{(j_A-1)n_B+j_B,(l_A-1)r_B+l_B} = A_{j_A,l_A} B_{j_B,l_B}$. This relationship can also be written as

$$\mathbf{C} = \begin{bmatrix} A_{1,1}\mathbf{B} & A_{1,2}\mathbf{B} & \cdots & A_{1,r_A}\mathbf{B} \\ A_{2,1}\mathbf{B} & A_{2,2}\mathbf{B} & \cdots & A_{2,r_A}\mathbf{B} \\ \vdots & \vdots & & \vdots \\ A_{n_A,1}\mathbf{B} & A_{n_A,2}\mathbf{B} & \cdots & A_{n_A,r_A}\mathbf{B} \end{bmatrix} \tag{8.29}$$

Note that the Kronecker product is much more general than matrix multiplication, and requires no restrictions on the dimensions of the arrays being multiplied. One can easily verify that the definition gives

$$(\mathbf{A} + \mathbf{B}) \otimes (\mathbf{C} + \mathbf{D}) = \mathbf{A} \otimes \mathbf{C} + \mathbf{A} \otimes \mathbf{D} + \mathbf{B} \otimes \mathbf{C} + \mathbf{B} \otimes \mathbf{D} \tag{8.30}$$

and

$$\mathbf{A} \otimes (\mathbf{B} \otimes \mathbf{C}) = (\mathbf{A} \otimes \mathbf{B}) \otimes \mathbf{C} \tag{8.31}$$

for any arrays $\mathbf{A}$, $\mathbf{B}$, $\mathbf{C}$ and $\mathbf{D}$, showing that the operation is distributive and associative. A simplified notation is used for the Kronecker product of a matrix with itself. In particular we will define Kronecker powers as $\mathbf{A}^{[2]} = \mathbf{A} \otimes \mathbf{A}$ and $\mathbf{A}^{[j]} = \mathbf{A} \otimes \mathbf{A}^{[j-1]}$ for $j > 2$. Note that in the particular case in which $\mathbf{A}$ and $\mathbf{B}$ each have only one column, we find that $\mathbf{A} \otimes \mathbf{B}$ also has only one column. That is, the Kronecker product of two vectors is a vector and the Kronecker power of a vector is a vector. This situation will be of particular interest in our state-space analysis. Since our equations of motion contain matrix products, we will also need a general relationship involving the combination of matrix products and Kronecker products. This formula is

$$(\mathbf{AB}) \otimes (\mathbf{CD}) = (\mathbf{A} \otimes \mathbf{C})(\mathbf{B} \otimes \mathbf{D}) \tag{8.32}$$

provided that the dimensions of the arrays are such that the matrix products $\mathbf{AB}$ and $\mathbf{CD}$ exist.

Even though we found the matrix formulation to be very convenient for the equations governing the second moments of the state variables, we will now rewrite those expressions using the Kronecker notation to illustrate its application. The array of second moment functions will be given by the vector $E[\vec{Y}^{[2]}(t)] = E[\vec{Y}(t) \otimes \vec{Y}(t)]$. Note that this vector contains every element of the matrix $E[\vec{Y}(t)\vec{Y}^T(t)]$ which we previously considered. The only difference is that the elements are now arranged as a vector of length n_Y^2 instead of in a matrix of dimension $n_Y \times n_Y$. The derivative of the Kronecker power is given by

$$\frac{d}{dt} E[\vec{Y}^{[2]}(t)] = E[\dot{\vec{Y}}(t) \otimes \vec{Y}(t)] + E[\vec{Y}(t) \otimes \dot{\vec{Y}}(t)]$$

so that the differential equation governing the second moments can be obtained by using Eq. 8.30 in taking the Kronecker product of $\vec{Y}(t)$ with the equation of motion, in the form of Eq. 8.5, then taking the expected value. That is,

$$E[\dot{\vec{Y}}(t) \otimes \vec{Y}(t)] + E[\mathbf{A}\vec{Y}(t) \otimes \vec{Y}(t)] = E[\vec{Q}(t) \otimes \vec{Y}(t)]$$

and

$$E[\vec{Y}(t) \otimes \dot{\vec{Y}}(t)] + E[\vec{Y}(t) \otimes \mathbf{A}\vec{Y}(t)] = E[\vec{Y}(t) \otimes \vec{Q}(t)]$$

and adding these expressions gives

$$\frac{d}{dt} E[\vec{Y}^{[2]}(t)] + E\Big(\vec{Y}(t) \otimes [\mathbf{A}\vec{Y}(t)]\Big) + E\Big([\mathbf{A}\vec{Y}(t)] \otimes \vec{Y}(t)\Big) = E[\vec{Y}(t) \otimes \vec{Q}(t)] + E[\vec{Q}(t) \otimes \vec{Y}(t)] \tag{8.33}$$

We can simplify this relationship by noting that $\vec{Y}(t) \otimes [\mathbf{A}\vec{Y}(t)] \equiv [\mathbf{I}\vec{Y}(t)] \otimes [\mathbf{A}\vec{Y}(t)]$, in which $\mathbf{I}$ denotes the identity matrix of dimension $n_Y \times n_Y$. This then allows us to use Eq. 8.32 in writing $\vec{Y}(t) \otimes [\mathbf{A}\vec{Y}(t)] = (\mathbf{I} \otimes \mathbf{A})\vec{Y}^{[2]}(t)$. Similarly, $[\mathbf{A}\vec{Y}(t)] \otimes \vec{Y}(t) = (\mathbf{A} \otimes \mathbf{I})\vec{Y}^{[2]}(t)$ so that Eq. 8.33 becomes

$$\frac{d}{dt} E[\vec{Y}^{[2]}(t)] + [(\mathbf{I} \otimes \mathbf{A}) + (\mathbf{A} \otimes \mathbf{I})] E[\vec{Y}^{[2]}(t)] = E[\vec{Y}(t) \otimes \vec{Q}(t)] + E[\vec{Q}(t) \otimes \vec{Y}(t)] \tag{8.34}$$

This expression governing the second moments of the state variables is exactly equivalent to Eq. 8.9. In fact, the (j,l) component of Eq. 8.9 is identical to the $(n_Y - 1)j + l$ component of Eq. 8.34.

As noted before, symmetry causes the (l,j) component of Eq. 8.9 to be identical to the (j,l) component. Similarly, the $(n_Y - 1)l + j$ and the $(n_Y - 1)j + l$ components of Eq. 8.34 are identical.

Since Eqs. 8.34 and 8.9 have identical components, there is really no advantage in using the Kronecker notation for the second moment equations. On the other hand, this new notation is easily extended to higher dimensions, whereas the matrix notation is not so easily extended. The array of third moments of the state variables, for example, can be written as $E[\vec{Y}^{[3]}(t)]$ and its derivative is

$$\frac{d}{dt}E[\vec{Y}^{[3]}(t)] = E[\vec{Y}(t) \otimes \vec{Y}(t) \otimes \dot{\vec{Y}}(t)] + E[\vec{Y}(t) \otimes \dot{\vec{Y}}(t) \otimes \vec{Y}(t)] + E[\dot{\vec{Y}}(t) \otimes \vec{Y}(t) \otimes \vec{Y}(t)]$$

Substituting for $\dot{\vec{Y}}(t)$ from Eq. 8.5, arranging terms, and simplifying according to Eq. 8.32 then gives

$$\frac{d}{dt}E[\vec{Y}^{[3]}(t)] + [(\mathbf{I} \otimes \mathbf{I} \otimes \mathbf{A}) + (\mathbf{I} \otimes \mathbf{A} \otimes \mathbf{I}) + (\mathbf{A} \otimes \mathbf{I} \otimes \mathbf{I})]E[\vec{Y}^{[3]}(t)] =$$
$$E[\vec{Y}(t) \otimes \vec{Y}(t) \otimes \vec{Q}(t)] + E[\vec{Y}(t) \otimes \vec{Q}(t) \otimes \vec{Y}(t)] + E[\vec{Q}(t) \otimes \vec{Y}(t) \otimes \vec{Y}(t)]$$

or

$$\frac{d}{dt}E[\vec{Y}^{[3]}(t)] + [(\mathbf{I}^{[2]} \otimes \mathbf{A}) + (\mathbf{I} \otimes \mathbf{A} \otimes \mathbf{I}) + (\mathbf{A} \otimes \mathbf{I}^{[2]})]E[\vec{Y}^{[3]}(t)] =$$
$$E[\vec{Y}^{[2]}(t) \otimes \vec{Q}(t)] + E[\vec{Y}(t) \otimes \vec{Q}(t) \otimes \vec{Y}(t)] + E[\vec{Q}(t) \otimes \vec{Y}^{[2]}(t)]$$

In this way one can write the general expression for the jth moments as

$$\frac{d}{dt}E[\vec{Y}^{[j]}(t)] + \mathbf{B}E[\vec{Y}^{[j]}(t)] = \vec{\psi}(t) \tag{8.35}$$

in which

$$\mathbf{B} = \sum_{l=1}^{j} \underbrace{\mathbf{I} \otimes \cdots \otimes \mathbf{I}}_{l-1} \otimes \mathbf{A} \otimes \underbrace{\mathbf{I} \otimes \cdots \otimes \mathbf{I}}_{j-l} = \sum_{l=1}^{j} \mathbf{I}^{[l-1]} \otimes \mathbf{A} \otimes \mathbf{I}^{[j-l]} \tag{8.36}$$

and

$$\vec{\psi}(t) = \sum_{l=1}^{j} E[\underbrace{\vec{Y}(t) \otimes \cdots \otimes \vec{Y}(t)}_{l-1} \otimes \vec{Q}(t) \otimes \underbrace{\vec{Y}(t) \otimes \cdots \otimes \vec{Y}(t)}_{j-l}] = \sum_{l=1}^{j} E[\vec{Y}^{[l-1]}(t) \otimes \vec{Q}(t) \otimes \vec{Y}^{[j-l]}(t)]$$

$$\tag{8.37}$$

The equations for higher-order cumulants are essentially identical to those of Eq. 8.35 for higher moments. In fact, the only complication in going from moments to cumulants is in the definition of a cumulant notation corresponding to the Kronecker products and powers. This can be done, though, by analogy with the product terms. In particular, we will define an array of cumulant terms which is organized in exactly the same way as the elements of the corresponding Kronecker product moments. For the general situation in which $\mathbf{A}_1, \mathbf{A}_2, \cdots, \mathbf{A}_j$ are matrices of random variables, we note that any particular component of the jth-order moment matrix $E(\mathbf{A}_1 \otimes \mathbf{A}_2 \otimes \cdots \otimes \mathbf{A}_j)$ is a cross-product of one element each from the matrices $\mathbf{A}_1, \mathbf{A}_2, \cdots, \mathbf{A}_j$. We now define the corresponding component of the matrix $\kappa_j^{\otimes}(\mathbf{A}_1, \mathbf{A}_2, \cdots, \mathbf{A}_j)$ to be the jth-order joint cumulant of those same j elements, one each from $\mathbf{A}_1, \mathbf{A}_2, \cdots, \mathbf{A}_j$. The linearity property of the cumulants, as given in Eq. 8.3, then allows us to rewrite Eq. 8.35 as

$$\frac{d}{dt}\kappa_j^{\otimes}[\vec{Y}(t), \cdots, \vec{Y}(t)] + \mathbf{B}\kappa_j^{\otimes}[\vec{Y}(t), \cdots, \vec{Y}(t)] = \sum_{l=1}^{j} \kappa_j^{\otimes}[\underbrace{\vec{Y}(t), \cdots, \vec{Y}(t)}_{l-1}, \vec{Q}(t), \underbrace{\vec{Y}(t), \cdots, \vec{Y}(t)}_{j-l}] \quad (8.38)$$

in which the matrix $\mathbf{B}$ is as defined in Eq. 8.36.

Let us now consider the special case in which the excitation vector process $\{\vec{Q}(t)\}$ is delta-correlated. By this we mean that

$$\kappa_j^{\otimes}[\vec{Q}(t_1), \cdots, \vec{Q}(t_j)] = (2\pi)^{j-1}\vec{S}_j(t_j)\delta(t_1 - t_j) \cdots \delta(t_{j-1} - t_j) \quad (8.39)$$

in which $\vec{S}_j(t)$ is a vector of dimension n_Y^j. Its components are spectral density terms which give the nonstationary intensity of the jth cumulants of the excitation. The elements of $\vec{S}_2(t)$ are ordinary autospectral and cross-spectral densities, the elements of $\vec{S}_3(t)$ are bispectral densities, etc. Using Eq. 8.39 one can write a generalization of Eq. 8.16 as

$$\int_{t_0}^{t} \cdots \int_{t_0}^{t} \kappa^{\otimes}[\vec{Q}(t_1), \cdots, \vec{Q}(t_j)]dt_1 \cdots dt_j = (2\pi)^{j-1}\int_{t_0}^{t} \vec{S}_j(v)\,dv$$

for any $t_0 < t$. The derivative with respect to t of this expression is

$$\sum_{k=1}^{j}\int_{t_0}^{t} \cdots \int_{t_0}^{t} \kappa_j^{\otimes}[\vec{Q}(t_1), \cdots, \vec{Q}(t_{k-1}), \vec{Q}(t), \vec{Q}(t_{k+1}), \cdots, \vec{Q}(t_j)]dt_1 \cdots dt_{k-1}dt_{k+1} \cdots dt_j = (2\pi)^{j-1}\vec{S}_j(t)$$

and substituting the form of Eq. 8.12 for the $\vec{Y}(t)$ response terms in the right-hand side of Eq. 8.38 gives exactly this same expression. Thus, Eq. 8.38 becomes

$$\frac{d}{dt}\kappa_j^{\otimes}[\vec{Y}(t), \cdots, \vec{Y}(t)] + \mathbf{B}\kappa_j^{\otimes}[\vec{Y}(t), \cdots, \vec{Y}(t)] = (2\pi)^{j-1}\vec{S}_j(t) \quad (8.40)$$

This generalization of Eq. 8.18 describes all jth-order cumulants of the response process $\{\vec{Y}(t)\}$ resulting from the delta-correlated excitation process $\{\vec{Q}(t)\}$.

The general solution of Eq. 8.40, of course, can be written as

$$\kappa_j^{\otimes}[\vec{Y}(t), \cdots, \vec{Y}(t)] = (2\pi)^{j-1}\int_{-\infty}^{t} e^{-(t-u)\mathbf{B}}\vec{S}_j(u)\,du \quad (8.41)$$

and the corresponding array of conditional cumulants given $\vec{Y}(t_0) = \vec{y}$ at some particular time t_0 is

$$\kappa_j^{\otimes}[\vec{Y}(t),\cdots,\vec{Y}(t)|\vec{Y}(t_0) = \vec{y}] = (2\pi)^{j-1}\int_{t_0}^{t} e^{-(t-u)\mathbf{B}}\,\vec{S}_j(u)\,du \quad \text{for} \quad j \geq 2 \tag{8.42}$$

Note that Eq. 8.42 gives the conditional cumulants as being zero at time $t = t_0$, in agreement with the fact that all cumulant components of order higher than unity are zero for any deterministic vector, such as $\vec{y}$.

The matrix exponential in the integrands of Eqs. 8.41 and 8.42 can be reduced to scalar exponentials by using the eigenvectors and eigenvalues of $\mathbf{B}$. Since the dimension of $\mathbf{B}$ is n_Y^j, the problem of finding the eigenvectors and eigenvalues seems rather formidable if n_Y and j are not small. On the other hand if the eigenvector matrix $\boldsymbol{\theta}_A$ and the eigenvalue matrix $\boldsymbol{\lambda}_A$ are known, such that $\mathbf{A}\boldsymbol{\theta}_A = \boldsymbol{\theta}_A\boldsymbol{\lambda}_A$, then the eigenvectors for $\mathbf{B}$ are the columns of the matrix $\boldsymbol{\theta}_B = \boldsymbol{\theta}_A^{[j]}$ (Papadimitriou, 1994). This is verified by noting that Eq. 8.32 gives

$$\mathbf{B}\boldsymbol{\theta}_B = \left(\sum_{l=1}^{j}\underbrace{\mathbf{I}\otimes\cdots\otimes\mathbf{I}}_{l-1}\otimes\mathbf{A}\otimes\underbrace{\mathbf{I}\otimes\cdots\otimes\mathbf{I}}_{j-l}\right)\left(\underbrace{\boldsymbol{\theta}_A\otimes\cdots\otimes\boldsymbol{\theta}_A}_{j}\right) = \sum_{l=1}^{j}\underbrace{\boldsymbol{\theta}_A\otimes\cdots\otimes\boldsymbol{\theta}_A}_{l-1}\otimes(\mathbf{A}\boldsymbol{\theta}_A)\otimes\underbrace{\boldsymbol{\theta}_A\otimes\cdots\otimes\boldsymbol{\theta}_A}_{j-l}$$

so that

$$\mathbf{B}\boldsymbol{\theta}_B = \sum_{l=1}^{j}\underbrace{\boldsymbol{\theta}_A\otimes\cdots\otimes\boldsymbol{\theta}_A}_{l-1}\otimes(\boldsymbol{\theta}_A\boldsymbol{\lambda}_A)\otimes\underbrace{\boldsymbol{\theta}_A\otimes\cdots\otimes\boldsymbol{\theta}_A}_{j-l} = \left(\underbrace{\boldsymbol{\theta}_A\otimes\cdots\otimes\boldsymbol{\theta}_A}_{j}\right)\left(\sum_{l=1}^{j}\underbrace{\mathbf{I}\otimes\cdots\otimes\mathbf{I}}_{l-1}\otimes\boldsymbol{\lambda}_A\otimes\underbrace{\mathbf{I}\otimes\cdots\otimes\mathbf{I}}_{j-l}\right)$$

and this is exactly $\boldsymbol{\theta}_B\boldsymbol{\lambda}_B$ with

$$\boldsymbol{\lambda}_B = \sum_{l=1}^{j}\mathbf{I}^{[(l-1)]}\otimes\boldsymbol{\lambda}_A\otimes\mathbf{I}^{[(j-l)]} \tag{8.43}$$

giving the eigenvalues of $\mathbf{B}$. Thus, one can do the eigenanalysis on the $\mathbf{A}$ matrix, which has dimension n_Y, rather than on $\mathbf{B}$ which has dimension n_Y^j.

Note that the dimension of the vectors $E[\vec{Y}^{[j]}(t)]$ and $\kappa_j^{\otimes}[\vec{Y}(t),\cdots,\vec{Y}(t)]$ of unknowns in Eqs. 8.35 and 8.40, respectively, is n_Y^j, but the number of distinct moments or cumulants of order j is much less than this. For example $\vec{Y}^{[j]}(t)$ includes all $j!$ permutations of the terms in $Y_{l_1}Y_{l_2}\cdots Y_{l_j}$ for distinct values of $l_1,\cdots,l_j$, but all of these permutations are identical. Exactly the same symmetry applies to the cumulants. Of course, the $j!$ equations governing these different representations of the same term are also identical. Thus, Eqs. 8.35 and 8.40 contain much redundant information. This is exactly the same situation as was found in Eqs. 8.9 and 8.10 for the state-space equations for the autocorrelation and autocovariance matrices. The symmetry of these matrices of unknowns resulted in all the off-diagonal terms being included twice in Eqs. 8.9 and 8.10. For higher-order moments or cumulants the amount of redundancy is much greater, though. The actual number of unique moments or cumulants of order j is given by the binomial coefficient $(n_Y + j - 1)!/[(n_Y - 1)!j!]$. One particular way to take account of this symmetry in order to consider only unique cumulants of a given order was presented by Lutes and Chen (1992), but its implementation is somewhat cumbersome.

It should be noted that higher-order moment equations are much less convenient than cumulant equations for delta-correlated excitations. Some indication of this fact was given in Eq. 8.19, using the autocorrelation matrix notation for the second moments of response. Some of the additional difficulties which result for higher-order moments can be demonstrated by consideration of the $\vec{\psi}(t)$ vector in Eq. 8.37 for $j = 3$:

$$\vec{\psi}(t) = E[\vec{Q}(t) \otimes \vec{Y}(t) \otimes \vec{Y}(t)] + E[\vec{Y}(t) \otimes \vec{Q}(t) \otimes \vec{Y}(t)] + E[\vec{Y}(t) \otimes \vec{Y}(t) \otimes \vec{Q}(t)]$$

Substituting Eq. 8.12 for $\vec{Y}(t)$ in this expression expands each of these three terms into nine terms, giving a total of 27. For example,

$$E[\vec{Y}(t) \otimes \vec{Q}(t) \otimes \vec{Y}(t)] = E[\vec{Y}(t_0) \otimes \vec{Q}(t) \otimes \vec{Y}(t_0)] + E[\vec{Y}(t_0) \otimes \vec{Q}(t) \otimes \int_{t_0}^t \vec{Q}(t_3)dt_3]$$

$$- E[\vec{Y}(t_0) \otimes \vec{Q}(t) \otimes \mathbf{A}\int_{t_0}^t \vec{Y}(t_3)dt_3] + E[\int_{t_0}^t \vec{Q}(t_1)dt_1 \otimes \vec{Q}(t) \otimes \vec{Y}(t_0)]$$

$$+ E[\int_{t_0}^t \vec{Q}(t_1)dt_1 \otimes \vec{Q}(t) \otimes \int_{t_0}^t \vec{Q}(t_3)dt_3] - E[\int_{t_0}^t \vec{Q}(t_1)dt_1 \otimes \vec{Q}(t) \otimes \mathbf{A}\int_{t_0}^t \vec{Y}(t_3)dt_3]$$

$$- E[\mathbf{A}\int_{t_0}^t \vec{Y}(t_1)dt_1 \otimes \vec{Q}(t) \otimes \vec{Y}(t_0)] - E[\mathbf{A}\int_{t_0}^t \vec{Y}(t_1)dt_1 \otimes \vec{Q}(t) \otimes \int_{t_0}^t \vec{Q}(t_3)dt_3]$$

$$+ E[\mathbf{A}\int_{t_0}^t \vec{Y}(t_1)dt_1 \otimes \vec{Q}(t) \otimes \mathbf{A}\int_{t_0}^t \vec{Y}(t_3)dt_3]$$

In the corresponding cumulant equation, the independence of $\vec{Q}(t)$ and $\vec{Q}(\tau)$ for $\tau < t$ causes all the terms but one to be zero, giving

$$\kappa_3^{\otimes}[\vec{Y}(t), \vec{Q}(t), \vec{Y}(t)] = \int_{t_0}^t \int_{t_0}^t \kappa_3^{\otimes}[\vec{Q}(t_1), \vec{Q}(t), \vec{Q}(t_3)]dt_1 dt_3$$

In the moment equation, though, each of the terms makes a contribution, and some of them are awkward to handle. For example, consider the one term having no integral. Independence of $\vec{Q}(t)$ and $\vec{Y}(t_0)$ can be used to give $E[\vec{Y}(t_0) \otimes \vec{Q}(t) \otimes \vec{Y}(t_0)] = E[\vec{Y}(t_0) \otimes \vec{\mu}_Q(t) \otimes \vec{Y}(t_0)]$. Each component of this array is made up of the product of one component from $\vec{\mu}_Q(t)$ and one component from $E[\vec{Y}(t_0) \otimes \vec{Y}(t_0)]$, but the ordering of the terms, due to $\vec{Q}(t)$ being in the middle of the Kronecker product, precludes direct factoring of the expression. A Kronecker algebra operation to handle such difficulties by using a so-called permutant matrix (Di Paola et al., 1992) does exist, but this further complicates the equation. Overall, it seems preferable to find the response cumulants from solution of Eq. 8.40 and then find any needed response moments from these quantities, rather than working directly with higher-order moment equations.

One may note that the results presented thus far in this chapter have all been limited to situations with delta-correlated excitations. The general form of Eq. 8.35 does apply for any form of stochastic excitation, as do the special cases of Eqs. 8.9, 8.10, and 8.34. However, the right-hand side of each of those equations contains terms which depend on the unknown dynamic response as well as on the known excitation. It is only for the delta-correlated situation that these equations are reduced to a form which can be solved. This limitation of state-space analysis can be eased if the excitation can be modeled as a filtered delta-correlated process. That is, for many problems it is possible to model the $\{\vec{Q}(t)\}$ process as the output from a linear filter which has a delta-correlated input. If this can be done for the excitation of Eq. 8.5, then it is possible to apply state-space moment or cumulant analysis to the system of interest, as well as to the linear filter. The procedure is to define a composite linear system which includes both the filter and the system of interest. The excitation of this composite system is the input to the filter, and the response includes $\{\vec{Q}(t)\}$ as well as $\{\vec{Y}(t)\}$.

****

Example 8.10: Consider the third-order cumulants of the response of the SDF oscillator with a stationary, delta-correlated excitation. Find expressions for the stationary values of the response cumulants and for the eigenvalues of **B** which would appear in a nonstationary solution.

The state-space description of the system is as in Example 8.4, with $n_Y = 2$,

$$\vec{Y}(t) = \begin{bmatrix} X(t) \\ \dot{X}(t) \end{bmatrix}, \qquad \mathbf{A} = \begin{bmatrix} 0 & -1 \\ k/m & c/m \end{bmatrix}, \qquad \vec{Q}(t) = \frac{1}{m}\begin{bmatrix} 0 \\ F(t) \end{bmatrix}$$

The Kronecker representation of the third-order cumulants is $\vec{V}(t) \equiv \kappa_3^{\otimes}[\vec{Y}(t),\vec{Y}(t),\vec{Y}(t)]$, which can be written out explicitly as

$$\vec{V}(t) = \Big(\kappa[X(t),X(t),X(t)], \kappa[X(t),X(t),\dot{X}(t)], \kappa[X(t),\dot{X}(t),X(t)], \kappa[X(t),\dot{X}(t),\dot{X}(t)],$$

$$\kappa[\dot{X}(t),X(t),X(t)], \kappa[\dot{X}(t),X(t),\dot{X}(t)], \kappa[\dot{X}(t),\dot{X}(t),X(t)], \kappa[\dot{X}(t),\dot{X}(t),\dot{X}(t)]\Big)^T$$

The right-hand side of the $\dot{\vec{V}}(t)+\mathbf{B}\vec{V}(t) = \vec{\psi}$ equation is very simple. In fact it is easily shown that $\vec{\psi} \equiv (2\pi)^2 \vec{S}_3$ has only one non-zero component, and that is $\psi_8 = (2\pi)^2 S_3 / m^3$, in which the scalar S_3 represents the bispectrum of the $\{F(t)\}$ process:

$$\kappa[F(t_1),F(t_2),F(t_3)] = (2\pi)^2 S_3 \delta(t_1 - t_3)\,\delta(t_2 - t_3)$$

Expansion of the matrix $\mathbf{B} = \mathbf{A}\otimes\mathbf{I}\otimes\mathbf{I} + \mathbf{I}\otimes\mathbf{A}\otimes\mathbf{I} + \mathbf{I}\otimes\mathbf{I}\otimes\mathbf{A}$ gives

$$\mathbf{B} = \begin{bmatrix} 0 & -1 & -1 & 0 & -1 & 0 & 0 & 0 \\ k/m & c/m & 0 & -1 & 0 & -1 & 0 & 0 \\ k/m & 0 & c/m & -1 & 0 & 0 & -1 & 0 \\ 0 & k/m & k/m & 2c/m & 0 & 0 & 0 & -1 \\ k/m & 0 & 0 & 0 & c/m & -1 & -1 & 0 \\ 0 & k/m & 0 & 0 & k/m & 2c/m & 0 & -1 \\ 0 & 0 & k/m & 0 & k/m & 0 & 2c/m & -1 \\ 0 & 0 & 0 & k/m & 0 & k/m & k/m & 3c/m \end{bmatrix}$$

Solving $\mathbf{B}\vec{V} = \vec{\psi}$ then gives the third cumulants of the stationary response as

$$\vec{V} = \frac{(2\pi)^2 S_3}{m^3}\left[\frac{2m^3}{6c^2 k + 3k^2 m}, 0, 0, \frac{m^2}{6c^2 + 3km}, 0, \frac{m^2}{6c^2 + 3km}, \frac{m^2}{6c^2 + 3km}, \frac{2cm^2}{6c^2 + 3km}\right]^T$$

Note the fact that the fourth, sixth, and seventh components of $\vec{V}$ are identical. We could tell that this must be true, though, by the fact that the definition of the vector $\vec{V}$ gives each of these components to be the joint cumulant of $X(t)$, $\dot{X}(t)$, and $\dot{X}(t)$. Similarly, the second, third, and fifth components of $\vec{V}$ are all equal to the joint cumulant of $X(t)$, $X(t)$, and $\dot{X}(t)$. These components are not only identical, but also are identically zero in the stationary situation. This follows from the fact that

$$\kappa[X(t), X(t), \dot{X}(t)] = \frac{1}{3}\frac{d}{dt}\kappa[X(t), X(t), X(t)]$$

just as

$$E[X^2(t)\dot{X}(t)] = \frac{1}{3}\frac{d}{dt}E[X^3(t)]$$

One can find the diagonal matrix λ_B by doing eigenanalysis of the matrix $\mathbf{B}$ as given, or by using Eq. 8.43 and the two eigenvalues of $\mathbf{A}$. Using the common notation of $\omega_0^2 = k/m$, $2\zeta\omega_0 = c/m$, and $\omega_d^2 = \omega_0^2(1-\zeta^2)$, we can write the non-zero elements of λ_A as $[\lambda_A]_{1,1} = \zeta\omega_0 + i\omega_d$ and $[\lambda_A]_{2,2} = \zeta\omega_0 - i\omega_d$. Using Eq. 8.43 then gives the non-zero elements of λ_B as: $[\lambda_B]_{1,1} = 3\zeta\omega_0 + 3i\omega_d$, $[\lambda_B]_{2,2} = 3\zeta\omega_0 + i\omega_d$, $[\lambda_B]_{3,3} = 3\zeta\omega_0 + i\omega_d$, $[\lambda_B]_{4,4} = 3\zeta\omega_0 - i\omega_d$, $[\lambda_B]_{5,5} = 3\zeta\omega_0 + i\omega_d$, $[\lambda_B]_{6,6} = 3\zeta\omega_0 - i\omega_d$, $[\lambda_B]_{7,7} = 3\zeta\omega_0 - i\omega_d$, and $[\lambda_B]_{8,8} = 3\zeta\omega_0 - 3i\omega_d$. We note the same redundancy in these eigenvalues as we previously found in the elements of the stationary solution. Namely the second, third and fifth elements are identical, and the fourth, sixth and seventh elements are identical. We could also easily write out the matrix of eigenvectors of $\mathbf{B}$ as $\theta_B = \theta_A^{[j]}$. The primary difficulty in writing out corresponding general solutions to nonstationary problems is simply the number of terms involved.

****

8.7   State-Space Equations for Stationary Autocovariance

We have derived equations governing the behavior of the moments and cumulants of $\vec{Y}(t)$ at any time t by considering joint moment or joint cumulants, respectively, of $\vec{Y}(t)$ and the equation of motion at time t. These equations usually constitute the most useful form of state-space analysis, but Spanos (1983) has demonstrated that the method can also be generalized to give equations governing autocorrelation or autocovariance functions. These more general relationships can be derived by considering joint moments or cumulants of either $\vec{Y}(s)$ or $\vec{Q}(s)$ and the equation of motion at time t, with no restriction of s being equal to t. We will illustrate this idea only for one of the simplest situations involving the second moment autocorrelation and cross-correlation functions. In principle the method can also be applied to higher-order moments or cumulants, but the calculations will become more complex. In order to obtain an equation for a cross-correlation function we simply multiply Eq. 8.5 on the right by $\vec{Q}^T(s)$ and then take the expected value, namely

$$\boldsymbol{\phi}_{\dot{Y}Q}(t,s) + \mathbf{A}\boldsymbol{\phi}_{YQ}(t,s) = \boldsymbol{\phi}_{QQ}(t,s) \tag{8.44}$$

Since the first term in this equation represents a partial derivative of $\boldsymbol{\phi}_{YQ}(t,s)$, this is a partial differential equation governing the behavior of this cross-correlation function between the $\{\vec{Y}(t)\}$ and $\{\vec{Q}(t)\}$ processes. Rather than pursue this general relationship, though, we will simplify the presentation by restricting our attention to the situation with a stationary cross-product $\boldsymbol{\phi}_{YQ}(t,s) = \mathbf{R}_{YQ}(t-s)$. In this case we have the ordinary differential equation

$$\mathbf{R'}_{YQ}(\tau) + \mathbf{A}\mathbf{R}_{YQ}(\tau) = \mathbf{R}_{QQ}(\tau) \tag{8.45}$$

in which the prime denotes a derivative with respect to τ. The solution can be written as

$$\mathbf{R}_{YQ}(\tau) = \int_{-\infty}^{\tau} \exp[-(\tau - u)\mathbf{A}]\mathbf{R}_{QQ}(u)\,du \tag{8.46}$$

Similarly, when we multiply the equation of motion at time t by $\vec{Y}^T(s)$ we obtain an equation for the autocorrelation function as

$$\mathbf{R'}_{YY}(\tau) + \mathbf{A}\mathbf{R}_{YY}(\tau) = \mathbf{R}_{QY}(\tau) \tag{8.47}$$

and the solution can be written as

$$\mathbf{R}_{YY}(\tau) = \int_{-\infty}^{\tau} \exp[-(\tau - u)\mathbf{A}]\mathbf{R}_{QY}(u)\,du \tag{8.48}$$

The "excitation" in this expression, though, is almost the same as the term given in Eq. 8.46. In particular, noting that $\mathbf{R}_{QY}(\tau) = \mathbf{R}_{YQ}^T(-\tau)$ and making a change of variables of $v = -u$ in the integral allows us to rewrite Eq. 8.46 as

$$\mathbf{R}_{QY}(\tau) = \int_{\tau}^{\infty} \mathbf{R}_{QQ}(v) \exp[(\tau - v)\mathbf{A}^T] \, dv \tag{8.49}$$

Substitution of Eq. 8.49 into 8.48 then gives

$$\mathbf{R}_{YY}(\tau) = \int_{-\infty}^{\tau} \int_{u}^{\infty} \exp[-(\tau - u)\mathbf{A}]\mathbf{R}_{QQ}(v) \exp[-(v - u)\mathbf{A}^T] \, dv \, du \tag{8.50}$$

Alternatively, one could rewrite Eq. 8.47 to obtain $\mathbf{R}_{YQ}(\tau) = -\mathbf{R}'_{YY}(\tau) + \mathbf{R}_{YY}(\tau)\mathbf{A}^T$ and substitute this into Eq. 8.45 to give a second-order differential equation of

$$\mathbf{R}''_{YY}(\tau) + \mathbf{A}\mathbf{R}'_{YY}(\tau) - \mathbf{R}'_{YY}(\tau)\mathbf{A}^T - \mathbf{A}\mathbf{R}_{YY}(\tau)\mathbf{A}^T = -\mathbf{R}_{QQ}(\tau)$$

then verify that this equation is satisfied by Eq. 8.50.

One can, of course, also rewrite all these expressions in terms of cross-covariance rather than cross-product terms and obtain the results corresponding to Eqs. 8.49 and 8.50 as

$$\mathbf{G}_{QY}(\tau) = \int_{\tau}^{\infty} \mathbf{G}_{QQ}(v) \exp[(\tau - v)\mathbf{A}^T] \, dv \tag{8.51}$$

and

$$\mathbf{G}_{YY}(\tau) = \int_{-\infty}^{\tau} \int_{u}^{\infty} \exp[-(\tau - u)\mathbf{A}]\mathbf{G}_{QQ}(v) \exp[-(v - u)\mathbf{A}^T] \, dv \, du \tag{8.52}$$

Consider now the special case with $\{\vec{Q}(t)\}$ being a stationary white noise process with

$$\mathbf{G}_{QQ}(\tau) = 2\pi \mathbf{S}_0 \delta(\tau)$$

Substituting this expression into Eqs. 8.51 and 8.52 gives

$$\mathbf{G}_{QY}(\tau) = 2\pi \mathbf{S}_0 \exp(\tau \mathbf{A}^T) \, U(-\tau) \tag{8.53}$$

and

$$\mathbf{G}_{YY}(\tau) = 2\pi \exp(-\tau \mathbf{A}) \int_{-\infty}^{\min(0,\tau)} \exp(u \mathbf{A}) \, \mathbf{S}_0 \exp(u \mathbf{A}^T) \, du \tag{8.54}$$

These results are consistent with those obtained in Section 8.4 for the special case of $\tau = 0$. In particular, the discontinuity in Eq. 8.53 at $\tau = 0$ leaves some doubt about the value of $\mathbf{K}_{QY}(t,t) \equiv \mathbf{G}_{QY}(0)$, but Eq. 8.54 allows evaluation of $\mathbf{G}_{YY}(0)$ without ambiguity

Note that the results in this section require knowledge only of the autocorrelation or autocovariance function for the excitation $\{\vec{Q}(t)\}$. Thus, for example, the results in Eqs. 8.53 and 8.54 are based only on the white noise property that $\vec{Q}(t)$ and $\vec{Q}(s)$ are uncorrelated for $s \neq t$. If we wish to obtain similar results for higher-order cumulant functions, then we will need to consider the corresponding higher-order moment or cumulant functions of the excitation. In particular, results comparable to Eqs. 8.53 and 8.54 would require that $\{\vec{Q}(t)\}$ satisfy other portions of the general delta-correlated relationship. We will not investigate these more general and complicated problems, though. Note that any higher-order cumulant function will depend on more than one time argument, so the formulation would involve partial differential equations. Only the stationary second moment and second cumulant (autocorrelation and autocovariance) involve only a single time argument, so are governed by ordinary differential equations.

It may also be noted that the results in this section are more general than those in the earlier portions of the chapter inasmuch as they include solutions for excitations which are not delta-correlated. In particular, Eqs. 8.46 and 8.50 describe characteristics of the response to an excitation which has the autocorrelation function $R_{QQ}(\tau)$, rather than being delta-correlated.

****

Example 8.11: Use Eq. 8.54 to find the autocovariance functions of $\{X(t)\}$ and $\{\dot{X}(t)\}$ for the response of a SDF oscillator excited by a stationary delta-correlated force with autospectral density of S_0.

Using the eigenvectors and eigenvalues of $\mathbf{A}$, we first rewrite Eq. 8.54 as

$$\mathbf{G}_{YY}(\tau) = 2\pi\,\boldsymbol{\theta}\exp(-\tau\boldsymbol{\lambda})\int_{-\infty}^{\min(0,\tau)}\exp(u\boldsymbol{\lambda})\boldsymbol{\theta}^{-1}\,\mathbf{S}_0\,\boldsymbol{\theta}^{-1^T}\exp(u\boldsymbol{\lambda})\boldsymbol{\theta}^T\,du$$

and write out the expression for the (j,l) term as

$$[\mathbf{G}_{YY}(\tau)]_{j,l} = 2\pi\sum_{r_1}\sum_{r_2}\sum_{r_3}\sum_{r_4}\theta_{j,r_1}e^{-\lambda_{r_1,r_1}\tau}\theta_{r_1,r_2}^{-1}[\mathbf{S}_0]_{r_2,r_3}\theta_{r_4,r_3}^{-1}\theta_{l,r_4}\int_{-\infty}^{\min(0,\tau)}e^{(\lambda_{r_1,r_1}+\lambda_{r_4,r_4})u}\,du$$

The state space formulation of the SDF oscillator as given in Example 8.4 has $\vec{Q}(t) = [0, F(t)/m]^T$ and this tells us that the autospectral density matrix for $\{\vec{Q}(t)\}$ is

$$\mathbf{S}_0 = \frac{2\pi S_0}{m^2}\begin{bmatrix} 0 & 0 \\ 0 & 1 \end{bmatrix}$$

Thus, the only contributions to $[\mathbf{G}_{YY}(\tau)]_{j,l}$ come from terms with $r_2 = r_3 = 2$. Using this fact, performing the integration with respect to u, and simplifying gives

$$[\mathbf{G}_{YY}(\tau)]_{j,l} = \frac{2\pi S_0}{m^2}\sum_{r_1=1}^{2}\sum_{r_4=1}^{2}\theta_{j,r_1}\theta_{r_1,2}^{-1}\theta_{r_4,2}^{-1}\theta_{l,r_4}\frac{e^{\lambda_{r_1,r_1}\min(0,-\tau)}e^{\lambda_{r_4,r_4}\min(0,\tau)}}{\lambda_{r_1,r_1}+\lambda_{r_4,r_4}}$$

The eigenvector matrix for $\mathbf{A}$ and its inverse are as given in Example 8.5:

$$\boldsymbol{\theta} = \begin{bmatrix} 1 & 1 \\ -\zeta\omega_0 - i\omega_d & -\zeta\omega_0 + i\omega_d \end{bmatrix}, \quad \boldsymbol{\theta}^{-1} = \frac{1}{2}\begin{bmatrix} 1 + i\zeta\omega_0/\omega_d & i/\omega_d \\ 1 - i\zeta\omega_0/\omega_d & -i/\omega_d \end{bmatrix}$$

and the non-zero elements of $\boldsymbol{\lambda}$ are the eigenvalues of $\mathbf{A}$: $\lambda_{1,1} = \zeta\omega_0 + i\omega_d$ and $\lambda_{2,2} = \zeta\omega_0 - i\omega_d$, in which ω_0, ζ, and ω_d are defined in the usual way. We then find the desired covariance functions by performing the summation over r_1 and r_4. The results from the diagonal elements are

$$G_{XX}(\tau) = [\mathbf{G}_{YY}(\tau)]_{1,1} = \frac{\pi S_0}{2m^2\zeta\omega_0^3}e^{-\zeta\omega_0|\tau|}\left(\cos(\omega_d\tau) + \frac{\zeta\omega_0}{\omega_d}\sin(\omega_d|\tau|)\right)$$

and

$$G_{\dot{X}\dot{X}}(\tau) = [\mathbf{G}_{YY}(\tau)]_{2,2} = \frac{\pi S_0}{2m^2\zeta\omega_0}e^{-\zeta\omega_0|\tau|}\left(\cos(\omega_d\tau) - \frac{\zeta\omega_0}{\omega_d}\sin(\omega_d|\tau|)\right)$$

These two expressions are identical to the results in Eqs. 3.54 and 3.56, obtained using the Duhamel convolution integral. Similarly, the $[\mathbf{G}_{YY}(\tau)]_{1,2}$ element can be shown to be identical to the $G_{X\dot{X}}(\tau)$ function given in Eq. 3.55.

****

8.8   Fokker-Planck Equation

Note the general approach which has been used in deriving the various state-space moment and cumulant equations. We chose a moment or cumulant of interest, took its derivative with respect to time, then used this information in deriving a differential equation which governs the evolution over time of the particular quantity. This same general procedure will now be used in deriving a differential equation which governs the evolution of the probability density of a nonstationary process. We will give the full derivation of the equations for a scalar process $\{Y(t)\}$, since that significantly simplifies the mathematical expressions, then indicate the generalization to a vector process $\{\vec{Y}(t)\}$.

Formally we can write the derivative of a nonstationary probability density function as

$$\frac{\partial}{\partial t}p_{Y(t)}(u) = \lim_{\Delta t \to 0}\frac{1}{\Delta t}\left[p_{Y(t+\Delta t)}(u) - p_{Y(t)}(u)\right] \tag{8.55}$$

and we can write the probability density for $Y(t+\Delta t)$ which appears in this expression as the marginal probability integral of the joint probability density of $Y(t)$ and $Y(t+\Delta t)$:

$$p_{Y(t+\Delta t)}(u) = \int_{-\infty}^{\infty}p_{Y(t),Y(t+\Delta t)}(v,u)\,dv = \int_{-\infty}^{\infty}p_{Y(t)}(v)p_{Y(t+\Delta t)}(u|Y(t)=v)\,dv \tag{8.56}$$

We will now denote the change in the $\{Y(t)\}$ process from time t to time $t+\Delta t$ as $\Delta Y(t) \equiv$

$Y(t + \Delta t) - Y(t)$ and write a conditional characteristic function for this increment as

$$M_{\Delta Y}(\theta, v, t) = E(e^{i\theta \Delta Y} | X(t) = v) = \int_{-\infty}^{\infty} e^{i\theta(u-v)} p_{Y(t+\Delta t)}(u | Y(t) = v) \, du$$

Expanding this function as a Maclaurin power series about the point $\theta = 0$ (see Section A.12 in Appendix A) gives

$$M_{\Delta Y}(\theta, v, t) = \sum_{j=0}^{\infty} \frac{\theta^j}{j!} \left[\frac{\partial^j}{\partial \theta^j} M_{\Delta Y}(\theta, v, t) \right]_{\theta=0} = \sum_{j=0}^{\infty} \frac{(i\theta)^j}{j!} E[(\Delta Y)^j | Y(t) = v]$$

and taking the Fourier transform of this expression gives the conditional probability density function as

$$p_{Y(t+\Delta t)}(u | Y(t) = v) = \frac{1}{2\pi} \int_{-\infty}^{\infty} e^{-i\theta(u-v)} \sum_{j=0}^{\infty} \frac{(i\theta)^j}{j!} E[(\Delta Y)^j | Y(t) = v] \, d\theta \qquad (8.57)$$

Substituting Eq. 8.57 into Eq. 8.56 allows us to write the unconditional probability density for $Y(t + \Delta t)$ as

$$p_{Y(t+\Delta t)}(u) = p_{Y(t)}(u) + \sum_{j=1}^{\infty} \frac{1}{2\pi} \int_{-\infty}^{\infty} p_{Y(t)}(v) \int_{-\infty}^{\infty} e^{-i\theta(u-v)} \frac{(i\theta)^j}{j!} E[(\Delta Y)^j | Y(t) = v] \, d\theta \, dv$$

in which the term with $j = 0$ has been separated from the summation and simplified by noting that

$$\frac{1}{2\pi} \int_{-\infty}^{\infty} e^{-i\theta(u-v)} \, d\theta = \delta(u - v) \qquad (8.58)$$

The reason for separating this one term is so that Eq. 8.55 can be simplified as

$$\frac{\partial}{\partial t} p_{Y(t)}(u) = \sum_{j=1}^{\infty} \frac{1}{2\pi} \int_{-\infty}^{\infty} p_{Y(t)}(v) \int_{-\infty}^{\infty} e^{-i\theta(u-v)} \frac{(i\theta)^j}{j!} C^{(j)}(v, t) \, d\theta \, dv \qquad (8.59)$$

in which we have introduced a new notation of

$$C^{(j)}(v, t) \equiv \lim_{\Delta t \to 0} \frac{1}{\Delta t} E[(\Delta Y)^j | Y(t) = v] \qquad (8.60)$$

presuming that these limits exist. Integration by parts with respect to v now allows us to rewrite Eq. 8.59 as

$$\frac{\partial}{\partial t} p_{Y(t)}(u) = \sum_{j=1}^{\infty} \frac{1}{2\pi} \int_{-\infty}^{\infty} \frac{(-1)^j}{j!} \int_{-\infty}^{\infty} e^{-i\theta(u-v)} \frac{\partial^j}{\partial v^j} [C^{(j)}(v, t) p_{Y(t)}(v)] \, d\theta \, dv$$

and using Eq. 8.58 for the integration with respect to θ gives a Dirac delta function in the argument so that the integration with respect to v is almost trivial and gives

$$\frac{\partial}{\partial t} p_{Y(t)}(u) = \sum_{j=1}^{\infty} \frac{(-1)^j}{j!} \frac{\partial^j}{\partial u^j} [C^{(j)}(u,t) p_{Y(t)}(u)] \tag{8.61}$$

This is the Fokker-Planck equation for the scalar process $\{Y(t)\}$. It is also sometimes called the Kolmogorov forward equation.[1]

An alternate derivation of Eq. 8.61 avoids the use of characteristic functions but uses an arbitrary, though well behaved, function $R(u)$ which can be expanded as

$$R(u) = \sum_{j=0}^{\infty} \frac{(u-v)^j}{j!} R^{(j)}(v) \quad \text{with} \quad R^{(j)}(v) \equiv \left[\frac{d^j R(u)}{du^j} \right]_{u=v}$$

This expansion is used in a combination of Eqs. 8.55 and 8.56 to give

$$\int_{-\infty}^{\infty} R(u) \frac{\partial}{\partial t} p_{Y(t)}(u)\, du = \lim_{\Delta t \to 0} \frac{1}{\Delta t} \left[-\int_{-\infty}^{\infty} R(u) p_{Y(t)}(u)\, du \right.$$
$$\left. + \int_{-\infty}^{\infty} \sum_{j=0}^{\infty} \frac{(u-v)^j}{j!} R^{(j)}(v) \int_{-\infty}^{\infty} p_{Y(t)}(v) p_{Y(t+\Delta t)}(u|\, Y(t) = v)\, dv\, du \right]$$

Noting that the $j = 0$ term in the summation cancels the first term in the brackets gives

$$\int_{-\infty}^{\infty} R(u) \frac{\partial}{\partial t} p_{Y(t)}(u)\, du = \lim_{\Delta t \to 0} \frac{1}{\Delta t} \sum_{j=1}^{\infty} \int_{-\infty}^{\infty} \int_{-\infty}^{\infty} \frac{(u-v)^j}{j!} R^{(j)}(v) p_{Y(t)}(v) p_{Y(t+\Delta t)}(u|\, Y(t) = v)\, dv\, du$$

Performing the integration with respect to u allows this to be rewritten as

$$\int_{-\infty}^{\infty} R(u) \frac{\partial}{\partial t} p_{Y(t)}(u)\, du = \sum_{j=1}^{\infty} \int_{-\infty}^{\infty} \frac{1}{j!} R^{(j)}(v) p_{Y(t)}(v) C^{(j)}(v,t)\, dv$$

in which $C^{(j)}(v,t)$ is as defined in Eq. 8.60. We now require that $R(v)$ and its derivatives be sufficiently well behaved when $|v|$ is infinite that we can integrate by parts to obtain

---

[1] The qualifier in this latter terminology is to distinguish this equation from a similar partial differential equation called the Kolmogorov backward equation. The backward equation involves derivatives with respect to t_0 and u_0 of a conditional probability density function, $p_{Y(t)}[u|\, Y(t_0) = u_0]$.

$$\int_{-\infty}^{\infty} R(u) \frac{\partial}{\partial t} p_{Y(t)}(u)\, du = \sum_{j=1}^{\infty} \int_{-\infty}^{\infty} \frac{(-1)^j}{j!} R(v) \frac{\partial}{\partial v^j}[p_{Y(t)}(v)C^{(j)}(v,t)]\, dv$$

Noting that both u and v in this expression are simply dummy variables of integration, we can rewrite this as

$$\int_{-\infty}^{\infty} R(u)\left[\frac{\partial}{\partial t} p_{Y(t)}(u) - \sum_{j=1}^{\infty} \frac{(-1)^j}{j!} \frac{\partial}{\partial u^j}[p_{Y(t)}(u)C^{(j)}(u,t)]\right] du = 0$$

but in order for this to be true for a general range of $R(u)$ functions it is necessary that Eq. 8.61 be satisfied.

Note that Eq. 8.61 is very general, applying to the probability density function of virtually any scalar stochastic process. In particular, we have not used any equation of motion in the derivation of Eq. 8.61. If we can somehow find the values of the $C^{(j)}(u,t)$ terms which are appropriate for a particular $\{Y(t)\}$ process, then we will have a partial differential equation which must be satisfied by $p_{Y(t)}(u)$. For an important class of dynamic problems it turns out that we can use the equation of motion in deriving the $C^{(j)}(u,t)$ terms. This is quite logical, since the definition of $C^{(j)}(u,t)$ in Eq. 8.60 shows that these terms are moments of the increment of the $\{Y(t)\}$ process, and we can expect that a procedure similar to state space moment analysis should yield values for such moments.[2]

For the scalar equation of motion of

$$\dot{Y}(t) + AY(t) = Q(t)$$

we can see that

$$\Delta Y = Y(t + \Delta t) - Y(t) = \int_t^{t+\Delta t} \dot{Y}(s)\, ds = \int_t^{t+\Delta t} Q(s)\, ds - A\int_t^{t+\Delta t} Y(s)\, ds$$

so that

$$C^{(1)}(u,t) = \lim_{\Delta t \to 0} \frac{1}{\Delta t}\left(\int_t^{t+\Delta t} E[Q(s)|Y(t) = u]\, ds - A\int_t^{t+\Delta t} E[Y(s)|Y(t) = u]\, ds\right)$$

Provided that $Y(t)$ is continuous in the vicinity of time t, we can say that $E[Y(s)|Y(t) = u] \approx u$ in the final integral. Similarly, we will presume that $E[Q(s)|Y(t) = u]$ is continuous in the first integral, giving

$$C^{(1)}(u,t) = E[Q(t)|Y(t) = u] - Au \qquad (8.62)$$

---

[2]   One term that is used for these coefficients is "derivate moments." Lin (1967) attributes this terminology to Moyal in 1949. Stratonovich (1963, translation of an earlier Russian work) calls them "intensity functions."

In a similar fashion

$$C^{(2)}(u,t) = \lim_{\Delta t \to 0} \frac{1}{\Delta t} \left(\int_t^{t+\Delta t} \int_t^{t+\Delta t} E[Q(s_1)Q(s_2)|Y(t)=u]ds_1 ds_2 \right.$$
$$-2\int_t^{t+\Delta t} \int_t^{t+\Delta t} E[Q(s_1)Y(s_2)|Y(t)=u]ds_1 ds_2$$
$$\left. +A^2\int_t^{t+\Delta t} \int_t^{t+\Delta t} E[Y(s_1)Y(s_2)|Y(t)=u]ds_1 ds_2 \right)$$

The double integrals in this expression will all be of order $(\Delta t)^2$, so that they will contribute nothing to $C^{(2)}(u,t)$ unless their integrands are infinite. Thus, the last term will contribute nothing provided that $E[Y(s_1)Y(s_2)|Y(t)=u]$ is finite, and the second term will contribute nothing provided that $E[Q(s_1)Y(s_2)|Y(t)=u]$ is finite. We have shown in Section 8.4, though, that $\phi_{YY}(s_1,s_2)$ and $\phi_{QY}(s_1,s_2)$ are finite even when the excitation $\{Q(t)\}$ is delta-correlated. Thus, we find that only the first term may make a contribution. Exactly the same arguments can be made for the higher-order integrals which occur in $C^{(j)}(u,t)$ for $j > 2$, with the result that

$$C^{(j)}(u,t) = \lim_{\Delta t \to 0} \frac{1}{\Delta t} \int_t^{t+\Delta t} \cdots \int_t^{t+\Delta t} E[Q(s_1)\cdots Q(s_j)|Y(t)=u]ds_1 \cdots ds_j \quad \text{for} \quad j \geq 2 \quad (8.63)$$

Thus, we see that Eq. 8.61 depends on the coefficient A from the equation of motion only through its effect on the $C^{(1)}(u,t)$ coefficient.

The derivation of a Fokker-Planck equation for a vector process $\{\vec{Y}(t)\}$ can be accomplished by either of the methods given for the scalar situation of Eq. 8.61. We will not repeat the details of either derivation, but will point out the key differences between the scalar and vector situations in the first derivation. The first significant change is in the power series expansion of the characteristic function, which now is

$$M_{\Delta \vec{Y}}(\vec{\theta},\vec{v},t) = \sum_{j_1=0}^{\infty} \cdots \sum_{j_{n_Y}=0}^{\infty} \frac{(i\theta_1)^{j_1}\cdots(i\theta_n)^{j_{n_Y}}}{j_1!\cdots j_{n_Y}!} E[(\Delta Y_1)^{j_1}\cdots(\Delta Y_{n_Y})^{j_{n_Y}}|\vec{Y}(t)=\vec{v}]$$

involving moments of the increments of various components of $\vec{Y}(t)$. After this term has been used in the Fourier transform relationship for $p_{\vec{Y}(t+\Delta t)}(\vec{u}|\vec{Y}(t)=\vec{v})$, similar to Eq. 8.57, it is only the one term with $j_1 = \cdots = j_{n_Y} = 0$ from the multiple summation which is separated out as $p_{\vec{Y}(t)}(\vec{u})$. Thus, all the other terms remain in the final equation, which can be written as

$$\frac{\partial}{\partial t}p_{\vec{Y}(t)}(\vec{u}) = \underbrace{\sum_{j_1=0}^{\infty} \cdots \sum_{j_{n_Y}=0}^{\infty}}_{(\text{except } j_1=\cdots=j_{n_Y}=0)} \frac{(-1)^{j_1+\cdots+j_{n_Y}}}{j_1!\cdots j_{n_Y}!} \frac{\partial^{j_1+\cdots+j_{n_Y}}}{\partial u_1^{j_1}\cdots \partial u_{n_Y}^{j_{n_Y}}}[C^{(j_1,\cdots,j_{n_Y})}(\vec{u},t)p_{\vec{Y}(t)}(\vec{u})] \quad (8.64)$$

with

$$C^{(j_1, \cdots, j_{n_Y})}(\vec{u}, t) = \lim_{\Delta t \to 0} \frac{1}{\Delta t} E[(\Delta Y_1)^{j_1} \cdots (\Delta Y_{n_Y})^{j_{n_Y}} | \vec{Y}(t) = \vec{u}] \tag{8.65}$$

This is the general form for the Fokker-Planck equation for a vector process. Of course, Eqs. 8.60 and 8.61 are simply the special cases of Eqs. 8.65 and 8.64, respectively, with $n_Y = 1$.

For a system governed by the vector state space equation of motion

$$\dot{\vec{Y}}(t) + \mathbf{A}\vec{Y}(t) = \vec{Q}(t)$$

one can derive the analogs of Eqs. 8.62 and 8.63 as

$$\vec{C}^{(1)}(\vec{u}, t) = E[\vec{Q}(t) | \vec{Y}(t) = \vec{u}] - \mathbf{A}\vec{u} \tag{8.66}$$

and

$$C^{(j_1, \cdots, j_{n_Y})}(\vec{u}, t) = \lim_{\Delta t \to 0} \frac{1}{\Delta t} \int_t^{t+\Delta t} \cdots \int_t^{t+\Delta t} E[Q_1(s_{1,1}) \cdots Q_1(s_{1,j_1}) \cdots Q_{n_Y}(s_{n_Y,1}) \cdots Q_n(s_{n_Y, j_{n_Y}}) | \vec{Y}(t) = \vec{u}]$$

$$ds_{1,1} \cdots ds_{1,j_1} \cdots ds_{n_Y,1} \cdots ds_{n_Y, j_{n_Y}} \tag{8.67}$$

One uses Eq. 8.67 whenever $j_1 + \cdots + j_{n_Y} > 1$, and the lth component of $\vec{C}^{(1)}(u, t)$ gives $C^{(j_1, \cdots, j_{n_Y})}(\vec{u}, t)$ when $j_l = 1$ is the only nonzero j value.

The development up to this point has been very general, but practical application of the partial differential equation in Eq. 8.64 generally is limited to the special case in which $\{\vec{Y}(t)\}$ is a so-called Markov process. The defining property of a Markov process can be written as

$$p_{\vec{Y}(t)}(\vec{v} | \vec{Y}(s_1) = \vec{u}_1, \cdots, \vec{Y}(s_l) = \vec{u}_l) = p_{\vec{Y}(t)}(\vec{v} | \vec{Y}(s_l) = \vec{u}_l)$$

if $s_1 \leq s_2 \leq \cdots \leq s_l \leq t$. That is, knowing several past values $\vec{Y}(s_1) = \vec{u}_1, \cdots, \vec{Y}(s_l) = \vec{u}_l$ of the process gives no more information about likely future values than does knowing only the most recent value $\vec{Y}(s_l) = \vec{u}_l$. This property can also be given by the statement that the future of the process is conditionally independent of the past, given the present value. The response of our causal linear state space equation does have the Markov property in the special case when the excitation is delta-correlated. In particular, $\vec{Y}(t)$ for $t > t_0$ can be written as in Eq. 7.59 as a function of $\vec{Y}(t_0)$ plus an integral involving $\vec{Q}(s)$ for $s \geq t_0$. If $\{\vec{Q}(t)\}$ were not delta-correlated, then knowing additional values of $\vec{Y}(\tau)$ for $\tau < t$ might give us information about the probable values of $\vec{Y}(t)$ by giving us information about probable values of $\vec{Q}(s)$ for $s \geq t_0$. When $\{\vec{Q}(t)\}$ is delta-correlated, though, $\vec{Q}(s)$ for $s \geq t_0$ is independent of $\vec{Q}(\tau)$ for $\tau < t$, and therefore it is conditionally independent of $\vec{Y}(\tau)$ for $\tau < t$. Thus, we would learn nothing more about the probable values of $\vec{Y}(t)$ by knowing additional values of $\vec{Y}(\tau)$ for $\tau < t_0$. This is the Markov property. For such a Markov process one can eliminate the conditioning in Eq. 8.65, giving the coefficients for Eq. 8.64 as

$$C^{(j_1,\cdots,j_{n_Y})}(\vec{u},t) = \lim_{\Delta t \to 0} \frac{1}{\Delta t} E[(\Delta Y_1)^{j_1} \cdots (\Delta Y_{n_Y})^{j_{n_Y}}] \tag{8.68}$$

since $\Delta \vec{Y}$ is independent of $\vec{Y}(t)$.

Clearly, the Markov property allows us to eliminate the conditioning from Eqs. 8.66 and 8.67 for the response of a linear system to a delta-correlated excitation, giving

$$\vec{C}^{(1)}(\vec{u},t) = \vec{\mu}_Q(t) - \mathbf{A}\vec{u} \tag{8.69}$$

and

$$C^{(j_1,\cdots,j_{n_Y})}(\vec{u},t) = \lim_{\Delta t \to 0} \frac{1}{\Delta t} \int_t^{t+\Delta t} \cdots \int_t^{t+\Delta t} E[Q_1(s_{1,1})\cdots Q_1(s_{1,j_1})\cdots Q_{n_Y}(s_{n_Y,1})\cdots Q_{n_Y}(s_{n_Y,j_{n_Y}})]$$
$$ds_1 \cdots ds_{1,j_1} \cdots ds_{n_Y} \cdots ds_{n_Y,j_{n_Y}} \quad \text{for} \quad j \geq 2 \tag{8.70}$$

In addition, though, the property of delta-correlation also allows us to evaluate the integral in Eq. 8.70 so that we know all the coefficients in Eq. 8.64. In this situation there is the possibility of finding the nonstationary probability density function $p_{\vec{Y}(t)}(\vec{u})$ by solving that partial differential equation.

In order to describe precisely the delta-correlated excitation, we will use the notation

$$\kappa_J[Q_1(s_{1,1}),\cdots,Q_1(s_{1,j_1}),\cdots,Q_{n_Y}(s_{n_Y,1}),\cdots,Q_{n_Y}(s_{n_Y,j_{n_Y}})] = (2\pi)^{J-1} S_{j_1,\cdots,j_{n_Y}}(s_{n_Y,j_{n_Y}})$$
$$\delta(s_{1,1} - s_{n_Y,j_{n_Y}})\cdots\delta(s_{1,j_1} - s_{n_Y,j_{n_Y}})\cdots\delta(s_{n_Y,1} - s_{n_Y,j_{n_Y}})\cdots\delta(s_{n_Y,j_{n_Y}-1} - s_{n_Y,j_{n_Y}}) \tag{8.71}$$

for the Jth-order cross-cumulant, with $J = j_1 + \cdots + j_{n_Y}$. This is a component version of the higher-order spectral density representation given in Kronecker notation in Eq. 8.39. Now we must consider the relationship between the Jth-order moments in Eq. 8.70 and the cumulants in Eq. 8.71. In principle one can always write a Jth-order moment for any set of random variables as a function of cumulants up to that order, but these general relationships are not simple when J is large. In our situation, though, the delta-correlation property allows us to solve the problem in a simple way. In particular, we note that the integral in Eq. 8.70 is over a volume of size $(\Delta t)^J$, and the only way that an integral over this volume can be linear in Δt is if the integrand behaves like a product of $J - 1$ Dirac delta functions. Thus, when we relate the Jth-order moment in Eq. 8.70 to cumulants we only need consider terms which have a product of $J - 1$ Dirac delta functions coming from Eq. 8.71. Fortunately, there is only one such term in Eq. 8.70, and it is exactly the Jth-order cumulant of the terms in the Jth-order moment. To illustrate this fact in a relatively simple situation, consider the special case of Eq. 8.70 with $j_1 = J$ and $j_2 = \cdots = j_{n_Y} = 0$. The first few terms in the expansion of the Jth-order moment in terms of cumulants are

$$E[Q_1(s_1)\cdots Q_1(s_J)] = \kappa_J[Q_1(s_1),\cdots,Q_1(s_J)]$$

$$+ b_1 \sum_l E[Q_1(s_l)]\kappa_{J-1}[Q_1(s_1),\cdots,Q_1(s_{l-1}),Q_1(s_{l+1}),\cdots,Q_1(s_J)]$$

$$+ b_2 \sum_{l_1}\sum_{l_2} \kappa_2[Q_1(s_{l_1}),Q_1(s_{l_2})]\kappa_{J-2}[\text{other } Qs]$$

$$+ b_3 \sum_{l_1}\sum_{l_2}\sum_{l_3} \kappa_3[Q_1(s_{l_1}),Q_1(s_{l_2}),Q_1(s_{l_3})]\kappa_{J-3}[\text{other } Qs] + \cdots$$

The first term on the right hand side of the equation does include a product of $J - 1$ Dirac delta functions, but each of the other terms shown includes only $J - 2$ Dirac delta functions, so contributes only a term of order $(\Delta t)^2$ to the integral of Eq. 8.70. Other terms not written out contribute even less significant terms. For example, there are terms which correspond to subdividing $\{Q_1(s_1),\cdots,Q_1(s_J)\}$ into three subgroups, then taking the cumulant of each of these subgroups. Each of these terms gives a contribution of order $(\Delta t)^3$ to the integral.

Using the fact that only the Jth-order cumulant term makes an order Δt contribution to Eq. 8.70, and using Eq. 8.71 in evaluating this contribution gives

$$C^{(j_1,\cdots,j_{n_Y})}(\vec{u},t) = (2\pi)^{J-1} S_{j_1,\cdots,j_{n_Y}}(t) \tag{8.72}$$

for the higher-order coefficients in the Fokker-Planck equation for the response of the linear system to a nonstationary delta-correlated excitation.

The situation is considerably simplified in the special case when the excitation $\{\vec{Q}(t)\}$ is Gaussian. The fact that all cumulants beyond the second order are identically zero for a Gaussian process means that most of the terms given by Eq. 8.72 are also zero. In particular, one only needs use Eq. 8.72 for second-order coefficients in this situation. Furthermore, Eq. 8.72 shows that the second-order coefficients are simply related to the components of the autospectral density matrix for the excitation process as

$$C^{(\overbrace{0,\cdots,0}^{j-1},1,0,\cdots 0,\overbrace{1,0,\cdots,0}^{n_Y-l})}(\vec{u},t) = 2\pi[\mathbf{S}_0(t)]_{j,l} \quad \text{for} \quad j \neq l \tag{8.73}$$

and

$$C^{(\overbrace{0,\cdots,0}^{j-1},2,\overbrace{0,\cdots,0}^{n_Y-j})}(\vec{u},t) = 2\pi[\mathbf{S}_0(t)]_{j,j} \tag{8.74}$$

with the usual definition of the autospectral density matrix such that $\mathbf{K}_{QQ}(t,s) = 2\pi\mathbf{S}_0(t)\delta(t-s)$ for the delta-correlated process. The Fokker-Planck relationship of Eq. 8.64 can then be written as

$$\frac{\partial}{\partial t} p_{\vec{Y}(t)}(\vec{u}) = -\sum_{j=1}^{n_Y} \frac{\partial}{\partial u_j}[C_j^{(1)}(\vec{u},t)p_{\vec{Y}(t)}(\vec{u})] + \pi \sum_{j=1}^{n_Y}\sum_{l=1}^{n_Y}[S_0(t)]_{j,l}\frac{\partial^2}{\partial u_j \partial u_l}p_{\vec{Y}(t)}(\vec{u}) \qquad (8.75)$$

with $C_j^{(1)}(\vec{u},t)$ given by Eq. 8.69.[3] Finding solutions of the problem, of course, also requires consideration of appropriate initial conditions and boundary conditions.

Using the formulas given it is relatively easy to find the coefficients in the Fokker-Planck equation for the situations with a delta-correlated excitation. The problem of solving this partial differential equation may be quite difficult, though, particularly if the excitation is not Gaussian, so that higher-order coefficients are non-zero. The problem is simpler if the excitation is Gaussian, but there is also less reason to use the Fokker-Planck approach in that situation. We already know that the response is Gaussian for a linear system with a Gaussian excitation, and a Gaussian distribution is completely defined in terms of its first and second moments or cumulants. Thus, for a Gaussian problem it is usually easier to use state-space analysis of mean and variance and then substitute these values into a Gaussian probability density function, rather than solving the Fokker-Planck equation to find that probability density function directly. Examples 8.12 - 8.14 illustrate the derivation of the Fokker-Planck coefficients in a few situations, and the verification of the Gaussian solution for simple cases with Gaussian excitations.

It should be noted that it is also possible to derive state-space moment or cumulant equations from the Fokker-Planck equation. This is illustrated in Example 8.15 for the second moment of a very simple system. There seems to be no obvious advantage to derivation of the moment or cumulant equations in this way though, since they can be derived by the methods of Sections 8.3 and 8.4 without consideration of a partial differential equation.

As with state-space moment or cumulant analysis, practical application of the Fokker-Planck equation is typically limited to problems with delta-correlated excitations. In the case of the Fokker-Planck equation there is no such limitation on the equation itself, but evaluation of the coefficients required the Markov property, and that property is dependent on delta-correlation of the excitation. Just as in moment or cumulant analysis, though, the results can be extended to include any problem in which the excitation can be modeled as a filtered delta-correlated process.

****

Example 8.12: Let $\{X(t)\}$ denote the response of a dashpot with coefficient c subjected to a nonstationary Gaussian white noise force $\{F(t)\}$ with mean $\mu_F(t)$ and autospectral density $S_0(t)$. Find the coefficients in the

---

[3] It may be noted that Stratonovich (1963) reserved the term Fokker-Planck equation for the Gaussian situation of Eq. 8.75, rather than using it for the more general forward Kolmogorov equation.

Fokker-Planck equation and find its solution for the situation with $X(0) = 0$.

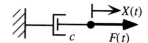

The equation of motion can be written as $\dot{X}(t) = F(t)/c$, which is the special case of $\dot{\vec{Y}}(t) + \mathbf{A}\vec{Y}(t) = \vec{Q}(t)$ with $n_Y = 1$, $\vec{Y}(t) = X(t)$, $\mathbf{A} = 0$, and $\vec{Q}(t) = F(t)/m$. Since the excitation is Gaussian, we can use Eq. 8.75 for the Fokker-Planck equation. The first-order coefficient in the Fokker-Planck equation is found from Eq. 8.69 as $C^{(1)}(u,t) = \mu_F(t)/c$. Using $[S_0(t)] = S_0(t)/c^2$, the second-order coefficient is $C^{(2)}(u,t) = 2\pi S_0(t)/c^2$. Thus, the Fokker-Planck equation is

$$\frac{\partial}{\partial t}p_{X(t)}(u) = -\frac{\mu_F(t)}{c}\frac{\partial}{\partial u}p_{X(t)}(u) + \frac{\pi S_0(t)}{c^2}\frac{\partial^2}{\partial u^2}p_{X(t)}(u)$$

Note that this equation is particularly simple because $C^{(1)}(u,t)$, as well as $C^{(2)}(u,t)$, is independent of u.

Rather than seeking a solution of this partial differential equation in a direct manner, we will take the simpler approach of using our prior knowledge that the response should be Gaussian and simply verify that the equation is solved by the Gaussian probability density function. That is, we will substitute

$$p_{X(t)}(u) = \frac{1}{\sqrt{2\pi}\sigma_X(t)}\exp\left[-\frac{1}{2}\left(\frac{u-\mu_X(t)}{\sigma_X(t)}\right)^2\right]$$

into the Fokker-Planck equation, and verify that the equation is satisfied if $\mu_X(t)$ and $\sigma_X(t)$ are chosen properly. For the scalar Gaussian probability density function, the three needed derivative terms can be written as

$$\frac{\partial}{\partial u}p_{X(t)}(u) = \frac{-1}{\sigma_X(t)}\left(\frac{u-\mu_X(t)}{\sigma_X(t)}\right)p_{X(t)}(u)$$

$$\frac{\partial^2}{\partial u^2}p_{X(t)}(u) = \frac{1}{\sigma_X^2(t)}\left[\left(\frac{u-\mu_X(t)}{\sigma_X(t)}\right)^2 - 1\right]p_{X(t)}(u)$$

and

$$\frac{\partial}{\partial t}p_{X(t)}(u) = \left(\frac{\dot{\sigma}_X(t)}{\sigma_X(t)}\left[\left(\frac{u-\mu_X(t)}{\sigma_X(t)}\right)^2 - 1\right] + \frac{\dot{\mu}_X(t)}{\sigma_X(t)}\left(\frac{u-\mu_X(t)}{\sigma_X(t)}\right)\right)p_{X(t)}(u)$$

$$= \frac{\partial}{\partial u}\left[\sigma_X(t)\dot{\sigma}_X(t)\frac{\partial}{\partial u}p_{X(t)}(u) - \dot{\mu}_X(t)p_{X(t)}(u)\right]$$

Substitution of these relationships into the Fokker-Planck equation gives

$$\frac{\partial}{\partial u}\left[\sigma_X(t)\dot{\sigma}_X(t)\frac{\partial}{\partial u}p_{X(t)}(u) - \dot{\mu}_X(t)p_{X(t)}(u) + \frac{\mu_F(t)}{c}p_{X(t)}(u) - \frac{\pi S_0(t)}{c^2}\frac{\partial}{\partial u}p_{X(t)}(u)\right] = 0$$

or

$$\frac{\partial}{\partial u}\left[\left(\frac{1}{2}\frac{d}{dt}\sigma_X^2(t) - \frac{\pi S_0(t)}{c^2}\right)\frac{\partial}{\partial u}p_{X(t)}(u) - \left(\dot{\mu}_X(t) - \frac{\mu_F(t)}{c}\right)p_{X(t)}(u)\right] = 0$$

The term in the brackets is zero for $|u| = \infty$, due to the nature of the probability density function. Thus, it must be zero everywhere since its derivative is zero everywhere. Furthermore, the only way that the term can be zero for all u values is for the coefficients of both $p_{X(t)}(u)$ and its derivative to be zero. Thus, we find that $\mu_X(t)$ and $\sigma_X^2(t)$ are governed by the equations

$$\frac{d}{dt}\mu_X(t) = \frac{\mu_F(t)}{c}, \qquad \frac{d}{dt}\sigma_X^2(t) = \frac{2\pi S_0(t)}{c^2}$$

The solutions of these ordinary differential equations, of course, depend on the initial conditions on the problem. For our stated initial condition of $X(0) = 0$, we know that $\mu_X(0) = 0$ and $\sigma_X^2(0) = 0$, so that the solutions are

$$\mu_X(t) = \frac{1}{c}\int_0^t \mu_F(s)\,ds, \qquad \sigma_X^2(t) = \frac{2\pi}{c^2}\int_0^t S_0(s)\,ds$$

Note that we have only verified that the Gaussian probability density function with these parameter values is one solution of the Fokker-Planck equation with the given initial conditions. In fact, it is the unique solution of the problem for the given initial conditions. The values of the nonstationary mean and variance of $\{X(t)\}$, of course, can be confirmed by other methods of analysis.
****

Example 8.13: Let $\{X(t)\}$ denote the response of the system shown when subjected to a nonstationary Gaussian white noise force $\{F(t)\}$ with mean $\mu_F(t)$ and autospectral density $S_0(t)$. Find the coefficients in the Fokker-Planck equation, verify that it has a Gaussian solution, and find the equations governing the evolution of the mean and variance in that Gaussian solution.

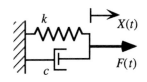

Proceeding in the same way as in Example 8.12, the equation of motion is $\dot{\vec{Y}}(t) + \mathbf{A}\vec{Y}(t) = \vec{Q}(t)$ with $n_Y = 1$, $\vec{Y}(t) = X(t)$, $\mathbf{A} = k/c$, and $Q(t) = F(t)/m$. The coefficients in the Fokker-Planck equation are found from Eqs. 8.69 and 8.72 as

$$C^{(1)}(u,t) = \frac{\mu_F(t)}{c} - \frac{k}{c}u, \qquad C^{(2)}(u,t) = \frac{2\pi S_0(t)}{c^2}$$

Thus, the Fokker-Planck relationship from Eq. 8.75 is

$$\frac{\partial}{\partial t} p_{X(t)}(u) = -\frac{\partial}{\partial u}\left[\left(\frac{\mu_F(t)}{c} - \frac{k}{c}u\right)p_{X(t)}(u)\right] + \frac{\pi S_0(t)}{c^2}\frac{\partial^2}{\partial u^2} p_{X(t)}(u)$$

Substituting the relationships given in Example 8.12 for the derivatives of a Gaussian probability density function converts this equation into

$$\frac{\partial}{\partial u}\left[\sigma_X(t)\dot\sigma_X(t)\frac{\partial}{\partial u}p_{X(t)}(u) - \dot\mu_X(t)p_{X(t)}(u) + \left(\frac{\mu_F(t)}{c} - \frac{k}{c}u\right)p_{X(t)}(u) - \frac{\pi S_0(t)}{c^2}\frac{\partial}{\partial u}p_{X(t)}(u)\right] = 0$$

This requires that

$$\frac{-1}{\sigma_X(t)}\left(\frac{u - \mu_X(t)}{\sigma_X(t)}\right)\left(\sigma_X(t)\dot\sigma_X(t) - \frac{\pi S_0(t)}{c^2}\right) - \dot\mu_X(t) + \frac{\mu_F(t)}{c} - \frac{k}{c}u = 0$$

This expression includes only terms that are linear in u and terms that are independent of u. Obviously both must be zero, in order that the equation be satisfied for all u values. Thus, we have

$$\frac{-1}{\sigma_X^2(t)}\left(\sigma_X(t)\dot\sigma_X(t) - \frac{\pi S_0(t)}{c^2}\right) = \frac{k}{c}$$

and

$$\frac{\mu_X(t)}{\sigma_X^2(t)}\left(\sigma_X(t)\dot\sigma_X(t) - \frac{\pi S_0(t)}{c^2}\right) - \dot\mu_X(t) = -\frac{\mu_F(t)}{c}$$

Combining these two equations gives

$$\frac{d}{dt}\mu_X(t) + \frac{k}{c}\mu_X(t) = \frac{\mu_F(t)}{c}$$

as the ordinary differential equation governing the evolution of the mean value function in the Gaussian probability density function. Rearranging the first of the two equations gives

$$\frac{d}{dt}\sigma_X^2(t) + 2\frac{k}{c}\sigma_X^2(t) = \frac{2\pi S_0(t)}{c^2}$$

as the corresponding equation for the variance. Note that the Gaussian assumption used here includes the implicit assumption that the given initial condition on $p_{X(t)}(u)$ can be described by a Gaussian probability density function (possibly with zero initial variance, as in Example 8.12). The Fokker-Planck equation will be the same for other initial conditions as well, but the solution will be different than the one given here.
****

Example 8.14: Give the complete Fokker-Planck equation for the 2DF
building model shown, when the excitation is a base acceleration $\{\ddot{Z}(t)\}$
which is a nonstationary shot noise. The terms k_1 and k_2 represent the total
stiffness of all columns in the stories, and $X_1(t)$ and $X_2(t)$ are motions
relative to the base of the structure.

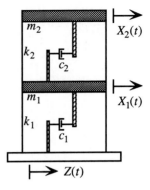

The equation of motion is $\dot{\vec{Y}}(t) + \mathbf{A}\vec{Y}(t) = \vec{Q}(t)$ with $n_Y = 4$, and

$$\vec{Y}(t) = \begin{bmatrix} X_1(t) \\ X_2(t) \\ \dot{X}_1(t) \\ \dot{X}_2(t) \end{bmatrix}, \quad \mathbf{A} = \begin{bmatrix} 0 & 0 & -1 & 0 \\ 0 & 0 & 0 & -1 \\ \dfrac{k_1+k_2}{m_1} & \dfrac{-k_2}{m_1} & \dfrac{c_1+c_2}{m_1} & \dfrac{-c_2}{m_1} \\ \dfrac{-k_2}{m_2} & \dfrac{k_2}{m_2} & \dfrac{-c_2}{m_2} & \dfrac{c_2}{m_2} \end{bmatrix}, \quad \vec{Q}(t) = \begin{bmatrix} 0 \\ 0 \\ -\ddot{Z}(t) \\ -\ddot{Z}(t) \end{bmatrix}$$

The first-order coefficients in the Fokker-Planck equation are found from Eq. 8.69, which can be written as

$$C_j^{(1)} = \mu_{Q_j}(t) - \sum_{l=1}^{4} A_{j,l} u_l$$

giving: $C_1^{(1)} = -u_3$, $C_2^{(1)} = -u_4$,

$$C_3^{(1)} = -\mu_{\ddot{Z}}(t) + \frac{k_1+k_2}{m_1} u_1 - \frac{k_2}{m_1} u_2 + \frac{c_1+c_2}{m_1} u_3 - \frac{c_2}{m_1} u_4$$

and

$$C_4^{(1)} = -\mu_{\ddot{Z}}(t) + \frac{k_2}{m_2} u_1 - \frac{k_2}{m_2} u_2 + \frac{c_2}{m_2} u_3 - \frac{c_2}{m_2} u_4$$

which can be further simplified by noting that $\mu_{\ddot{Z}}(t) = \mu_F \dot{\mu}_N(t)$ for the nonstationary shot noise, in which F
denotes a typical pulse of the shot noise, and $\dot{\mu}_N(t)$ is the expected arrival rate for the pulses (see Example
3.6). Since the excitation is a non-Gaussian delta-correlated process, the non-zero higher-order coefficients
are found from Eq. 8.72 with $S_{j_1,j_2,j_3,j_4}(t)$ being the $(j_3 + j_4)$-order autospectral density of $\{-\ddot{Z}(t)\}$ if
$j_1 = j_2 = 0$, and being zero otherwise. In Example 5.7 we found that the Jth-order autospectral density of
the shot noise is $E(F^J)\dot{\mu}_N(t) / (2\pi)^{J-1}$. Thus, Eq. 8.72 gives

$$\begin{aligned} C^{(j_1,j_2,j_3,j_4)} &= E[(-F)^{j_3+j_4}]\dot{\mu}_N(t) \quad \text{for} \quad j_1 = j_2 = 0 \\ &= 0 \qquad\qquad\qquad\qquad \text{otherwise.} \end{aligned}$$

Using the simplified notation of $p(\vec{u})$ for $p_{X_1(t),X_2(t),\dot{X}_1(t),\dot{X}_2(t)}(u_1,u_2,u_3,u_4)$, the Fokker-Planck relationship
from Eq. 8.64 is

$$\frac{\partial}{\partial t} p(\vec{u}) = -\frac{\partial}{\partial u_1}[-u_3 p(\vec{u})] - \frac{\partial}{\partial u_2}[-u_4 p(\vec{u})]$$

$$-\frac{\partial}{\partial u_3}\left[\left(-\mu_F \dot{\mu}_N(t) + \frac{k_1+k_2}{m_1} u_1 - \frac{k_2}{m_1} u_2 + \frac{c_1+c_2}{m_1} u_3 - \frac{c_2}{m_1} u_4\right) p(\vec{u})\right]$$

$$-\frac{\partial}{\partial u_4}\left[\left(-\mu_F \dot{\mu}_N(t) + \frac{k_2}{m_2} u_1 - \frac{k_2}{m_2} u_2 + \frac{c_2}{m_2} u_3 - \frac{c_2}{m_2} u_4\right) p(\vec{u})\right]$$

$$+E(F^2)\dot{\mu}_N(t)\left[\frac{1}{2}\frac{\partial^2}{\partial u_3^2} p(\vec{u}) + \frac{\partial^2}{\partial u_3 \partial u_4} p(\vec{u}) + \frac{1}{2}\frac{\partial^2}{\partial u_4^2} p(\vec{u})\right]$$

$$+E(F^3)\dot{\mu}_N(t)\left[\frac{1}{6}\frac{\partial^3}{\partial u_3^3} p(\vec{u}) + \frac{1}{2}\frac{\partial^3}{\partial u_3^2 \partial u_4} p(\vec{u}) + \frac{1}{2}\frac{\partial^3}{\partial u_3 \partial u_4^2} p(\vec{u}) + \frac{1}{6}\frac{\partial^3}{\partial u_4^3} p(\vec{u})\right]+\cdots$$

or

$$\frac{\partial}{\partial t} p(\vec{u}) = u_3 \frac{\partial}{\partial u_1} p(\vec{u}) + u_4 \frac{\partial}{\partial u_2} p(\vec{u}) - \frac{c_1+c_2}{m_1} p(\vec{u}) + \frac{c_2}{m_2} p(\vec{u})$$

$$-\left(-\mu_F \dot{\mu}_N(t) + \frac{k_1+k_2}{m_1} u_1 - \frac{k_2}{m_1} u_2 + \frac{c_1+c_2}{m_1} u_3 - \frac{c_2}{m_1} u_4\right)\frac{\partial}{\partial u_3} p(\vec{u})$$

$$-\left(-\mu_F \dot{\mu}_N(t) + \frac{k_2}{m_2} u_1 - \frac{k_2}{m_2} u_2 + \frac{c_2}{m_2} u_3 - \frac{c_2}{m_2} u_4\right)\frac{\partial}{\partial u_4} p(\vec{u})$$

$$+\dot{\mu}_N(t)\sum_{J=2}^{\infty} E(F^J)\sum_{l=0}^{J} \frac{J!}{(J-l)!l!} \frac{\partial^2}{\partial u_3^l \partial u_4^{J-l}} p(\vec{u})$$

****

Example 8.15: For the system of Example 8.13, but with $\mu_F(t) = 0$, derive the state-space equation for the mean squared value of $X(t)$ by using the Fokker-Planck equation.

Setting $\mu_F(t) = 0$ in the Fokker-Planck equation we found in Example 8.13 gives

$$\frac{\partial}{\partial t} p_{X(t)}(u) = -\frac{\partial}{\partial u}\left[-\frac{k}{c} u\, p_{X(t)}(u)\right] + \frac{\pi S_0(t)}{c^2} \frac{\partial^2}{\partial u^2} p_{X(t)}(u)$$

Multiplying this equation by u^2 and integrating over all possible u values gives

$$\int_{-\infty}^{\infty} u^2 \frac{\partial}{\partial t} p_{X(t)}(u)\, du = -\int_{-\infty}^{\infty} u^2 \frac{\partial}{\partial u}\left[-\frac{k}{c} u\, p_{X(t)}(u)\right] du + \frac{\pi S_0(t)}{c^2} \int_{-\infty}^{\infty} u^2 \frac{\partial^2}{\partial u^2} p_{X(t)}(u)\, du$$

The left-hand side of the equation is easily simplified by reversing the order of integration and differentiation, to give an integral which is exactly the derivative with respect to time of $E[X^2(t)]$. The other two integrals can be evaluated by integrating by parts — once in the first integral on the right-hand side of the equation, and twice in the other term. This gives

$$\frac{d}{dt}E[X^2(t)] = -2\frac{k}{c}\int_{-\infty}^{\infty}u^2 p_{X(t)}(u)\,du + 2\frac{\pi S_0(t)}{c^2}\int_{-\infty}^{\infty} p_{X(t)}(u)\,du$$

which can be rewritten as

$$\frac{d}{dt}E[X^2(t)] + 2\frac{k}{c}E[X^2(t)] = \frac{2\pi S_0(t)}{c^2}$$

Note that this is identical in form to the variance equation obtained in Example 8.3 directly from the equation of motion.

****

It should be noted that this introduction to Fokker-Planck analysis is quite elementary in nature and omits many advanced topics. More detailed analyses, such as those given by Soong and Grigoriu (1993) and Lin and Cai (1995), particularly emphasize its relationship to so-called "Itô calculus." This form of analysis is particularly important for the study of dynamics problems in which the parameters of the system vary stochastically (Ibrahim 1985), in which case one must be quite careful in interpreting the meaning of a differential equation. If the parametric excitation is white noise, for example, the two common interpretations of the differential of the response process (Itô and Stratonovich) differ by a quantity called the Wong-Zakai correction term. These issues do not arise for the problems considered here since we consider the stochastic excitation always to be applied externally, which is also referred to as an additive excitation (in distinction to the parametric problem which is considered to have an internal or multiplicative excitation). For the external excitation there is only one interpretation of the stochastic differential and the simplified procedures presented here are applicable.

Exercises
****
First and Second Moments and Covariance
****

8.1 Consider a building subjected to a wind force $\{F(t)\}$. The building is modeled as a linear SDF system:

$$m\ddot{X}(t) + c\dot{X}(t) + kX(t) = F(t)$$

with $m = 200,000\,\text{kg}$, $c = 8\,\text{kN}\cdot\text{s}/\text{m}$, and $k = 3,200\,\text{kN}/\text{m}$. The force $\{F(t)\}$ is modeled as a stationary process having a mean value of $\mu_F = 20$ kN and an autospectral density of $S_{FF}(\omega) = 0.5$ $(\text{kN})^2/(\text{rad}/\text{s})$ for all ω.

Solve the appropriate state-space equations in order to find:

(a) The stationary mean value of the displacement response, μ_X.

(b) The stationary variance of the displacement response, σ_X^2.

****

8.2 Perform second moment state-space analysis of the system shown. The excitation is a mean zero, stationary white noise with autospectral density of $S_{FF}(\omega) = S_0$ for all ω.

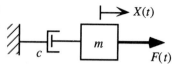

(a) Using $n_Y = 2$ and $\vec{Y}(t) = [X(t), \dot{X}(t)]^T$, solve the state-space equation to find the stationary variance of velocity, $\sigma_{\dot{X}}^2$.
(b) Show that the equation of part (a) fails to give a stationary solution for the variance of displacement, σ_X^2.
(c) Show that one can also solve for $\sigma_{\dot{X}}^2$ by using the state-space equation with $n_Y = 1$ and $\vec{Y}(t) = \dot{X}(t)$.

****

8.3 Consider a linear system governed by the following third-order differential equation:
$$\dddot{X}(t) + a\ddot{X}(t) + b\dot{X}(t) + cX(t) = F(t)$$
with $a > 0$, $b > 0$, $c > 0$ and $ab > c$, where each overdot denotes a derivative with respect to time. Let $\{F(t)\}$ be a stationary white noise process with $E[F(t)] = 0$, and $E[F(t)F(s)] = 2\pi S_0 \delta(t-s)$.
(a) Using a state vector of $\vec{Y}(t) = [X(t), \dot{X}(t), \ddot{X}(t)]^T$, formulate the state-space moment equations for the second moments of the response.
(b) Find the stationary values of $E(X^2)$, $E(\dot{X}^2)$, and $E(\ddot{X}^2)$ by solving the state-space moment equations.

****

8.4 Consider a linear system governed by the following fourth-order differential equation:
$$\ddddot{X}(t) + a_3\dddot{X}(t) + a_2\ddot{X}(t) + a_1\dot{X}(t) + a_0X(t) = F(t)$$
with $a_0 > 0$, $a_1 > 0$, $a_2 > 0$, $a_3 > 0$, and $a_1 a_2 a_3 > a_1^2 + a_0 a_3^2$, where each overdot denotes a derivative with respect to time. Let $\{F(t)\}$ be a stationary white noise process with $E[F(t)] = 0$, and $E[F(t)F(s)] = 2\pi S_0 \delta(t-s)$.
(a) Using a state vector of $\vec{Y}(t) = [X(t), \dot{X}(t), \ddot{X}(t), \dddot{X}(t)]^T$, formulate the state-space moment equations for the second moments of the response.
(b) Find the stationary values of $E(X^2)$, $E(\dot{X}^2)$, $E(\ddot{X}^2)$, and $E(\dddot{X}^2)$ by solving the state-space moment equations.

****

8.5 Consider the state-space cumulant equations derived in Example 8.2 for the third-order system with $m\ddot{X}_1(t) + c_1\dot{X}_1(t) + k_1X_1(t) + k_2X_2(t) = F(t)$, and $k_2X_2(t) = c_2[\dot{X}_1(t) - \dot{X}_2(t)]$
Let the excitation be a mean zero stationary white noise with autocovariance function $K_{FF}(t+\tau,t) = 2\pi S_0(t)\delta(\tau)$.
(a) Evaluate the $\vec{\psi}(t)$ vector defined in Example 8.2.
(b) Solve the stationary state-space equations to find expressions for all the second moment quantities included in the vector $\vec{V}(t)$ of Example 8.2.

****

8.6 For the structure and excitation of Exercise 8.1, solve the appropriate state-space equations in order to find:

(a) The conditional mean of $\{X(t)\}$ given an initial condition of $X(0) = 10\,\text{mm}$.

(b) The conditional variance of $\{X(t)\}$ given an initial condition of $X(0) = 10\,\text{mm}$.

****

8.7 Consider a building subjected to an earthquake ground motion $\{Z(t)\}$. The building is modeled as a linear SDF system:

$$m\ddot{X}(t) + c\dot{X}(t) + kX(t) = -m\ddot{Z}(t)$$

with $m = 200{,}000\,\text{kg}$, $c = 32\,\text{kN}\cdot\text{s}/\text{m}$, and $k = 3{,}200\,\text{kN}/\text{m}$. The ground acceleration $\{\ddot{Z}(t)\}$ is modeled as a nonstationary, mean zero, delta-correlated process having an autocovariance function of

$$K_{\ddot{Z}\ddot{Z}}(t+\tau,t) = 0.04\,\text{m}^2/(\text{rad}\cdot\text{s}^3)(e^{-0.2t} - e^{-0.25t})U(t)\delta(\tau)$$

(a) Use state-space cumulant equations to find the response standard deviations $\sigma_X(t)$ and $\sigma_{\dot{X}}(t)$.

(b) Compare the exact value of $\sigma_X(t)$ from part (a) with the "pseudo-stationary" result of $[\pi S_0(t)m^2/(kc)]^{1/2}$, in which $S_0(t)$ is defined by $K_{FF}(t+\tau,t) = 2\pi S_0(t)\delta(\tau)$ (see Example 8.5).

****

Energy Balance
****

8.8 Consider the system of Exercise 8.2.

(a) Use the variance result to find the rate of energy dissipation in the system.

(b) Confirm that this rate of energy dissipation equals the rate of energy addition calculated from Eq. 8.25.

****

8.9 Let the excitation processes $\{F_1(t)\}$ and $\{F_2(t)\}$ in Example 8.6 both be mean zero.

(a) Use the variance and covariance results to find the rate of energy dissipation in the system.

(b) Confirm that this rate of energy dissipation equals the rate of energy addition calculated from Eq. 8.25.

****

8.10 You wish to approximate the response levels at the top of a multi-story building. Past experience shows that this response is usually dominated by the fundamental mode, so you want to develop an equivalent SDF model based only on that mode. For your structure, the fundamental mode has a natural frequency ω_0 rad/sec, damping ζ, and an approximately linear mode shape:

$$X(t) = X_T(t)\frac{y}{L}$$

in which X is the displacement at height y, L is the height of the building, and X_T is the top displacement. Assume that the total mass M is uniformly distributed over the height (M/L per meter).

(a) Find an expression for the kinetic energy (KE) in the structure in terms of the top velocity $\dot{X}_T$.

(b) Assume that an earthquake gives the base of the building a mean zero white noise acceleration with an autospectral density of S_0. Find the expected value of the rate at which energy is added to the structure: $E(PA\ )$.

(c) Let PD denote the rate at which energy is dissipated by the structure, and assume that in stationary response $E(PD) = 4\zeta\omega_0 E(KE)$ (as for a SDF system). Using the results of (a) and (b), find $E(\dot{X}_T^2)$ and $E(X_T^2)$ for stationary response.
****

Kronecker Formulation
****

8.11 Write out all four component equations of expression 8.34 for the second moments of the response of a SDF oscillator, using the **A** matrix given in Example 8.4. Verify that these expressions are identical to those from Eq. 8.9.
****

8.12 Write out all nine component equations of expression 8.34 for second moments using the **A** matrix for the third-order system of Example 8.2 and Exercise 8.5. Verify that these expressions are identical to those from Eq. 8.9.
****

Fokker-Planck equation
****

8.13 Let $\{X(t)\}$ be the response of a SDF oscillator with an excitation which is a mean zero, Gaussian, stationary white noise with an autospectral density value of S_0.
(a) Give the complete Fokker-Planck equation for the system.
(b) Verify that this equation is satisfied by a stationary Gaussian probability density function with second moments given by Eq. 3.58 of Chapter 3.
****

8.14 Let $\{X(t)\}$ be the response of the third-order system of Exercise 8.3 with an excitation which is a mean zero, Gaussian, stationary white noise with an autospectral density value of S_0.
(a) Give the complete Fokker-Planck equation for the system.
(b) Verify that this equation is satisfied by a stationary Gaussian probability density function with the second moment values found in Exercise 8.3.
****

8.15 Give the complete Fokker-Planck equation for the fourth-order system of Exercise 8.4 with a stationary Gaussian white noise excitation which has a mean value of μ_F.
****

8.16 Give the complete Fokker-Planck equation for the third-order system of Example 8.2 and Exercises 8.5 and 8.12 in the situation in which the stationary, mean zero, white noise excitation is also Gaussian.
****

Chapter 9

Introduction to Nonlinear Stochastic Vibration

9.1 Approaches to the Problem

Exact analytical solutions have been found for only relatively few problems of nonlinear stochastic dynamics. Thus, much of the analysis of such problems relies on various approximate techniques. The existing analytical solutions are important, though, in at least two ways. Most obvious is the situation in which some important nonlinear system can be adequately approximated by a different nonlinear model for which an analytical solution is known, so that the solution can be used directly. Systems with exact analytical solutions can also be used in validating the assumptions of some approximate technique prior to its application to another problem for which no analytical solution is known.

In this brief introduction to nonlinear stochastic vibration, we will focus on two aspects of the problem. We will review some exact analytical state-space formulations of nonlinear problems, along with the corresponding solutions for some significant special cases. In addition, we will present the most commonly used category of approximate methods, called equivalent linearization. Even though it involves the most difficult equations of any of our methods, Fokker-Planck analysis will be presented first. The reason for this seemingly odd choice is that Fokker-Planck analysis provides exact analytical solutions to certain problems. These relatively limited exact results can then be used as a basis for judging the value and the shortcomings of simpler approximate solutions.[1] Before presenting the details of either exact or approximate solutions, though, we will discuss some basic ideas of nonlinear analysis.

Probably the greatest obstacle to analysis of nonlinear problems is the fact that superposition is generally not applicable. Since almost all of our linear analysis techniques are based on the concept of finding a general solution by superimposing particular solutions, it is necessary that we find alternate formulations for nonlinear problems. The time domain and frequency domain analysis methods of Chapters 3 and 4 are particularly limited for nonlinear problems because they both invoke the concept of superposition at the very outset of the analysis procedure. For example, the Duhamel convolution integral of Chapter 3 can be viewed as giving the $X(t)$ response at time t as a superposition of the responses at that time due to $F(u)$ excitation contributions at all earlier times. Similarly, the fundamental idea in Chapter 4 is that for any given frequency ω the Fourier transform $\tilde{X}(\omega)$ of the response is the same as the response to the Fourier transform $\tilde{F}(\omega)$ of the excitation, and this result is valid only if superposition applies.

---

[1] There is no difficulty in proceeding directly from Section 9.1 to Section 9.3, in case the reader wishes to bypass the Fokker-Planck analysis of nonlinear systems. This is certainly recommended for anyone who has not become familiar with the material in Section 8.8.

The derivation of state-space equations as introduced in Chapter 8, on the other hand, does not depend on superposition or any other property associated with linearity. Thus, one can also derive these equations for nonlinear systems. In this regard, though, it is important to distinguish between the derivation and the solution of the state-space equations. For example, the state-space moment and cumulant equations that we derived in Chapter 8 were linear, so we used superposition, in the form of convolution integrals, in writing their solutions. For nonlinear systems, though, we will find that the state-space moment and cumulant equations are also nonlinear. Thus, the form of the solutions of these equations will be different than in Chapter 8, even though the method of derivation is unchanged. The situation is somewhat different for the Fokker-Planck equation, since it has variable coefficients which may be nonlinear functions of the state variables, even for a linear system. Nonetheless, the increased complexity associated with nonlinear analysis still applies. For example, a nonlinear system with a Gaussian excitation has a non-Gaussian response, and the simple Fokker-Planck solutions demonstrated in Chapter 8 were only for Gaussian processes.

The most commonly used approach for seeking approximate answers to nonlinear dynamic problems is to somehow replace the nonlinear system with a linear system. We will consider the term "equivalent linearization" to apply to all such methods, regardless of the technique used to choose the form or parameters of the linear substitute system. "Statistical linearization" is the most commonly used term to describe techniques for choosing the linearized system on the basis of minimum mean squared error, but the term can also be applied to minimization of some other probabilistic measure of discrepancy between the nonlinear and linear systems.

The primary reason for using equivalent linearization techniques, of course, is simplicity. The contents of this book up until this point are an example of the extensive literature which exists on the stochastic dynamics of linear systems. If a nonlinear system can be replaced by a linear one, then all the techniques for linear systems can be applied to give approximate predictions regarding the response of the nonlinear problem. Interpretation, as well as prediction, though, can be a basis for considering linearization. Even if we have in some way determined the true response levels for a nonlinear system, we may find it useful to seek a linear system with similar response behavior, so that we can interpret our results based on our experience with linear dynamics.

It should be emphasized that the concept of "equivalence" between any linear and nonlinear systems is always limited in scope. For example, we can sometimes find a linear system which matches certain mean and variance values of a nonlinear system, but there are always other statistics which are not the same for the two systems. Probably the simplest and most significant difference between linear and nonlinear models has to do with the Gaussian probability distribution. As previously noted, any linear system with a Gaussian excitation also has a Gaussian response. This allows the prediction of various response probability values based only on knowledge of the mean and variance of the response process. For a nonlinear system this relationship no longer holds. That is, a nonlinear system with a Gaussian excitation has a non-Gaussian response. Thus, even if an equivalent linearization scheme gives us good approximations of the mean and variance of the response of a nonlinear system, this will generally not allow us to compute accurately the probability of some rare event. The techniques of Chapter 5 can be used to obtain improved estimates of such probabilities, but this requires knowledge of more than mean and variance. Thus, one must always be careful to remember the limitations of any linearization and not to make erroneous conclusions based on inappropriate analogies with linear response. In spite of these limitations, though, equivalent linearization can often give very useful approximations of the response of a nonlinear system.

The state-space and equivalent linearization methods presented here are probably the most commonly used techniques for nonlinear stochastic analysis. Nonetheless, it should be kept in mind that there are a number of other approaches, and they continue to be developed. For example, perturbation provides a quite straightforward and general approach for a system that deviates only slightly from linear behavior (for example, see Lin, 1967; Nigam, 1983). We have chosen to emphasize linearization rather than perturbation, though, for several reasons. These include the following facts: first-order perturbation usually agrees with linearization, it is difficult to find higher-order perturbation terms, perturbation sequences do not necessarily converge for finite nonlinearities, and interpretation is generally simpler for linearization than for perturbation models.

Stochastic averaging is another important technique which is not considered here. Although not a fundamentally different approach to the problem, this technique is quite useful inasmuch as it reduces the order of a problem by using time averages of certain slowly varying quantities (Lin and Cai, 1995; Roberts and Spanos, 1986). Mention was also made at the beginning of this section of the possibility of seeking an equivalence between a given nonlinear system and a substitute nonlinear system for which the state-space solution is known. This method has been used with some success, even when implemented with relatively crude concepts of equivalence (for example, by Chen and Lutes, 1994), and some work has been done on techniques for improving the equivalent nonlinearization procedure by using mean square error minimization (Caughey, 1986; Cai and Lin, 1988; Zhu and Yu, 1989; To and Li, 1991). The usefulness of the method is limited in practice, though, by the relatively small number of models for which exact solutions are known [see Lin and Cai (1995) for a summary]. Another approach that has received considerable attention in recent years extends the time and frequency domain integrals of Chapters 3 and 4 by using the Volterra series (Schetzen, 1980; Rugh, 1981). Truncating this series has given useful approximate results for some problems, particularly those involving polynomial nonlinearities (Choi et al., 1985; Naess and Ness, 1992; Winterstein et al., 1994). The approach has also been combined with equivalent nonlinearization to give a technique of first replacing the problem of interest with one having a polynomial nonlinearity, then using the approximate solution for the substitute system (Donley and Spanos, 1990; Kareem et al., 1995; Li et al., 1995). Further developments can certainly be expected in the future.

9.2   Fokker-Planck Equation

The general Fokker-Planck equation for the probability density of any vector process $\{\vec{Y}(t)\}$ was given in Eq. 8.64 as

$$\frac{\partial}{\partial t} p_{\vec{Y}(t)}(\vec{u}) = \underbrace{\sum_{j_1=0}^{\infty} \cdots \sum_{j_{n_Y}=0}^{\infty}}_{(\text{except } j_1=\cdots=j_{n_Y}=0)} \frac{(-1)^{j_1+\cdots+j_{n_Y}}}{j_1!\cdots j_{n_Y}!} \frac{\partial^{j_1+\cdots+j_{n_Y}}}{\partial u_1^{j_1}\cdots\partial u_{n_Y}^{j_{n_Y}}} [C^{(j_1,\cdots,j_{n_Y})}(\vec{u},t) p_{\vec{Y}(t)}(\vec{u})] \qquad (9.1)$$

This equation applies equally well to the response of a nonlinear system and a linear system. Stated in another way, the equation applies to the probability density function of the vector process quite independently of whether that process is related to any dynamics problem. Thus, we may also find the equation useful in analyzing nonlinear vibration problems if we can determine the $C^{(j_1,\cdots,j_{n_Y})}(\vec{u},t)$ coefficients for the situation in which $\vec{Y}(t)$ is the vector of state variables for a nonlinear system.

We will now consider a very general formulation of nonlinear dynamics problems with a state vector $\vec{Y}(t)$ described by

$$\dot{\vec{Y}}(t) + \vec{g}[\vec{Y}(t)] = \vec{Q}(t) \qquad (9.2)$$

Note that this equation is identical to Eq. 8.5, except that the linear restoring force $\mathbf{A}\vec{Y}(t)$ has now been replaced by the nonlinear vector function $\vec{g}[\vec{Y}(t)]$. In particular, each component of $\vec{g}[\vec{Y}(t)]$ is a scalar nonlinear function of the state variables: $g_j[\vec{Y}(t)] \equiv g_j[Y_1(t),\cdots,Y_{n_Y}(t)]$ for $j = 1, \cdots, n_Y$.

Using the same procedure as in Section 8.8, we can say that

$$\Delta Y_j = \int_t^{t+\Delta t} \dot{Y}_j(s)\,ds = \int_t^{t+\Delta t} Q_j(s)\,ds - \int_t^{t+\Delta t} g_j[\vec{Y}(s)]\,ds$$

so that the vector of first-order coefficients, corresponding to Eq. 8.66, is

$$\vec{C}^{(1)}(u,t) = \lim_{\Delta t \to 0} \frac{1}{\Delta t} E[\Delta \vec{Y} \mid \vec{Y}(t) = \vec{u}] = E[\vec{Q}(t) \mid \vec{Y}(t) = \vec{u}] - \vec{g}(\vec{u})$$

Provided that $E\big(Q_j(s_1)g_l[Y(s_2)]\mid\vec{Y}(t) = \vec{u}\big)$ is finite for all choices of j and l, we can say that the coefficients based on higher-order moments of the increments of $\{\vec{Y}(t)\}$ must depend only on terms like the integral of $Q_j(s)$ appearing on the right-hand side of Eq. 9.2. Furthermore, these integrals are identical to those for a linear system, so that these higher-order coefficients are identical to those given by Eq. 8.67 for the linear system.

If we now restrict the excitation process $\{\vec{Q}(t)\}$ to be delta-correlated, so that $\vec{Q}(t)$ is independent of $\vec{Q}(s)$ for $s \neq t$, we can say that $\{\vec{Y}(t)\}$ has the Markov property and the Fokker-Planck coefficients are given by

$$\vec{C}^{(1)}(\vec{u},t) = \vec{\mu}_Q(t) - \vec{g}(\vec{u}) \tag{9.3}$$

and

$$C^{(j_1,\cdots,j_{n_Y})}(\vec{u},t) \equiv \lim_{\Delta t \to 0} \frac{1}{\Delta t} E[(\Delta Y_1)^{j_1}\cdots(\Delta Y_{n_Y})^{j_{n_Y}}|\vec{Y}(t) = \vec{u}]$$

$$= (2\pi)^{J-1} S_{j_1,\cdots,j_{n_Y}}(t) \tag{9.4}$$

in which $J = j_1 + \cdots + j_{n_Y}$ and $S_{j_1,\cdots,j_{n_Y}}(t)$ is a nonstationary spectral density of order J for the delta-correlated process . Clearly Eq. 9.3 is the obvious nonlinear generalization of Eq. 8.69, and Eq. 9.4 is actually identical to Eq. 8.72. These two equations then give the coefficients in the Fokker-Planck relationship of Eq. 9.1 for a nonlinear system with a delta-correlated excitation. In the special case in which the excitation is also Gaussian we know that the higher-order cumulants of the excitation are zero so that $C^{(j_1,\cdots,j_{n_Y})}(\vec{u},t) = 0$ for $J > 2$ and

$$C^{(\overbrace{0,\cdots,0}^{j-1},1,0,\cdots 0,1,\overbrace{0,\cdots,0}^{n_Y-l})}(\vec{u},t) = 2\pi[\mathbf{S}_0(t)]_{j,l} \qquad \text{for } j \neq l \tag{9.5}$$

and

$$C^{(\overbrace{0,\cdots,0}^{j-1},2,\overbrace{0,\cdots,0}^{n_Y-j})}(\vec{u},t) = 2\pi[\mathbf{S}_0(t)]_{j,j} \tag{9.6}$$

in which $\mathbf{S}_0(t)$ is defined by the covariance relationship $\mathbf{K}_{QQ}(t,s) = 2\pi\mathbf{S}_0(t)\delta(t-s)$.

We will now turn our attention to certain special cases for which solutions of Eq. 9.1 have been found for nonlinear systems. We will start with the simplest situation of a system with a first-order differential equation of motion:

$$\dot{X}(t) + f'[X(t)] = F(t) \tag{9.7}$$

in which $f'(u)$ is the derivative of a nonlinear function $f(u)$ satisfying $f(u) \to \infty$ as $|u| \to \infty$, and the excitation $\{F(t)\}$ is a mean zero, stationary Gaussian white noise with autospectral density S_0. This equation could describe the behavior of a linear dashpot and a nonlinear spring attached in parallel, or a mass attached to a nonlinear dashpot, although the latter model would require that $\{X(t)\}$ represent the velocity of the system.

From Eqs. 9.3 and 9.6 we have the first-order and second-order coefficients in the Fokker-Planck equation for the system of Eq. 9.7 as

$$C^{(1)}(u,t) = -g(u) = -f'(u)$$

and

$$C^{(2)}(u,t) = 2\pi\mathbf{S}_0(t)$$

Substituting these expressions into the Fokker-Planck equation gives

$$\frac{\partial}{\partial t} p_{X(t)}(u) = -\frac{\partial}{\partial u}[-f'(u)\,p_{X(t)}(u)] + \pi S_0 \frac{\partial^2}{\partial u^2} p_{X(t)}(u) = \frac{\partial}{\partial u}\left[f'(u)\,p_{X(t)}(u) + \pi S_0 \frac{\partial}{\partial u} p_{X(t)}(u)\right]$$

It is easy to verify that a stationary solution of this equation is

$$p_{X(t)}(u) = B \exp\!\left(\frac{-f(u)}{\pi S_0}\right) \tag{9.8}$$

so that

$$\frac{\partial}{\partial u} p_{X(t)}(u) = -\frac{f'(u)}{\pi S_0} p_{X(t)}(u)$$

and the term in the brackets of the Fokker-Planck equation is identically zero, giving

$$\frac{\partial}{\partial t} p_{X(t)}(u) = 0$$

as it must be for a stationary process. The definition of $p_{X(t)}(u)$ as given in Eq. 9.8 is completed by evaluating the constant B from

$$B^{-1} = \int_{-\infty}^{\infty} \exp\!\left(\frac{-f(u)}{\pi S_0}\right) du \tag{9.9}$$

****

Example 9.1: Let $\{X(t)\}$ denote the response of the system with the equation of motion

$$\dot{X}(t) + k_1 X(t) + k_3 X^3(t) = F(t)$$

with $k_1 \geq 0$, $k_3 \geq 0$, and $\{F(t)\}$ being a mean zero, stationary Gaussian white noise with autospectral density S_0. Find the probability distribution and the variance of the stationary $\{X(t)\}$ response.

Using $f'(u) = k_1 u + k_3 u^3$ gives $f(u) = k_1 u^2/2 + k_3 u^4/4$ so that Eqs. 9.8 and 9.9 give

$$p_{X(t)}(u) = \frac{\exp\!\left[\dfrac{-1}{\pi S_0}\left(\dfrac{k_1 u^2}{2} + \dfrac{k_3 u^4}{4}\right)\right]}{\displaystyle\int_{-\infty}^{\infty} \exp\!\left[\dfrac{-1}{\pi S_0}\left(\dfrac{k_1 v^2}{2} + \dfrac{k_3 v^4}{4}\right)\right] dv}$$

It is not convenient to find a closed form expression for the integral in the denominator of this expression, or of the integral of $u^2\, p_{X(t)}(u)$, which is needed in the evaluation of the variance σ_X^2. Thus, numerical integration will be used to provide sample results. First, though, it is convenient to introduce a normalizing constant of $\sigma_0^2 = \pi S_0 / k_1$, which is the variance of $X(t)$ for the limiting linear case of $k_3 = 0$. Using a change of variables of $u = w\sigma_0$ then gives the probability density function as

$$p_{X(t)}(w\sigma_0) = \frac{\exp\left(-\dfrac{w^2}{2} - \alpha\dfrac{w^4}{4}\right)}{\displaystyle\int_{-\infty}^{\infty}\exp\left(-\dfrac{v^2}{2} - \alpha\dfrac{v^4}{4}\right)dv}$$

in which $\alpha = (k_3\sigma_0^2 / k_1)$. A sketch is given showing how the variance of $X(t)$ depends on the dimensionless parameter α. A sketch is also given showing a common measure of the non-Gaussianity of the response of the nonlinear system. In particular, we have numerically evaluated the fourth moment $E(X^4)$ so that we can plot $kurtosis = E(X^4) / \sigma_X^4$. Note that the fourth moment and kurtosis have been used to illustrate non-Gaussianity since the third moment and skewness are zero due to the symmetry of the problem. The kurtosis, of course, is 3.0 for a Gaussian process, and this limit is reached at $k_3 = 0$.

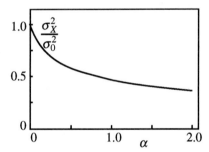
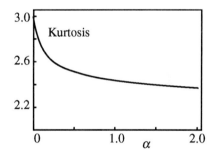

Next we show the probability density function for $X(t)$ for two particular values of the α nonlinearity parameter. In each case a Gaussian probability density function with the same variance as the non-Gaussian one is shown for comparison.

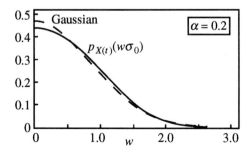
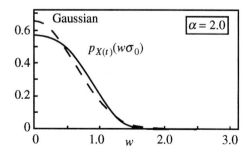

As previously noted, the limiting case of $k_3 = 0$ gives a Gaussian response with variance σ_0^2. For the other limiting case of $k_1 = 0$ it is also possible to evaluate the integrals in closed form. In particular, the integrals of interest can all be written in terms of gamma functions (see Example A.31 of Appendix A) as

$$\int_0^\infty \exp(-au^4)\,du = \frac{\Gamma(0.25)}{4a^{0.25}} = 0.9064\,a^{-0.25}$$

$$\int_0^\infty u^2 \exp(-au^4)\,du = \frac{\Gamma(0.75)}{4a^{0.75}} = 0.3064\,a^{-0.75}$$

and

$$\int_0^\infty u^4 \exp(-au^4)\,du = \frac{\Gamma(1.25)}{4a^{1.25}} = \frac{\Gamma(0.25)}{16a^{1.25}} = 0.2266\,a^{-1.25}$$

Substitution of these values gives $\sigma_X^2 = 0.6760(\pi S_0 / k_3)^{1/2}$ and *kurtosis* = 2.188.

****

Example 9.2: Let $\{X(t)\}$ denote the response of the system with the equation of motion

$$\dot{X}(t) + k\,|X(t)|^b\,\mathrm{sgn}[X(t)] = F(t)$$

with $\{F(t)\}$ being a mean zero, stationary Gaussian white noise with autospectral density S_0. Find the probability distribution and the variance of the stationary $\{X(t)\}$ response.

Using $f'(u) = k|u|^b\,\mathrm{sgn}(u) = k|u|^{(b-1)}u$ gives $f(u) = k|u|^{b+1}/(b+1)$ so that Eqs. 9.8 and 9.9 give

$$p_{X(t)}(u) = B\exp\left[\frac{-k|u|^{b+1}}{(b+1)\pi S_0}\right]$$

Using the change of variables $w = ku^{n+1}/[\pi S_0(n+1)]$ allows B to be evaluated as follows:

$$B^{-1} = 2\int_0^\infty \exp\left[\frac{-ku^{b+1}}{(b+1)\pi S_0}\right]du = 2\int_0^\infty e^{-w}\left[\frac{w^{-b/(b+1)}}{b+1}\left(\frac{(b+1)\pi S_0}{k}\right)^{1/(b+1)}\right]dw$$

and

$$B = \left(\frac{k}{(b+1)\pi S_0}\right)^{1/(b+1)}\left(\frac{b+1}{2}\right)\left[\Gamma\left(\frac{1}{b+1}\right)\right]^{-1}$$

The variance of $\{X(t)\}$ is then given by

$$\sigma_X^2 = E(X^2) = 2B\int_0^\infty u^2 \exp\left[\frac{-ku^{b+1}}{(b+1)\pi S_0}\right]du = 2B\int_0^\infty e^{-w}\left[\frac{w^{(j-b)/(b+1)}}{b+1}\left(\frac{(b+1)\pi S_0}{k}\right)^{(j+1)/(b+1)}\right]dw$$

or

$$\sigma_X^2 = \left(\frac{\pi S_0}{k}\right)^{2/(b+1)}(b+1)^{2/(b+1)}\Gamma\left(\frac{3}{b+1}\right)\left[\Gamma\left(\frac{1}{b+1}\right)\right]^{-1}$$

****

One of the more general second-order differential equations for which a solution is known involves a stochastic process $\{X(t)\}$ with an equation of motion of

$$m\ddot{X}(t) + f_2'\big(H[X(t), \dot{X}(t)]\big)\dot{X}(t) + f_1'[X(t)] = F(t) \tag{9.10}$$

in which $f_1'(\cdot)$ and $f_2'(\cdot)$ are the derivatives of nonlinear functions satisfying $uf_1'(u) \geq 0$ for all u, $f_1(u) \to \infty$ for $|u| \to \infty$, $f_2'(u) \geq 0$ for all $u \geq 0$, $f_2(u) \to \infty$ for $u \to \infty$, and H is a non-negative term which represents the energy in the system. Note that the $f_2'\big(H[X(t), \dot{X}(t)]\big)\dot{X}(t)$ term of Eq. 9.10 is a nonlinear damping term, since it corresponds to a force which opposes the direction of the velocity of the mass. Similarly, $f_1'[X(t)]$ is a nonlinear spring term, giving a restoring force which opposes the displacement of the mass from equilibrium.

The term $f_1(u)$ is exactly the potential energy in the nonlinear system, since it is the integral of the nonlinear restoring force. Thus, the total energy in the system is

$$H[X(t), \dot{X}(t)] = f_1[X(t)] + \frac{m}{2}\dot{X}^2(t)$$

Using $\vec{Y}(t) = [X(t), \dot{X}(t)]^T$ as the state vector gives the terms in Eq. 9.2 as

$$\vec{g}[\vec{Y}(t)] = \begin{bmatrix} -\dot{X}(t) \\ f_1'[X(t)]/m + f_2'\big(H[X(t), \dot{X}(t)]\big)\dot{X}(t)/m \end{bmatrix}$$

and $\vec{Q}(t) = [0, F(t)]^T$. We will restrict the excitation $\{F(t)\}$ to be mean zero, stationary, Gaussian white noise, so that it is delta-correlated and the first-order coefficients in the Fokker-Planck equation are

$$\vec{C}^{(1)}(\vec{u}, t) = -\vec{g}(\vec{u}) = \begin{bmatrix} u_2 \\ -f_1'(u_1)/m - f_2'[H(u_1, u_2)]u_2/m \end{bmatrix} \tag{9.11}$$

Note that $\mathbf{K}_{QQ}(t, s) = 2\pi \mathbf{S}_0 \delta(t - s)$ gives only one non-zero element in the $\mathbf{S}_0$ matrix; that element is the autospectral density of $\{F(t)/m\}$, which we will denote by the scalar S_0/m^2. This gives the only non-zero higher-order coefficient in the Fokker-Planck equation as

$$C^{(0,2)}(\vec{u}, t) = 2\pi[\mathbf{S}_0]_{2,2} = 2\pi S_0/m^2 \tag{9.12}$$

Substituting Eqs. 9.11 and 9.12 into Eq. 9.1 gives the Fokker-Planck equation for the system of Eq. 9.10 as

$$\frac{\partial}{\partial t} p_{\vec{Y}(t)}(\vec{u}) = -\frac{\partial}{\partial u_1}[u_2\, p_{\vec{Y}(t)}(\vec{u})] - \frac{1}{m}\frac{\partial}{\partial u_2}\left[\left(-f_1'(u_1) - f_2'\,[H(u_1,u_2)]u_2\right) p_{\vec{Y}(t)}(\vec{u})\right]$$

$$+ \frac{\pi S_0}{m^2}\frac{\partial^2}{\partial u_2^2} p_{\vec{Y}(t)}(\vec{u}) \tag{9.13}$$

No exact nonstationary solution of this equation has been found, but Caughey (1964) demonstrated that it has a stationary solution of

$$p_{\vec{Y}(t)}(\vec{u}) = B \exp\left[\frac{-1}{\pi S_0} f_2\,[H(u_1,u_2)]\right] \tag{9.14}$$

in which

$$B^{-1} = \int_{-\infty}^{\infty}\int_{-\infty}^{\infty} \exp\left[\frac{-1}{\pi S_0} f_2\,[H(u_1,u_2)]\right] du_1\, du_2$$

so that the integral of $p_{\vec{Y}(t)}(\vec{u})$ is unity.

In order to verify that Eq. 9.14 satisfies the Fokker-Planck equation, we can note that

$$\frac{\partial^2}{\partial u_2^2} p_{\vec{Y}(t)}(\vec{u}) = \frac{\partial}{\partial u_2}\left(p_{\vec{Y}(t)}(\vec{u})\left[\frac{-1}{\pi S_0} f_2'[H(u_1,u_2)]\frac{\partial\, H(u_1,u_2)}{\partial u_2}\right]\right)$$

$$= \frac{\partial}{\partial u_2}\left(p_{\vec{Y}(t)}(\vec{u})\left[\frac{-mu_2}{\pi S_0} f_2'[H(u_1,u_2)]\right]\right)$$

and substituting this into Eq. 9.13 gives

$$\frac{\partial}{\partial t} p_{\vec{Y}(t)}(\vec{u}) = -\frac{\partial}{\partial u_1}[u_2\, p_{\vec{Y}(t)}(\vec{u})] + \frac{1}{m}\frac{\partial}{\partial u_2}[f_1'(u_1) p_{\vec{Y}(t)}(\vec{u})]$$

Evaluation of the two derivatives on the right-hand side of this expression then gives

$$\frac{\partial}{\partial t} p_{\vec{Y}(t)}(\vec{u}) = 0$$

confirming that Eq. 9.14 is a stationary solution of Eq. 9.13.

Note that the $f_1'[X(t)]$ nonlinear spring term in Eq. 9.10 has a very natural form, and describes many cases which may be of practical interest, several of which had been discovered prior to Caughey's presentation of the general solution. Simple examples include odd polynomials, such as the cubic function in the common Duffing oscillator. The nonlinear damping term of $f_2'\left(H[X(t),\dot{X}(t)]\right)\dot{X}(t)$, on the other hand, seems rather unnatural. Clearly the form was chosen to give a Fokker-Planck equation which could be solved, rather than on the basis of direct modeling of common physical problems. The model does serve the purpose, though, of giving exact solutions for a class of examples having nonlinear damping. Furthermore, a model with the damping coefficient f_2' varying as a function of the energy in the system may be equally as reasonable as either of the more obvious choices of having the damping force (and also the damping coefficient) depend only on the $\dot{X}(t)$ value or having the damping coefficient depend on $|X(t)|$. In many physical problems we have so little information about the true nature of the energy dissipation that we would have difficulty in determining which of these models would be more appropriate. For narrowband motion with a given dominant frequency, we could use the model of Eq. 9.10 or either of the other two models suggested to represent any reasonable nonlinear relationship between the energy loss per cycle and the amplitude of the motion.

An important special case of Eq. 9.10 is when $f_2'[H(u_1, u_2)]$ is a constant, which we will denote by c. In this situation Eq. 9.14 gives

$$p_{\vec{Y}(t)}(\vec{u}) = B \exp\left[\frac{-c\,H(u_1, u_2)}{\pi S_0}\right] = A \exp\left[\frac{-c\,f_1(u_1)}{\pi S_0}\right] \exp\left[\frac{-mc\,u_2^2}{2\pi S_0}\right] \tag{9.15}$$

The fact that this stationary probability density can be factored into the product of a function of u_1 and a function of u_2 shows that $X(t)$ and $\dot{X}(t)$ are independent random variables, for any given t value. Furthermore, the form of the dependence on u_2, which is the dummy variable for $Y_2(t) \equiv \dot{X}(t)$, shows that $\dot{X}(t)$ is a Gaussian random variable with a variance of

$$\sigma_{\dot{X}}^2(t) = \frac{\pi S_0}{mc} \tag{9.16}$$

One can easily verify that this value is identical to that obtained in Chapters 3 and 4 for the response of a linear SDF system to white noise excitation. Of course this result could have been predicted, since Eq. 9.10 with $f_2'[H(u_1, u_2)] = c$ includes the linear SDF system as the special case when $f_1'(u_1) = ku_1$. Nonetheless, it seems somewhat surprising that the probability distribution and the variance of $\dot{X}(t)$ are not affected by the presence of nonlinearity in the spring of the SDF oscillator.

It should be noted that $\{\dot{X}(t)\}$ is not a Gaussian process even though $\dot{X}(t)$ is a Gaussian random variable for every value of t for the system with $f_2'[H(u_1, u_2)] = c$. This is easily proved as follows. If $\{\dot{X}(t)\}$ were a Gaussian process, then its integral $\{X(t)\}$ would also be a Gaussian process, as shown in Chapter 5. However, Eq. 9.15 shows that $p_{X(t)}(u)$ has the form

$$p_{X(t)}(u) = \hat{B} \exp\left[\frac{-c\,f_1(u_1)}{\pi S_0}\right] \tag{9.17}$$

with $\hat{B} = B\sqrt{2\pi}\sigma_{\dot{X}}$, and this clearly is not a Gaussian distribution unless $f_1'(u_1) = ku_1$. Thus,

$\{X(t)\}$ is not a Gaussian process, and this proves that $\{\dot{X}(t)\}$ is not a Gaussian process. Even though $\dot{X}(t)$ is a Gaussian random variable for every value of t in the situation with linear damping, it appears that random variables such as $\dot{X}(t)$ and $\dot{X}(s)$ are not jointly Gaussian, as is required for a Gaussian process.

****

Example 9.3: Let $\{X(t)\}$ denote the response of the nonlinear oscillator with the equation of motion

$$m\ddot{X}(t) + c\dot{X}(t) + k_1 X(t) + k_3 X^3(t) = F(t)$$

with $k_1 \geq 0$, $k_3 \geq 0$, and $\{F(t)\}$ being a mean zero, stationary Gaussian white noise with autospectral density S_0. Find the probability distribution and the variance of the stationary $\{X(t)\}$ and $\{\dot{X}(t)\}$ responses.

Using $f_2'(u) = c$ and $f_1'(u) = k_1 u + k_3 u^3$ gives

$$H(u_1, u_2) = \frac{k_1 u_1^2}{2} + \frac{k_3 u_1^4}{4} + \frac{m u_2^2}{2}$$

and

$$p_{\vec{Y}(t)}(\vec{u}) = B \exp\left[\frac{-c}{\pi S_0}\left(\frac{k_1 u_1^2}{2} + \frac{k_3 u_1^4}{4} + \frac{m u_2^2}{2} \right) \right]$$

Factoring this expression, as in Eqs. 9.15 and 9.17, gives $\dot{X}(t)$ as being Gaussian with variance $\sigma_{\dot{X}}^2 = \pi S_0 / (cm)$, and $X(t)$ as having a non-Gaussian probability density function of

$$p_{X(t)}(u_1) = B\sqrt{2\pi}\,\sigma_{\dot{X}} \exp\left[\frac{-c}{\pi S_0}\left(\frac{k_1 u_1^2}{2} + \frac{k_3 u_1^4}{4} \right) \right]$$

One may note that this probability distribution has the same form as that of $X(t)$ in Example 9.1. The only difference in the exponent is the presence of the c term in the present problem. The multiplier in front of the exponential, of course, is simply the value which assures that the integral of the probability density function is unity. If we let σ_0^2 denote the response variance for the limiting linear case of $k_3 = 0$:

$$\sigma_0^2 = \frac{\pi S_0}{c k_1}$$

then the change of variables of $u = w\sigma_0$ gives the probability density function as

$$p_{X(t)}(w\sigma_0) = \frac{\exp\left(-\dfrac{w^2}{2} - \alpha \dfrac{w^4}{4} \right)}{\displaystyle\int_{-\infty}^{\infty} \exp\left(-\dfrac{v^2}{2} - \alpha \dfrac{v^4}{4} \right) dv}$$

with $\alpha = (k_3 \sigma_0^2 / k_1)$. This expression is identical to that found in Example 9.1. Thus, the plots of probability density function, variance and kurtosis shown in Example 9.1 also apply to the $X(t)$ response in this oscillator problem.

****

Example 9.4: Let $\{X(t)\}$ denote the response of the nonlinear oscillator with the equation of motion

$$m\ddot{X}(t) + c_1[\dot{X}^3(t) + (k/m)X^2(t)\dot{X}(t)] + kX(t) = F(t)$$

with$\{F(t)\}$ being a mean zero, stationary Gaussian white noise with autospectral density S_0. Find the probability distribution and the variance of the stationary $\{X(t)\}$ and $\{\dot{X}(t)\}$ responses.

Since this system has a linear spring the energy term is simply $H[X(t),\dot{X}(t)] = [m\dot{X}^2(t) + kX^2(t)]/2$. Thus, the equation of motion agrees with Eq. 9.10 if we choose $f_1'(u) = ku$ and $f_2'(u) = 2c_1 u / m$. This gives $f_2(u) = c_1 u^2 / m$ so that Eq. 9.14 becomes

$$p_{X(t),\dot{X}(t)}(u_1, u_2) = B\exp\left(\frac{-c_1(mu_2^2 + ku_1^2)^2}{4m\pi S_0}\right)$$

This joint probability density function does not factor into a function of u_1 multiplied by a function of u_2, which is an indication that $X(t)$ and $\dot{X}(t)$ are not independent random variables for this problem. The most direct way to find the marginal probability density functions for $X(t)$ and $\dot{X}(t)$ is to integrate over all u_1 and u_2 values, respectively. For example

$$p_{X(t)}(u_1) = B\int_{-\infty}^{\infty} \exp\left(\frac{-c_1(mu_2^2 + ku_1^2)^2}{4m\pi S_0}\right) du_2$$

with B evaluated from the fact that the integral of $p_{X(t)}(u_1)$ with respect to u_1 must be unity. Further integration then gives the variance values for $X(t)$ as

$$\sigma_X^2 = B\int_{-\infty}^{\infty} u_1^2 \, p_{X(t)}(u_1) \, du_1 = B\int_{-\infty}^{\infty}\int_{-\infty}^{\infty} u_1^2 \exp\left(\frac{-c_1(mu_2^2 + ku_1^2)^2}{4m\pi S_0}\right) du_2 \, du_1$$

However, one can make some conclusions about the relative magnitudes and the probability distributions of $X(t)$ and $\dot{X}(t)$ without any integration. In particular, if a normalized version of $\dot{X}(t)$ is introduced as $Z(t) = \dot{X}(t)/\omega_0$ with $\omega_0 = (k/m)^{1/2}$, then we have

$$p_{X(t),Z(t)}(u_1, u_2) = B\omega_0 \exp\left(\frac{-c_1 k^2(u_2^2 + u_1^2)^2}{4m\pi S_0}\right)$$

The symmetry of this relationship shows that $p_{X(t)}(u) = p_{Z(t)}(u) = \omega_0 p_{\dot{X}(t)}(\omega_0 u)$. Thus, the probability distribution of $Z(t) = \dot{X}(t) / \omega_0$ is identical to that of $X(t)$. Among other things, this shows that $\sigma_{\dot{X}}^2 = \omega_0^2 \sigma_X^2$. It also shows that the kurtosis of $\dot{X}(t)$ is the same as that of $X(t)$.

For this particular problem one can obtain exact results for the value of B, the variance of the response, etc. by converting the double integral with respect to u_1 and u_2 into polar coordinates. Specifically, if we let $u_1 = r\cos\theta$ and $u_2 = r\sin\theta$ then

$$B^{-1} = \omega_0 \int_{-\infty}^{\infty}\int_{-\infty}^{\infty} \exp\left(\frac{-c_1 k^2 (u_2^2 + u_1^2)^2}{4m\pi S_0}\right) du_1\, du_2 = \omega_0 \int_0^{2\pi}\int_0^{\infty} \exp\left(\frac{-c_1 k^2 r^4}{4m\pi S_0}\right) r\, dr_1\, d\theta$$

which can be evaluated as

$$B^{-1} = 2\pi\,\omega_0 \int_0^{\infty} \exp\left(\frac{-c_1 k^2 r^4}{4m\pi S_0}\right) r\, dr = 2\pi\,\omega_0 \frac{\Gamma(1/2)}{4}\sqrt{\frac{4m\pi S_0}{c_1 k^2}} = \pi^2 \sqrt{\frac{S_0}{kc_1}}$$

since $\Gamma(1/2) = \pi^{1/2}$. Similarly

$$\sigma_X^2 = B\omega_0 \int_{-\infty}^{\infty}\int_{-\infty}^{\infty} u_1^2 \exp\left(\frac{-c_1 k^2 (u_2^2 + u_1^2)^2}{4m\pi S_0}\right) du_1\, du_2 = \omega_0 \int_0^{2\pi}\int_0^{\infty} \exp\left(\frac{-c_1 k^2 r^4}{4m\pi S_0}\right) r^3 \cos^2(\theta)\, dr_1\, d\theta$$

or

$$\sigma_X^2 = \pi\,B\omega_0 \int_0^{\infty} \exp\left(\frac{-c_1 k^2 r^4}{4m\pi S_0}\right) r^3\, dr = \pi\,B\omega_0 \frac{m\pi S_0}{c_1 k^2} = \frac{1}{k}\sqrt{\frac{mS_0}{c_1}}$$

From the fact that $\sigma_{\dot{X}}^2 = \omega_0^2 \sigma_X^2$, we can also write

$$\sigma_{\dot{X}}^2 = \sqrt{\frac{S_0}{mc_1}}$$

The sketch shows numerical values for the marginal probability density function of $X(t)$, since there is not a simple analytical solution for this function. A Gaussian distribution is also shown for comparison.

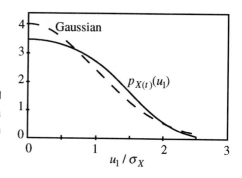

It should be kept in mind that this discussion of Fokker-Planck analysis of nonlinear systems is very introductory in nature, and is included largely for the purpose of presenting exact solutions to a few nonlinear problems. Stochastic averaging, as mentioned in Section 8.8, is often used to reduce the order of problems and other special techniques have been developed for investigating the stability and bifurcation of solutions. A summary of some of these results, including references to experimental and numerical simulation studies, has been given by Ibrahim (1995).

9.3   Statistical Linearization

The simplest linearization situation involves finding the coefficients in a linear function such that it approximates a given nonlinear function $g[X(t)]$. The usual approach for doing this in a stochastic problem involves minimizing the mean squared error. Specifically, we will write the linear function as $a_0 + a_1[X(t) - \mu_X(t)]$, and choose a_0 and a_1 to minimize $E([\hat{E}^2(t)])$ in which the error $\hat{E}(t)$ is defined as

$$\hat{E}(t) = a_0 + a_1[X(t) - \mu_X(t)] - g[X(t)]$$

Setting the derivatives with respect to a_0 and a_1 equal to zero gives the two equations

$$0 = \frac{d}{da_0}E([\hat{E}^2(t)]) = 2E\left(\hat{E}(t)\frac{d\hat{E}(t)}{da_0}\right) = 2E[\hat{E}(t)] = 2[a_0 - E(g[X(t)])]$$

and

$$0 = \frac{d}{da_1}E([\hat{E}^2(t)]) = 2E\left(\hat{E}(t)\frac{d\hat{E}(t)}{da_1}\right) = 2E\left(\hat{E}(t)[X(t) - \mu_X(t)]\right) = 2\left[a_1\sigma_X^2(t) - K_{X(t),g[X(t)]}\right]$$

The general solution is[2]

$$a_0 = E(g[X(t)]) \tag{9.18}$$

and

$$a_1 = \frac{K_{X(t),g[X(t)]}}{\sigma_X^2(t)} \tag{9.19}$$

If $\{X(t)\}$ is mean zero, which can always be achieved by proper definition of the process, then Eq. 9.19 can be rewritten as

$$a_1 = \frac{E(X(t)g[X(t)])}{E[X^2(t)]} \qquad \text{for} \quad \mu_X(t) = 0 \tag{9.20}$$

If one knows the probability distribution of $X(t)$ then it is possible, at least in principle, to solve Eq. 9.18 and either Eq. 9.19 or 9.20 to find the optimal values of a_0 and a_1. One difficulty of using this procedure in a dynamics problem, though, is the fact that one may not know the necessary probability distribution of $X(t)$. For example, if $\{X(t)\}$ is the response of a nonlinear dynamic system, in which $g[X(t)]$ is a nonlinear restoring force, then the probability distribution of $X(t)$ will be unknown. In this situation it is necessary to make some approximation. Another limitation of the procedure regards its accuracy. Even if one can exactly satisfy Eqs. 9.18 and 9.19, that does not assure that the linear function is a good approximation of $g[X(t)]$. The fitting procedure has only minimized the second moment of the $\hat{E}(t)$ error term, and has not even considered other measures of error. In spite of these facts, it is often possible to get very useful results from statistical linearization.

---

[2] Determining a_0 and a_1 is precisely a problem in classical linear regression, which is discussed briefly in Section A.8 of Appendix A.

The Gaussian distribution is most commonly used to approximate an unknown probability distribution of $\{X(t)\}$. One justification for this approach is based on the common assumption that many naturally occurring excitations are approximately Gaussian, along with the consequences of linearization. Specifically, if the excitation of a nonlinear system is Gaussian, then the dynamic response of a linearized substitute system is also Gaussian; this is used as the approximation of the distribution of $\{X(t)\}$. Stated another way, if the excitation is Gaussian and the system nonlinearity is not very severe, then linearization should work well and the response of the nonlinear system is expected to be approximately Gaussian. In most applications of linearization, though, it is not feasible to demonstrate that the Gaussian assumption is valid. Fortunately, the choice of the linearization parameters a_0 and a_1 only requires estimates of certain first and second moment terms, as shown in Eqs. 9.18 and 9.19. It is quite possible for the Gaussian approximation to give acceptable estimates of these moments even when it does not accurately match the tails of the distribution of $\{X(t)\}$.

Now we will generalize the statistical linearization procedure to include the case of a nonlinear function $g[\vec{X}(t)]$ of a number of random variables arranged as the components of the vector process $\{\vec{X}(t)\}$. In this case we will write the linear approximation as

$$g[\vec{X}(t)] \approx a_0 + \sum_{j=1}^{n} a_j [X_j(t) - \mu_{X_j}(t)] \tag{9.21}$$

and choose the values of the coefficients a_0, a_1, $\cdots$, a_n so as to minimize the mean squared value of the error written as

$$\hat{E}(t) = a_0 + \sum_{j=1}^{n} a_j [X_j(t) - \mu_{X_j}(t)] - g[\vec{X}(t)]$$

The result can be written as

$$a_0 = E\big(g[\vec{X}(t)]\big) \tag{9.22}$$

and

$$\vec{a} = \mathbf{K}_{XX}^{-1}(t,t) Cov\big(\vec{X}(t), g[\vec{X}(t)]\big) = \mathbf{K}_{XX}^{-1}(t,t) E\big([\vec{X}(t) - \vec{\mu}_X(t)] g[\vec{X}(t)]\big) \tag{9.23}$$

in which $\vec{a} = [a_1, \cdots . a_n]^T$ and $Cov\big(\vec{X}(t), g[\vec{X}(t)]\big)$ is a vector containing the covariance of $g[\vec{X}(t)]$ with each component of $\vec{X}(t)$. These equations, of course, include Eqs. 9.18 and 9.19 as the special case with $n = 1$.

When it is presumed that $\{\vec{X}(t)\}$ has the Gaussian distribution, it is possible to rewrite Eqs. 9.22 and 9.23 in an alternate form which is sometimes more convenient to use. In particular, we will use the jointly Gaussian distribution to write

$$E\big([\vec{X}(t) - \vec{\mu}_X(t)]g[\vec{X}(t)]\big) = \int_{-\infty}^{\infty} \cdots \int_{-\infty}^{\infty} [\vec{u} - \vec{\mu}_X(t)]g(\vec{u})$$

$$\frac{\exp\left(-\frac{1}{2}[\vec{u} - \vec{\mu}_X(t)]^T \mathbf{K}_{XX}^{-1}(t,t)[\vec{u} - \vec{\mu}_X(t)]\right)}{(2\pi)^{n/2}|\mathbf{K}_{XX}(t,t)|^{1/2}} du_1 \cdots du_n$$

The jth component of $\vec{a}$ can then be written as

$$a_j = \sum_{k=1}^{n} [\mathbf{K}_{XX}^{-1}(t,t)]_{j,k} E\big([X_k(t) - \mu_{X_k}(t)]g[\vec{X}(t)]\big)$$

or

$$a_j = \sum_{k=1}^{n} [\mathbf{K}_{XX}^{-1}(t,t)]_{j,k} \int_{-\infty}^{\infty} \cdots \int_{-\infty}^{\infty} [u_k - \mu_{X_k}(t)]g(\vec{u})$$

$$\frac{\exp\left(-\frac{1}{2}[\vec{u} - \vec{\mu}_X(t)]^T \mathbf{K}_{XX}^{-1}(t,t)[\vec{u} - \vec{\mu}_X(t)]\right)}{(2\pi)^{n/2}|\mathbf{K}_{XX}(t,t)|^{1/2}} du_1 \cdots du_n$$

Noting that

$$\sum_{k=1}^{n} [\mathbf{K}_{XX}^{-1}(t,t)]_{j,k} [u_k - \mu_{X_k}(t)] \exp\left(-\frac{1}{2}[\vec{u} - \vec{\mu}_X(t)]^T \mathbf{K}_{XX}^{-1}(t,t)[\vec{u} - \vec{\mu}_X(t)]\right)$$

$$= -\frac{\partial}{\partial u_j} \exp\left(-\frac{1}{2}[\vec{u} - \vec{\mu}_X(t)]^T \mathbf{K}_{XX}^{-1}(t,t)[\vec{u} - \vec{\mu}_X(t)]\right)$$

now allows us to integrate by parts with respect to u_j, giving

$$a_j = \int_{-\infty}^{\infty} \cdots \int_{-\infty}^{\infty} \frac{\partial g(\vec{u})}{\partial u_j} \frac{\exp\left(-\frac{1}{2}[\vec{u} - \vec{\mu}_X(t)]^T \mathbf{K}_{XX}^{-1}(t,t)[\vec{u} - \vec{\mu}_X(t)]\right)}{(2\pi)^{n/2}|\mathbf{K}_{XX}(t,t)|^{1/2}} du_1 \cdots du_n$$

which is simply[3]

$$a_j = E\left(\frac{\partial g[\vec{X}(t)]}{\partial X_j(t)}\right) \tag{9.24}$$

For $n \geq 2$ the use of Eq. 9.24 will usually be somewhat simpler than Eq. 9.23. Each formulation requires the evaluation of n expectations of nonlinear functions of $\vec{X}(t)$. However, the

---

[3] This Gaussian linearization result was apparently first introduced by Kazakov in 1965, but it is often associated with the names of Atalik and Utku, who first applied it to the analysis of stochastic vibration in 1976.

expectations in Eq. 9.23 directly give the components in $\mathbf{K}_{XX}(t,t)\vec{a}$, so determination of $\vec{a}$ generally requires the solution of simultaneous equations. On the other hand, the expectations in Eq. 9.24 directly give the components of $\vec{a}$. Equation 9.24, of course, also applies for the special case of $n = 1$. In that case, though, it is not necessarily any easier to use Eq. 9.24 than Eq. 9.19 for finding the value of a_1.

9.4   Linearization of Dynamics Problems

We will now demonstrate the use of statistical linearization for some relatively simple dynamics problems which contain either scalar or vector nonlinear terms $g[X(t)]$, $g[\vec{X}(t)]$, or $\vec{g}[\vec{X}(t)]$, in which $\{X(t)\}$ or $\{\vec{X}(t)\}$ denotes the response process of the system. The basic approach in each situation is quite straightforward. Each nonlinear term is replaced by its linear approximation, according to the formulas in Section 9.3. This gives the linearization parameters as functions of various expectations which depend, in turn, on the statistics of the response process. The response of the linearized system is then found from the methods of linear analysis, using time domain integration, frequency domain integration, or state-space methods, according to the preference of the analyst. This step gives the response parameters as functions of the linearization parameters. Thus, we generally now have simultaneous equations, some giving the linearization parameters as functions of the response levels, and others giving the response levels as functions of the linearization parameters. The final step is to solve these equations to find the response levels as functions of the excitation and the parameters of the nonlinear system. In most situations this solution can only be found by iteration. Using assumed values of the linearization parameters gives initial estimates of the response levels and using the estimated response levels gives improved estimates of the linearization parameters. In some situations, though, it is possible to solve the simultaneous equations by analytical methods. We will illustrate a few of these situations, and then we will give an example in which the iterative approach is used. Finally, we will outline the application of the method to general multi-degree-of-freedom problems with nonstationary response, although we will not devote the space necessary to work such advanced examples.[4]

We will begin with the nonlinear first-order system described by

$$\dot{X}(t) + g[X(t)] = F(t) \tag{9.25}$$

which is identical to Eq. 9.7 with the nonlinear term rewritten as $g(\cdot)$ instead of $f'(\cdot)$. For simplicity we will take $g(\cdot)$ to be an odd function and the $\{F(t)\}$ excitation process to be a mean zero, stationary, Gaussian white noise with an autocorrelation function of $R_{FF}(\tau) = 2\pi S_0 \delta(\tau)$.

Based on the antisymmetry of $g[X(t)]$ we can see that $\mu_X = 0$ and $E(g[X(t)]) = 0$. Thus, using Eqs. 9.18 and 9.20 for the linearization of $g[X(t)]$ gives Eq. 9.25 as being replaced by

---

[4]   Much more extensive investigations of statistical linearization are given by Roberts and Spanos (1990).

$$\dot{X}(t) + a_1 X(t) = F(t) \tag{9.26}$$

with the coefficient given by Eq. 9.20. Thus, Eq. 9.20 gives the dependence of the linearization parameter a_1 on the response levels of the process. That is, if the probability distribution of $X(t)$ were known, then Eq. 9.20 would give us the value of a_1. The converse relationship giving the dependence of the statistics of $X(t)$ on a_1 is found by linear stochastic analysis of Eq. 9.26. In particular, one can use the methods of Chapters 3, 4 or 8 to show that the stationary solution of Eq. 9.26 has

$$E[X^2(t)] = \frac{\pi S_0}{a_1} \tag{9.27}$$

Eliminating a_1 between Eqs. 9.20 and 9.27 gives

$$E\big(X(t)g[X(t)]\big) = \pi S_0 \tag{9.28}$$

as a necessary condition of linearization. One can then use this equation to find $\sigma_X^2 \equiv E[X^2(t)]$, provided that one can relate $E\big(X(t)g[X(t)]\big)$ to σ_X^2 for the given $g(\cdot)$ function. As noted in the preceding section, this is usually done by assuming that $X(t)$ is Gaussian. The assumption that $X(t)$ is Gaussian also allows us the alternative of using Eq. 9.24 in place of Eq. 9.20 for relating a_1 to the response values. This gives

$$E\big(g'[X(t)]\big) = \frac{\pi S_0}{E[X^2(t)]} \tag{9.29}$$

as being equivalent to Eq. 9.28. In this form it appears that Eq. 9.28 is somewhat simpler than Eq. 9.29.

****

Example 9.5: Let $\{X(t)\}$ denote the response of the system of Example 9.1 with an equation of motion of

$$\dot{X}(t) + k_1 X(t) + k_3 X^3(t) = F(t)$$

in which $\{F(t)\}$ is a mean zero, stationary, Gaussian, white noise with autospectral density S_0. Using statistical linearization, find the value of the parameter a_1 of Eqs. 9.26 and 9.27, and estimate the value of σ_X^2 for stationary response.

Using the approximation that $X(t)$ is Gaussian and mean zero gives $E\big(g[X(t)]\big) = 0$ and

$$E\big(X(t)g[X(t)]\big) = k_1 E[X^2(t)] + k_3 E[X^4(t)] = k_1 \sigma_X^2 + 3k_3 \sigma_X^4$$

so that Eq. 9.18 confirms that $a_0 = 0$ and Eq. 9.20 gives

$$a_1 = \frac{E\big(X(t)g[X(t)]\big)}{E[X^2(t)]} = k_1 + 3k_3 \sigma_X^2$$

Equivalently, we could have obtained this latter result by using Eq. 9.24 to write

$$a_1 = E\left(\frac{\partial g[X(t)]}{\partial X(t)}\right) = E[k_1 + 3k_3 X^2(t)] = k_1 + 3k_3 \sigma_X^2$$

Either Eq. 9.28 or 9.29 shows that the result of using this expression in conjunction with Eq. 9.27 gives

$$k_1 \sigma_X^2 + 3k_3 \sigma_X^4 = \pi S_0$$

which has a solution of

$$\sigma_X^2 = \frac{k_1}{6k_3}\left(\sqrt{1 + \frac{12 k_3 \pi S_0}{k_1^2}} - 1\right)$$

Using the notation of $\sigma_0^2 = \pi S_0 / k_1$ for the variance of the system in the limiting case of $k_3 = 0$, as was done in Example 9.1, this can be rewritten as

$$\frac{\sigma_X^2}{\sigma_0^2} = \frac{k_1}{6k_3 \sigma_0^2}\left(\sqrt{1 + \frac{12 k_3 \sigma_0^2}{k_1}} - 1\right)$$

The sketch compares this result with the exact solution obtained in Example 9.1, using $\alpha = (k_3 \sigma_0^2 / k_1)$. It is obvious that statistical linearization gives a very good approximation of the response variance for this nonlinear problem.

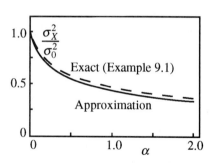

Recall that it was shown in Example 9.1 that the kurtosis of $X(t)$ varies from 3.0 down to less than 2.4 for the range of k_3 values shown in the sketch. Thus, this example shows a situation in which statistical linearization gives a good variance estimate even though the response is significantly non-Gaussian. It must be kept in mind, though, that this statistical linearization procedure does not give us any estimate of the non-Gaussianity of the response, and using a Gaussian approximation for $X(t)$ could give very poor estimates of the probability of occurrence of rare events. For example, letting $\alpha = 2.0$ in the exact solution of Example 9.1 gives $P[X(t) > 3\sigma_X(t)] = 3.21 \times 10^{-5}$ whereas a Gaussian distribution gives this probability as 1.35×10^{-3}, or almost two orders of magnitude larger than the exact value. Statistical linearization is surprisingly accurate in predicting the variance, but caution is necessary in using the linearization results for other purposes.

****

Example 9.6: Let $\{X(t)\}$ denote the response of the system of Example 9.2 with the equation of motion

$$\dot{X}(t) + k|X(t)|^b \text{sgn}[X(t)] = F(t)$$

with $\{F(t)\}$ being a mean zero, stationary Gaussian white noise with autospectral density S_0. Using statistical linearization, find the value of the parameter a_1 of Eqs. 9.26 and 9.27, and estimate the value of σ_X^2 for stationary response.

Using the approximation that $X(t)$ is Gaussian and mean zero gives $E\big(g[X(t)]\big) = 0$ and

$$E(X(t)g[X(t)]) = kE\big(|X(t)|^{b+1}\big) = 2k\int_0^\infty \frac{u^{b+1}\exp[-u^2/(2\sigma_X^2)]}{\sqrt{2\pi}\,\sigma_X}\,du = 2k\frac{\sigma_X}{\sqrt{2\pi}}\int_0^\infty (2\sigma_X^2 w)^{b/2}e^{-w}\,dw$$

so that Eq. 9.18 confirms that $a_0 = 0$ and Eq. 9.20 gives

$$a_1 = \frac{E(X(t)g[X(t)])}{E[X^2(t)]} = k\sigma_X^{b-1}\frac{2^{(b+2)/2}}{\sqrt{2\pi}}\Gamma\left(\frac{b}{2}+1\right) = kb\sigma_X^{b-1}\frac{2^{b/2}}{\sqrt{2\pi}}\Gamma\left(\frac{b}{2}\right)$$

Equivalently, we could have obtained this latter result by using Eq. 9.24 to write

$$a_1 = E\left(\frac{\partial\,g[X(t)]}{\partial\,X(t)}\right) = kbE[|X(t)|^{b-1}] = 2kb\int_0^\infty \frac{u^{b-1}\exp[-u^2/(2\sigma_X^2)]}{\sqrt{2\pi}\,\sigma_X}\,du = kb\sigma_X^{n-1}\frac{2^{b/2}}{\sqrt{2\pi}}\Gamma\left(\frac{b}{2}\right)$$

Either Eq. 9.28 or 9.29 shows that the result of using this expression in conjunction with Eq. 9.27 gives

$$kb\frac{2^{b/2}}{\sqrt{2\pi}}\Gamma\left(\frac{b}{2}\right)\sigma_X^{b+1} = \pi S_0$$

which has a solution of

$$\sigma_X^2 = \left(\frac{\pi S_0}{k}\right)^{2/(b+1)}\left(\frac{\sqrt{2\pi}}{2^{b/2}b\Gamma(b/2)}\right)^{2/(b+1)}$$

The sketch compares this result with the exact solution obtained in Example 9.2 by giving values of the ratio

$$R = \frac{\left(\sigma_X^2\right)_{equiv.\,lin.}}{\left(\sigma_X^2\right)_{exact}}$$

For this system it is clear that statistical linearization gives a good approximation of the response variance only when b is relatively near unity.

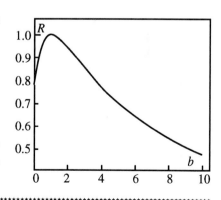

* *

Next we will consider the linearization of an oscillator with a nonlinear restoring force, which may depend on both $X(t)$ and $\dot{X}(t)$:

$$m\ddot{X}(t) + f[X(t), \dot{X}(t)] = F(t) \tag{9.30}$$

The nonlinear function is then linearized according to Eq. 9.21 to give

$$m\ddot{X}(t) + a_2\dot{X}(t) + a_1X(t) + a_0 = F(t) \tag{9.31}$$

with a_0 being found from Eq. 9.22 as

$$a_0 = E\big(f[X(t), \dot{X}(t)]\big)$$

The a_1 and a_2 coefficients can be found from Eq. 9.23, which gives two simultaneous equations:

$$a_1\sigma_X^2(t) + a_2\rho(t)\sigma_X(t)\sigma_{\dot{X}}(t) = Cov\big(X(t), f[X(t), \dot{X}(t)]\big) \tag{9.32}$$

and

$$a_1\rho(t)\sigma_X(t)\sigma_{\dot{X}}(t) + a_2\sigma_{\dot{X}}^2(t) = Cov\big(\dot{X}(t), f[X(t), \dot{X}(t)]\big) \tag{9.33}$$

in which $\rho(t)\sigma_X(t)\sigma_{\dot{X}}(t)$ denotes the covariance of $X(t)$ and $\dot{X}(t)$.

If we restrict our attention to the special case in which $\{X(t)\}$ is a stationary process, then we know that $\rho(t) = 0$ and $\mu_{\dot{X}}(t) = 0$, so that Eqs. 9.32 and 9.33 uncouple to give

$$a_1 = \frac{Cov\big(X(t), f[X(t), \dot{X}(t)]\big)}{\sigma_X^2(t)} \tag{9.34}$$

and

$$a_2 = \frac{Cov\big(\dot{X}(t), f[X(t), \dot{X}(t)]\big)}{\sigma_{\dot{X}}^2(t)} = \frac{E\big(\dot{X}(t)f[X(t), \dot{X}(t)]\big)}{E[\dot{X}^2(t)]} \tag{9.35}$$

If we make the further restriction that $\mu_X(t) = 0$, then we can rewrite Eq. 9.34 as

$$a_1 = \frac{E\big(X(t)f[X(t), \dot{X}(t)]\big)}{E[X^2(t)]}$$

One situation which will lead to this special case of $X(t)$ being mean zero is for $\{F(t)\}$ to have a probability distribution which is symmetric about zero, and $f(\cdot, \cdot)$ to be odd in the sense that $f(u_1, u_2) = -f(-u_1, -u_2)$.

Let us now consider the special case in which $\{F(t)\}$ is mean zero, stationary, white noise with autospectral density S_0. For this excitation we know that the response of the linearized SDF oscillator described by Eq. 9.31 is stationary, has a mean value of $\mu_X = -a_0/a_1$, and variance values of

$$\sigma_X^2 = \frac{\pi S_0}{a_1 a_2}$$

and

$$\sigma_{\dot{X}}^2 = E[\dot{X}^2] = \frac{\pi S_0}{m a_2}$$

One can now use these two equations in conjunction with Eqs. 9.34 and 9.35, and eliminate a_1 and a_2 to obtain necessary conditions for stationary response as

$$Cov\left(X(t), f[X(t), \dot{X}(t)]\right) = m\sigma_{\dot{X}}^2(t) = mE[\dot{X}^2(t)] \tag{9.36}$$

and

$$Cov\left(\dot{X}(t), f[X(t), \dot{X}(t)]\right) = E\left(\dot{X}(t) f[X(t), \dot{X}(t)]\right) = \pi S_0 / m \tag{9.37}$$

If the $\{F(t)\}$ process is Gaussian, then we also have the choice of using Eq. 9.24 in place of Eq. 9.23. This gives

$$a_1 = E\left(\frac{\partial}{\partial X(t)} f[X(t), \dot{X}(t)]\right)$$

and

$$a_2 = E\left(\frac{\partial}{\partial \dot{X}(t)} f[X(t), \dot{X}(t)]\right)$$

so that Eqs. 9.34 and 9.35 give necessary equations for the stationary response as

$$E\left(\frac{\partial}{\partial X(t)} f[X(t), \dot{X}(t)]\right) = \frac{m\sigma_{\dot{X}}^2}{\sigma_X^2} \tag{9.38}$$

and

$$E\left(\frac{\partial}{\partial \dot{X}(t)} f[X(t), \dot{X}(t)]\right) = \frac{\pi S_0}{m\sigma_{\dot{X}}^2}$$

For the case of Gaussian response these expressions are equivalent to Eqs. 9.36 and 9.37, but they are not necessarily any simpler than those equations.

It is also instructive to consider the special case in which the nonlinearity of Eq. 9.30 corresponds to a linear damping and a nonlinear spring. Following the notation of Eqs. 9.15 - 9.17, we will write this as

$$f[X(t), \dot{X}(t)] = c\dot{X}(t) + f_1'[X(t)]$$

Now we can use the fact that $Cov[X(t), \dot{X}(t)] = E[X(t)\dot{X}(t)] = 0$ for stationary response to obtain

$$Cov\left(X(t), f[X(t), \dot{X}(t)]\right) = Cov\left(X(t), f_1'[X(t)]\right) \tag{9.39}$$

Similarly we can use the fact that

$$E\left(\dot{X}(t)\, f_1'[X(t)]\right) \equiv \frac{d}{dt} E\left(f_1[X(t)]\right)$$

must be zero for stationary response to say that

$$E\left(\dot{X}(t) f[X(t),\dot{X}(t)]\right) = cE[\dot{X}^2] = c\sigma_{\dot{X}}^2 \tag{9.40}$$

From Eqs. 9.34 and 9.35 we now find that

$$a_1 = \frac{Cov\left(X(t),\, f_1'[X(t)]\right)}{\sigma_X^2} \tag{9.41}$$

and

$$a_2 = c \tag{9.42}$$

We also know that

$$a_0 = E\left(f_1'[X(t)]\right) \tag{9.43}$$

since $\dot{X}(t)$ is mean zero for stationary response.

In fact we could have obtained the results in Eqs. 9.41 - 9.43 by a slightly simpler approach. Since the $c\dot{X}(t)$ term in our $f[X(t),\dot{X}(t)]$ function is linear, there really is no need to include it in our linearization procedure. That is, we can say that our equation of motion for the oscillator is

$$m\ddot{X}(t) + c\dot{X}(t) + f_1'[X(t)] = F(t) \tag{9.44}$$

and the only linearization required is of the term $f_1'[X(t)] \approx a_0 + a_1 X(t)$. This one-dimensional linearization then gives exactly Eqs. 9.41 and 9.43, respectively, for a_1 and a_0, and the fact that Eq. 9.42 gives $a_2 = c$ demonstrates that either of these formulations of statistical linearization gives the same linearized system. This illustrates a general property of statistical linearization. If a linear term is somehow included within the nonlinear function which is being linearized, then that linear term will appear unchanged in the results of the linearization.

Substituting Eq. 9.40 into Eq. 9.37 now gives us the linearization estimate of $\sigma_{\dot{X}}^2$ as $\pi S_0 / (cm)$. This result of the approximate method, though, is identical to the exact result presented in Eq. 9.16 for the special case when the white noise excitation is also Gaussian. Thus, if the excitation is Gaussian, then statistical linearization gives an exact value for $\sigma_{\dot{X}}^2$ for the system with linear damping and a nonlinear spring. A formula related to the corresponding estimate of σ_X^2 is obtained by substituting Eq. 9.39 and σ_X^2 into Eq. 9.36, giving

$$Cov\left(X(t),\, f_1'[X(t)]\right) = \pi S_0 / c \tag{9.45}$$

Actual determination of σ_X^2 from this formula requires an assumption about the probability distribution of $X(t)$. If $X(t)$ is assumed to be Gaussian and mean zero, then one also has the alternative of using Eq. 9.38 to write

$$\sigma_X^2 E\big(f_1''[X(t)]\big) = \pi S_0 / c \tag{9.46}$$

which is then equivalent to Eq. 9.45.

****

Example 9.7: Let $\{X(t)\}$ denote the response of the oscillator of Example 9.3 with an equation of motion of

$$m\ddot{X}(t) + c\dot{X}(t) + k_1 X(t) + k_3 X^3(t) = F(t)$$

in which $\{F(t)\}$ is a mean zero, stationary Gaussian white noise with autospectral density S_0. Using statistical linearization, find the value of the a_1 and a_2 parameters of Eqs. 9.31, 9.34 and 9.35, and estimate the value of σ_X^2 and $\sigma_{\dot{X}}^2$ for stationary response.

Since this equation of motion is a special case of Eq. 9.44, we know that $a_2 = c$ and $\sigma_{\dot{X}}^2 = \pi S_0 / (cm)$. Using the notation of Eq. 9.44 we can then write the nonlinear term as $f_1'(u_1) = k_1 u_1 + k_3 u_1^3$, so that Eq. 9.41 gives

$$a_1 = \frac{E\big(X(t) f_1'[X(t)]\big)}{\sigma_X^2} = \frac{k_1 E[X^2] + k_3 E[X^4]}{\sigma_X^2} = k_1 + k_3 E[X^4]/\sigma_X^2$$

Using the assumption that $X(t)$ is Gaussian now gives

$$a_1 = k_1 + 3k_3 \sigma_X^2$$

Note that this expression for a_1 is identical with the results of Eq. 9.24, involving the expected value of the partial derivative of $f_1'[X(t)]$.

Using Eq. 9.45 now gives

$$3k_3 \sigma_X^4 + k_1 \sigma_X^2 = m\sigma_{\dot{X}}^2 = \frac{\pi S_0}{c}$$

The variance of $X(t)$ is found by solving this quadratic equation to give

$$\sigma_X^2 = \frac{k_1}{6k_3}\left(\sqrt{1 + \frac{12k_3 \pi S_0}{ck_1^2}} - 1 \right)$$

Using the notation $\sigma_0^2 = \pi S_0 / (ck_1)$ for the variance in the limiting case of $k_3 = 0$, as was done in Example

9.3, allows the σ_X^2 result to be written in normalized form as

$$\frac{\sigma_X^2}{\sigma_0^2} = \frac{k_1}{6k_3\sigma_0^2}\left(\sqrt{1 + \frac{12k_3\sigma_0^2}{k_1}} - 1\right)$$

Note, though, that this expression is identical to the one obtained and plotted in Example 9.5. Thus, the plot shown in that example also demonstrates that statistical linearization gives good variance approximations for $X(t)$ in the current problem.

****

Example 9.8: Let $\{X(t)\}$ denote the response of the oscillator with an equation of motion of

$$m\ddot{X}(t) + c_1\dot{X}^3(t) + c_2 X^2(t)\dot{X}(t) + kX(t) = F(t)$$

in which $\{F(t)\}$ is a mean zero, stationary Gaussian white noise with autospectral density S_0. Using statistical linearization, find the value of the linearization parameters and estimate the value of σ_X^2 and $\sigma_{\dot{X}}^2$ for stationary response.

As in Eqs. 9.44 - 9.46, there is no need to include a linear term in our linearization. Thus, we can write the problem as

$$m\ddot{X}(t) + f[X(t), \dot{X}(t)] + kX(t) = F(t)$$

with the linearization problem being the determination of a_1 and a_2 in

$$f(u_1, u_2) = c_1 u_2^3 + c_2 u_1^2 u_2 \approx a_1 u_1 + a_2 u_2$$

No a_0 term is included because the symmetry of the problem gives $E\big(f[X(t), \dot{X}(t)]\big) = 0$. Using Eq. 9.23 and noting that $X(t)$ and $\dot{X}(t)$ are uncorrelated for stationary response gives

$$a_1 = \frac{E\big(X(t)f[X(t), \dot{X}(t)]\big)}{E[X^2(t)]} = \frac{c_1 E[X(t)\dot{X}^3(t)] + c_2 E[X^3(t)\dot{X}(t)]}{\sigma_X^2} = \frac{c_1 E[X(t)\dot{X}^3(t)]}{\sigma_X^2}$$

and

$$a_2 = \frac{E\big(\dot{X}(t)f[X(t), \dot{X}(t)]\big)}{E[\dot{X}^2(t)]} = \frac{c_1 E[\dot{X}^4(t)] + c_2 E[X^2(t)\dot{X}^2(t)]}{\sigma_{\dot{X}}^2}$$

Note that the expression for a_1 has been simplified by using the information that $E[X^3(t)\dot{X}(t)] = 0$ for stationary response. The other expectations in these expressions are unknown, except that we can approximate them by using a simplifying assumption for the probability distribution of $X(t)$ and $\dot{X}(t)$. In particular, if we assume that $X(t)$ and $\dot{X}(t)$ are jointly Gaussian then they are also independent so that we have $E[X(t)\dot{X}^3(t)] = 0$, $E[\dot{X}^4(t)] = 3\sigma_{\dot{X}}^4$, and $E[X^2(t)\dot{X}^2(t)] = \sigma_X^2\sigma_{\dot{X}}^2$. This gives the linearization coefficients as $a_1 = 0$ and $a_2 = 3c_1\sigma_{\dot{X}}^2 + c_2\sigma_X^2$. Thus, the linearized system is

$$m\ddot{X}(t) + (3c_1\sigma_{\dot{X}}^2 + c_2\sigma_X^2)\dot{X}(t) + kX(t) = F(t)$$

We know that the response of this linearized system has

$$\sigma_X^2 = \frac{\pi S_0}{k[3c_1\sigma_{\dot{X}}^2 + c_2\sigma_X^2]}$$

and $\sigma_{\dot{X}}^2 = (k/m)\sigma_X^2$. Substituting this latter relationship into the former one and solving for σ_X^2 gives

$$\sigma_X^2 = \sqrt{\frac{\pi S_0}{k\left(\dfrac{3c_1 k}{m} + c_2\right)}}$$

For the special case in which $c_2 = c_1 k / m$, this nonlinear system is the same as the one studied in Example 9.4. Thus, we have an exact solution for that special case, and we can use it as a basis of comparison to determine the accuracy of the approximate result obtained here. For this particular value of c_2, the approximate result becomes

$$\sigma_X^2 = \frac{1}{2k}\sqrt{\frac{m\pi S_0}{c_1}}$$

and this is 11% lower than the exact variance value for this particular problem. This same error also applies to $\sigma_{\dot{X}}^2$, since the approximate result gives $\sigma_{\dot{X}}^2 / \sigma_X^2 = k/m$, just as was true in the exact solution. It should be kept in mind, as well, that this problem is quite significantly non-Gaussian, as was shown in Example 9.4.
****
Example 9.9: Let $\{X(t)\}$ denote the response of the nonlinear oscillator of Examples 9.3 and 9.7 with the equation of motion of

$$m\ddot{X}(t) + c\dot{X}(t) + k_1 X(t) + k_3 X^3(t) = F(t)$$

but with $\{F(t)\}$ being a mean zero, stationary Gaussian process with a non-white noise autospectral density of

$$S_{FF}(\omega) = S_0 U(0.7\omega_0 - |\omega|)$$

in which ω_0 is defined as $\omega_0 = (k_1/m)^{1/2}$. That is, $S_{FF}(\omega) = S_0$ for $|\omega| < 0.7\omega_0$ and equals zero otherwise. Using statistical linearization, estimate the variance of the stationary $\{X(t)\}$ and $\{\dot{X}(t)\}$ responses for the special case of $k_3 = 2ck_1^2 / (\pi S_0)$ and $\zeta_0 \equiv c/(2m\omega_0) = 0.05$.

For this excitation we will not be able to use simple closed form analytical solutions of the linearized system. Thus, we will use iteration in finding a solution. We can directly use the linearized model found in Example 9.7:

$$m\ddot{X}(t) + c\dot{X}(t) + a_1 X(t) = F(t)$$

with $a_1 = k_1 + k_3 E(X^4)/\sigma_X^2$. We will also use the assumption that $X(t)$ is approximately Gaussian to convert this into $a_1 = k_1 + 3k_3\sigma_X^2$. Since we do not have a closed form solution for the response variance for this linearized problem, we will use the harmonic transfer function of $H_x(\omega) = [a_1 - m\omega^2 + 2ic\omega]^{-1}$ and integrate this to obtain

$$\sigma_X^2 = \int_{-\infty}^{\infty} S_{XX}(\omega)\,d\omega = S_0\int_{-0.7\omega_0}^{0.7\omega_0} |H_x(\omega)|^2\,d\omega = 2S_0\int_0^{0.7\omega_0} \frac{d\omega}{(a_1 - m\omega^2)^2 + (2c\omega)^2}$$

For convenience, we will use a nondimensional frequency of $\eta = \omega/\omega_0$ to obtain

$$\sigma_X^2 = \frac{2S_0\omega_0}{a_1^2}\int_0^{0.7}\left[\left(1 - \frac{k_1\eta^2}{a_1}\right)^2 + \left(\frac{2k_1\xi_0\eta}{a_1}\right)^2\right]^{-1} d\eta$$

To begin the iterative process we will simply neglect the k_3 term, taking $a_1 = k_1$ and numerically evaluating the integral to obtain $\sigma_X^2 = (2S_0\omega_0/k_1^2)(1.114)$. Using this value along with the given value of k_3 gives an estimate of $a_1 = k_1 + 3k_3\sigma_X^2 = k_1(1.142)$. Using this value in a new evaluation of the frequency integral gives an improved estimate of the response variance as $\sigma_X^2 = (2S_0\omega_0/k_1^2)(1.029)$. This, in turn, gives an improved estimate of the linearization parameter of $a_1 = k_1 + 3k_3\sigma_X^2 = k_1(1.131)$. Carrying out two more steps of the iteration gives $\sigma_X^2 = (2S_0\omega_0/k_1^2)(1.034)$ and $a_1 = k_1 + 3k_3\sigma_X^2 = k_1(1.132)$ then $\sigma_X^2 = (2S_0\omega_0/k_1^2)(1.033)$ and $a_1 = k_1 + 3k_3\sigma_X^2 = k_1(1.132)$. Since this represents convergence, our estimate of the response variance is

$$\sigma_X^2 = \frac{2S_0\omega_0}{k_1^2}(1.033) = \frac{2.067\,S_0}{(k_1^3 m)^{1/2}}$$

For this linear system we know that $\sigma_{\dot{X}}^2 = (a_1/m)\sigma_X^2$, so our estimate of $\sigma_{\dot{X}}^2$ is

$$\sigma_{\dot{X}}^2 = \frac{1.132k_1}{m}\frac{2.067\,S_0}{(k_1^3 m)^{1/2}} = \frac{2.340\,S_0}{(k_1 m^3)^{1/2}}$$

****

We can also use statistical linearization for the general formulation of the nonlinear dynamics problem given in Eq. 9.2. In particular, if the equation of motion is

$$\dot{\vec{Y}}(t) + \vec{g}[\vec{Y}(t)] = \vec{Q}(t)$$

then the linearization will consist of using an approximation of the form

$$\vec{g}[\vec{Y}(t)] \approx \vec{b}(t) + \mathbf{A}(t)[\vec{Y}(t) - \vec{\mu}_Y(t)] \tag{9.47}$$

The linearization parameters in the vector $\vec{b}(t)$ and the matrix $\mathbf{A}(t)$ of this relationship are found from

Eqs. 9.22 and 9.23 to be given by

$$\vec{b}(t) = E\big(\vec{g}[\vec{Y}(t)]\big) \tag{9.48}$$

and the solution of

$$\mathbf{K}_{YY}(t,t)\mathbf{A}^{T}(t) = \mathbf{K}_{Yg}(t,t) \tag{9.49}$$

in which $\mathbf{K}_{Yg}(t,t)$ denotes the cross-covariance matrix $Cov\big(\vec{Y}(t), \vec{g}[\vec{Y}(t)]\big)$. That is,

$$\mathbf{K}_{Yg}(t,t) = E\big(\vec{Y}(t)\vec{g}^{T}[\vec{Y}(t)]\big) - \vec{\mu}_{Y}(t)E\big(\vec{g}^{T}[\vec{Y}(t)]\big) \tag{9.50}$$

Using the linearization of Eq. 9.47 gives the substitute linear equation of motion as

$$\dot{\vec{Y}}(t) + \vec{b}(t) + \mathbf{A}(t)[\vec{Y}(t) - \vec{\mu}_{Y}(t)] = \vec{Q}(t)$$

This relationship can be put into a more useful form by noting that the expected value of the original nonlinear equation of motion gives

$$\vec{\mu}_{\dot{Y}}(t) + E\big(\vec{g}[\vec{Y}(t)]\big) = \vec{\mu}_{Q}(t)$$

and combining this with Eq. 9.48 gives

$$\vec{b}(t) = \vec{\mu}_{Q}(t) - \vec{\mu}_{\dot{Y}}(t) \tag{9.51}$$

This allows the linearized equation of motion to be rewritten as

$$[\dot{\vec{Y}}(t) - \vec{\mu}_{\dot{Y}}(t)] + \mathbf{A}(t)[\vec{Y}(t) - \vec{\mu}_{Y}(t)] = [\vec{Q}(t) - \vec{\mu}_{Q}(t)] \tag{9.52}$$

Note that this equation only involves the deviation of $\vec{Y}(t)$ away from its mean value. Thus, it can tell us about the covariance and higher-order cumulants of the response, but it does not govern the behavior of $\vec{\mu}_{Y}(t)$. This first moment behavior is given by Eq. 9.51, which can be rewritten as

$$\vec{\mu}_{\dot{Y}}(t) = \vec{\mu}_{Q}(t) - \vec{b}(t) \tag{9.53}$$

One can analyze Eq. 9.52 by using any method appropriate for linear systems. In particular, one can use state-space analysis to write cumulant equations as in Chapter 8. The equation for the covariance of the response, in particular, is given by Eq. 8.10:

$$\frac{d}{dt}\mathbf{K}_{YY}(t,t) + \mathbf{A}(t)\mathbf{K}_{YY}(t,t) + \mathbf{K}_{YY}(t,t)\mathbf{A}^{T}(t) = \mathbf{K}_{QY}(t,t) + \mathbf{K}_{YQ}(t,t) \tag{9.54}$$

and if the excitation is delta-correlated this can be reduced to

$$\frac{d}{dt}\mathbf{K}_{YY}(t,t) + \mathbf{A}(t)\mathbf{K}_{YY}(t,t) + \mathbf{K}_{YY}(t,t)\mathbf{A}^T(t) = 2\pi\mathbf{S}_0(t) \qquad (9.55)$$

as in Eq. 8.18. It should also be noted that if the excitation $\{\vec{Q}(t)\}$ is a Gaussian process, then the response $\{\vec{Y}(t)\}$ from Eq. 9.54 will also be Gaussian. Furthermore, if this result is anticipated in the evaluation of the linearization parameters, then one can use Eq. 9.24 in place of Eq. 9.23, giving

$$A_{jl}(t) = E\left(\frac{\partial g_j[\vec{Y}(t)]}{\partial Y_l(t)}\right)$$

as being equivalent to Eq. 9.49. This formulation can sometimes simplify the evaluation of the linearization coefficients.

By switching to Kronecker notation (as in Section 8.6), we could also write the expressions for higher-order cumulants in the same fashion. Note that if the excitation and response processes are stationary, then $\vec{b}$ and $\mathbf{A}$ are independent of t so that our linearized cumulant equations have constant coefficients as in the simpler examples considered earlier in this section.

9.5   Linearization of Hysteretic Systems

Oscillators having a hysteretic restoring force represent a special category of nonlinear systems which are often of practical importance. By a hysteretic force we mean that the force at a particular time $t = t_1$ depends not only on the values of state variables such as $X(t_1)$ and $\dot{X}(t_1)$ at that instant of time, but also on the time history of $\{X(t)\}$ for $t < t_1$. We will use a notation of $g[\{X(t)\}]$ for such a hysteretic term. Some of the simplest hysteretic models are obtained by introducing idealized Coulomb friction in conjunction with spring elements. For example, simply placing the Coulomb slider in series with a linear spring as in Fig. 9.1(a) gives the hysteretic restoring force shown in Fig. 9.1(b). In addition to representing the system with friction, this model which is always bounded by constant values, designated here as $\pm k\,x_y$, approximates the force-deflection behavior for a uniaxially loaded member made of a material with an elastoplastic stress-strain relationship. Because of this latter fact, the model is commonly called the elastoplastic nonlinearity. The time history behavior is very simple, with the slope of the force-deflection relationship being k whenever $|g[\{X(t)\}]| < k\,x_y$.

By adding another spring in parallel with the elastoplastic element of Fig. 9.1, we can obtain a system with so-called bilinear hysteretic behavior. When we add a mass and a dashpot, we obtain the bilinear hysteretic oscillator shown in Fig. 9.2. There are two important reasons for adding the parallel spring, designated as k_2 in Fig. 9.2, to the bilinear hysteretic oscillator. The first reason is based on the physical fact that most real systems do not lose all their stiffness at the first onset of yielding. That is, the restoring force in a real system usually does continue to increase somewhat after the inception of yielding. The second reason for giving k_2 a non-zero value is that the variance

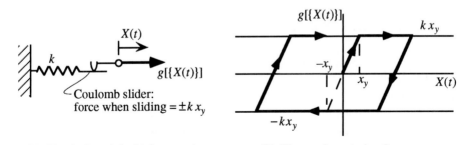

(a) Physical model with hysteresis   (b) Hysteretic restoring force

Figure 9.1 System with elastoplastic hysteresis

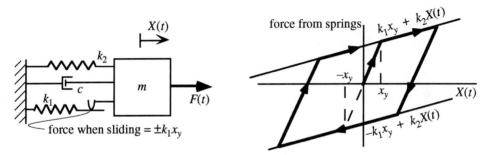

Figure 9.2 Oscillator with bilinear hysteresis

of $X(t)$ grows without bound as t increases for the oscillator with $k_2 = 0$. Thus, the system with $k_2 = 0$ presents some mathematical difficulties since it never has a stationary response to a stationary excitation. In a sense, these two justifications for $k_2 \neq 0$ can be related by the argument that at least one reason that designers avoid the possibility of having a system which is completely elastoplastic is its inherent lack of stability; this is also what leads to the absence of a stationary stochastic solution.

Real physical systems, of course, often have much more complicated hysteretic nonlinearities than the bilinear behavior represented by Fig. 9.1. The linearization methods which we will present here can be applied to a system with any form of hysteresis loop, but we will limit our attention to non-deteriorating systems, in the sense that a periodic time history for $X(t)$ leads to a periodic time history for $g[\{X(t)\}]$.

We will now consider the dynamics of an oscillator including a hysteretic nonlinear element in series with a linear spring and dashpot. This hysteretic element can be elastoplastic as shown in Fig. 9.1 or of some other form. The equation of motion for the oscillator is

$$m\ddot{X}(t) + c\dot{X}(t) + k_2 X(t) + g[\{X(t)\}] = F(t) \tag{9.56}$$

The first method that we will present follows quite closely the general procedure used in Section 9.3. In particular, we will use a linear approximation of

$$g[\{X(t)\}] \approx a_0 + a_1 X(t) + a_2 \dot{X}(t) \tag{9.57}$$

Note that the inclusion of $\dot{X}(t)$ in this linearization may seem somewhat arbitrary, since it does not appear explicitly in the nonlinear $g[\{X(t)\}]$ term. There are several ways that one can justify the inclusion of the $\dot{X}(t)$ term in Eq. 9.57, but the consideration of energy dissipation is probably the simplest. We know that the area of the hysteresis loop of a nonlinear spring represents an amount of energy dissipation during one cycle of motion across the spring element. Since energy dissipation has a very important influence on the level of dynamic response in most stochastic vibration problems, it is clear that we should not use a linearized model which lacks the ability to account for this hysteretic energy dissipation. Including the $a_2 \dot{X}(t)$ term in Eq. 9.57 amounts to introducing an additional viscous damping term into the equation of motion, which does provide for possible matching of the energy dissipation in the original nonlinear system.

The mean squared error in the approximation of Eq. 9.57 is minimized by using Eqs. 9.22 and 9.23 to find the linearization coefficients, just as in the nonhysteretic situation. We will simplify this, though, by limiting our attention to the situation in which the excitation and the restoring force have symmetry such that $X(t)$ and $g[\{X(t)\}]$ are mean zero. This then gives $a_0 = 0$, from Eq. 9.22, and the other coefficients are found from Eq. 9.23 as

$$a_1 = \frac{E(X(t)g[\{X(t)\}])}{E[X^2(t)]} \tag{9.58}$$

and

$$a_2 = \frac{E(\dot{X}(t)g[\{X(t)\}])}{E[\dot{X}^2(t)]} \tag{9.59}$$

The hysteretic nature of the nonlinearity complicates the evaluation of the terms in the numerators of Eqs. 9.58 and 9.59. We can make an assumption about the probability distribution of $X(t)$, just as we did in Sections 9.3 and 9.4, but that is still not enough information to allow evaluation of $E(X(t)g[\{X(t)\}])$ and $E(\dot{X}(t)g[\{X(t)\}])$. The difficulty, of course, is that $g[\{X(t)\}]$ also depends on the past time history of $X(t)$, so that one also needs some assumption about the nature of the possible time histories. The usual approach is to assume that $\{X(t)\}$ is a narrowband process. In particular, we will assume that we can write

$$X(t) = A(t)\cos[\omega_a t + \theta(t)] \tag{9.60}$$

with $\theta(t)$ being uniformly distributed on the interval from zero to 2π, and $\{A(t)\}$ and $\{\theta(t)\}$ being independent processes which vary slowly. That is, we assume that $\dot{\theta}(t)$ is generally small compared to the average frequency ω_a, and $\dot{A}(t)$ is generally small compared to $\omega_a A(t)$. With these assumptions we can always say that $A(t)$ and $\theta(t)$ have been at approximately their present values throughout the previous cycle. If this is true, then we know that $g[\{X(t)\}]$ lies on the perimeter of the hysteresis loop with the amplitude $A(t)$. Furthermore, we can find the distribution of $g[\{X(t)\}]$ around that perimeter by using the uniform probability distribution of $\theta(t)$. In particular, we can write the conditional expectation of $X(t)\,g[\{X(t)\}]$ as

$$E\big(X(t)g[\{X(t)\}]|\,A(t)=u\big)=ug_c(u)/2$$

with

$$g_c(u)=\frac{1}{\pi}\int_0^{2\pi}\cos(\psi)\,g[u\cos(\psi)]d\psi \tag{9.61}$$

in which the integral is evaluated with $g[u\cos(\psi)]$ following the perimeter of the hysteresis loop of amplitude u. Taking the unconditional expectation by using an appropriate probability distribution for $A(t)$ then gives

$$E\big(X(t)g[\{X(t)\}]\big)=\frac{E\big(A(t)g_c[A(t)]\big)}{2}=\frac{1}{2}\int_0^{\infty}u\,g_c(u)\,p_{A(t)}(u)\,du \tag{9.62}$$

One can also use Eq. 9.60 to rewrite the denominator of Eq. 9.58 as $E[X^2(t)]=E[A^2(t)]/2$, so that

$$a_1=\frac{E\big(A(t)g_c[A(t)]\big)}{E[A^2(t)]} \tag{9.63}$$

The expected value of $\dot{X}(t)g[\{X(t)\}]$ is evaluated in a directly parallel fashion. In particular, the derivative of the narrowband process is approximated as

$$\dot{X}(t)=-\omega_a A(t)\sin[\omega_a t+\theta(t)] \tag{9.64}$$

which gives $E[\dot{X}^2(t)]=\omega_a^2 E[A^2(t)]/2$. Again presuming that $g[\{X(t)\}]$ is on the perimeter of the hysteresis loop of amplitude $A(t)$ gives[5]

$$E\big(\dot{X}(t)g[\{X(t)\}]|\,A(t)=u\big)=-\omega_a u g_s(u)/2$$

with

$$g_s(u)=\frac{1}{\pi}\int_0^{2\pi}\sin(\psi)\,g[u\cos(\psi)]d\psi \tag{9.65}$$

so that one has

---

[5]   The linearization presented here is essentially the same as that given by Caughey (1960b), although his justification of Eqs. 9.61 and 9.65 is slightly different. In particular, he considers these integrals to represent a time average over one cycle of oscillator response, as in the Krylov-Bogoliubov method for linearization of deterministic problems. This interpretation, which can be considered an application of stochastic averaging, makes Eqs. 9.62 and 9.66 become combination averages, which are expected values of time averages.

$$E\big(\dot{X}(t)g[\{X(t)\}]\big) = \frac{-\omega_a}{2}E\big(A(t)g_s[A(t)]\big) = \frac{-\omega_a}{2}\int_0^\infty u\,g_s(u)\,p_{A(t)}(u)\,du \qquad (9.66)$$

and

$$a_2 = -\frac{1}{\omega_a}\frac{E\big(A(t)g_s[A(t)]\big)}{E[A^2(t)]} \qquad (9.67)$$

One can also simplify this result a little by noting that a change of variables of $w = u\cos(\psi)$ in eq, 9.65 gives

$$g_s(u) = \frac{-1}{\pi u}\int_{loop} g(w)\,dw = \frac{-1}{\pi u}\times(\text{Area of Hysteresis Loop with Amplitude } u)$$

It is now necessary to make some assumption about the probability distribution of the amplitude $A(t)$ in order to evaluate the expectations in Eqs. 9.62 and 9.66, and thus obtain the values of the a_1 and a_2 linearization coefficients from Eqs. 9.63 and 9.67. The usual assumption is to say that the narrowband $\{X(t)\}$ process is nearly Gaussian, and this implies that the Rayleigh distribution can be used for $A(t)$. This is essentially the same argument as was used in Section 5.3, where $A(t)$ for a Gaussian process had exactly the Rayleigh probability distribution when it was defined such as to exactly satisfy Eqs. 9.60 and 9.64. Using the Rayleigh probability distribution (or some other known function) for $p_{A(t)}(u)$ then gives a_1 and a_2 as functions of σ_X. Overall, it may be seen that when the narrowband assumption is used for $\{X(t)\}$, the difficulty in linearizing a hysteretic function of $\{X(t)\}$ is only slightly greater than in linearizing a nonhysteretic function of $X(t)$ and $\dot{X}(t)$.

After the linearization coefficients have been evaluated, one can substitute $a_1 X(t) + a_2 \dot{X}(t)$ in place of $g[\{X(t)\}]$ in Eq. 9.56. This gives the approximate equation of motion as

$$m\ddot{X}(t) + (c + a_2)\dot{X}(t) + (k_2 + a_1)X(t) = F(t)$$

from which it is obvious that a_1 represents an additional stiffness term and a_2 represents additional damping. That is, this linearization procedure replaces the elastoplastic element with a spring and a dashpot in parallel. If this linearized system does have a narrowband response, then its average frequency must be approximately

$$\omega_a = \sqrt{\frac{k_2 + a_1}{m}} \qquad (9.68)$$

since that is the resonant frequency of the new equation of motion. Furthermore, we can use the usual expression for the variance of the response of the linear SDF oscillator to say that

$$\sigma_X^2 = \frac{\pi S_0}{(c + a_2)(k_2 + a_1)} \qquad (9.69)$$

The variance of $\dot{X}(t)$, of course, is approximated by $\omega_a^2 \sigma_X^2$. Numerical results are obtained by simultaneously solving Eqs. 9.63, 9.67, 9.68, and 9.69.

Note that it is reasonable to expect the accuracy of this linearization procedure to depend on the adequacy of two particular approximations: the narrowband assumption used in obtaining Eqs. 9.61 and 9.65, and the probability distribution assumed for the amplitude $A(t)$ of the narrowband response. If the system is nearly linear then the narrowband assumption should be appropriate. Also, if the excitation of the system is Gaussian and the nonlinearity is small, then the response should be nearly Gaussian so that a Rayleigh approximation for the amplitude is justified.

****

Example 9.10: Find the linearization parameters and the mean squared response levels for the bilinear hysteretic oscillator shown in Fig. 9.2 when the excitation $F(t)$ is a stationary, mean zero, Gaussian white noise with autospectral density S_0.

The system is described by Eq. 9.56 with $g[\{X(t)\}]$ as given in Fig. 9.1. Recall that the symmetry of the situation gives $a_0 = 0$. In order to evaluate a_1 and a_2 we need to evaluate the $g_c(u)$ and $g_s(u)$ functions defined by Eqs. 9.61 and 9.65. First we note that for $u < x_y$ the elastoplastic term is $g[u\cos(\psi)] = k_1 u \cos(\psi)$, so that

$$g_c(u) = k_1 u, \qquad g_s(u) = 0 \quad \text{for} \quad u < x_y$$

For $u > x_y$, we split the integration with respect to ψ in Eq. 9.61 into four parts, corresponding to the four straight segments of the hysteresis loop. The result is

$$g_c(u) = \frac{1}{\pi}\left[\int_0^{\psi^*} \cos(\psi)k_1[x_y - u(1-\cos\psi)]d\psi + \int_{\psi^*}^{\pi} \cos(\psi)(-k_1 x_y)d\psi \right.$$
$$\left. + \int_{\pi}^{\pi+\psi^*} \cos(\psi)k_1[-x_y + u(1-\cos\psi)]d\psi + \int_{\pi+\psi^*}^{2\pi} \cos(\psi)(k_1 x_y)d\psi \right]$$

in which

$$\psi^* = \cos^{-1}\left(\frac{u-2x_y}{u}\right)$$

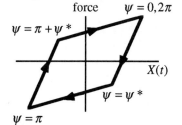

as shown in the sketch. After simplification, this becomes

$$g_c(u) = \frac{k_1 u \psi^*}{\pi} - \frac{2k_1}{\pi}\left(\frac{u-2x_y}{u}\right)\sqrt{x_y(u-x_y)} \quad \text{for} \quad u > x_y$$

One can obtain $g_s(u)$ either by performing a similar integration around the loop or by finding the area of the hysteresis loop in Fig. 9.1. The result is

$$g_s(u) = -\frac{4k_1}{\pi}\frac{x_y(u-x_y)}{u} \quad \text{for} \quad u > x_y$$

Presuming that the response is narrowband and Gaussian, we now use the Rayleigh probability density function for $A(t)$:

$$p_{A(t)}(u) = \frac{u}{\sigma_X^2}\exp\left(\frac{-u^2}{2\sigma_X^2}\right)$$

and evaluate the expectations of Eqs. 9.63 and 9.67. The integral in Eq. 9.63 has the form of

$$a_1 = \frac{2}{\sigma_X^2}\left[\int_0^{x_y}k_1u^2\frac{u}{\sigma_X^2}\exp\left(\frac{-u^2}{2\sigma_X^2}\right)du\right.$$
$$\left.+\int_{x_y}^{\infty}\left[\frac{k_1u}{\pi}\cos^{-1}\left(\frac{u-2x_y}{u}\right)-\frac{2k_1}{\pi}\left(\frac{u-2x_y}{u}\right)\sqrt{x_y(u-x_y)}\right]\frac{u}{\sigma_X^2}\exp\left(\frac{-u^2}{2\sigma_X^2}\right)du\right]$$

and Caughey (1960b) showed that it can be converted into

$$a_1 = k_1\left[1-\frac{8}{\pi}\int_1^{\infty}\left(\frac{1}{z^3}+\frac{x_y^2}{2\sigma_X^2 z}\right)\sqrt{z-1}\exp\left(\frac{-x_y^2 z^2}{2\sigma_X^2}\right)dz\right]$$

No closed form has been found for this integral, but it can be easily evaluated numerically if one first chooses a value for the dimensionless ratio σ_X/x_y. Note that σ_X/x_y can viewed as an rms ductility ratio, since it represents the rms response displacement divided by the yield displacement. The evaluation of a_2 is somewhat easier. Writing out the integral for the expected value in Eq. 9.67 gives

$$a_2 = \frac{8k_1}{\pi\omega_a\sigma_X^2}\int_{x_y}^{\infty}x_y(u-x_y)\frac{u}{\sigma_X^2}\exp\left(\frac{-u^2}{2\sigma_X^2}\right)du$$

and this reduces to

$$a_2 = \sqrt{\frac{2}{\pi}}\frac{k_1 x_y}{\omega_a\sigma_X}\left[1-\Phi\left(\frac{x_y}{\sigma_X}\right)\right]$$

in which $\Phi(\cdot)$ denotes the cumulative distribution function for a standardized Gaussian random variable, and is related to the error function (see Eq. B.4 of Appendix B). The sketches show normalized values of a_1 and a_2 for a fairly wide range of σ_X/x_y values.

The decay of a_1 from its initial value of k_1 for $\sigma_X << x_y$ to its final value of zero for $\sigma_X >> x_y$ corresponds to ω_a^2 varying from $(k_1+k_2)/m$ to k_2/m. If the excitation is very small then $\sigma_X << x_y$ and the elastoplastic element experiences little or no yielding so that the k_1 spring does contribute effectively to the system stiffness. If the excitation is very large so that $\sigma_X \gg x_y$, then the force added by the elastoplastic element is relatively small so that the total stiffness approaches k_2. The damping effect of the elastoplastic element has a rather different characteristic. The dashpot effect is the greatest when $\sigma_X \approx 1.5x_y$, and is near zero for both very large and very small values of σ_X / x_y. Of course the energy dissipated per cycle does continue to grow as σ_X increases beyond $1.5x_y$, even though the value of the "equivalent" dashpot coefficient a_2 decreases.

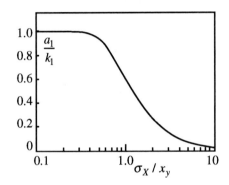 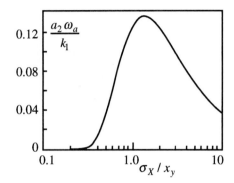

One can use an iterative procedure to find the solution of Eqs. 9.63, 9.67, 9.68, and 9.69 for given values of the system parameters m, c, k_2, k_1, and x_y, and the excitation level S_0. Alternatively, one can consider fixed values of some parameters and plot the results versus other parameters. We have prepared three plots of this form, in which σ_X / σ_F^* is plotted versus σ_F^* / x_y, with σ_F^* defined as

$$\sigma_F^{*\,2} = \frac{2 S_0 \omega_0}{(k_1 + k_2)^2}$$

with $\omega_0 = \sqrt{(k_1 + k_2)/m}$ = resonant frequency for the linear system which results when x_y is infinite. For each curve the values of k_2 / k_1 and $\zeta_0 \equiv c / [2\sqrt{(k_1 + k_2)m}]$ are held constant. Note that for given system parameters, σ_F^* is a measure of the level of the excitation. Thus, any curve of the type given can be considered to represent response normalized by the excitation, plotted versus the excitation normalized by the yield level. The curves were obtained by choosing a value of σ_X / x_y, evaluating a_1 / k_1 and $\omega_a a_2 / k_1$ as for the preceding plots, using Eq. 9.68 for the ω_a term in the latter of these expressions, and finally using Eq. 9.69 to give the response variance.

The plots includes three different values of k_2 / k_1, to represent three different situations. For $k_2 / k_1 = 9$ it is reasonable to consider the nonlinearity to be small, so that the assumptions used in the linearization procedure should be appropriate. For $k_2 / k_1 = 1$ the system can lose up to half of its stiffness due to yielding and there is considerable hysteretic energy dissipation, so it is questionable whether the narrowband and Gaussian assumptions are appropriate for $\{X(t)\}$. The situation with $k_2 / k_1 = 0.05$ goes far beyond the range in which it is reasonable to expect the assumptions to be valid. Numerical data for the response levels of the bilinear hysteretic oscillator with $k_2 / k_1 = 1$ and $k_2 / k_1 = 0.05$ were presented by Iwan and Lutes (1968), and some of these data are reproduced in the plots. In particular, the plots include data for $\zeta_0 = 0$ and $\zeta_0 = 0.01$ for $k_2 / k_1 = 1$ and for $\zeta_0 = 0$ and $\zeta_0 = 0.05$ for $k_2 / k_1 = 0.05$. Somewhat surprisingly, the data show that the linearization results are generally quite good for $k_2 / k_1 = 1$, even though this is hardly a situation with a small amount of nonlinearity. For $k_2 / k_1 = 0.05$, on the other hand, there are sometimes significant discrepancies between the results from the bilinear hysteretic system and those from linearization.

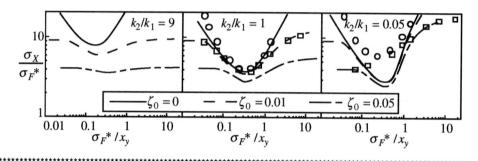

For some hysteretic systems, it is possible to approach the problem of linearization in a different way which eliminates the need for the narrowband approximation used in obtaining Eqs. 9.62 and 9.66. This approach involves the introduction of one or more additional state variables such that the $g[\{X(t)\}]$ restoring force can be written as a nonhysteretic function of the new set of state variables. This idea is easily illustrated for the elastoplastic element which contributes the hysteresis to the bilinear oscillator of Example 9.10.[6] An appropriate additional state variable for this example is shown as $Z(t)$ in Fig. 9.3. Using this variable allows one to express the force across the elastoplastic element as

$$\hat{g}[X(t), \dot{X}(t), Z(t)] = k\, Z(t) \tag{9.70}$$

One notes that the state variables $X(t)$ and $\dot{X}(t)$ do not appear explicitly in this equation for the force, but they are necessary if one is to determine the rate of change of the force:

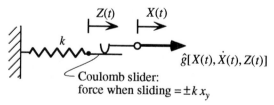

Figure 9.3 Nonhysteretic description of the elastoplastic force

---

[6]   A linearization of this type for the bilinear hysteretic element was used by Kaul and Penzien in 1974, although their formulation of the linearization problem involved the Fokker-Planck equation rather than the approach used here.

$$\dot{\hat{g}}[X(t), \dot{X}(t), Z(t)] = k\,\dot{Z}(t)$$

with

$$\dot{Z}(t) = \dot{X}(t)\Big(1 - U[Z(t) - x_y]U[\dot{X}(t)] - U[-Z(t) - x_y]U[-\dot{X}(t)]\Big) \tag{9.71}$$

That is, when $Z(t)$ reaches the level $\pm x_y$, we know that $\dot{Z}(t)$ changes from $\dot{X}(t)$ to zero, so that $\hat{g}[X(t), \dot{X}(t), Z(t)]$ remains unchanged until $\dot{X}(t)$ reverses its sign. When this sign reversal occurs for $\dot{X}(t)$, the value of $\hat{g}[X(t), \dot{X}(t), Z(t)]$ begins to move back from $\pm k x_y$ toward zero.

Since Eq. 9.70 is already linear, there is no reason to apply statistical linearization to it, but definition of an appropriate linear model in this formulation involves replacing Eq. 9.71 with a linear relationship of the form

$$\dot{Z}(t) \approx a_0 + a_1 X(t) + a_2 \dot{X}(t) + a_3 Z(t) \tag{9.72}$$

We will now make the major assumption that $\{X(t)\}$ and $\{Z(t)\}$ can be approximated as being mean zero and jointly Gaussian processes. This makes $X(t)$, $\dot{X}(t)$, and $Z(t)$ jointly Gaussian random variables, so we can conveniently use Eq. 9.24 in evaluating the coefficients of linearization. The results are

$$a_0 = E[\dot{Z}(t)] = 0 \tag{9.73}$$

$$a_1 = E\left(\frac{\partial \dot{Z}(t)}{\partial X(t)}\right) = 0 \tag{9.74}$$

$$a_2 = E\left(\frac{\partial \dot{Z}(t)}{\partial \dot{X}(t)}\right) = E\Big(1 - U[Z(t) - x_y]U[\dot{X}(t)]\Big) - E\Big(U[-Z(t) - x_y]U[-\dot{X}(t)]\Big)$$

$$= 1 - 2\int_{x_y}^{\infty}\int_{0}^{\infty} p_{\dot{X}(t),Z(t)}(v,w)\,dv\,dw \tag{9.75}$$

and

$$a_3 = E\left(\frac{\partial \dot{Z}(t)}{\partial Z(t)}\right) = -E\Big(\dot{X}(t)\,\delta[Z(t) - x_y]U[\dot{X}(t)]\Big) + E\Big(\dot{X}(t)\,\delta[-Z(t) - x_y]U[-\dot{X}(t)]\Big)$$

$$= -2\int_{0}^{\infty} v\, p_{\dot{X}(t),Z(t)}(v, x_y)\,dv \tag{9.76}$$

In addition to Eqs. 9.73 - 9.76 relating the values of the linearization coefficients to the response levels, it is necessary to have other equations relating the response levels to the linearization parameters. These equations come from stochastic analysis of the equation of motion of a system containing this linearized elastoplastic element. This can be illustrated by considering the bilinear hysteretic system of Example 9.10. Using Eqs. 9.70 to linearize the elastoplastic element in this system gives an equation of motion of

$$m\ddot{X}(t) + c\dot{X}(t) + k_2 X(t) + k_1 Z(t) = F(t) \tag{9.77}$$

and the fact that a_0 and a_1 are zero allows the auxiliary relationship of Eq. 9.72 to be written as

$$\dot{Z}(t) = a_2 \dot{X}(t) + a_3 Z(t) \tag{9.78}$$

One can find the response levels of this system as functions of the parameters m, c, k_1, k_2, a_2, and a_3 by any of the methods appropriate for a linear system. If the excitation $F(t)$ is stationary white noise, then it is particularly convenient to use the state-space moment or cumulant methods of Chapter 8 to derive simultaneous algebraic equations giving the response variances and covariances. In particular, one can take the state vector as $\vec{Y}(t) = [X(t), \dot{X}(t), Z(t)]^T$, so that the response variances and covariances are given by the components of a symmetric matrix of dimension 3. One of the six simultaneous equations turns out to be trivial, simply giving the well-known fact that the covariance of $X(t)$ and $\dot{X}(t)$ is zero. The other five equations must be solved simultaneously with Eqs. 9.75 and 9.76 in order to find not only the five remaining terms describing the response levels, but also the a_2 and a_3 coefficients.

****

Example 9.11: Find the linearization parameters and the mean squared response levels predicted by Eqs. 9.70 - 9.76 for stationary response of the bilinear hysteretic oscillator shown in Fig. 9.2 with an excitation $F(t)$ which is a stationary, mean zero, Gaussian white noise with autospectral density S_0.

As previously noted, the system is described by Eqs. 9.77 and 9.78 and the linearization parameters a_2 and a_3 are related to the response levels by Eqs. 9.75 and 9.76. Writing these two expressions as integrals of the jointly Gaussian probability density function gives

$$a_2 = 1 - 2\int_{x_y/\sigma_Z}^{\infty} \frac{e^{-r^2/2}}{\sqrt{2\pi}} \Phi\left(\frac{\rho_{\dot{X},Z}r}{\sqrt{1-\rho_{\dot{X},Z}^2}}\right) dr$$

and

$$a_3 = -2\frac{\sigma_{\dot{X}}}{\sigma_Z}\hat{a}_3$$

with

$$\hat{a}_3 = \frac{\rho_{\dot{X},Z}x_y}{\sqrt{2\pi}\sigma_Z}\exp\left(\frac{-x_y^2}{2\sigma_Z^2}\right)\Phi\left(\frac{\rho_{\dot{X},Z}x_y}{\sigma_Z\sqrt{1-\rho_{\dot{X},Z}^2}}\right) + \frac{1}{2\pi}\sqrt{1-\rho_{\dot{X},Z}^2}\exp\left(\frac{-x_y^2}{2\sigma_Z^2(1-\rho_{\dot{X},Z}^2)}\right)$$

Note that a_2 and $\hat{a}_3$ are dimensionless quantities which depend only on the two parameters $\rho_{\dot{X},Z}$ and x_y/σ_Z, whereas a_3 has units of frequency and also depends on the value $\sigma_{\dot{X}}/\sigma_Z$.

In order to evaluate parameters and response levels, we must use expressions giving the response levels of the linearized system of Eqs. 9.77 and 9.78. We will do this by using the state-space formulation of the covariance equations, as given in Section 8.4. Thus, we rewrite the equations of motion in the standard form of $\vec{Y}(t) + \mathbf{A}\vec{Y}(t) = \vec{Q}(t)$, with $\vec{Y}(t) = [X(t), \dot{X}(t), Z(t)]^T$, $\vec{Q}(t) = [0, F(t)/m, 0]^T$, and

$$\mathbf{A} = \begin{bmatrix} 0 & -1 & 0 \\ k_2/m & c/m & k_1/m \\ 0 & -a_2 & -a_3 \end{bmatrix}$$

Since we are only interested in stationary response, we can simplify Eq. 8.18 to be

$$\mathbf{A}\mathbf{K}_{YY} + \mathbf{K}_{YY}\mathbf{A}^T = 2\pi\mathbf{S}_0$$

with the only non-zero element of the $\mathbf{S}_0$ matrix being the scalar S_0 in the $(2,2)$ position. After some very minor simplification, the six distinct components of this relationship can be written as

$$K_{X\dot{X}} = 0$$

$$cK_{X\dot{X}} + k_1 K_{XZ} + k_2 K_{XX} - mK_{\dot{X}\dot{X}} = 0$$

$$a_2 K_{X\dot{X}} + a_3 K_{XZ} + K_{\dot{X}Z} = 0$$

$$k_2 K_{X\dot{X}} + cK_{\dot{X}\dot{X}} + k_1 K_{\dot{X}Z} = \pi S_0 / m$$

$$k_2 K_{XZ} + (c - a_3 m) K_{\dot{X}Z} - a_2 m K_{\dot{X}\dot{X}} + k_1 K_{ZZ} = 0$$

and

$$-a_2 K_{\dot{X}Z} - a_3 K_{ZZ} = 0$$

The first of these equations, $K_{X\dot{X}} = 0$, is simply a well-known property of any stationary process, and this can be used to give a slight simplification in the second, third, and fourth equations. Similarly, Eq. 9.78 shows that the sixth equation is the corresponding condition that $K_{\dot{Z}Z} = 0$. Using the linearization condition of $a_3 = -2(\sigma_{\dot{X}} / \sigma_Z)\hat{a}_3$ along with this stationarity condition gives

$$-a_2 \rho_{\dot{X}Z} + 2\hat{a}_3 = 0$$

as a necessary condition for the joint probability density function of $\dot{X}(t)$ and $Z(t)$. One can use this relationship along with the expressions for a_2 and $\hat{a}_3$ to derive a unique relationship between $\rho_{\dot{X}Z}$ and x_y / σ_Z. In particular, for a selected value of x_y / σ_Z one can use an initial guess for $\rho_{\dot{X}Z}$, evaluate estimates of a_2 and $\hat{a}_3$, reevaluate $\rho_{\dot{X}Z}$, etc. After finding these values for the selected x_y / σ_Z value, one can use the third response equation to write

$$K_{XZ} = -K_{\dot{X}Z} / a_3 = \rho_{\dot{X}Z}\sigma_Z^2 / (2\hat{a}_3) = \sigma_Z^2 / a_2$$

Rewriting $K_{\dot{X}\dot{X}}$ and K_{ZZ} as $\sigma_{\dot{X}}^2$ and σ_Z^2, respectively, allows relatively simple solution of the fourth and fifth response equations to find these quantities. The second response equation then gives the value of $K_{XX} \equiv \sigma_X^2$. Although there are multiple solution branches for these equations, there seems to be only one solution which gives real positive values of σ_X^2, $\sigma_{\dot{X}}^2$, and σ_Z^2 for any given value of x_y / σ_Z. The sketches show the results of this linearization for the situations with $k_2 = k_1$ and $k_2 = 0.05k_1$. The form of the plots is the same as in Example 9.10. For $k_2 = 0.05k_1$ and intermediate values of σ_F^* / x_y, it is observed that the fit of this linearization to the numerical data is significantly better than for the linearization of Example 9.10. For larger values of σ_F^* / x_y, this linearization is less accurate than the one of Example 9.10, for either value of k_2 / k_1.

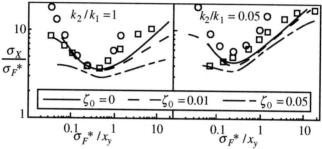

Another version of the linearization of the elastoplastic element was offered by Asano and Iwan (1984) who noted that Eq. 9.70 allows the possibility of the absolute value of the restoring force exceeding kx_y, whereas that should be impossible. To avoid this situation they replaced Eq. 9.70 with an expression which truncates the restoring force at the levels $\pm kx_y$. This can be written as

$$\hat{g}[X(t),\dot{X}(t),Z(t)] = k\,Z(t)\big(1 - U[Z(t) - x_y]U[\dot{X}(t)] - U[-Z(t) - x_y]U[-\dot{X}(t)]\big)$$
$$+k\,x_y\big(U[Z(t) - x_y]U[\dot{X}(t)] - U[-Z(t) - x_y]U[-\dot{X}(t)]\big) \qquad (9.79)$$

Note that Eq. 9.79 is identical to Eq. 9.70 whenever $|Z(t)| \le x_y$. Furthermore, the behavior of the truly elastoplastic element never allows the system to enter the region of $|Z(t)| > x_y$, where Eqs. 9.70 and 9.79 differ from each other. Thus, both Eq. 9.70 and 9.79 exactly describe truly elastoplastic behavior. When one linearizes the equations in the usual way, though, the results are not identical. In particular, one must consider Eq. 9.79 to give a nonlinear relationship between the elastoplastic force and the state variables. Linearization of this model, then, involves not only using Eqs. 9.78 in place of 9.71, but also using an approximation of

$$\hat{g}[X(t),\dot{X}(t),Z(t)] \approx b_0 + b_1 X(t) + b_2 \dot{X}(t) + b_3 Z(t) \qquad (9.80)$$

in place of Eq. 9.79. Using the jointly Gaussian assumption for the state variables gives the coefficients as

$$b_0 = E\big(\hat{g}[X(t),\dot{X}(t),Z(t)]\big) = 0 \qquad (9.81)$$

$$b_1 = E\left(\frac{\partial \hat{g}[X(t), \dot{X}(t), Z(t)]}{\partial X(t)}\right) = 0 \tag{9.82}$$

$$b_2 = E\left(\frac{\partial \hat{g}[X(t), \dot{X}(t), Z(t)]}{\partial \dot{X}(t)}\right) = k E\Big[Z(t) \delta[\dot{X}(t)]\big(-U[Z(t) - x_y] + U[-Z(t) - x_y]\big)\Big]$$

$$+ k x_y E\Big[\delta[\dot{X}(t)]\big(U[Z(t) - x_y] + U[-Z(t) - x_y]\big)\Big]$$

$$= -2k \int_{x_y}^{\infty} (w - x_y)\, p_{\dot{X}(t), Z(t)}(0, w)\, dw \tag{9.83}$$

$$b_3 = E\left(\frac{\partial \hat{g}[X(t), \dot{X}(t), Z(t)]}{\partial Z(t)}\right) = k E\big(1 - U[Z(t) - x_y] U[\dot{X}(t)] - U[-Z(t) - x_y] U[-\dot{X}(t)]\big)$$

$$= k\left[1 - 2\int_{x_y}^{\infty}\int_0^{\infty} p_{\dot{X}(t), Z(t)}(v, w)\, dv\, dw\right] \tag{9.84}$$

One may also note that Eqs. 9.84 and 9.75 give b_3 as being identical to $k a_2$.

It is somewhat disturbing to find that Eqs. 9.70 and 9.79 are both exact descriptions of the elastoplastic nonlinearity, but they give different results when analyzed by statistical linearization. For this system there is not a unique answer even when one adopts the standard procedure of minimizing the mean squared error and uses the standard assumption that the response quantities have a jointly Gaussian probability distribution. By using a different linearization procedure, or assuming a different probability distribution, of course, it is possible to obtain any number of other results, but there is some ambiguity even within the standard procedure. In this particular instance, the additional ambiguity arose because the new state variable $Z(t)$ was bounded, so that it was undefined on certain regions of the state space. The Gaussian assumption, though, does assign probability to the entire state space, so that the linearization results are affected by assumptions about behavior in the regions which really should have zero probability. One way to avoid this particular difficulty is to analyze only models which have a finite probability density everywhere. This is one of the motivations for the popularity in recent years of a smooth hysteresis model in which the new state variable $Z(t)$ is proportional to the hysteretic portion of the restoring force, as in Eq. 9.77, and its rate of change is governed by the differential equation

$$\dot{Z}(t) = -c_1 |\dot{X}(t)| Z^{c_3-1}(t) |Z(t) - c_2| Z^{c_3}(t)| \dot{X}(t)$$

for some parameters c_1, c_2, and c_3.[7] This gives $Z(t)$ as unbounded, so that there is not any region of zero probability. Wen (1976) achieved generally good results using statistical linearization for this hysteretic system, and the model seems to be capable of approximating many physical problems.

---

[7] This form is a generalization by Wen (1976) of a model use by Bouc (1968) with $c_3 = 1$.

In some problems it is not as obvious how to avoid having regions of zero probability in the state space. For example, when modeling damaged structures it may be desirable to consider the tangential stiffness at time t to be a function of the maximum distortion of the structure at any time in the past. This can be accomodated by defining a state variable $Y(t)$ which is the maximum absolute value of the displacement $X(t)$ at any prior time (as in Section 6.5). This new variable $Y(t)$ is not bounded, but there should be zero probability on the regions $X(t) > Y(t)$ and $X(t) < -Y(t)$. Statistical linearization has been used for such problems (Senthilnathan and Lutes, 1991) but the difficulties and ambiguities involved are similar to those for elastoplastic behavior.

****

Example 9.12: Find the linearization parameters and the mean squared response levels predicted by Eqs. 9.80 - 9.84 for stationary response of the bilinear hysteretic oscillator shown in Fig. 9.2 with an excitation $F(t)$ which is a stationary, mean zero, Gaussian white noise with autospectral density S_0.

The procedure is the same as in Example 9.11. In addition to the a_2 and a_3 parameters evaluated there, the linearization now also uses $b_3 = k_1 a_2$, and the parameter b_2 from Eq. 9.83, which can be rewritten as

$$b_2 = -2k_1 \frac{\sigma_Z}{\sigma_{\dot{X}}} \hat{b}_2$$

with

$$\hat{b}_2 = \frac{1}{2\pi} \sqrt{1-\rho_{\dot{X},Z}^2} \exp\left(\frac{-x_y^2}{2\sigma_Z^2(1-\rho_{\dot{X},Z}^2)}\right) - \frac{x_y}{\sqrt{2\pi}\,\sigma_Z} \Phi\left(\frac{-x_y}{\sigma_Z\sqrt{1-\rho_{\dot{X},Z}^2}}\right)$$

Note that $\hat{b}_2$, like a_2 and $\hat{a}_3$, is a function only of $\rho_{\dot{X},Z}$ and x_y / σ_Z, so can be evaluated independently from the analysis of the dynamics of the oscillator.

Substituting Eq. 9.80, with $b_0 = 0$, $b_1 = 0$, and $b_3 = k_1 a_2$ into Eq. 9.56 gives the equation of motion for the linearized system as

$$m\ddot{X}(t) + (c + b_2)\dot{X}(t) + k_2 X(t) + k_1 a_2 Z(t) = F(t)$$

This equation must be solved in conjunction with Eq. 9.78, giving the rate of change of $Z(t)$. Note that the new equation of motion has the same form as Eq. 9.77, but it has different stiffness and damping values. In particular, the presence of a_2, which is a number between zero and unity, results in a reduced stiffness for this linearization of the problem. Similarly, b_2 is a negative quantity, so it causes a reduction in the damping of the system. Thus, we can anticipate that the response of this linear model will be greater than that for Eq. 9.70. The details of the state-space analysis to find the response quantities as functions of the linearization parameters will be omitted, since the procedure is fundamentally the same as in Example 9.11. The σ_X / σ_F^* results are shown in the plot. It is observed that this linearization does a better job of fitting the numerical simulation data than does the method of either Example 9.10 or 9.11. The difference between the linearization shown here and that of Example 9.11 is most obvious for relatively high values of σ_F^* / x_y, since this is the situation where Eqs. 9.70 and 9.79 are most significantly different.

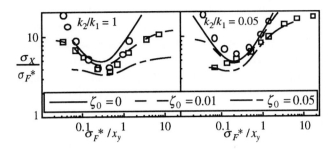

* *

9.6 State-Space Moment and Cumulant Equations

We will now consider how the state-space methods of Chapter 8 can be applied to nonlinear problems. For this purpose we will presume that the nonlinear equation of motion is written as in Eq. 9.2:

$$\dot{\vec{Y}}(t) + \vec{g}[\vec{Y}(t)] = \vec{Q}(t) \qquad (9.85)$$

This general equation is adequate for systems with nonlinear systems without hysteresis. However, with the introduction of additional state variables, as illustrated by $Z(t)$ in Eqs. 9.70 and 9.71, it can also be applied to many problems which are generally classified as hysteretic.

The procedure of Section 8.3 can now be used to derive state-space mean and covariance equations. The mean value equation of

$$\dot{\vec{\mu}}_Y(t) + E\left(\vec{g}[\vec{Y}(t)]\right) = \vec{\mu}_Q(t) \qquad (9.86)$$

results from simply taking the expectation of Eq. 9.85. Recalling that

$$\frac{d}{dt}\mathbf{K}_{YY}(t,t) = \mathbf{K}_{\dot{Y}Y}(t,t) + \mathbf{K}_{Y\dot{Y}}(t,t)$$

we can take cross-covariances involving terms of Eq. 9.85 and the state vector to give

$$\frac{d}{dt}\mathbf{K}_{YY}(t,t) + \mathbf{K}_{gY}(t,t) + \mathbf{K}_{Yg}(t,t) = \mathbf{K}_{QY}(t,t) + \mathbf{K}_{YQ}(t,t) \qquad (9.87)$$

in which $\mathbf{K}_{Yg}(t,t)$ is as defined in Eq. 9.50, and $\mathbf{K}_{gY}(t,t)$ is its transpose.

One use of the exact mean and covariance relationships given in Eqs. 9.86 and 9.87 is to provide valuable information about the assumptions of the statistical linearization procedure introduced in Section 9.4. In particular, a comparison of Eq. 9.86 and the approximate linearization result given in Eq. 9.53 shows that integrating the two expressions will give exactly the same $\vec{\mu}_Y(t)$ values if $\vec{b}(t)$ is identically the same as $E\left(\vec{g}[\vec{Y}(t)]\right)$. This latter relationship, though, is exactly the linearization condition given in Eq. 9.48. Similarly, one can compare Eqs. 9.54 and 9.87, and note that they are identical if $\mathbf{A}(t)\mathbf{K}_{YY}(t,t) = \mathbf{K}_{gY}(t,t)$, which is exactly equivalent to the linearization condition given in Eq. 9.49. Thus, if one can manage to choose $\vec{b}(t)$ and $\mathbf{A}(t)$ in such a way that the conditions of Eqs. 9.48 and 9.49 are exactly satisfied, then statistical linearization will give exact values of the mean and covariance of response of the nonlinear system. In one sense, this provides a justification for use of the statistical linearization procedure which is much stronger than any argument that we were able to make in Section 9.4. On the other hand, one must realize that it is generally impossible to satisfy Eqs. 9.48 and 9.49. In order to satisfy these linearization conditions it would be necessary to know the probability distribution of the response vector $\vec{Y}(t)$, but this probability distribution is precisely the unknown quantity in the problem. Nonetheless, it is useful to know that any mean and covariance error of statistical linearization lies entirely in the evaluation of the linearization coefficients, and not in the difference between the linearized equation and the original nonlinear equation. For higher-order cumulants, of course, the situation is more complicated. For example, if $\{\vec{Q}(t)\}$ is Gaussian then the linearized system will give $\{\vec{Y}(t)\}$ as also being Gaussian, so that its higher-order cumulants are zero. Obviously, this is not true for the nonlinear problem.

Returning to consideration of the nonlinear state-space equations, note that the right-hand side of Eq. 9.87 is identical to that of Eq. 8.10. That is, the right-hand side of the equation is not affected by the nonlinearity which has been introduced into the system. Furthermore, the methods used in Section 8.4 to simplify this right-hand side are not in any way dependent on the system being linear. Thus, if the excitation vector $\vec{Q}(t)$ is delta-correlated with a nonstationary covariance of $K_{QQ}(t,s) = 2\pi \mathbf{S}_0(t)\delta(t-s)$, then one can write the nonlinear version of Eq. 8.18 as

$$\frac{d}{dt}\mathbf{K}_{YY}(t,t) + \mathbf{K}_{gY}(t,t) + \mathbf{K}_{Yg}(t,t) = 2\pi \mathbf{S}_0(t) \tag{9.88}$$

By switching to Kronecker notation we can also give the corresponding equations for higher-order cumulants. The general relationship which is analogous to Eq. 8.38 for the linear problem is

$$\frac{d}{dt}\kappa_j^{\otimes}[\vec{Y}(t),\cdots,\vec{Y}(t)] + \sum_{l=1}^{j}\kappa_j^{\otimes}[\underbrace{\vec{Y}(t),\cdots,\vec{Y}(t)}_{l-1},\vec{g}[Y(t)],\underbrace{\vec{Y}(t),\cdots,\vec{Y}(t)}_{j-l}]$$

$$= \sum_{l=1}^{j}\kappa_j^{\otimes}[\underbrace{\vec{Y}(t),\cdots,\vec{Y}(t)}_{l-1},\vec{Q}(t),\underbrace{\vec{Y}(t),\cdots,\vec{Y}(t)}_{j-l}] \tag{9.89}$$

For the special case of a delta-correlated excitation we will, as in Section 8.6, write the jth-order cumulant array for the excitation as $\kappa_j^{\otimes}[\vec{Q}(t_1),\cdots\vec{Q}(t_j)] = (2\pi)^{j-1}\vec{S}_j(t_j)\delta(t_1-t_j)\cdots\delta(t_{j-1}-t_j)$, which allows Eq. 9.89 to be rewritten as

$$\frac{d}{dt}\kappa_j^\otimes[\vec{Y}(t),\cdots,\vec{Y}(t)]+\sum_{l=1}^{j}\kappa_j^\otimes[\underbrace{\vec{Y}(t),\cdots,\vec{Y}(t)}_{l-1},\vec{g}[Y(t)],\underbrace{\vec{Y}(t),\cdots,\vec{Y}(t)}_{j-l}]=(2\pi)^{j-1}\vec{S}_j(t) \qquad (9.90)$$

The state-space cumulant relationships of Eqs. 9.87 and 9.89 give exact descriptions of the general situation in which the excitation of the nonlinear system is not delta-correlated, but they are less useful than the expressions in Eqs. 9.88 and 9.90 for delta-correlated excitations. This, of course, is because of the difficulty in evaluating the terms on the right-hand side of Eq. 9.89 for a general stochastic excitation. If the excitation can be modeled as a filtered delta-correlated process, then useful cumulant equations can be formulated by using the technique of analyzing the filter and the nonlinear system of interest as one composite system, as was mentioned in Chapter 8 for linear problems.

The difficulty in using Eq. 9.86, 9.88, or 9.90 to estimate response levels is in finding adequate approximations for cumulants involving $\vec{Y}(t)$ and the nonlinear function $g[\vec{Y}(t)]$. Not surprisingly, if $\{\vec{Q}(t)\}$ is a Gaussian process then it is common to approximate $\{\vec{Y}(t)\}$ as also being Gaussian. This is called **Gaussian closure.**[8] With this assumption the mean and cross-covariance terms involving $g[\vec{Y}(t)]$ can be written as functions of the mean and covariance of $\vec{Y}(t)$, so that it is possible to find solutions to the equations. In fact, these mean and covariance solutions obtained by considering $\vec{Y}(t)$ to be Gaussian are identical to those obtained by statistical linearization with the same assumption about the distribution of $\vec{Y}(t)$. For example, in state-space analysis the value of $E(\vec{g}[Y(t)])$ is taken as the coefficient in Eq. 9.86, whereas in statistical linearization it is used as the value of the vector $\vec{b}(t)$ of linearization coefficients, but using this value for $\vec{b}(t)$ makes Eq. 9.53 identical to Eq. 9.86. Similarly, the $\mathbf{K}_{gY}$ and $\mathbf{K}_{Yg}$ matrices can be taken either as the coefficients in the state-space formulation of Eq. 9.88, or as the values of $\mathbf{A}(t)\mathbf{K}_{YY}(t,t)$ and $\mathbf{K}_{YY}(t,t)\mathbf{A}^T(t)$, which are the corresponding coefficients in Eq. 9.55 for statistical linearization. Thus, state-space cumulant analysis and statistical linearization are completely equivalent when both are applied with the Gaussian assumption. This is in spite of the fact that there seem to be more assumptions involved in the statistical linearization approach.

In many nonlinear problems it is possible to write, at least approximately, the nonlinear function as a polynomial of the state variables. For example, this may be done by writing a power series expansion for the nonlinearity. Such a polynomial nonlinearity leads to an infinite set of coupled state-space equations governing the response cumulants or moments. To illustrate this situation in the simplest possible situation, we will consider the moment equations when the state vector of Eq. 9.85 is a scalar and the nonlinear function $g[Y(t)]$ is a polynomial. The equation of motion is then

$$\dot{Y}(t)+\sum_{k=0}^{K}a_k Y^k(t)=Q(t) \qquad (9.91)$$

---

[8]   The term "closure" comes from consideration of systems with polynomial nonlinearities, which are discussed in the following paragraphs.

and the moment equations can be found by multiplying this equation by powers of $Y(t)$ and then taking the expected value. In particular, multiplying by $Y^{j-1}(t)$ gives

$$E[Y^{j-1}(t)\dot{Y}(t)] + \sum_{k=0}^{K} a_k E[Y^{k+j-1}(t)] = E[Q(t)Y^{j-1}(t)]$$

which can be rewritten as

$$\frac{1}{j}\frac{d}{dt}E[Y^j(t)] + \sum_{k=0}^{K} a_k E[Y^{k+j-1}(t)] = E[Q(t)Y^{j-1}(t)] \qquad (9.92)$$

The derivative term in Eq. 9.92 demonstrates that this equation governs the evolution of the jth moment of $Y(t)$. The summation, though, includes moments up to the order $(K + j - 1)$, where K is the highest-order term included in the polynomial of Eq. 9.91. Taking $j = 1$ shows that the behavior of the mean value term $E[Y(t)]$ depends on the values of $E[Y^2(t)], \cdots, E[Y^K(t)]$. Similarly, using $j = 2$ shows that the evolution of $E[Y^2(t)]$ depends on the values of $E[Y(t)]$, $E[Y^3(t)], \cdots, E[Y^{K+1}(t)]$, etc. Since the evolution of any moment term depends on the values of still higher moment terms, the set of coupled equations continues without limit. Methods for truncating this infinite set of equations are commonly called **closure methods**.

Gaussian closure is the most common technique for truncating the set of equations which result from consideration of a polynomial nonlinearity. For the scalar situation of Eq. 9.92 this involves assuming that any moment of $Y(t)$ can be written in terms of the mean and variance in the same way as for a Gaussian random variable. In particular, it is assumed that

$$E[Y^m(t)] = E\left[\left(\mu_Y(t) + [Y(t) - \mu_Y(t)]\right)^m\right] = \sum_{r=0}^{m} \frac{m!}{r!(m-r)!}\mu_Y^{m-r}(t)E\left([Y(t) - \mu_Y(t)]^r\right)$$

or

$$E[Y^m(t)] = \sum_{r=0,2,4}^{[m]} \frac{m!}{r!(m-r)!}(1)(3)\cdots(r-1)\mu_Y^{m-r}(t)\sigma_Y^r(t) \qquad (9.93)$$

in which $[m]$ denotes either m or $m - 1$, depending on which is an even integer (see Eq. A.60 and Example A.32 of Appendix A). With this assumption, $\mu_Y(t)$ and $\sigma_Y(t)$ are the only unknowns in Eq. 9.92. Furthermore, by choosing two particular values of j in this equation one can obtain two simultaneous equations from which, in principle, one can solve for the values of $\mu_Y(t)$ and $\sigma_Y(t)$. One must expect that the values obtained for the unknowns will depend on the choices of j, since the other moment equations will generally not be satisfied. The usual procedure in applying such a closure method is to use the lowest-order moment equations which will serve the purpose. Thus, Gaussian closure for the scalar problem involves using Eq. 9.92 with $j = 1$ and with $j = 2$, giving the equations which directly govern the evolution of $\mu_Y(t)$ and $E[Y^2(t)] \equiv \mu_Y^2(t) + \sigma_Y^2(t)$. If symmetry dictates that $\mu_Y(t)$ is zero for the problem of interest, then the equation for $j = 1$ is trivial and the equation for $j = 2$ gives the evolution of the variance.

It may be noted that Gaussian closure of the state-space moment equations gives exactly the same results as simply assuming that $\{Y(t)\}$ is a Gaussian process, even though the formulation of the assumption is slightly different. In particular, Gaussian closure is often presented as an assumption that certain moments of $Y(t)$ are the same as they would be if $Y(t)$ were Gaussian, rather than as an assumption that $Y(t)$ truly is Gaussian, but the results are the same. Furthermore, we have previously demonstrated that the nonlinear state-space moment or cumulant equations are exactly the same as those for the system obtained by statistical linearization when $\{Y(t)\}$ is assumed to be Gaussian in both approaches. Thus, we see that the most straightforward implementations of Gaussian closure and statistical linearization make the two methods exactly equivalent.

A more general method to close the infinite set of moment equations for a problem with a polynomial nonlinearity is to assume that all cumulants of $Y(t)$ beyond some specified order are so small that they can be neglected (Crandall, 1980; Wu and Lin, 1984). If one chooses to consider only the first J cumulants of $Y(t)$, for example, then it is only necessary to consider the moment expressions from Eq. 9.92 for $j = 1$ to J, since there are only J unknowns to be evaluated. When J is chosen to be 2, this general cumulant neglect closure approach reduces to Gaussian closure. Cumulant neglect closure with $J > 2$ generally provides better approximations of the the mean and the variance of the response, and also gives at least crude approximations of other cumulants up to order J. On the other hand, the method does not necessarily converge as J is made larger. The reason seems to be related to the fact that it is not possible for a stochastic process to have zero values for all cumulants beyond order J unless $J = 1$ or 2 .[9] Gaussian closure can be viewed as approximating $\{Y(t)\}$ by a Gaussian process, but it is not possible to view any higher-order cumulant neglect closure scheme as corresponding to use of such a substitute process.

The generalization to a vector process of the polynomial nonlinearity of Eq. 9.91 has the form

$$\dot{\vec{Y}}(t) + \sum_{k=0}^{K} \mathbf{A}_k \vec{Y}^{[k]}(t) = \vec{Q}(t)$$

in which $\vec{Y}^{[k]}(t)$ is a Kronecker power of the vector and $\mathbf{A}_k$ is a matrix of dimension $n_Y \times n_Y^k$. The moment equation then becomes

$$\frac{d}{dt} E[\vec{Y}^{[j]}(t)] + \sum_{l=1}^{j} E[\vec{Y}^{[l-1]}(t) \otimes \sum_{k=0}^{K} \mathbf{A}_k \vec{Y}^{[k]}(t) \otimes \vec{Y}^{[j-l]}(t)] = \sum_{l=1}^{j} E[\vec{Y}^{[l-1]}(t) \otimes Q(t) \otimes \vec{Y}^{[j-l]}(t)]$$

Using the property of Eq. 8.32 this can also be rewritten as

---

[9]   Kendall and Stuart (1977) attribute this result to Marcinkiewicz in 1938.

$$\frac{d}{dt}E[\vec{Y}^{[j]}(t)] + \sum_{l=1}^{j}\sum_{k=0}^{K}(I^{[l-1]}\otimes \mathbf{A}_k \otimes I^{[j-l]})E[\vec{Y}^{[k+j-1]}(t)] = \sum_{l=1}^{j}E[\vec{Y}^{[l-1]}(t)\otimes Q(t) \otimes \vec{Y}^{[j-l]}(t)$$

(9.94)

Use of cumulant neglect closure with this formulation is somewhat complicated by the fact that the equation is written for moments rather than cumulants, and the relationship between the two becomes somewhat complicated for higher-order terms. Another difficulty is that the right-hand side of these equations is quite complicated, even if the excitation $\{Q(t)\}$ is delta-correlated. Both of these particular difficulties are alleviated by using the cumulant form of the equations. In this case, the right-hand side of the equation for a delta-correlated excitation is the same as in Eq. 9.90:

$$\frac{d}{dt}\kappa_j^{\otimes}[\vec{Y}(t),\cdots,\vec{Y}(t)] + \sum_{l=1}^{j}\sum_{k=0}^{K}(I^{[l-1]}\otimes \mathbf{A}_k \otimes I^{[j-l]})\kappa_j^{\otimes}[\underbrace{\vec{Y}(t),\cdots,\vec{Y}(t)}_{l-1}, \vec{Y}^{[k]}(t),\underbrace{\vec{Y}(t),\cdots,\vec{Y}(t)}_{j-l}]$$

$$= (2\pi)^{j-1}\vec{S}_j(t) \quad (9.95)$$

The primary difficulty in using Eq. 9.95 concerns the fact that the Kronecker cumulant term on the left-hand side of the equation includes the Kronecker product $\vec{Y}^{[k]}(t)$, in addition to $\vec{Y}(t)$ terms. In spite of the difficulties, procedures have been developed for the implementation of cumulant neglect closure for both the moment analysis of Eq. 9.94 (Di Paola and Muscolino, 1990; Di Paola et al., 1992; Di Paola and Falsone, 1993) and the cumulant analysis of Eq. 9.95 (Papadimitriou and Lutes, 1994; Papadimitriou, 1995).

At least one other approach has also been used to find approximate solutions to Eq. 9.90 or 9.95. This approach is based on the fact that one can often anticipate not only that certain state variables will have non-Gaussian distributions, but also the general sense in which they will differ from Gaussian. In particular, one can often predict that some variable will have a greater than Gaussian or a smaller than Gaussian probability of large excursions. In such a situation, it may be possible to assume a particular non-Gaussian form for the probability distribution of the state variable. One of the easiest ways to do this is to assume a Gaussian distribution for a new variable which is defined as a nonlinear function of the non-Gaussian state variable. This method was used by Iyengar and Dash (1978), who rewrote a state variable $Z(t)$ which was bounded by $\pm b$ as $Z(t) = (2b/\pi)\tan^{-1}[\hat{Z}(t)]$, and then assumed that $\hat{Z}(t)$ was Gaussian. In essence this amounts to rewriting the equation of motion for the system in terms of a new state variable $\hat{Z}(t)$ in place of $Z(t)$, then using Gaussian closure. In fact, Iyengar and Dash referred to the technique as Gaussian closure, even though it can also be considered to be a method of non-Gaussian closure for the problem formulated in terms of the original state variables. The method seems to have considerable potential for use, and it avoids the major difficulty of cumulant neglect closure schemes. That is, the new state variable, as illustrated by $\hat{Z}(t)$, is a physically meaningful quantity while a random variable with only J non-zero cumulants generally is not physically meaningful. No general procedures have as yet been formulated, though, for achieving non-Gaussian closure by using nonlinear transformations in conjunction with Gaussian closure.

****

Example 9.13: Let $\{X(t)\}$ denote the response of the system of Examples 9.1 and 9.5 with an equation of motion of

$$\dot{X}(t) + k_1 X(t) + k_3 X^3(t) = F(t)$$

in which $\{F(t)\}$ is a mean zero, stationary, delta-correlated process with autospectral density S_0. Write the state-space equation for the second moment of the system response, verify that the result agrees with that of statistical linearization, and apply Gaussian closure to simplify the equation.

This scalar system is described by Eq. 9.85 with $Y(t) = X(t)$, $g[X(t)] = k_1 X(t) + k_3 X^3(t)$ and $Q(t) = F(t)$. Since the system is symmetric and the excitation has a mean value of zero, the response gives both $X(t)$ and $g[X(t)]$ as also being mean zero. The state-space covariance relationship of Eq. 9.88 can then be written as

$$\frac{d}{dt} E[X^2(t)] + 2E\big(X(t)g[X(t)]\big) = 2\pi S_0$$

or

$$\frac{d}{dt} E[X^2(t)] + 2k_1 E[X^2(t)] + 2k_3 E[X^4(t)] = 2\pi S_0$$

One may note that this equation is also identical to the result of simply multiplying $X(t)$ times each term of the equation of motion, and then taking the expected value, since

$$\frac{d}{dt} E[X^2(t)] = 2E[X(t)\dot{X}(t)]$$

and $E[F(t)X(t)] = \pi S_0$. One always has the option of deriving state-space moment or cumulant equations by this direct procedure. The advantage of Eq. 9.88 is that it may help to clarify exactly which expressions are needed for the evaluation of given moments or cumulants in a more complicated problem.

Recall that in Example 9.5 we introduced a term a_1 which can be written as

$$a_1 = \frac{E\big(X(t)g[X(t)]\big)}{E[X^2(t)]} = k_1 + k_3 \frac{E[X^4(t)]}{E[X^2(t)]}$$

If we now introduce that term into our current analysis of the nonlinear system, the moment equation takes the form

$$\frac{d}{dt} E[X^2(t)] + 2a_1 E[X^2(t)] = 2\pi S_0$$

This relationship, though, is identical to the second moment state-space equation for the linear system described by $\dot{X}(t) + a_1 X(t) = F(t)$, which is the linearized model in Example 9.5. Since no approximations have been used in deriving this state-space relationship for the nonlinear problem, it exactly describes the second moment of the response. Thus, this example confirms the fact that statistical linearization also gave an exact relationship for the second moment of the nonlinear response. The difficulty in practice, of course, is that one cannot exactly evaluate the linearization coefficient a_1, since it involves another moment of the response. The approximate stationary solution based on a Gaussian assumption was given in Example 9.5.

In order to implement Gaussian closure, we now evaluate all the terms in the state-space moment equations in terms of the unknown variance (or second moment) of the response. We do this by assuming that all expected values are the same as they would be if $X(t)$ were Gaussian. In our state-space moment equations for the current problem, $E[X^4(t)]$ is the only term which is not already written in terms of the second moment of the response. Thus, we use the Gaussian assumption to obtain

$$E[X^4(t)] = \int_{-\infty}^{\infty} u^4 \, p_{X(t)}(u) \, du = \int_{-\infty}^{\infty} u^4 \frac{e^{-u^2/(2\sigma_X^2)}}{\sqrt{2\pi}\,\sigma_X} \, du = 3\sigma_X^4 = 3\left(E[X^2(t)]\right)^2$$

With this result, the state-space moment equation becomes

$$\frac{d}{dt} E[X^2(t)] + 2k_1 E[X^2(t)] + 6k_3 \left(E[X^2(t)]\right)^2 = 2\pi S_0$$

One could numerically solve this nonlinear differential equation to find an approximation of the nonstationary $E[X^2(t)]$. For the special case of stationary response, the equation becomes algebraic and the solution is as found in Example 9.5.
****

Example 9.14: Let $\{X(t)\}$ denote the response of the oscillator of Example 9.8 with an equation of motion of

$$m\ddot{X}(t) + c_1\dot{X}^3(t) + c_2 X^2(t)\dot{X}(t) + kX(t) = F(t)$$

in which $\{F(t)\}$ is a mean zero, delta-correlated process with autospectral density S_0. Write the general state-space equations for the second moments of the system response, verify that the results agree with those of nonstationary statistical linearization, and apply Gaussian closure to simplify the equations.

Using the usual state vector of $\vec{Y}(t) = [X(t), \dot{X}(t)]^T$, the component equations of motion are

$$\dot{Y}_1(t) - Y_2(t) = 0$$

and

$$m\dot{Y}_2(t) + c_1 Y_2^3(t) + c_2 Y_1^2(t)Y_2(t) + kY_1(t) = F(t)$$

This agrees with Eq. 9.85 if we write the nonlinear restoring force as

$$\vec{g}[\vec{Y}(t)] = \begin{bmatrix} -Y_2(t) \\ (c_1/m)Y_2^3(t) + (c_2/m)Y_1^2(t)Y_2(t) + (k/m)Y_1(t) \end{bmatrix}$$

and the excitation as $\vec{Q}(t) = [0, F(t)/m]^T$. Since the mean values are equal to zero, the cross-covariances are the same as cross-products, and the matrix $\mathbf{K}_{Yg}(t,t)$ in Eq. 9.88 is

$$\mathbf{K}_{Yg}(t,t) = \begin{bmatrix} -E[Y_1(t)Y_2(t)] & (c_1/m)E[Y_1(t)Y_2^3(t)] + (c_2/m)\,E[Y_1^3(t)Y_2(t)] + (k/m)E[Y_1^2(t)] \\ -E[Y_2^2(t)] & (c_1/m)E[Y_2^4(t)] + (c_2/m)\,E[Y_1^2(t)Y_2^2(t)] + (k/m)E[Y_1(t)Y_2(t)] \end{bmatrix}$$

Of course, $\mathbf{K}_{gY}(t,t)$ is the transpose of this expression. The $\mathbf{S}_0(t)$ matrix is simply

$$\mathbf{S}_0(t) = \begin{bmatrix} 0 & 0 \\ 0 & S_0/m^2 \end{bmatrix}$$

Thus, the three distinct component equations from Eq. 9.88 can be written as

$$\frac{d}{dt}E[X^2(t)] - 2E[X(t)\dot{X}(t)] = 0$$

$$\frac{d}{dt}E[X(t)\dot{X}(t)] + \frac{c_1}{m}E[X(t)\dot{X}^3(t)] + \frac{c_2}{m}E[X^3(t)\dot{X}(t)] + \frac{k}{m}E[X^2(t)] - E[\dot{X}^2(t)] = 0$$

and

$$\frac{d}{dt}E[\dot{X}^2(t)] + 2\frac{c_1}{m}E[\dot{X}^4(t)] + 2\frac{c_2}{m}E[X^2(t)\dot{X}^2(t)] + 2\frac{k}{m}E[X(t)\dot{X}(t)] = \frac{2\pi S_0}{m^2}$$

Note that the first of these three state-space moment equations is trivial, since it only states a well-known property of the derivative. Furthermore, the second and third equations could have been derived in an alternative direct manner. In particular, we can obtain two related equations by multiplying the original equation of motion by $X(t)$ and $\dot{X}(t)$, respectively, and then taking the expected value of the products. These equations are

$$mE[X(t)\ddot{X}(t)] + c_1E[X(t)\dot{X}^3(t)] + c_2\,E[X^3(t)\dot{X}(t)] + kE[X^2(t)] = 0$$

and

$$mE[\dot{X}(t)\ddot{X}(t)] + c_1E[\dot{X}^4(t)] + c_2\,E[X^2(t)\dot{X}^2(t)] + kE[X(t)\dot{X}(t)] = \frac{2\pi S_0}{m}$$

and noting that

$$\frac{d}{dt}E[X(t)\dot{X}(t)] = E[X(t)\ddot{X}(t)] + E[\dot{X}^2(t)] \quad \text{and} \quad \frac{d}{dt}E[\dot{X}^2(t)] = 2E[\dot{X}(t)\ddot{X}(t)]$$

allows these expressions to be converted into the two state-space moment equations already obtained. The right-hand sides of the equations, of course, have been simplified by using the relationships given in Section 8.4 for delta-correlated excitations.

As in Example 9.8, the linearized equation can be written as $m\ddot{X}(t) + a_2 \dot{X}(t) + (a_1+k)X(t) = F(t)$. The values of a_1 and a_2 for stationary response were obtained in Example 9.8, but more general results are required for comparison with the general nonstationary state-space equations. From Eq. 9.23, the general linearization relationship of $g[\vec{X}(t)] = c_1 \dot{X}^3(t) + c_2 X^2(t)\dot{X}(t)$ can be written as

$$\mathbf{K}_{XX}\vec{a} = Cov\left(\vec{X}(t), g[\vec{X}(t)]\right) \equiv \begin{bmatrix} c_1 K_{X,\dot{X}^3}(t,t) + c_2 K_{X,X^2\dot{X}}(t,t) \\ c_1 K_{\dot{X},\dot{X}^3}(t,t) + c_2 K_{\dot{X},X^2\dot{X}}(t,t) \end{bmatrix}$$

Since the processes are mean zero, though, the component equations can be written as

$$a_1 E[X^2(t)] + a_2 E[X(t)\dot{X}(t)] = c_1 E[X(t)\dot{X}^3(t)] + c_2 E[X^3(t)\dot{X}(t)]$$

and

$$a_1 E[X(t)\dot{X}(t)] + a_2 E[\dot{X}^2(t)] = c_1 E[\dot{X}^4(t)] + c_2 E[X^2(t)\dot{X}^2(t)]$$

The state-space equations for the linearized system are

$$\frac{d}{dt}E[X(t)\dot{X}(t)] - E[\dot{X}^2(t)] + \frac{a_2}{m}E[X(t)\dot{X}(t)] + \frac{a_1+k}{m}E[X^2(t)] = 0$$

and

$$\frac{d}{dt}E[\dot{X}^2(t)] + 2\frac{a_2}{m}E[\dot{X}^2(t)] + 2\frac{a_1+k}{m}E[X(t)\dot{X}(t)] = \frac{2\pi S_0}{m^2}$$

and substituting from the equations for a_1 and a_2 makes these equations identical to the component equations derived for the nonlinear system.

As in Example 9.13, we note that no approximations have been used in deriving these state-space relationships of the nonlinear problem, so they exactly describe the moments of the response. Thus, this example also confirms the fact that statistical linearization gives exact relationships for the second moments of the nonlinear response, but with the practical limitation that one cannot exactly evaluate the linearization coefficients a_1 and a_2. The approximate stationary solution based on a Gaussian assumption was given in Example 9.8.

To obtain the Gaussian closure approximation of the nonstationary nonlinear state-space moment equations, we also use relationships which would be true if $X(t)$ and $\dot{X}(t)$ were mean zero and jointly Gaussian. In particular, this gives

$$E[X(t)\dot{X}^3(t)] = 3E[X(t)\dot{X}(t)]E[\dot{X}^2(t)]$$

$$E[X^3(t)\dot{X}(t)] = 3E[X(t)\dot{X}(t)]E[X^2(t)]$$

$$E[\dot{X}^4(t)] = 3\left(E[\dot{X}^2(t)]\right)^2$$

and

$$E[X^2(t)\dot{X}^2(t)] = E[X^2(t)]E[\dot{X}^2(t)] + 2\left(E[X(t)\dot{X}(t)]\right)^2$$

Substituting these relationships gives the state-space equations as

$$\frac{d}{dt}E[X^2(t)] - 2E[X(t)\dot{X}(t)] = 0$$

$$\frac{d}{dt}E[X(t)\dot{X}(t)] + \frac{3c_1}{m}E[X(t)\dot{X}(t)]E[\dot{X}^2(t)]$$
$$+ \frac{3c_2}{m}E[X(t)\dot{X}(t)]E[X^2(t)] + \frac{k}{m}E[X^2(t)] - E[\dot{X}^2(t)] = 0$$

and

$$\frac{d}{dt}E[\dot{X}^2(t)] + \frac{6c_1}{m}\left(E[\dot{X}^2(t)]\right)^2 + \frac{2c_2}{m}E[X^2(t)]E[\dot{X}^2(t)]$$
$$+ \frac{4c_2}{m}\left(E[X(t)\dot{X}(t)]\right)^2 + 2\frac{k}{m}E[X(t)\dot{X}(t)] = \frac{2\pi S_0}{m^2}$$

In principle, one can now solve these three simultaneous differential equations for the three unknown second moment terms: $E[X^2(t)]$, $E[\dot{X}^2(t)]$, and $E[X(t)\dot{X}(t)]$. The problem is made somewhat complicated, of course, by the fact that two of the equations are nonlinear.

Restricting attention to the special case of stationary response simplifies the situation by making all the derivative terms in the equations be zero, and gives $E[X(t)\dot{X}(t)] = 0$. The values of $E[X^2(t)]$ and $E[\dot{X}^2(t)]$ are then exactly as found in Example 9.8.

****

Example 9.15: Consider the nonlinear oscillator described by

$$\ddot{X}(t) + c\dot{X}(t)\exp[-bX^2(t)] + kX^3(t)\dot{X}^2(t) = F(t)$$

in which $\{F(t)\}$ is a mean zero, stationary delta-correlated process with autospectral density S_0. Write the general state-space moment equations for the system, and find the approximate stationary solution which results from Gaussian closure.

Using the alternative approach presented in Example 9.14, we can obtain appropriate state-space moment equations by multiplying the equation of motion by the components of the $\vec{Y}(t) \equiv [X(t), \dot{X}(t)]^T$ state vector, then taking the expected values. The resulting equations are

$$E[X(t)\ddot{X}(t)] + cE\left(X(t)\dot{X}(t)\exp[-bX^2(t)]\right) + kE[X^4(t)\dot{X}^2(t)] = E[X(t)F(t)] = 0$$

and

$$E[\dot{X}(t)\ddot{X}(t)] + cE\left(\dot{X}^2(t)\exp[-bX^2(t)]\right) + kE[X^3(t)\dot{X}^3(t)] = E[\dot{X}(t)F(t)] = \pi S_0$$

For stationary response, though, we can use the simplifications that $E[\dot{X}(t)\ddot{X}(t)] = 0$, $E[X(t)\ddot{X}(t)] = -E[\dot{X}^2(t)]$, and $E\left(X(t)\dot{X}(t)\exp[-bX^2(t)]\right) = 0$. The last of these three relationships follows from the fact that the term has the form of $E\left(\dot{X}(t)f'[X(t)]\right)$, which is exactly the derivative with respect to t of $E(f[X(t)])$. Using these conditions gives the exact state-space equations for stationary response as

$$-E[\dot{X}^2(t)] + kE[X^4(t)\dot{X}^2(t)] = 0$$

and

$$cE\left(\dot{X}^2(t)\exp[-bX^2(t)]\right) + kE[X^3(t)\dot{X}^3(t)] = \pi S_0$$

Now we introduce the Gaussian assumption in order to further simplify the equations. In particular, we presume that $X(t)$ and $\dot{X}(t)$ are jointly Gaussian, which requires that they are also independent since they are uncorrelated. This gives

$$E[X^4(t)\dot{X}^2(t)] = E[X^4(t)]E[\dot{X}^2(t)] = 3\sigma_X^4\sigma_{\dot{X}}^2$$

$$E[X^3(t)\dot{X}^3(t)] = E[X^3(t)]E[\dot{X}^3(t)] = 0$$

and

$$E\left(\dot{X}^2(t)\exp[-bX^2(t)]\right) = E[\dot{X}^2(t)]E\left(\exp[-bX^2(t)]\right)$$

and the Gaussian distribution allows the final term to be evaluated as

$$E\left(\exp[-bX^2(t)]\right) = \frac{1}{\sqrt{2\pi}\,\sigma_X}\int_{-\infty}^{\infty}\exp\left[-u^2\left(\frac{2\sigma_X^2 b^2 + 1}{2\sigma_X^2}\right)\right]du$$

$$= \frac{1}{\sqrt{2\pi}\sqrt{2\sigma_X^2 b^2 + 1}}\int_{-\infty}^{\infty}e^{-w^2/2}\,dw = \frac{1}{\sqrt{2\sigma_X^2 b^2 + 1}}$$

Thus, the state-space equations become

$$-\sigma_{\dot{X}}^2 + 3k\sigma_X^4\sigma_{\dot{X}}^2 = 0$$

and

$$\frac{c\sigma_{\dot{X}}^2}{\sqrt{2\sigma_X^2 b^2 + 1}} = \pi S_0$$

From the first of these equations, we determine that

$$\sigma_X^2 = \frac{1}{\sqrt{3k}}$$

and the second equation then gives

$$\sigma_{\dot{X}}^2 = \frac{\pi S_0}{c}\sqrt{2\sigma_X^2 b^2 + 1} = \frac{\pi S_0}{c}\sqrt{\frac{2b^2}{\sqrt{3k}} + 1}$$

****

Exercises

****

Fokker-Planck Equation

****

9.1 Consider the response of the nonlinear system governed by the first-order differential equation

$$\dot{X}(t) + b\frac{X(t)}{|X(t)|} = F(t)$$

in which $\{F(t)\}$ is a mean zero, Gaussian, white noise process with an autospectral density of S_0.

(a) Give the Fokker-Planck equation governing the probability density of the first-order Markov process $\{X(t)\}$ and evaluate the two non-zero coefficients in that equation:

$$C^{(1)}(u,t) = \lim_{\Delta t \to 0}\frac{1}{\Delta t}E[\Delta X | X(t) = u]$$

$$C^{(2)}(u) = \lim_{\Delta t \to 0}\frac{1}{\Delta t}E\big[(\Delta X)^2 | X(t) = u\big]$$

(b) Verify that the Fokker-Planck equation is satisfied by the stationary probability density function

$$p_{X(t)}(u) = B\,\exp\!\left(-\frac{b|u|}{\pi S_0}\right)$$

and evaluate B for this density function.

(c) Evaluate the variance σ_X^2 and the kurtosis, $E[X^4(t)]/\sigma_X^4$, for this stationary response process.

****

9.2 Consider the response of the nonlinear system governed by the first-order differential equation

$$\dot{X}(t) + bX^5(t) = F(t)$$

in which $\{F(t)\}$ is a mean zero, Gaussian, white noise process with an autospectral density of S_0.

(a) Give the Fokker-Planck equation governing the probability density of the first-order Markov process $\{X(t)\}$ and evaluate the two non-zero coefficients, $C^{(1)}(u)$ and $C^{(2)}(u)$, in that equation.

(b) Verify that the Fokker-Planck equation has a stationary solution of

$$p_{X(t)}(u) = B\exp\!\left(-\frac{bu^6}{6\pi S_0}\right)$$

and evaluate B in that expression.

(c) Evaluate the variance σ_X^2 and the kurtosis, $E[X^4(t)]/\sigma_X^4$, for this stationary response process.

****

9.3 Consider the response of the nonlinear system governed by the first-order differential equation

$$\dot{X}(t) + b\frac{X(t)}{\sqrt{|X(t)|}} = F(t)$$

in which $\{F(t)\}$ is a mean zero, Gaussian, white noise process with an autospectral density of S_0.

(a) Give the Fokker-Planck equation governing the probability density of the first-order Markov process $\{X(t)\}$ and evaluate the two non-zero coefficients, $C^{(1)}(u)$ and $C^{(2)}(u)$, in that equation.

(b) Verify that the Fokker-Planck equation has a stationary solution of

$$p_{X(t)}(u) = B \exp\left(-\frac{2b|u|^{3/2}}{3\pi S_0}\right)$$

and evaluate B in that expression.

(c) Evaluate the variance σ_X^2 and the kurtosis, $E[X^4(t)]/\sigma_X^4$, for this stationary response process.

****

9.4 Consider the response of the nonlinear system governed by the first-order differential equation

$$\dot{X}(t) + k_1 X(t) + k_5 X^5(t) = F(t)$$

in which $\{F(t)\}$ is a mean zero, Gaussian, white noise process with an autocorrelation function of
$$E[F(t)F(s)] = 2\pi S_0 \delta(t-s)$$

(a) Give the Fokker-Planck equation governing the probability density of the first-order Markov process $\{X(t)\}$ and evaluate the two non-zero coefficients in that equation: $C^{(1)}(u)$ and $C^{(2)}(u)$.

(b) Verify that the Fokker-Planck equation is satisfied by the stationary probability density function

$$p_{X(t)}(u) = B \exp\left(-\frac{1}{\pi S_0}\left[k_1\frac{u^2}{2} + k_5\frac{u^6}{6}\right]\right)$$

in which B is a constant chosen to make the integral of the expression with respect to u be unity.

****

9.5 Consider the nonlinear system governed by the second-order differential equation

$$m\ddot{X}(t) + c\dot{X}(t) + k\frac{X(t)}{|X(t)|} = F(t)$$

in which $\{F(t)\}$ is a mean zero, Gaussian, white noise process with an autocorrelation function of
$$E[F(t)F(s)] = 2\pi S_0 \delta(t-s)$$

(a) Using the state vector $\vec{Y}(t) = [X(t), \dot{X}(t)]^T$, give the Fokker-Planck equation governing the probability density $p_{\vec{Y}(t)}(\vec{u})$ for the vector Markov process $\{\vec{Y}(t)\}$ and evaluate the first- and second-order coefficients in that equation:

$$C_1^{(1)}(\vec{u}) = \lim_{\Delta t \to 0} \frac{1}{\Delta t} E\left[\Delta Y_1 \mid \vec{Y}(t) = \vec{u}\right], \qquad C_2^{(1)}(\vec{u}) = \lim_{\Delta t \to 0} \frac{1}{\Delta t} E\left[\Delta Y_2 \mid \vec{Y}(t) = \vec{u}\right]$$

$$C^{(2,0)}(\vec{u}) = \lim_{\Delta t \to 0} \frac{1}{\Delta t} E\left[(\Delta Y_1)^2 \mid \vec{Y}(t) = \vec{u}\right], \qquad C^{(0,2)}(\vec{u}) = \lim_{\Delta t \to 0} \frac{1}{\Delta t} E\left[(\Delta Y_2)^2 \mid \vec{Y}(t) = \vec{u}\right]$$

$$C^{(1,1)}(\vec{u}) = \lim_{\Delta t \to 0} \frac{1}{\Delta t} E\left[\Delta Y_1 \Delta Y_2 \mid \vec{Y}(t) = \vec{u}\right]$$

(b) Verify that the Fokker-Planck equation has a stationary solution of

$$p_{\vec{Y}(t)}(\vec{u}) = B \exp\left[-\frac{c}{\pi S_0}\left(k|u_1| + \frac{mu_2^2}{2}\right)\right]$$

and evaluate B in that expression.

(c) Find stationary values of σ_X^2 and $\sigma_{\dot{X}}^2$.

(d) Find stationary values of the kurtosis of $X(t)$ and $\dot{X}(t)$.

****

9.6 Consider the nonlinear system governed by the second-order differential equation

$$m\ddot{X}(t) + c\dot{X}(t) + k|X(t)|X(t) = F(t)$$

in which $\{F(t)\}$ is a mean zero, Gaussian, white noise process with an autospectral density of S_0.

(a) Using the state vector $\vec{Y}(t) = [X(t), \dot{X}(t)]^T$, give the Fokker-Planck equation governing the probability density $p_{\vec{Y}(t)}(\vec{u})$ for the vector Markov process $\{\vec{Y}(t)\}$ and evaluate the first- and second-order coefficients in that equation.

(b) Verify that the Fokker-Planck equation has a stationary solution of

$$p_{\vec{Y}(t)}(\vec{u}) = B \; \exp\left[-\frac{c}{\pi S_0}\left(\frac{k|u_1^3|}{3} + \frac{mu_2^2}{2}\right)\right]$$

and evaluate B in that expression.

(c) Find stationary values of σ_X^2 and $\sigma_{\dot{X}}^2$.

(d) Find stationary values of the kurtosis of $X(t)$ and $\dot{X}(t)$.

****

9.7 Consider the nonlinear system governed by the second-order differential equation

$$m\ddot{X}(t) + c_2\left|\dot{X}(t)\right|\dot{X}(t) + k_1 X(t) + k_5 X^5(t) = F(t)$$

in which $\{F(t)\}$ is a mean zero, Gaussian, white noise process with autospectral density of S_0.

Using the state vector $\vec{Y}(t) = [X(t), \dot{X}(t)]^T$, give the Fokker-Planck equation governing the probability density $p_{\vec{Y}(t)}(\vec{u})$ for the vector Markov process $\{\vec{Y}(t)\}$ and evaluate the first- and second-order coefficients in that equation.

****

9.8 Consider the nonlinear system governed by the second-order differential equation

$$\ddot{X}(t) + c\dot{X}^3(t) + k|X(t)|^{1/2} X(t) = F(t)$$

in which $\{F(t)\}$ is a mean zero, Gaussian, white noise process with an autospectral density of S_0.

Using the state vector $\vec{Y}(t) = [X(t), \dot{X}(t)]^T$, give the Fokker-Planck equation governing the probability density $p_{\vec{Y}(t)}(\vec{u})$ for the vector Markov process $\{\vec{Y}(t)\}$ and evaluate the first- and second-order coefficients in that equation.

****

9.9 The Fokker-Planck equation for a certain vector Markov process $\{\vec{Y}(t)\}$ has the form

$$\frac{\partial}{\partial t}p_{\vec{Y}(t)}(\vec{u}) + u_2\frac{\partial}{\partial u_1}p_{\vec{Y}(t)}(\vec{u}) - \left(2bu_1u_2 + 3cu_1^2u_2^2\right)p_{\vec{Y}(t)}(\vec{u})$$

$$-\left(bu_1u_2^2 + cu_1^2u_2^3\right)\frac{\partial}{\partial u_2}p_{\vec{Y}(t)}(\vec{u}) = \pi S_0\frac{\partial^2}{\partial u_2^2}p_{\vec{Y}(t)}(\vec{u})$$

Furthermore, this $\{\vec{Y}(t)\}$ process is the state vector $\vec{Y}(t) = [X(t), \dot{X}(t)]^T$ for the solution of a differential equation $\ddot{X}(t) + g[X(t), \dot{X}(t)] = F(t)$ in which $\{F(t)\}$ is a mean zero, Gaussian, white noise process with an autospectral density of S_0.

Find the appropriate $g[X(t), \dot{X}(t)]$ function in this equation of motion.

****

Statistical Linearization
****
9.10 Let $\{X(t)\}$ denote the response of the system of Exercise 9.1 with an equation of motion of
$$\dot{X}(t) + b\frac{X(t)}{|X(t)|} = F(t)$$
in which $\{F(t)\}$ is a mean zero, Gaussian, white noise process with an autospectral density of S_0.
(a) Using statistical linearization, find an expression for the parameter a_1 of Eqs. 9.26 and 9.27.
(b) Evaluate a_1 in terms of σ_X^2 using the assumption that $\{X(t)\}$ is a stationary Gaussian process.
(c) Verify that this stationary value of a_1 agrees with the results of Eq. 9.24.
(d) Estimate the value of σ_X^2 for stationary response.
****
9.11 Let $\{X(t)\}$ denote the response of the system of Exercise 9.2 with an equation of motion of
$$\dot{X}(t) + bX^5(t) = F(t)$$
in which $\{F(t)\}$ is a mean zero, Gaussian, white noise process with an autospectral density of S_0.
(a) Using statistical linearization, find an expression for the parameter a_1 of Eqs. 9.26 and 9.27.
(b) Evaluate a_1 in terms of σ_X^2 using the assumption that $\{X(t)\}$ is a stationary Gaussian process.
(c) Verify that this stationary value of a_1 agrees with the results of Eq. 9.24.
(d) Estimate the value of σ_X^2 for stationary response.
****
9.12 Let $\{X(t)\}$ denote the response of the system of Exercise 9.3 with an equation of motion of
$$\dot{X}(t) + b\frac{X(t)}{\sqrt{|X(t)|}} = F(t)$$
in which $\{F(t)\}$ is a mean zero, Gaussian, white noise process with an autospectral density of S_0.
(a) Using statistical linearization, find an expression for the parameter a_1 of Eqs. 9.26 and 9.27.
(b) Evaluate a_1 in terms of σ_X^2 using the assumption that $\{X(t)\}$ is a stationary Gaussian process.
(c) Verify that this stationary value of a_1 agrees with the results of Eq. 9.24.
(d) Estimate the value of σ_X^2 for stationary response.
****
9.13 Let $\{X(t)\}$ denote the response of the system of Exercise 9.4 with an equation of motion of
$$\dot{X}(t) + k_1 X(t) + k_5 X^5(t) = F(t)$$
in which $\{F(t)\}$ is a mean zero, Gaussian, white noise process with an autospectral density of S_0.
(a) Using statistical linearization, find an expression for the parameter a_1 of Eqs. 9.26 and 9.27.
(b) Evaluate a_1 in terms of σ_X^2 using the assumption that $\{X(t)\}$ is a stationary Gaussian process.
(c) Verify that this stationary value of a_1 agrees with the results of Eq. 9.24.
(d) Find a cubic algebraic equation which could be solved to obtain an estimate of the value of σ_X^2 for stationary response.
****
9.14 Let $\{X(t)\}$ denote the response of the system of Exercise 9.5 with an equation of motion of
$$m\ddot{X}(t) + c\dot{X}(t) + k\frac{X(t)}{|X(t)|} = F(t)$$
in which $\{F(t)\}$ is a mean zero, Gaussian, white noise process with an autospectral density of S_0.
(a) Using statistical linearization, find expressions for the a_1 and a_2 parameters of Eqs. 9.31, 9.34,

and 9.35.

(b) Evaluate a_1 and a_2 in terms of σ_X^2 and $\sigma_{\dot{X}}^2$ using the assumption that $\{X(t)\}$ is a stationary Gaussian process.

(c) Verify that the stationary values of a_1 and a_2 agree with the results of Eq. 9.24.

(d) Estimate the values of σ_X^2 and $\sigma_{\dot{X}}^2$ for stationary response.

****

9.15 Let $\{X(t)\}$ denote the response of the system of Exercise 9.6 with an equation of motion of
$$m\ddot{X}(t) + c\dot{X}(t) + k|X(t)|X(t) = F(t)$$
in which $\{F(t)\}$ is a mean zero, Gaussian, white noise process with an autospectral density of S_0.

(a) Using statistical linearization, find expressions for the a_1 and a_2 parameters of Eqs. 9.31, 9.34, and 9.35.

(b) Evaluate a_1 and a_2 in terms of σ_X^2 and $\sigma_{\dot{X}}^2$ using the assumption that $\{X(t)\}$ is a stationary Gaussian process.

(c) Verify that the stationary values of a_1 and a_2 agree with the results of Eq. 9.24.

(d) Estimate the values of σ_X^2 and $\sigma_{\dot{X}}^2$ for stationary response.

****

9.16 Let $\{X(t)\}$ denote the response of the system of Exercise 9.7 with an equation of motion of
$$m\ddot{X}(t) + c_2|\dot{X}(t)|\dot{X}(t) + k_1 X(t) + k_5 X^5(t) = F(t)$$
in which $\{F(t)\}$ is a mean zero, Gaussian, white noise process with autospectral density of S_0.

(a) Using statistical linearization, find expressions for the a_1 and a_2 parameters of Eqs. 9.31, 9.34, and 9.35.

(b) Evaluate a_1 and a_2 in terms of σ_X^2 and $\sigma_{\dot{X}}^2$ using the assumption that $\{X(t)\}$ is a stationary Gaussian process.

(c) Verify that the stationary values of a_1 and a_2 agree with the results of Eq. 9.24.

(d) Estimate the value of $\sigma_{\dot{X}}^2$ for stationary response, and find a cubic algebraic equation which could be solved to obtain a corresponding estimate of the value of σ_X^2.

****

9.17 Let $\{X(t)\}$ denote the response of the system of Exercise 9.8 with an equation of motion of
$$\ddot{X}(t) + c\dot{X}^3(t) + k|X(t)|^{1/2} X(t) = F(t)$$
in which $\{F(t)\}$ is a mean zero, Gaussian, white noise process with an autospectral density of S_0.

(a) Using statistical linearization, find expressions for the a_1 and a_2 parameters of Eqs.9.31, 9.34, and 9.35.

(b) Evaluate a_1 and a_2 in terms of σ_X^2 and $\sigma_{\dot{X}}^2$ using the assumption that $\{X(t)\}$ is a stationary Gaussian process.

(c) Verify that the stationary values of a_1 and a_2 agree with the results of Eq. 9.24.

(d) Estimate the values of σ_X^2 and $\sigma_{\dot{X}}^2$ for stationary response.

****

State-Space Moment Equations

****

9.18 Let $\{X(t)\}$ denote the response of the system of Exercises 9.1 and 9.10 with an equation of motion of

$$\dot{X}(t) + b\frac{X(t)}{|X(t)|} = F(t)$$

in which $\{F(t)\}$ is a mean zero, Gaussian, white noise process with an autospectral density of S_0.

(a) Derive an exact state-space equation for the second moment of the response.

(b) Verify that the state-space moment equation in part (a) is the same as for the linearized system of Exercise 9.10.

(c) Use Gaussian closure to estimate the value of σ_X^2 for stationary response.

****

9.19 Let $\{X(t)\}$ denote the response of the system of Exercises 9.2 and 9.11 with an equation of motion of

$$\dot{X}(t) + bX^5(t) = F(t)$$

in which $\{F(t)\}$ is a mean zero, Gaussian, white noise process with an autospectral density of S_0.

(a) Derive an exact state-space equation for the second moment of the response.

(b) Verify that the state-space moment equation in part (a) is the same as for the linearized system of Exercise 9.11.

(c) Use Gaussian closure to estimate the value of σ_X^2 for stationary response.

****

9.20 Let $\{X(t)\}$ denote the response of the system of Exercises 9.3 and 9.12 with an equation of motion of

$$\dot{X}(t) + b\frac{X(t)}{\sqrt{|X(t)|}} = F(t)$$

in which $\{F(t)\}$ is a mean zero, Gaussian, white noise process with an autospectral density of S_0.

(a) Derive an exact state-space equation for the second moment of the response.

(b) Verify that the state-space moment equation in part (a) is the same as for the linearized system of Exercise 9.12.

(c) Use Gaussian closure to estimate the value of σ_X^2 for stationary response.

****

9.21 Let $\{X(t)\}$ denote the response of the system of Exercises 9.4 and 9.13 with an equation of motion of

$$\dot{X}(t) + k_1 X(t) + k_5 X^5(t) = F(t)$$

in which $\{F(t)\}$ is a mean zero, Gaussian, white noise process with an autospectral density of S_0.

(a) Derive an exact state-space equation for the second moment of the response.

(b) Verify that the state-space moment equation in part (a) is the same as for the linearized system of Exercise 9.13.

(c) Use Gaussian closure to find a cubic algebraic equation which could be solved to obtain an estimate of the value of σ_X^2 for stationary response.

****

9.22 Let $\{X(t)\}$ denote the response of the system of Exercises 9.5 and 9.14 with an equation of motion of

$$m\ddot{X}(t) + c\dot{X}(t) + k\frac{X(t)}{|X(t)|} = F(t)$$

in which $\{F(t)\}$ is a mean zero, Gaussian, white noise process with an autospectral density of S_0.

(a) Derive exact state-space equations for the second moments of the response of the system.

(b) Verify that the state-space moment equations in part (a) are the same as for the linearized system of Exercise 9.14.

(c) Use Gaussian closure to estimate the values of σ_X^2 and $\sigma_{\dot{X}}^2$ for stationary response.

****

9.23 Let $\{X(t)\}$ denote the response of the system of Exercises 9.6 and 9.15 with an equation of motion of

$$m\ddot{X}(t) + c\dot{X}(t) + k|X(t)|X(t) = F(t)$$

in which $\{F(t)\}$ is a mean zero, Gaussian, white noise process with an autospectral density of S_0.

(a) Derive exact state-space equations for the second moments of the response of the system.

(b) Verify that the state-space moment equations in part (a) are the same as for the linearized system of Exercise 9.15.

(c) Use Gaussian closure to estimate the values of σ_X^2 and $\sigma_{\dot{X}}^2$ for stationary response.

****

9.24 Let $\{X(t)\}$ denote the response of the system of Exercises 9.7 and 9.16 with an equation of motion of

$$m\ddot{X}(t) + c_2|\dot{X}(t)|\dot{X}(t) + k_1 X(t) + k_5 X^5(t) = F(t)$$

in which $\{F(t)\}$ is a mean zero, Gaussian, white noise process with autospectral density of S_0.

(a) Derive exact state-space equations for the second moments of the response of the system.

(b) Verify that the state-space moment equations in part (a) are the same as for the linearized system of Exercise 9.16.

(c) Use Gaussian closure to estimate the value of $\sigma_{\dot{X}}^2$ for stationary response, and to find a cubic algebraic equation which could be solved to obtain an estimate of the corresponding value of σ_X^2.

****

9.25 Let $\{X(t)\}$ denote the response of the system of Exercises 9.8 and 9.17 with an equation of motion of

$$\ddot{X}(t) + c\dot{X}^3(t) + k|X(t)|^{1/2} X(t) = F(t)$$

in which $\{F(t)\}$ is a mean zero, Gaussian, white noise process with an autospectral density of S_0.

(a) Derive exact state-space equations for the second moments of the response of the system.

(b) Verify that the state-space moment equations in part (a) are the same as for the linearized system of Exercise 9.17.

(c) Use Gaussian closure to estimate the values of σ_X^2 and $\sigma_{\dot{X}}^2$ for stationary response.

****

Chapter 10

Stochastic Analysis of Fatigue Damage

10.1 Fundamentals of Structural Fatigue

Structural or mechanical fatigue can be defined as the process of accumulation of damage due to application of a time-varying strain. It can be expected to occur whenever a structure is subjected to time-varying loads, and in many situations it may govern the design of the structure. Each time a load cycle is applied, an incremental amount of damage occurs. This damage is cumulative in nature and the accumulation of damage continues until failure occurs. If fatigue cracks are detected early enough, then repair may be possible, but such detection is not always possible or feasible. The results can be disastrous failures. Dramatic examples have included aircraft, offshore oil platforms, and highway bridges.

The concern here will be with fatigue due to loads which vary in such an erratic manner that they are appropriately modeled as stochastic processes. Such behavior is typical of many environmental loadings which cause fatigue. Even though the details of the time histories of loadings or responses which will occur in the future cannot be predicted, it is presumed that sufficient consistency exists that certain statistics (or averages) of those future time histories can be predicted. This allows estimations of probabilities of failure, mean time to failure, etc., although the failure time for any single structure or element remains unknown.

The fact of damage accumulation is readily observable in some situations in which one or more cracks can be observed to be propagating. However, prior to the existence of a visible crack there may be a significant period during which damage is accumulating within the structure. Some analysts consider this initial phase simply to be the microscopic portion of the crack growth, while others make a greater distinction between the crack initiation and the crack propagation phases. In this introductory discussion we will use the term "fatigue" to describe both of the mentioned situations, and the concept of accumulated damage measured by a damage function $D(t)$ will be used to denote progress toward failure whether or not that progress is observable.

The damage function $D(t)$ is presumed to start at or very near zero for a new structure, and is normalized to be unity when failure occurs. Furthermore, it is a non-decreasing function of time. One should note, though, that this is a very vague specification of the function. For example, there are many definitions of the term "failure" in common usage. These vary from "appearance of a visible crack" to "complete fracture." Even for a given definition of failure, though, there are many possibilities for the damage function. For example, if some function $f(t)$ goes monotonically from zero to unity as t goes from zero to failure time (T), then $f(t)$ is one possible choice for the damage function $D(t)$. However $f^n(t)$, for any positive exponent n, also satisfies the conditions for a damage function. Even in a simple situation like the propagation of a one-dimensional crack,

377

the definition of damage is not unique. The length of the crack may be observable, but there is no rule that the damage function must be taken to be proportional to the length of the crack. Thus, the challenge is to identify a useful damage model, rather than to find the true damage model. The models discussed here are quite simple, but have been found to be very useful.

In order to apply stochastic process analysis to fatigue prediction, it is first necessary to have a model relating fatigue damage to time history characteristics for deterministic problems. Thus, we will begin with a brief investigation of the concept of damage accumulation under deterministic loadings, before proceeding to simple stochastic approaches. The simplest of all the deterministic fatigue loadings is periodic in nature, which is classified as **constant amplitude** in the usual nomenclature of fatigue. Empirical data from such constant amplitude testing forms the basis for all of our methods for predicting fatigue life under more complicated loadings.

Constant amplitude fatigue behavior is determined experimentally from tests in which a load or deflection is controlled and varied in a simple periodic (possibly harmonic) manner until failure occurs. In this situation, fatigue failure is usually found to depend significantly on only two characteristics of the time history of stress. These two characteristics can be taken as the minimum and maximum values of stress during the cycle. An equivalent formulation uses the **mean stress** value, defined as the average of the minimum and maximum stresses, and the **stress range** value, defined as the maximum stress minus the minimum stress. The exact shape of the cycle of periodic loading has been found not to be important for predicting the number of constant amplitude cycles until failure. Neither has the frequency of loading been found to be important except for either very high frequencies or situations of corrosion fatigue. Furthermore, the stress range effect is usually found to be considerably more important than the mean stress effect. Thus, the result of a constant amplitude fatigue test is often described by two bits of information: the stress range, which we will denote by S_r; and the number of cycles until failure, denoted by N_f. A typical experimental investigation of constant amplitude fatigue for specimens of a given configuration and material involves performing a large number of tests including a number of values of S_r, then plotting the (S_r, N_f) results. This is called an **S-N curve**.[1]

Since S_r is a scalar quantity, the typical S-N curve is an ordinary two-dimensional plot. Theoretically, though, one can extend the idea of an S-N relationship to more complicated periodic situations, even though the geometry needed to describe the results becomes more involved. For example, if $\vec{S}_r$ is a two-dimensional vector representing both stress range and mean stress, then N_f versus $\vec{S}_r$ is a surface rather than a curve. One could further refine the model by letting $\vec{S}_r$ include characteristics of several components of stress, such as the three stress invariants. The basic idea would remain unchanged, though, and the term S-N curve could still be used for the relationship between the observed life N_f and the characteristics of the applied loading which have been included in $\vec{S}_r$.

---

[1] It should be noted that S_r is a completely different concept than spectral density, although the symbol S is used for both. Hopefully the subscript r on the stress range will assist in avoiding confusion about the notation.

Ignoring, for the moment, any effect of mean stress, we will emphasize the dependence of fatigue life on stress range by writing $N_f(S_r)$ for the fatigue life observed for a given value S_r of the stress range. In principle, the S-N curve of $N_f(S_r)$ versus S_r could be any non-increasing curve, but experimental data commonly show that a large portion of that curve is well approximated by an equation of the form

$$N_f(S_r) = K S_r^{-m} \qquad\qquad (10.1)$$

in which K and m are positive constants whose values depend on both the material and the geometry of the specimen. This S-N form plots as a straight line on log-log paper (see Example 10.1). If S_r is given either very small or very large values, then the form of Eq. 10.1 will generally no longer be appropriate. For many materials, N_f seems to go to infinity when S_r is smaller than some particular value. This particular value is called the fatigue limit or endurance limit. At the other extreme, a sufficiently large S_r value may produce an unstable situation in which failure occurs rapidly. Fatigue analysis, though, usually consists of predicting failure due to moderately large loads, so that Eq. 10.1 is often quite useful.

****

Example 10.1: The table gives the fatigue results for a set of welded specimens. Find an S-N curve of the form of Eq. 10.1 to approximate the data.

The sketch shows the 16 data points plotted in the traditional way of $\log(S_r)$ versus $\log(N_f)$.

S_r MPa	N_f
667	18,360
667	21,880
667	16,610
667	15,340
667	14,400
533	31,550
533	39,170
533	32,774
267	432,000
267	356,400
267	417,000
267	284,480
267	212,350
267	716,600
267	386,840
267	202,814

Note that even for this relatively small data set there is significant scatter of the N_f values observed for loadings with identical values of S_r, particularly at the lowest S_r level, for which the largest N_f value is almost 3.5 times larger than the smallest one. Given this scatter, which is not at all unusual for fatigue results, it is necessary to use some statistical procedure to choose the best values of K and m. This is usually done by performing a so-called linear regression of $\log(N_f)$ on $\log(S_r)$.[2] This is simply the procedure of choosing K and m to minimize the value of

---

[2] It should be noted that this linear regression of data does not involve any use of a probability model, so is distinctly different from the probabilistic procedure with the same name, which is discussed briefly in Section A.8 of Appendix A.

$$\sum_{j=1}^{J}\Big(\log(N_f)_j - [\log(K) - m\log(S_r)_j]\Big)^2$$

in which $\log(S_r)_j$ and $\log(N_f)_j$ denote the jth data point, and J is the total number of data points. Computer algorithms for this calculation are readily available, and the result can also be written as

$$\log(K) = \frac{\sum_{j=1}^{J}[\log(S_r)_j]^2 \sum_{j=1}^{J}\log(N_f)_j - \sum_{j=1}^{J}\log(S_r)_j \sum_{j=1}^{J}[\log(S_r)_j \log(N_f)_j]}{J\sum_{j=1}^{J}[\log(S_r)_j]^2 - \left(\sum_{j=1}^{J}\log(S_r)_j\right)^2}$$

and

$$m = \frac{\sum_{j=1}^{J}\log(S_r)_j \sum_{j=1}^{J}\log(N_f)_j - J\sum_{j=1}^{J}[\log(S_r)_j \log(N_f)_j]}{J\sum_{j=1}^{J}[\log(S_r)_j]^2 - \left(\sum_{j=1}^{J}\log(S_r)_j\right)^2}$$

The data in the table give values of $m = 3.303$ and $\log(K) = 31.22$. One of the simplest ways to write the resulting S-N relationship is as

$$N_f = \left(\frac{12,700}{S_r}\right)^{3.303}$$

****

As noted, the second most important characteristic of the constant amplitude loading is usually considered to be the mean stress. Although mean stress is often neglected in fatigue prediction, there are simple empirical formulas which can be used to account for it in an approximate way. The simplest such approach is called the Goodman correction, and it postulates that the damage done by a loading $x(t)$ with mean stress x_m and stress range S_r is the same as would be done by another loading with mean zero and range S_e such that

$$\frac{S_r}{S_e} + \frac{x_m}{x_u} = 1$$

in which x_u denotes the ultimate stress capacity of the material. The Gerber formula is similar, and has often been determined to be in better agreement with empirical data:

$$\frac{S_r}{S_e} + \left(\frac{x_m}{x_u}\right)^2 = 1 \tag{10.2}$$

Clearly the Gerber equation typically gives fairly small differences between S_e and S_r. For example, choosing $x_m = 0.3x_u$ gives $S_e = 1.10S_r$. This appears to be a fairly minor change in the stress range, but it can still have a noticeable effect on the predicted fatigue life. The value of the exponent m in Eq. 10.1 may be as small as about 3 for a specimen with a weld or a sharp stress concentration, but it may exceed 10 for a very smooth specimen of homogeneous material. In these two particular cases, a 10% increase in the effective stress range gives 25% and 61% decreases in N_f, respectively. Note that use of the Gerber (or Goodman) formula allows one to include both mean stress and stress range in fatigue prediction while maintaining the simplicity of a one-dimensional S-N relationship.

****

Example 10.2: The mean stress was not zero in the tests reported in Example 10.1, but rather had the values shown in the table. The ultimate stress of the material was $x_u = 516$ MPa. Find an S-N curve of the form of Eq. 10.1 that uses an effective stress range based on the Gerber correction.

The Gerber formula of Eq. 10.2 gives the effective stress range values as $S_e = 314$ MPa and 542 MPa for $S_r = 267$ MPa and 533 MPa, respectively. For $S_r = 667$ MPa we have $S_e = S_r$, since there is no mean stress. The sketch shows the S-N data plotted using S_e, and linear regression gives the S-N curve shown, which is described by

$$N_f = \left(\frac{7,350}{S_e}\right)^{4.043}$$

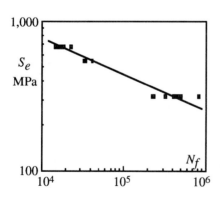

S_r MPa	x_m MPa	N_f
667	0	18,360
667	0	21,880
667	0	16,610
667	0	15,340
667	0	14,400
533	66.7	31,550
533	66.7	39,170
533	66.7	32,774
267	200	432,000
267	200	356,400
267	200	417,000
267	200	284,480
267	200	212,350
267	200	716,600
267	200	386,840
267	200	202,814

Note that the values of both m and K are changed significantly by the inclusion of the mean stress for this particular example. The change in m, in particular, is related to the fact that the mean stress is directly tied to S_r for this data set. If x_m were not correlated with S_r, one would expect it might cause a significant change in the position, but not the slope, of the S-N curve. This would amount to a change in K, but not in m.
****

10.2   Modeling Accumulated Damage

In order to develop a more precise model for the damage, let ΔD_j denote the increment of damage during cycle j, and $N(t)$ denote the number of applied cycles of load up to time t. Thus, we can say that

$$D(t) = \sum_{j=1}^{N(t)} \Delta D_j \tag{10.3}$$

Further, let T = failure time, so that $N_f = N(T)$ is the number of cycles to failure. This gives $D(T) = 1$, so that

$$1 = \sum_{j=1}^{N(T)} \Delta D_j = \sum_{j=1}^{N_f} \Delta D_j \tag{10.4}$$

For a periodic constant amplitude loading with stress range S_r, the upper limit of the summation is $N_f(S_r)$. If we then assume that the ΔD_j values are all the same throughout the test, then Eq. 10.4 gives the damage per cycle as

$$\Delta D_j = \frac{1}{N_f(S_r)} \tag{10.5}$$

which is called a linear damage assumption since it obviously gives $D(t) = N(t) / N_f$ growing linearly with the number of cycles.

Most engineering structures are, however, subjected to loadings which are much more complicated in form than are the periodic loadings used in laboratory fatigue testing. The basic problem of fatigue analysis is to use appropriately the S-N data from periodic tests to predict the fatigue life of an element or assembly which is subjected to a service load having a complicated time history. The concept of accumulated damage, $D(t)$, is basically an invention to help deal with this problem. In particular, Eq. 10.3 is presumed to apply to general non-periodic conditions, with $D(T) = 1$ still corresponding to failure. The problem, then, is to divide a complicated time history into cycles, to count those cycles [i.e., to find $N(t)$], and to determine an incremental damage ΔD_j for each cycle. Actually, half cycles are sometimes more conveniently identified than cycles but the concept is the same.

One of the most obvious cycle identification schemes is to consider the segment of a stress time history $x(t)$ between any two subsequent local extrema (from a peak to a valley or from a valley to a peak) to be a half cycle. In this scheme, the number of cycles is the same as the number of peaks. Another simple scheme ignores all but the largest extrema between any two subsequent upcrossings of zero, then proceeds as discussed. This generally gives a significantly smaller number of cycles — namely, the number of upcrossings of the level zero. Neither of these schemes, though, seems to give appropriate answers in some quite simple situations. In particular, consider the effect of adding a high-frequency component to a basically low-frequency time history, as shown in Fig. 10.1. Using the subsequent peaks and valleys identifies the half cycles shown in part (b) of the figure.

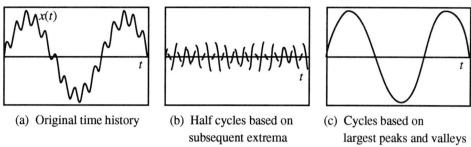

(a) Original time history   (b) Half cycles based on   (c) Cycles based on
                                       subsequent extrema   largest peaks and valleys
Figure 10.1 Inadequate cycle identification schemes

This gives a large number of cycles, but no large half cycles are identified, since there are many peaks and valleys between any high peak and any low valley of the original time history shown in part (a). Using only the largest peaks and valleys, on the other hand, does give large half cycles but it completely ignores most of the effect of the high-frequency component, as illustrated in part (c) of the figure. Thus, it becomes apparent that more elaborate techniques are needed for identifying cycles within a complicated time history.

Clearly, for identifying cycles within a complicated time history one would like to use a scheme which does count all the cycles, but does not lose the large cycles which happen to be interrupted by small cycles. The most commonly adopted cycle identification scheme of this type is called the "rainflow" method.[3] When a rainflow analysis is complete, every segment of $x(t)$ belongs to exactly one of the identified half cycles, and there is one half cycle terminating and another beginning at each extremum (local peak or local valley) of the time history. Furthermore, the peaks and valleys are paired in such a way as to give the largest possible half cycle, then the largest possible remaining half cycle, etc. If one assumes that the stress-strain behavior is slightly nonlinear and satisfies Masing's hypothesis,[4] then a rainflow cycle can be identified with a closed stress-strain hysteresis loop. Various other interpretations, and several algorithms, for rainflow analysis have been published. Figure 10.2 gives one algorithm due to Downing and Socie (1982). It is one of the simplest and most efficient forms available.

Figure 10.3 illustrates the cycles identified by the rainflow method for the time history of Fig. 10.1, and also illustrates the concept of relating rainflow cycles with closed hysteresis loops for a nonlinear stress-strain relationship. Although the plot shows a very pronounced nonlinearity, this is

---

[3] This method is generally attributed to Matsuishi and Endo in 1968. It was quite thoroughly studied by Dowling (1972) and is available in standard reference books such as the one by Fuchs and Stephens (1980).

[4] This hypothesis states that either half of any closed hysteresis loop has the same shape as the initial loading curve for the specimen, but it is magnified by a factor of two. Jennings (1964) refers to Tanabashi and Kaneta (1962) for the association of this hypothesis with the name Masing.

Downing and Socie Rainflow Algorithm (1982)
The origin of the stress time history has been taken to align with the highest peak, with the earlier portion placed at the end. The extrema are the elements of the sequence $\{x_1, x_2, \cdots, x_j, \cdots\}$, with $x_1 = x_{max}$. $\{q_1, q_2, \cdots, q_n\}$ is an array of variable dimension n.
Step 0. Initialize: $n = 1$, $q_1 = x_1 = x_{max}$
Step 1. $n \to n+1$, q_n = next extremum from $\{x_1, x_2, \cdots, x_j, \cdots\}$ Stop if there are no more data.

Step 2. If $n < 3$, then go to Step 1.
          Otherwise, establish temporary ranges R_1 and R_2:
          $R_1 = \|x_n - x_{n-1}\|$ (range under consideration)
          $R_2 = \|x_{n-1} - x_{n-2}\|$ (prior range adjacent to R_1)

Step 3. Compare R_1 and R_2:
          If $R_1 < R_2$ then go to Step 1.
          If $R_1 \geq R_2$ then go to Step 4.

Step 4. Identify R_2 as a rainflow range and remove its extrema from $\{q_1, q_2, \cdots, q_n\}$:
          $q_n \to q_{n-2}$, $n \to n-2$
          Go to Step 2.

Figure 10.2 Rainflow algorithm

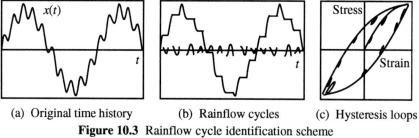

(a) Original time history   (b) Rainflow cycles   (c) Hysteresis loops

Figure 10.3 Rainflow cycle identification scheme

only for purposes of illustration. It is presumed that the material really behaves in an almost linear manner, but an accurate plot of this type would hide the hysteresis loops which we seek to illustrate. The hysteresis loops are shown with $x(t)$ being the stress time history, but for a material that is almost linear it makes little practical difference whether $x(t)$ is associated with the stress or the strain axis in the plot. Note that in the Downing and Socie algorithm, the initial portion of the time history of stress is moved to the end, such that the rearranged time history begins with the highest peak. This arrangement eliminates any initial loading curve from the hysteresis loops in Fig. 10.3(c), and makes every portion of the time history a part of a closed hysteresis loop.

Once a cycle identification scheme has been applied, one must choose the ΔD_j values giving the increment of damage during each cycle. This is usually done by the Palmgren-Miner hypothesis. This hypothesis, in practice, consists of using Eq. 10.5 for the damage increments in Eqs. 10.3 and 10.4 for the variable amplitude situation. The failure condition is then

$$1 = \sum_{j=1}^{N(T)} [N_f(S_{r,j})]^{-1} \tag{10.6}$$

in which T is the time to variable amplitude failure and $S_{r,j}$ is the stress range of cycle j. This equation would be exactly correct if the following two conditions were met: the damage $D(t)$ grew linearly in the constant amplitude tests, and any variable amplitude cycle of range S_r caused the same amount of damage as it would in a constant amplitude test. Actually these two conditions can be combined into one statement that the Palmgren-Miner hypothesis is correct if any cycle of given range S_r always causes the same amount of damage in any testing situation.

While these statements are sufficient conditions to justify the use of Eq. 10.6, they are not necessary. At least, they need not hold for any particular $D(t)$ damage function. For example, consider a situation in which the incremental damage ΔD_j depends on the accumulated damage $D(t)$ as well as on $S_{r,j}$, but with this dependence being separable into the product of two functions:

$$\Delta D_j = g[D(t)]h(S_{r,j}) \tag{10.7}$$

for some functions $g(\cdot)$ and $h(\cdot)$. Now define a new function $H(t)$ as

$$H(t) = \sum_{j=1}^{N(t)} h(S_{r,j}) = \sum_{j=1}^{N(t)} \frac{\Delta D_j}{g[D(t)]} \tag{10.8}$$

This gives a failure condition of

$$H(T) = \sum_{j=1}^{N(T)} \frac{\Delta D_j}{g[D(t)]} \approx \int_0^1 \frac{dD}{g(D)}$$

in which the final form is based on a presumption that $\Delta D_j \ll 1$, so that the summation is essentially the same as the integral. One can take $D^*(t) \equiv H(t) / H(T)$ as a new damage measure, and Eq. 10.8 shows that it grows linearly. Thus, if there exists a separable incremental damage measure of the form of Eq. 10.7, then there also exists a linear damage measure $D^*(t)$ so that Eq. 10.6 must be satisfied. Another terminology which has been used to refer to the separable damage function in Eq.

10.7 is that it is nonlinear but not stress-dependent (Kutt and Bieniek, 1983). This latter nomenclature refers to the fact that a plot of $D(t)$ versus t/T is the same for any S_r level in a constant amplitude test. If this function is denoted by $q(t/T)$, then the linear damage function $D*$ is simply the inverse function $q^{-1}(D)$.

Various crack propagation models provide examples of separable damage functions as hypothesized in Eq. 10.7. It is natural in such situations to define $D(t)$ as the ratio of crack length at time t to the crack length at failure, but empirical data show that this $D(t)$ does not grow linearly. Rather, the crack growth rate is greater when the crack length is greater. However, there are a number of analytical formulations in which the rate of crack growth is the product of a function of crack length and another function of the loading. One commonly used form is called the Paris crack growth law, and it approximates the rate of crack growth as proportional to a power m of the range of the stress intensity.[5] The range of the stress intensity, though, is proportional to the range of load multiplied by the square root of crack length. Thus, this model gives exactly Eq. 10.7 with $g(D) = D^{m/2}$ and $h(S_r)$ proportional to S_r^m. This gives a model in which $D(t)$ grows nonlinearly for constant amplitude tests, since ΔD_j depends strongly on $D(t)$. Nonetheless, Eq. 10.6 is appropriate for the model, since there exists an alternative damage function $D*$ which does grow linearly. In this particular example it is quite easy to derive the $D*$ function, but the more important point is that one does not need actually to identify it. In order to justify using the Palmgren-Miner approximation, one only needs to know that the linear damage function exists.

Even though crack propagation does provide examples of separable damage functions, it should be noted that not all crack propagation models are in this category. In particular, there are a number of models which include load sequence effects, through residual stress or crack retardation terms. These models do not satisfy Eq. 10.7, so use of the Palmgren-Miner approximation is not necessarily acceptable if one of these models is deemed appropriate.

As usually applied, the Palmgren-Miner summation for fatigue damage requires knowledge of at least two characteristics: the rate of occurrence of stress ranges, and the magnitudes of the stress ranges. In addition, an approximation of the mean stress effect may also be included through the use of an equivalent effective stress range, as in Eq. 10.2. For a deterministic time history, the rate of occurrence of stress ranges is usually taken to be the rate of occurrence of local peaks, although the rate of occurrence of upcrossings of the mean stress also has some use. Both of these quantities have been investigated for stochastic processes in Chapter 6. Determination of both the stress range and the mean stress of a cycle can be accomplished if one knows the values of the peak and valley which constitute the extrema of that cycle. Thus, the probabilistic description of all local extrema of a stochastic time history is necessary for a complete probabilistic description of stochastic fatigue. Complete solutions to this problem are not available and simplifying approximations are commonly adopted.

---

[5]   The term "stress intensity" is commonly used in fracture mechanics and is a measure of the magnitude of stress in the vicinity of a singular point for which linear elasticity theory predicts an infinite stress.

10.3 Stochastic Analysis of Fatigue

We now wish to apply probability theory to simple fatigue models in order to make probabilistic predictions of the time to failure for a structure or specimen subjected to a stochastic loading process. The quantity T denoting the failure time for the specimen will now be a random variable, and the stress time history will be given by the stochastic process $\{X(t)\}$. The goal is to characterize T as completely as possible. This could involve finding the mean μ_T, the variance σ_T^2, the probability density function $p_T(t)$, etc. In fact, though, we will not find exact solutions for any of these quantities. The damage models are formulated to allow computation of $D(t)$ for given values of t. It is generally difficult or impossible to find exact solutions of the inverse problem of characterizing the random variable T defined by $D(T) = 1$.

As noted in Eq. 10.3, the accumulated damage can be considered to be the sum of incremental damage per cycle values ΔD_j, and if we presume that $\{X(t)\}$ is stationary and D is a linear damage function, then all the ΔD_j increments are identically distributed. Furthermore, when $D(t)$ is approaching unity we expect the number $N(t)$ of cycles to be large, since it approaches the fatigue life $N(T)$ in the limit. Under these conditions it is generally true that $\sigma_D(t)/\mu_D(t)$ decays like $[N(t)]^{-1/2}$ as $N(t)$ becomes large. Precisely, the necessary condition is that the damage increments ΔD_j and ΔD_k become uncorrelated when j and k are well separated such that

$$\sum_{l=-\infty}^{\infty} Cov[\Delta D_j \Delta D_{j+l}] < \infty$$

which can also be stated as the requirement that the sequence of ΔD_j values is ergodic in mean value (see Section 2.7). Under these conditions, the variations of the random ΔD_j values tend to average out so that D has relatively less uncertainty than do the individual ΔD_j values. This has also been verified by much more rigorous techniques (see Crandall and Mark, 1963).

The fact that $\sigma_D(t)/\mu_D(t)$ becomes very small as t approaches T in most fatigue problems assures us that σ_T/μ_T is also very small, which allows us to use the approximation

$$E[D(\mu_T)] \approx 1 \tag{10.9}$$

That is, we say that mean time to failure, μ_T, is approximately the time at which the mean value of damage, $\mu_D(t)$, reaches unity. Furthermore, we have little need to estimate σ_T or $p_T(t)$ if $\sigma_T \ll \mu_T$. Any random variable with $\sigma \ll \mu$ is almost deterministic at the value μ. Thus, we will concentrate on estimating μ_T as approximated by Eq. 10.9.

A word of caution regarding uncertainty about the value of T is appropriate here. This discussion seems to suggest that we will have almost no uncertainty about the value of T in stochastic fatigue problems. Experimental data, on the other hand, usually lead to a conclusion that there is considerable uncertainty about T. It is not uncommon for supposedly identical specimens subjected to the same loading to give T values which differ by as much as a factor of ten (i.e., one T value is ten times the other). This apparent contradiction is, in fact, a reflection of other factors not included in the fatigue theory presented here. There is also considerable variation in the T or N_f values obtained from identical constant amplitude tests of supposedly identical specimens, as seen in the data in Examples 10.1 and 10.2. Supposedly the explanation is that the specimens really are not identical so that probability theory should also be used to describe the constant amplitude fatigue properties of any given (untested) specimen. This approach certainly is possible, but it is not the central problem of interest here. That is, it does not specifically concern the effect of stochastic time histories. The gist of the statement that $\sigma_T \ll \mu_T$ for stochastic time histories is that having a stochastic time history results in very little *increase* in the uncertainty about T, compared to a constant amplitude situation. This is also supported by experimental data. In fact, the data often suggest that there is less statistical scatter in stochastic fatigue than in deterministic fatigue.

Before exploring any particular stochastic fatigue theory, let us note the form that results from taking the expected value of the equation for damage accumulation. In particular, let us limit attention to the situation in which the stress process $\{X(t)\}$ is stationary so that the incremental damage values $\{\Delta D_1, \Delta D_2, \cdots, \Delta D_j, \cdots\}$ are identically distributed. Adding the assumption that the random variable $N(t)$ is independent of this sequence of damage values gives

$$E[D(t)] = E[N(t)]E(\Delta D)$$

and the Palmgren-Miner damage model of Eqs. 10.5 and 10.6 reduces this to

$$E[D(t)] = E[N(t)]E\left(\frac{1}{N_f(S_r)}\right) \tag{10.10}$$

in which S_r is now a random variable representing any of the identically distributed stress ranges $\{S_{r,1}, S_{r,2}, \cdots, S_{r,j}, \cdots\}$ in $\{X(t)\}$. Recall that $N_f(\cdot)$ is a deterministic function representing the constant amplitude S-N curve, but a deterministic function of a random argument is a random variable. Combining Eqs. 10.9 and 10.10 gives

$$\frac{1}{E[N(T)]} = E\left(\frac{1}{N_f(S_r)}\right) \tag{10.11}$$

which can be used to obtain an estimate of the expected number of cycles to failure, $E[N(T)]$. Converting this estimate of $E[N(T)]$ into an estimate of $E(T)$ requires knowledge of the mean rate of occurrence of cycles. This is generally not difficult, though, once a definition of "cycle" is chosen. Although the linear damage function of Eq. 10.5 was used in deriving this result, it should be noted that the result is also valid if the problem has a separable damage function, as in Eq. 10.7, since that assures the existence of a linear damage function.

One can also argue that the linear damage growth model is a much more restrictive assumption than is really needed to justify use of Eq. 10.10. Rather than saying that the damage is $[N_f(S_r)]^{-1}$ in each cycle of stress range S_r, it is only necessary to say that the damage in all cycles of stress range S_r is equal to $[N_f(S_r)]^{-1}$ multiplied by the number of such cycles. For example, if a complete fatigue test includes 900 cycles with S_r in the interval $[u, u + \Delta u]$, for some small value of Δu, then the key assumption is that all 900 of these cycles give a combined damage contribution of $900[N_f(u)]^{-1}$. This includes the possibility that each cycle causes a damage increment of $[N_f(u)]^{-1}$, but it also includes many other possibilities. For example, it includes the possibility that the damage increments are small at the beginning of the test and grow larger as failure is approached, as in Eq. 10.7. It is also compatible with some sequence effect models in which the damage in cycle j is affected by the stress range of cycles $j - 1, j - 2$, etc.

In order to clarify this argument, we will define an average damage increment $\overline{\Delta D}(u)$, in which the average is over all cycles with $u \leq S_r \leq u + \Delta u$ during a complete stochastic fatigue test. The Palmgren-Miner hypothesis is then justified if $\overline{\Delta D}(u) = [N_f(u)]^{-1}$. The separable damage function $D(t)$ of Eq. 10.7 is one example of a nonlinear damage function which does give this average value of the damage increment. The applicability of this idea to damage models including sequence effects can be illustrated by considering the simplest such situation, in which the ΔD_j damage increment depends only on $S_{r,j}$ and $S_{r,j-1}$. Now for $u \leq S_{r,j} \leq u + \Delta u$ we expect $S_{r,j-1}$ to have a variety of possible values, some greater than u and some smaller than u. Thus, we anticipate that the sequence effect may have one probability of reducing ΔD_j and another probability of increasing ΔD_j during any given cycle. Over the course of the time history, though, it would be quite possible for these effects to cancel out in such a way that $\overline{\Delta D}(u)$ was unchanged by the sequence effect. This would justify use of the Palmgren-Miner hypothesis. This is not intended as an argument that sequence effects are unimportant. Rather it is an explanation of why they are secondary effects. An obvious situation in which such a sequence effect might be important is when ΔD_j is a highly nonlinear function of $S_{r,j-1}$

If the S-N curve is taken to have the power law form of Eq. 10.1, then the Palmgren-Miner model of Eq. 10.11 reduces to

$$\frac{1}{E[N(T)]} \equiv E(\Delta D) = K^{-1}E(S_r^m) \tag{10.12}$$

so that the crucial step in estimating the expected value of the fatigue life is simply the evaluation of the mth moment of the stress range S_r. It should be kept in mind, though, that m may not be an integer, since it is determined on the basis of obtaining a best fit of experimental data. If the S-N curve does not have this simple shape, then the Palmgren-Miner model requires that one evaluate the expected value of the more general nonlinear function in Eq. 10.11.

As noted earlier, the rainflow method is generally considered one of the best methods for identifying the cycles within a variable amplitude time history. Thus, we would like to apply it to stochastic fatigue as well. That is, we would like to consider the random variables $N(t)$ and S_r to represent the number and amplitude, respectively, of rainflow cycles within the stationary stochastic time history of $\{X(t)\}$. Since the rainflow method gives one cycle for each peak of the time history, it is easy to calculate the rate of occurrence of rainflow cycles by using the formula for ν_P from Chapter 6. It is generally not easy, though, to find an analytical evaluation or approximation for $E(S_r^m)$ for rainflow cycles. Lindgren and Rychlik (1987) have derived the relationship between the probability distribution of the rainflow cycles and that of the sequence of extrema of the stress process. The difficulty, though, is in finding or estimating the joint probability distribution of the extrema. Formulas for the marginal distributions were given in Section 6.4, but the joint distributions are generally unknown. Rychlik (1989) has presented an approximate result based on the assumption that the extrema of the stress process form an n-step Markov chain with a finite number of states.[6] He has presented some numerical results for the cases of $n = 1$ and 2, but this involves using a numerical procedure to derive the needed conditional probability distributions of the extrema of a given time history.

Purely empirical attempts have also been made to find the probability distribution of rainflow ranges from simulation studies, but without great success. Corazao (see Lutes et al., 1984), in particular, simulated time histories for Gaussian processes corresponding to 85 different autospectral density curves. For each time history, he performed rainflow analysis and subsequently attempted to fit the resulting empirical cumulative distribution functions for the random variable S_r with each of six relatively simple distribution functions. In each distribution function, he chose all possible parameters on the basis of a best fit to the data. Even with this very flexible approach, he was unsuccessful in finding a form for the cumulative distribution function that would fit the empirical results, particularly for broadband processes. It seems likely that this attempt to describe the distribution of rainflow ranges failed largely because it was presumed that the shape of the probability distribution was determined solely by the irregularity factor IF introduced in Section 6.3. That is, it was presumed that two different autospectral density curves having the same IF value would also have the same cumulative distribution function for the random variable S_r. The justification for the use of this assumption for a Gaussian process was by analogy with the distribution of the peak random variable P, for which the shape of the cumulative distribution function does depend only on IF (see Example 6.9). Many other bandwidth parameters are possible, though, and it may be that some of them are more appropriate for describing rainflow ranges.

Thus, the problem of finding a closed-form solution or simple approximation for the probability distribution of rainflow ranges remains unsolved. Nonetheless, there are certain limiting cases in which one can evaluate $E(S_r^m)$ for rainflow cycles. The most notable of these is when the $\{X(t)\}$ process is Gaussian and very narrowband, and this situation will be investigated in the

---

[6]   This n-step Markov property can be defined as follows: The conditional distribution of the jth extremum, given the values of the n most recent extrema, is independent of all earlier extrema. This is a discrete version of the Markov property defined in Section 8.8.

following section. It should also be noted that one always has the option of studying stochastic fatigue by simulating long time histories of samples from $\{X(t)\}$, then performing deterministic rainflow analysis of these samples. This, however, is somewhat awkward for use as a routine design tool. More easily implemented approximate methods are discussed in the following two sections.

The relatively simple approach presented here, of course, is not the only method for the stochastic analysis of fatigue. Among the more mathematically sophisticated formulations is one in which $D(t)$ is modeled as a Markov process, such that the increment of damage during any "duty cycle" is a random variable which depends on $D(t)$ and the stress $X(t)$, but is independent of prior values (Bogdanoff and Kozin, 1985). This is a very general approach, but it has been found to be very difficult to identify the conditional probability distribution of the damage increment based on experimental data (Sobczyk and Spencer, 1992; Madsen et al., 1986). Other approaches focus exclusively on crack growth, which can be considered to be an observable $D(t)$ function. Experimental data have been used, for example, in developing models in which the rate of crack growth depends on the values of prior overloads, which affect the state of stress and strain at the crack tip (see Sobczyk and Spencer, 1992). In many practical situations, though, a specimen may have very little remaining fatigue life after the first appearance of a visible crack. Furthermore, it is not obvious whether the models which describe visible crack growth also apply to earlier parts of the fatigue process, and there is limited benefit in having an accurate model of visible crack growth if that is only a small fraction of the total fatigue life. Overall, it seems that most predictions of fatigue life in practical problems are based on the approach presented here, in which information about the accumulated damage is obtained from a constant amplitude S-N curve.

10.4 Rayleigh Approximation

Several analytical approximation techniques exist whereby one can estimate the stochastic fatigue life based only on the knowledge of the autospectral density function of the stress process. In particular, these methods typically make use of the values of certain spectral moments, as defined in Eq. 4.41. These techniques are classified as spectral methods. The first spectral method that we will discuss, called the Rayleigh approximation method, is one of the simplest and most widely used analytical techniques for stochastic fatigue analysis. This method was originally developed to predict the fatigue life under a narrowband Gaussian loading.

For a very narrowband $\{X(t)\}$ process, it is reasonable to say that the value of the stress range S_r is twice the value of either the peak or amplitude of the process. Of course, the peak and amplitude distributions are essentially identical for such a very narrowband process. Furthermore, the number of cycles per unit time of the process can be taken as either the rate of occurrence of peaks or the rate of occurrence of upcrossings of the mean value of $\{X(t)\}$. Calculation of the mth moment of S_r in this situation is particularly simple if $\{X(t)\}$ is also Gaussian, since the amplitude then has the Rayleigh distribution. Neglecting any effect of mean stress, one finds that

$$E(\Delta D) = K^{-1}E(S_r^m) = K^{-1}2^m \int_0^\infty \frac{u^{m+1}}{\sigma_X^2} \exp\left(\frac{-u^2}{2\sigma_X^2}\right) du = K^{-1}(2)^{3m/2}\sigma_X^m \Gamma\left(1+\frac{m}{2}\right)$$

or

$$E(\Delta D) = K^{-1}(2)^{3m/2}\lambda_0^{m/2}\Gamma\left(1+\frac{m}{2}\right) \qquad (10.13)$$

in which $\Gamma(\cdot)$ is the gamma function[7] and λ_0 denotes the zero-order spectral moment.[8] The simplest form for the appropriate rate of cycle occurrence for this mean zero narrowband process is given by the rate of upcrossings of the mean value, $v_X^+(\mu_X)$. Since the process is Gaussian (see Example 6.1) this is given by $\sigma_{\dot{X}}/(2\pi\sigma_X)$, so that the estimate of failure time can be written as

$$E(T) = \frac{2\pi\sigma_X}{E(\Delta D)\sigma_{\dot{X}}} = \frac{2\pi}{E(\Delta D)}\sqrt{\frac{\lambda_0}{\lambda_2}} \qquad (10.14)$$

If μ_X is not zero but is constant throughout the loading process, then one can use Eq. 10.2 to obtain an equivalent stress range, which merely introduces a constant scale factor for each range. In particular, it gives $E(\Delta D)$ as Eq. 10.13 divided by $[1-(\mu_X/x_u)^2]^m$.

The so-called Rayleigh approximation of fatigue damage consists of using Eqs. 10.13 and 10.14 to predict $E(T)$. Its simplicity is, no doubt, one of the strong motivations for its widespread use. In order to predict $E(T)$ by this method, one only needs to find σ_X and $v_X^+(\mu_X)$ for the $\{X(t)\}$ process. Furthermore, these two characteristics are easily evaluated either from an autospectral density curve, by using the λ_0 and λ_2 spectral moments, or from a representative sample time history. In addition, it should be noted that the results of the Rayleigh approximation tend to those of rainflow analysis in the limiting situation of a very narrowband Gaussian process. A common approach is to take the rainflow results as the basis of comparison for other approximate techniques, and on this basis the Rayleigh method becomes perfect in the limit as the bandwidth of a Gaussian stress process tends to zero.

Although Eqs. 10.13 and 10.14 were obtained from narrowband assumptions, they are also very commonly used to predict the fatigue life for stress processes which are not narrowband. In fact, the Rayleigh approximation is probably the most widely used analytical method for predicting fatigue life for any sort of stochastic stress process. It must be kept in mind, though, that the assumptions in the Rayleigh method are not necessarily appropriate for broadband processes, and simulation results confirm that Rayleigh predictions can differ significantly from rainflow predictions when stress processes are not narrowband. This is true for stress processes with broadband autospectral densities, but it is much more noticeable for a so-called bimodal autospectral density, having two narrowband peaks at different frequency values (Lutes and Larsen, 1990; Ortiz and Chen, 1987; Wirsching and Light, 1980).

---

[7] More information on the gamma function is given in Example A.31 of Appendix A.

[8] Lin (1967) attributes this widely used result to Miles in 1954.

Using Eqs. 10.13 and 10.14 for a stress process which is not narrowband may be viewed as finding the damage for an "equivalent narrowband process." That is, rather than attempting to define cycle ranges in $\{X(t)\}$, it is presumed that the damage done by $\{X(t)\}$ is the same as would be done by a narrowband Gaussian process with the same rate of upcrossings of its mean value. It is difficult to assess the validity of this choice of an "equivalent" process. Using the rate of mean crossings, $v_X^+(\mu_X)$, as the rate of cycle occurrence seems to neglect the effect of high-frequency components. On the other hand, using S_r as twice the value of the peaks of the narrowband process gives unduly large magnitudes of the ranges for many time histories, such as ones similar to those in Figs. 10.1 and 10.3. Thus, the two errors in the Rayleigh approximation at least partially cancel each other.

****

Example 10.3: Compare the Rayleigh and rainflow predictions of the fatigue life for the special case of a stationary, mean zero, Gaussian stress process and an S-N curve given by Eq. 10.1 with $m = 1$.

From Eqs. 10.13 and 10.14 the Rayleigh approximation for $m = 1$ can be written as

$$E(T) = \frac{2\pi\,\sigma_X}{K^{-1}(2)^{3/2}\sigma_X\,\Gamma\!\left(\frac{3}{2}\right)\sigma_{\dot{X}}} = \frac{\sqrt{2\pi}\,K}{\sigma_{\dot{X}}}$$

The situation with $m = 1$ is a special case in which we can also exactly evaluate the rainflow prediction of the fatigue life. In particular, Eqs. 10.1 and 10.6 give the failure condition as

$$1 = K^{-1}\sum_{j=1}^{N(T)} S_{r,j}$$

However, we can rewrite this summation of S_r values by noting the contributions to the summation from each time increment of length dt. In particular, there is an excursion $|\dot{X}(t)|dt$ during the time increment, and this increment of excursion becomes a part of some $S_{r,j}$ stress range. Thus, it adds directly to the summation of all those stress ranges, and we can say that

$$\sum_{j=1}^{N(T)} S_{r,j} = \frac{1}{2}\int_0^T |\dot{X}(t)|\,dt$$

in which the factor of 1/2 comes from the fact that a full cycle with range S_r corresponds to a total excursion of $2S_r$. Substituting this relationship and taking the expected value gives

$$2 = K^{-1}E(T)E(|\dot{X}(t)|)$$

For the Gaussian process, we find that

$$E(|\dot{X}(t)|) = 2\int_0^{\infty} \frac{u}{\sqrt{2\pi}\sigma_{\dot{X}}}\exp\left(\frac{-u^2}{2\sigma_{\dot{X}}^2}\right)du = \sqrt{\frac{2}{\pi}}\,\sigma_{\dot{X}}$$

and this gives

$$E(T) = \frac{2K}{E(|\dot{X}(t)|)} = \frac{\sqrt{2\pi}\,K}{\sigma_{\dot{X}}}$$

This, however, is identical to the result of the Rayleigh approximation. Note, also, that we have used no assumption that the stress is narrowband. This is the only situation in which the Rayleigh approximation is in perfect agreement with rainflow analysis regardless of the form of the autospectral density of the stress.

****

Example 10.4: A particular mechanical bracket has been subjected to constant amplitude fatigue tests at the single level of $S_r = 150$ MPa, and the observed fatigue life was $N_f = 10^6$ cycles. Based on experience with similar devices, it is estimated that the parameter m of the S-N curve is in the range of $3 \le m \le 5$. Based on this limited information, it is necessary to choose a level for the standard deviation σ_X of the narrowband, mean zero, Gaussian stress which will be applied to the bracket in actual service. Using the limiting values of $m = 3$ and $m = 5$ find the level of σ_X meeting each of two design situations: $E[N(T)] = 10^6$, and $E[N(T)] = 10^8$.

Beginning with $m = 3$ we find the value of K in the S-N curve of Eq. 10.1 such that $10^6 = K/(150)^3$, giving $K = 3.375 \times 10^{12}$. For $E[N(T)] = 10^6$, Eq. 10.13 then gives $10^{-6} = K^{-1}2^{4.5}\sigma_X^3(3\sqrt{\pi}/4)$. Solving this gives the answer of $\sigma_X = 48.2$ MPa. Similarly, $E[N(T)] = 10^8$ gives $\sigma_X = 10.4$ MPa.

For $m = 5$, we proceed in the same way and find that $K = 7.594 \times 10^{16}$. For $E[N(T)] = 10^6$, Eq. 10.13 becomes $10^{-6} = K^{-1}2^{7.5}\sigma_X^5(15\sqrt{\pi}/8)$, which gives $\sigma_X = 41.7$ MPa, and $E[N(T)] = 10^8$ gives $\sigma_X = 16.6$ MPa.

Given the uncertainty about the value of m, the safe choice is to use $\sigma_X = 41.7$ MPa for a design condition of $E[N(T)] = 10^6$, and $\sigma_X = 10.4$ MPa for a design condition of $E[N(T)] = 10^8$.

Note that one must be cautious in choosing conservative approximations for this type of problem. In particular, one cannot conclude that either choice of m value is generally more conservative than the other. For the design condition of $E[N(T)] = 10^6$ it is more conservative to assume that $m = 5$, since this gives the allowable stress as having $\sigma_X = 41.7$ MPa, which is 13% smaller than the value one would obtain by using $m = 3$. For the alternate design condition of $E[N(T)] = 10^8$, it is more conservative to assume that $m = 3$, since the resulting allowable value of $\sigma_X = 10.4$ MPa is 37% smaller than would be obtained by using $m = 5$. In general, one must consider the range of possible m values when dealing with such limited experimental data, rather than simply using the largest possible or smallest possible m value.

****

10.5   Other Spectral Methods

One attempt to improve on the Rayleigh approximation technique can be called the peak approximation method. The fundamental idea of this approach is to estimate the probability distribution of the stress ranges according to the probability distribution for the peaks of a Gaussian $\{X(t)\}$ process, as given in Example 6.9, and to calculate the number of cycles according to the rate of occurrence of peaks. For a process which is not very narrowband, these assumptions sound more plausible than the amplitude distribution and the mean crossing rate used in the Rayleigh approximation. Actually both of these peak approximation assumptions are slightly altered in practice, since some peaks of $\{X(t)\}$ are negative, and these peaks cannot correspond to values of the non-negative quantity S_r. The modification to neglect the negative peaks consists of using an appropriate conditional probability distribution. It was shown in Example 6.10 that the fraction of negative peaks for any continuous process $\{X(t)\}$ is $(1 - IF)/2$, in which IF is the irregularity factor, so the relevant conditional probability density function for positive peaks for a mean zero process becomes

$$p_{P(t)}(u \mid P(t) \geq 0) = \left(\frac{2}{1+IF}\right) p_{P(t)}(u) U(u) \tag{10.15}$$

One cannot generally perform a simple analytical evaluation of $E[P^m(t)]$ for this distribution because of the Gaussian cumulative distribution function which appears in Eq. 6.10, but numerical integration can be used fairly easily. Also, various analytical approximations can be obtained by curve fitting the results of numerical integration. One such simple approximation is given by (Lutes et al., 1984)

$$E[P^m(t) \mid P(t) \geq 0] \approx (2)^{m/2} \sigma_X^m \left(1 + \frac{IF(1-IF)}{4}\right)^{(-1+m/2)} \frac{\Gamma\left(\dfrac{m+1+IF}{2}\right)}{\Gamma\left(\dfrac{1+IF}{2}\right)} \tag{10.16}$$

This equation agrees exactly with the results of Eq. 10.15 for the special cases of $IF = 0$ and $IF = 1$, for any positive m value. Furthermore, it is within 3% of results from numerical integration for all IF values for m values varying from one to nine.

Since the peak approximation of stress ranges is based only on positive peaks, it is natural to use the rate of occurrence of positive peaks as the rate of occurrence of cycles in this approach. Thus, $(1 + IF) v_P / 2$ becomes the rate of occurrence of stress ranges in this approximation, and one can also use the fact that $v_P = v_X^+(\mu_X) / IF$. The final result of using these expressions in conjunction with $S_r = 2P(t)$ for non-negative peaks can be written as

$$E(T) \approx \frac{2\pi K}{2^m E[P^m(t) \mid P(t) \geq 0]} \frac{2(IF)}{(1+IF)} \frac{\sigma_X}{\sigma_{\dot{X}}} = \frac{4\pi K (IF)}{2^m (1+IF) E[P^m(t) \mid P(t) \geq 0]} \sqrt{\frac{\lambda_0}{\lambda_2}} \tag{10.17}$$

Equations 10.16 and 10.17 can then be used to obtain results for the peak approximation method.

The information required about $\{X(t)\}$ is the value of the irregularity factor, in addition to the σ_X and $\sigma_{\dot{X}}$ values which are needed for the Rayleigh approximation. In terms of spectral moments, one now needs λ_0, λ_2, and λ_4, since IF for a Gaussian process is the same as α_2, which is computed from these three spectral moments (see Example 6.7).

The basic concept of the peak approximation seems to be very sound for a narrowband process. Using the conditional distribution for $P(t)$ seems to be somewhat more reasonable than using the Rayleigh distribution, and using the rate of occurrence of positive peaks gives a correction for the inherent underestimation of the number of cycles when only crossings of the mean value are counted. Numerical results show, though, that these two corrections mostly cancel out for a narrowband process ($IF \approx 1$), so that $E(T)$ is changed very little. The more accurate probability distribution for S_r gives a reduced value for $E(S_r^m)$, and this almost offsets the effect of the increased rate of cycle occurrence. For a narrowband process the peak approximation method is reasonable, but it gives the same results as the simpler Rayleigh approximation. For some other problems, the peak approximation and the Rayleigh approximation may give very different results. Examples 10.6 and 10.7 will illustrate two such situations.

One of the more recent spectral techniques is called the single moment method (Lutes and Larsen, 1990). It is particularly easy to apply when one is given an autospectral density curve, because it depends on only one spectral moment. It has the general form of $E(T) = c(\lambda_a)^b$, in which a, b, and c are positive constants which may depend on the specimen, but not on the autospectral density. One can determine the appropriate values for a, b, and c on the basis of making this formula agree with the rainflow and Rayleigh methods for the limiting situation of a very narrowband process. In particular, consider an autospectral density of

$$S_{XX}(\omega) = \frac{\sigma_X^2}{2}[\delta(\omega + \omega_0) + \delta(\omega - \omega_0)]$$

so that the spectral moment λ_a is given by $\lambda_a = \sigma_X^2 \omega_0^a$, and the single moment approximation gives

$$E(T) = c\sigma_X^{2b}\omega_0^{ab}$$

For this autospectral density, Eqs. 10.13 and 10.14 give the Rayleigh approximation as

$$E(T) = \frac{2\pi}{\omega_0 E(\Delta D)} = \frac{2\pi K}{(2)^{3m/2}\Gamma\left(1 + \frac{m}{2}\right)\omega_0\sigma_X^m}$$

Thus, the single moment method agrees with the Rayleigh method in the narrowband situation only if

$$a = \frac{2}{m}, \qquad b = \frac{-m}{2}, \qquad c = \frac{2\pi K}{(2)^{3m/2} \Gamma\left(1 + \frac{m}{2}\right)}$$

With these parameter values, the single moment method can be written as

$$E(T) = \frac{2\pi K}{(2)^{3m/2} \Gamma\left(1 + \frac{m}{2}\right) (\lambda_{2/m})^{m/2}} \qquad (10.18)$$

The development of Eq. 10.18 assures that the single moment expression agrees with the Rayleigh approximation for a narrowband process, and we know that this also assures agreement with the rainflow method for this type of autospectral density. The rather surprising fact is that the results of Eq. 10.18 also are in quite good agreement with simulation results from the rainflow method for a great variety of autospectral densities. Some of the data which support this conclusion are included in Example 10.7. It should be noted that $2/m$ is generally not an integer, so that this method uses a noninteger spectral moment. Although, this may seem unusual, it is consistent with the fact that simulation data have indicated that rainflow damage seems to be more closely related to noninteger spectral moments than to integer spectral moments such as λ_1, λ_2, etc. (Lutes et al., 1984).

Ortiz and Chen (1987) have presented another spectral method using noninteger spectral moments, and demonstrated that it gives very good results for some spectral densities. It may be written as

$$E(T) = \frac{2\pi K (\lambda_{2+2/m})^{m/2}}{(2)^{3m/2} \Gamma\left(1 + \frac{m}{2}\right) (\lambda_{2/m})^{m/2} \lambda_2^{(m-1)/2} \lambda_4^{1/2}} \qquad (10.19)$$

which is slightly more complicated to use since it involves four spectral moments. Rainflow results from simulation have shown that this method is sometimes more accurate than the single moment method, but that in the worst case its error is substantially greater than for any situation with the single moment method. This will be illustrated in Examples 10.6 and 10.7.

Recall that the Rayleigh approximation is simple and widely known, and involves assumptions that are reasonable for a narrowband process. Thus, it may be helpful to characterize any other method for analyzing a broadband process in terms of how much its fatigue life prediction differs from that of the Rayleigh approximation. For this purpose, we define a Rayleigh ratio term as

$$RR = \frac{E(T) \text{ by alternate approximation}}{E(T) \text{ by Rayleigh approximation}} \qquad (10.20)$$

An idea of this type was apparently first introduced by Wirsching and Light (1980) who characterized the difference between the Rayleigh approximation and the rainflow method by using a correction factor of the form $CF = (RR)^{-1}$.

The Rayleigh ratio may be viewed as a method of interpreting results, but if this RR factor for the rainflow method were known as a function of some bandwidth parameter, then it would also provide a simple method of estimating the fatigue life. One could find the rainflow prediction of $E(T)$ by calculating the Rayleigh approximation and then multiplying by RR. In addition to the problem of predicting RR, though, there is no simple way to know what bandwidth parameter will work best for this purpose. Wirsching and Light obtained empirical correction factor values by simulating time histories of response, performing rainflow analysis, and plotting the results versus the irregularity factor IF. For their autospectral density curves and m values of 3, 4, 5, and 6, they found that their simulation results gave values of CF which could be approximated by

$$CF = a + (1-a)\left(1 - \sqrt{1 - (IF)^2}\right)^b \tag{10.21}$$

in which

$$a = 0.926 - 0.033m, \qquad b = 1.587m - 2.323 \tag{10.22}$$

This relationship implies that different spectral density curves having nearly the same value of IF give CF values which are approximately the same, but Wirsching and Light found that this approximation was not correct for $m = 10$. In particular, their CF values for $m = 10$ were found to show considerable scatter when plotted versus IF. Examples 10.6 and 10.7 demonstrate other situations in which the empirical relationship is not adequate.

For very narrowband processes, there is complete agreement between the results of all of the spectral prediction methods that we have discussed. This, of course, is appropriate. Our Palmgren-Miner approach involves making stochastic fatigue predictions based on the results of constant amplitude tests, and the very narrowband process is the stochastic process which most closely resembles a constant amplitude loading. It has been found, though, that there are sometimes significant differences between the results of the various fatigue prediction methods for stress processes which are not narrowband. The following three examples provide limited comparisons between the predictions of the various approximation schemes.

****

Example 10.5: Let the mean zero, Gaussian stress process $\{X(t)\}$ have an autospectral density function of $S_{XX}(\omega) = S_0 U(\omega_0 - |\omega|)$. For a specimen with $m = 4$, compare the fatigue lives predicted by the Rayleigh approximation, the peak approximation, the single moment method, the formula of Ortiz and Chen, and the formula of Wirsching and Light.

For this spectral density, we find that the spectral moments are given by

$$\lambda_j = \frac{2S_0\,\omega_0^{j+1}}{j+1}$$

From Eqs. 10.13 and 10.14 we obtain the Rayleigh approximation as

$$E(T_{Ray}) = 2\pi\sqrt{\frac{\lambda_0}{\lambda_2}}\,\frac{K}{2^6\lambda_0^2(2)} = \frac{\pi\,K}{64\lambda_0^{3/2}\lambda_2^{1/2}} = \frac{\pi\,K\sqrt{3}}{256\,S_0^2\,\omega_0^3} = \frac{K}{47.05\,S_0^2\,\omega_0^3}$$

The irregularity factor is $IF = \lambda_2/\sqrt{\lambda_0\lambda_4} = \sqrt{5}/3$, and Eqs. 10.16 and 10.17 then give the peak approximation as $E(T_{peak}) = K/(47.15 S_0^2\,\omega_0^3)$. Similarly, Eqs. 10.21 and 10.22 give $a = 0.794$, $b = 4.025$, and $RR = (CF)^{-1} = 1.256$. In order to evaluate the single moment approximation, we first find $\lambda_{1/2} = 4S_0\omega_0^{3/2}/3$ and then Eq. 10.18 gives $E(T_{sm}) = K/(36.22\,S_0^2\,\omega_0^3)$. Finally, we also evaluate $\lambda_{5/2} = 4S_0\omega_0^{7/2}/7$, and Eq. 10.19 gives the Ortiz and Chen approximation as $E(T_{oc}) = K/(38.18\,S_0^2\,\omega_0^3)$.

Comparing the results, we see that the peak approximation predicts a fatigue life which is 0.2% smaller than the Rayleigh approximation. The single moment method, the approximation of Ortiz and Chen, and the Wirsching and Light correction factor all predict a fatigue life that is greater than the Rayleigh prediction, with the increases being 30%, 23%, and 26%, respectively.
****

Example 10.6: Let the mean zero, Gaussian stress process $\{X(t)\}$ be the response of a SDF oscillator excited by white noise. For a specimen with $m = 4$, compare the fatigue lives predicted by the Rayleigh approximation, the peak approximation, the single moment method, the formula of Ortiz and Chen, and the formula of Wirsching and Light.

From the results in Section 4.7, we can write the autospectral density for $\{X(t)\}$ as

$$S_{XX}(\omega) = S_0 m^{-2}\left[\left(\omega_0^2 - \omega^2\right)^2 + \left(2\zeta\omega_0\omega\right)^2\right]^{-1}$$

in which ω_0 and ζ are the resonant frequency and the damping of the oscillator, and S_0 is the autospectral density of the white noise excitation. From the results in Chapter 3 or 4, we know the values of λ_0 and λ_2 to be

$$\lambda_0 \equiv \sigma_X^2 = \frac{\pi\,S_0}{2m^2\,\zeta\omega_0^3}, \qquad \lambda_2 \equiv \sigma_{\dot{X}}^2 = \frac{\pi\,S_0}{2m^2\,\zeta\omega_0}$$

Using these two spectral moments in Eqs. 10.13 and 10.14 with $m = 4$ gives the Rayleigh approximation result as

$$E(T) = \frac{2\pi\,K}{2^6\sigma_X^4\Gamma(3)\omega_0} = \frac{Km^4\zeta^2\omega_0^5}{16\pi\,S_0^2}$$

There is a problem in applying some of the other spectral methods to this particular problem. In particular, this $S_{XX}(\omega)$ gives $\lambda_4 = \infty$, as was noted in Example 4.8. Thus, we have $IF = 0$, for which Eq. 10.17 gives the peak approximation result as $E(T) = 0$, and Eq. 10.21 gives $RR = (CF)^{-1} = a^{-1} = 1.26$ for the approximation of Wirsching and Light. Similarly, Eq. 10.19 gives $E(T) = 0$ for the Ortiz and Chen approximation, as well, since $\lambda_4 = \infty$ and the other pertinent spectral moments are finite.

In order to apply the single moment method to a situation with $m = 4$, we need to calculate the $\lambda_{0.5}$ spectral moment. For our autospectral density, this can be written as

$$\lambda_{0.5} = 2\int_0^\infty \sqrt{\omega}\, S_{XX}(\omega)\, d\omega = \frac{2S_0}{m^2 \omega_0^{5/2}} \int_0^\infty \frac{\sqrt{\eta}\, d\eta}{(1-\eta^2)^2 + (2\zeta\eta)^2}$$

in which the dimensionless variable η is ω / ω_0. This integral can be evaluated numerically to obtain values which can be used in Eq. 10.18 to give single moment results. One convenient way to present these numerical results is as the Rayleigh ratio RR of Eq. 10.20. These RR values have been computed for ζ values varying from zero to unity, and it is found that they agree with $RR = 1 + \zeta$. This implies that

$$\lambda_{0.5} = \frac{\pi S_0}{2m^2 \omega_0^{5/2} \zeta \sqrt{1+\zeta}}$$

For $IF = 0$ the Wirsching and Light formulas give a value of RR which depends only on m, not on the other properties of the autospectral density function. Thus, this empirical approximation gives the rather unusual prediction that the rainflow fatigue life will be 26% greater than the Rayleigh approximation, regardless of the amount of damping in the oscillator. This value of RR agrees with the single moment result only for $\zeta = 0.26$.

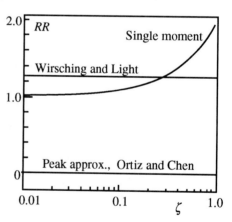

This is an example in which the single moment method is the only one of the alternative spectral methods which gives what one might consider to be a bandwidth correction for the Rayleigh approximation. That is, the bandwidth of the oscillator response is generally considered to depend on ζ, and only the single moment method gives an RR value which depends on ζ.

****

Example 10.7: Let the mean zero, Gaussian stress process $\{X(t)\}$ be the sum of two very narrowband processes, such that $\{X(t)\}$ has a bimodal autospectral density. For a specimen with $m = 3$, compare the fatigue lives predicted by the Rayleigh approximation, the peak approximation, the single moment method, the formula of Ortiz and Chen, and the formulas of Wirsching and Light.

We will write the stress process $\{X(t)\}$ as $X(t) = X_1(t) + X_2(t)$, with $\{X_1(t)\}$ and $\{X_2(t)\}$ being independent processes with autospectral densities of $S_1(\omega)$ and $S_2(\omega)$, respectively. For simplification we will also choose $S_2(\omega)$ to have the form $S_2(\omega) = (b/r)S_1(\omega/r)$, so that it has the same shape as $S_1(\omega)$.

The parameters r and b are the relative frequency and the relative variance of $\{X_2(t)\}$ as compared to $\{X_1(t)\}$. That is, if $\{X_1(t)\}$ is narrowband at a frequency of ω_0 and has variance σ_0^2, then $\{X_2(t)\}$ is narrowband at frequency $r\omega_0$ and has variance $b\sigma_0^2$. This gives any spectral moment of $\{X(t)\}$ as having the form $\lambda_j = (\lambda_j)_0(1 + br^j)$, in which $(\lambda_j)_0$ denotes the corresponding spectral moment of $\{X_1(t)\}$. The spectral method comparisons can be made even simpler by letting the autospectral density of $\{X_1(t)\}$ be the limiting narrowband function, giving

$$S_1(\omega) = \frac{\sigma_0^2}{2}[\delta(\omega + \omega_0) + \delta(\omega - \omega_0)], \qquad \lambda_j = \sigma_0^2 \omega_0^j (1 + br^j)$$

One can now use these spectral moments in finding all the spectral predictions of $E(T)$, and the Rayleigh ratio values for the other three methods.

Note that the bimodal autospectral density reduces to a narrowband unimodal shape if b is either very small or very large, and also if $r = 1$. As expected, one can show that the RR values do tend to unity in these limiting situations. Furthermore, we also know that the Rayleigh approximation will be a good approximation of the rainflow analysis for these narrowband situations. For more general values of b and r, for which the autospectral density is bimodal, there can be notable differences in the various predictions. The peak approximation method always predicts a shorter fatigue life than does the Rayleigh approximation. On the other hand, the single moment method and the Wirsching and Light formulas consistently predict that the fatigue life should be greater than the Rayleigh approximation. The Ortiz and Chen predictions are sometimes above and sometimes below the Rayleigh results. In fact, the primary difference between the results of Ortiz and Chen and those of the single moment method is that the RR values are sometimes less than unity for the former method. Although the Wirsching and Light and single moment results agree to the extent that they always give RR greater than unity, there are major differences in the numerical values. It is found that the differences between the various approximations become more pronounced when r takes on larger values.

The comparison of the various spectral methods, of course, does not necessarily demonstrate which method is most accurate. This assessment is usually based on the results of rainflow analysis of simulated time histories. One of the more extensive such studies (Larsen and Lutes, 1991) has included 180 bimodal autospectral densities, and we can compare Rayleigh ratio values from these results with those from the spectral approximations. We will do this, though,

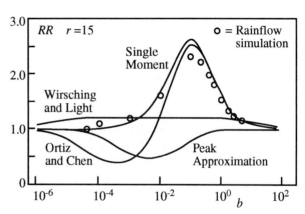

only for the largest r value included in the bimodal simulations, since this gives the greatest differences between the various predictions. Thus, for our situation with $m = 3$, we plot RR values versus b for $r = 15$. The plot includes RR data points from rainflow analysis, as well as curves for all the spectral methods and the Wirsching and Light approximation.

Clearly, the single moment predictions are in the best agreement with the rainflow values. In particular, RR values less than unity, as predicted by the peak approximation and the Ortiz and Chen approximation, are not found by the rainflow method. Also, the rainflow RR values do substantially exceed the maximum value of 1.20 predicted by Wirsching and Light.

****

   These three examples cover only a small fraction of the possible autocorrelation functions which one could investigate, but they do illustrate the similarities and critical differences which have been found between the various spectral methods. Example 10.5 is typical of problems in which the autospectral density is unimodal and limited to a finite frequency range. In such situations one generally finds that the predictions of the single moment method, Ortiz and Chen's formula, and Wirsching and Light's formula are in reasonably good agreement, predicting fatigue life values which are somewhat above the Rayleigh prediction, while the peak approximation is quite close to the Rayleigh value.[9] The autospectral densities in Examples 10.6 and 10.7 emphasize the differences between the various methods. In particular, Example 10.6 removes the limitation of a bounded frequency range, and Example 10.7 introduces a bimodal autospectral density. Both of these situations must be considered to be common occurrences, so that the differences revealed by these examples have serious implications regarding the reliability of the various spectral methods.

   The peak approximation differs significantly from the Rayleigh approximation only for IF values smaller than about 0.5, but the difference can be quite significant for these non-narrowband processes. However, the idea of using $S_r = 2P(t)$ with $P(t)$ representing every positive peak in a non-narrowband process is not very reasonable. For a stochastic time history resembling Fig. 10.1(a), for example, it is obvious that $IF \ll 1$, and most of the additional peaks correspond to fairly small oscillations in the time history, not to completely reversed cycles with $S_r = 2P(t)$. For such a process with a broadband or bimodal autospectral density, it is not surprising that the peak approximation method significantly overpredicts the rate of damage accumulation. Overall, the peak approximation has never been found to be an improvement over the Rayleigh method, and in some broadband situations it is much less satisfactory than the Rayleigh method.

   The major difficulty with the Ortiz and Chen procedure is that it sometimes predicts RR values which are significantly less than unity, while simulation has never demonstrated this to be true for the rainflow method. In Example 10.7 it was seen that the Ortiz and Chen prediction of $E(T)$ for $m = 3$ was sometimes only about 40% of the rainflow value for bimodal spectral densities, and Example 10.6 illustrated a situation in which the Ortiz and Chen prediction of $E(T)$ was zero. Thus, this formula can give major errors in the predicted fatigue life. The Wirsching and Light formula does

---

[9]   It should be noted that an example of this type is also given by Wirsching et al. (1995).

always give RR values greater than unity, but these values are sometimes in significant disagreement with rainflow results, as shown in Example 10.7. Furthermore, its RR prediction for the SDF system of Example 10.6 is independent of the damping in the oscillator, whereas the bandwidth of the response process is generally acknowledged to vary with damping. The Wirsching and Light prediction that $E(T)$ is 26% greater than the Rayleigh value even if the damping tends to zero is not only inaccurate, but also unconservative.

Example 10.6 showed that the single moment method is the only one of the spectral techniques which is capable of providing a consistent bandwidth correction to the Rayleigh approximation for the unimodal response of a SDF oscillator excited by white noise. Similarly, Example 10.7 showed that it is the only one which is consistent with rainflow analysis of bimodal time histories. Although there can be significant discrepancies between the single moment method and rainflow analysis, these errors are much smaller than the maximum error for any of the other spectral methods. Overall, the single moment method is particularly attractive, based on simplicity, general consistency with the idea of bandwidth, and agreement with rainflow analysis of simulated time histories.

Based on the available data, it appears that one should avoid using high-order spectral moments in any attempt to find a better spectral approximation method. In particular, a small high-frequency component in the autospectral density can cause a significant increase in the λ_4 spectral moment which enters into the Ortiz and Chen formulation. Furthermore, the major error of this method in Examples 10.6 and 10.7 does occur in situations in which there is such a small high-frequency component which is contributing substantially to λ_4, and thereby reducing the predicted value of $E(T)$. The rainflow fatigue calculations, on the other hand, are not very sensitive to small high-frequency components, and the single moment method, which uses only $\lambda_{2/m}$, also seems to avoid such difficulties. Since the irregularity factor also depends on λ_4, one can anticipate difficulty when using any method which predicts fatigue life based on the IF value.

10.6   Non-Gaussian Effects

Since fatigue damage is a significantly nonlinear function of stress, we anticipate that it may be sensitive to variations in the probability distribution of that stress. For example, if the stress process $\{X(t)\}$ has a greater than Gaussian probability of taking on large values, then this is likely to cause large stress ranges, and these may cause significantly accelerated fatigue damage. We will summarize one simple approach for approximating this non-Gaussian effect for a narrowband stress process.

In Section 5.4 we discussed the possibility of modeling a non-Gaussian process $\{X(t)\}$ by using

$$X(t) = \mu_X(t) + \sigma_X(t)g[Y(t)]$$

in which $\{Y(t)\}$ is a mean zero, Gaussian process and g($\cdot$) is a monotonic nonlinear function. We can now use this idea to estimate the fatigue damage accumulation for the $\{X(t)\}$ process. In particular, we will use this idea to generalize the Rayleigh approximation method. In the Rayleigh approximation, we calculate the expected fatigue damage per cycle by taking the stress range as $S_r = 2A(t)$, in which $\{A(t)\}$ is the amplitude of $\{X(t)\}$. Recall, though, that this involves the observation that the probability distribution of $A(t)$ is essentially the same as that of a peak $P(t)$ for a mean zero, narrowband process. If we now let $P_X(t)$ and $P_Y(t)$ designate the peaks of $\{X(t)\}$ and $\{Y(t)\}$, respectively, then we can say that

$$P_X(t) = \mu_X(t) + \sigma_X(t) g[P_Y(t)]$$

since $g(\cdot)$ is a monotonic function. Similarly the relationship between the valleys of the processes is

$$V_X(t) = \mu_X(t) + \sigma_X(t) g[V_Y(t)]$$

We now impose the condition that $\{Y(t)\}$ is narrowband, which gives $P_Y(t) = A_Y(t)$ and $V_Y(t) = -A_Y(t)$. Of course the peak and valley do not occur simultaneously, but the narrowband assumption implies that $A_Y(t)$ varies slowly, so that we can consider it to have the same value at the time of the peak and the valley. This gives

$$S_{r,X} = P_X(t) - V_X(t) = \sigma_X(t)\big(g[A_Y(t)] - g[-A_Y(t)]\big)$$

Since $\{Y(t)\}$ is Gaussian and narrowband, we know that its amplitude $A_Y(t)$ has the Rayleigh distribution, so that we can write the following integral for the expected damage per cycle:

$$E(\Delta D) = K^{-1}\sigma_X^m E\big([g(A_Y) - g(-A_Y)]^m\big)$$

$$= K^{-1}\lambda_0^{m/2} \int_0^\infty [g(u) - g(-u)]^m u \exp(-u^2/2)\,du \tag{10.23}$$

This can be used with Eq. 10.14 to give an estimate of the expected fatigue life. Consider now the special case in which the $g(\cdot)$ is antisymmetric, so that $g(-u) = -g(u)$. This further simplifies Eq. 10.23 to give

$$E(\Delta D) = 2^m K^{-1}\lambda_0^{m/2} \int_0^\infty g^m(u) u \exp(-u^2/2)\,du \tag{10.24}$$

There are many possible choices for the $g(\cdot)$ function (e.g., Sarkani et al., 1994; Lutes et al., 1984), but we will limit our attention to the cubic Hermite polynomial forms studied by Winterstein:

$$X(t) = \sigma_X \sum_{j=1}^{3} b_j H_{ej}[Y(t)] \qquad \text{for} \qquad \textit{kurtosis} > 3 \tag{10.25}$$

and

$$Y(t) = \sum_{j=1}^{3} c_j H_{ej}[X(t)/\sigma_X] \qquad \text{for} \qquad \textit{kurtosis} < 3 \tag{10.26}$$

which are the simplified forms of Eqs. 5.31 and 5.36 for the case of $\mu_X = 0$. Expressions for the b_j and c_j coefficients are given in Section 5.4. Of course, it is necessary to invert Eq. 10.26 before we can use it to calculate the expected value of the damage increment in Eq. 10.23. One form in which the inverse can be written is

$$g(u) = \frac{c_2}{3c_3} + \left(Q_2 + \sqrt{Q_2^2 - Q_1^3}\right)^{1/3} + \left(Q_2 - \sqrt{Q_2^2 - Q_1^3}\right)^{1/3}$$

with

$$Q_1 = 1 - \frac{c_1}{3c_3} + \left(\frac{c_2}{3c_3}\right)^2, \qquad Q_2 = \frac{u}{2c_3} + \frac{c_1 c_2}{6c_3^2} - \left(\frac{c_2}{3c_3}\right)^3$$

and numerical integration can then be used for the evaluation of $E(\Delta D)$.

In order to illustrate the effect of non-Gaussianity on fatigue, we introduce a Gaussian ratio term defined as

$$GR = \frac{E(T) \text{ for alternate distribution}}{E(T) \text{ for Gaussian distribution}} \tag{10.27}$$

Note that this is very similar in concept to the Rayleigh ratio RR introduced to present the effect of the autospectral density. Fig. 10.4 shows numerical values of GR for the symmetric situation with *skewness* $= 0$ and $c_2 = 0$. In addition to the results from numerical integration of Eq. 10.24, we include some results obtained by Hu from rainflow analysis of simulated time histories (Lutes et al., 1984). It should be noted that the simulation results represent the average of the results for several different stress processes, not all of which were very narrowband. Thus, the simulation data may show some effect of bandwidth as well as of kurtosis. Nonetheless, they do confirm that Eq. 10.24 gives the proper trend in predicting the non-Gaussian effect, particularly for *kurtosis* > 3.

Winterstein (1985, 1988) also offered a simplification whereby $(GR)^{-1}$ is approximated as a linear function of the kurtosis, matching the slope of the curve from Eq. 10.24 at the point *kurtosis* $= 3$. It has been found that this approximation seems to overestimate GR for small kurtosis values, and that the simulation data is better approximated by taking GR as a linear function of the kurtosis in that situation. Thus, a reasonable approximation seems to be

$$GR = \left[1 + \frac{m(m-1)(kurtosis - 3)}{24}\right]^{-1} \qquad \text{for} \qquad kurtosis > 3$$

$$GR = 1 - \frac{m(m-1)(kurtosis - 3)}{24} \qquad \text{for} \qquad kurtosis < 3 \tag{10.28}$$

Figure 10.5 shows that this approximation is in very good agreement with the simulation data. Rather surprisingly, it seems that these particular data are better fitted by Eq. 10.28 than by Eq.

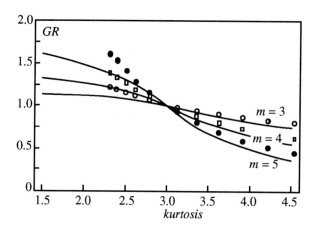

Figure 10.4 Non-Gaussian effect for narrowband process — Eq. 10.24

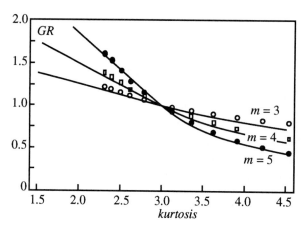

Figure 10.5 Non-Gaussian effect for narrowband process — Eq. 10.28

10.24. There is no obvious reason to expect this to be true in general, but it does appear that the simple formulas in Eq. 10.28 may be adequate for predicting the non-Gaussian effect for moderately non-Gaussian processes.

Practical interest in the non-Gaussian effect usually focuses on the situation with kurtosis greater than three, since that is the situation in which one might seriously overpredict the life of a structure by using the usual Gaussian assumption. If the true stress process has a smaller than Gaussian probability of large stress values, then it is generally acceptable to ignore non-Gaussian effects. If the opposite is true, though, then it is prudent to include this effect when predicting the fatigue life. This can be done by the approximate formulas given here, or by performing rainflow analysis of representative stress time histories. One practical situation in which stress processes are known to be non-Gaussian relates to the stresses in offshore structures subjected to wave loadings. The nonlinear relationship between water velocity and the hydrodynamic loading can cause the stress process to have significantly elevated kurtosis values, even if the water waves are assumed to be Gaussian (Lutes and Wang, 1993). The skewness of the stress process has also been investigated for this offshore problem, but it appears that skewness is much less significant than kurtosis for predicting fatigue values (Wang and Lutes, 1993). These offshore investigations have also provided some evidence that if the stress is neither Gaussian nor narrowband, it is sometimes acceptable to use correction factors for both the autospectral density effect and the non-Gaussian effect. This approximation gives the final estimate of $E(T)$ as $GR \times RR$ times the value from the Rayleigh approximation.

****

Example 10.8: Consider the fatigue life of a structural joint for which the S-N curve has been found to be $N = K S^{-4}$. In actual use, the joint will be subjected to a narrowband stochastic stress $\{X(t)\}$ with mean zero and standard deviation σ_X. This narrowband stress can be written as $X(t) = A(t)\cos[\omega_a t + \theta(t)]$, with $A(t)$ and $\theta(t)$ being independent, and $\theta(t)$ uniformly distributed on the set of possible values. Rather than being Rayleigh distributed, it is believed that $A(t)$ has the one-sided Gaussian distribution:

$$p_{A(t)}(u) = \frac{1}{\sqrt{\pi}\,\sigma_X}\exp\left(\frac{-u^2}{4\sigma_X^2}\right)U(u)$$

Compare the fatigue life predicted by the Rayleigh approximation, the use of $S_r = 2A(t)$ and $p_{A(t)}(u)$ in Eq. 10.13, and the use of Eq. 10.28.

Before starting on the fatigue computation, it is appropriate to check for consistency of the model. In particular, we find that

$$E[A^2(t)] = \int_0^\infty u^2\, p_{A(t)}(u)\, du = 2\sigma_X^2$$

so that the narrowband process gives $E[X^2(t)] = E[A^2(t)]/2 = \sigma_X^2$ (see Example 2.3 in Chapter 2). Thus, the model is consistent with σ_X^2 being the variance of the stress. Now we note that the Rayleigh approximation from Eqs. 10.13 and 10.14 is $E(T_{Ray}) = \pi K / (64\lambda_0^{3/2}\lambda_2^{1/2})$, as in Example 10.5. For this narrowband process we can say that $\lambda_2 = \omega_a^2\lambda_0$, and we also know that $\lambda_0 = \sigma_X^2$. This allows the Rayleigh approximation to be rewritten as $E(T_{Ray}) = \pi K / (64\omega_a\sigma_X^4)$.

Using $S_r = 2A(t)$ in Eq. 10.13 gives

$$E(\Delta D) = K^{-1}E(S_r^4) = 2^4 K^{-1}\int_0^\infty u^4 p_{A(t)}(u)\, du = 192 K^{-1}\sigma_X^4$$

and Eq. 10.14 then gives the expected fatigue life as $E(T) = \pi K / (96\omega_a\sigma_X^4)$. Thus, this gives a prediction of $E(T)$ which is 33% smaller than the Rayleigh approximation.

In order to use Eq. 10.28, we must find the kurtosis of the stress process. To do this, we first integrate over the possible values for $\theta(t)$ to find that

$$E[X^4(t)] = E[A^4(t)] \int_0^{2\pi} \frac{\cos^4(\psi)}{2\pi} d\psi = \frac{3}{8} E[A^4(t)]$$

From the integral used in the previous step, we find that $E[A^4(t)] = 12\sigma_X^4$, so that $E[X^4(t)] = 4.5\sigma_X^4$. Thus this process has $kurtosis = 4.5$, and Eq. 10.28 then gives

$$GR = \frac{24}{24 + 12(1.5)} = 0.57$$

which predicts a 43% reduction in fatigue life as compared to the Rayleigh approximation, which would be appropriate for a Gaussian process. This value is relatively consistent with the 33% reduction calculated in the previous step. To neglect the non-Gaussian effect would be a more serious error, since it would result in a fatigue life prediction which was significantly too large.

****

Exercises
****

Mean Stress and S-N Curve
****

10.1 Use the Gerber correction to find the values of the effective stress ranges for the data shown for an ultimate stress of $x_u = 516\,\text{MPa}$. Find the S-N curve of the form $N_f = K\,S_e^{-m}$.

[Note that the values of S_r, N_f, and x_u have been taken to be the same as in Examples 10.1 and 10.2. Furthermore, the average value of the mean stress x_m is approximately the same here as in Example 10.2, but x_m is not correlated with S_r, as it was in the example problem.]

S_r MPa	x_m MPa	N_f	S_r MPa	x_m MPa	N_f
667	141	18,360	267	170	432,000
667	176	21,880	267	145	356,400
667	24	16,610	267	144	417,000
667	104	15,340	267	201	284,480
667	26	14,400	267	38	212,350
533	92	31,550	267	41	716,600
533	61	39,170	267	3	386,840
533	113	32,774	267	43	202,814

****

Rayleigh Approximation
****

10.2 Use the Rayleigh approximation to estimate the expected number of cycles until failure, $E[N(T)]$, for a mean zero stress process $\{X(t)\}$ with $\sigma_X = 25$ MPa, using:
(a) the S-N curve found in Example 10.1.
(b) the S-N curve found in Example 10.2.
(c) the S-N curve found in Exercise 10.1

[Note: Properties of the gamma function are given in Example A.31 of Appendix A.]

****

10.3 For a material with an ultimate stress of $x_u = 516$ MPa, use the Rayleigh approximation to estimate the expected number of cycles until failure, $E[N(T)]$, for a stress process $\{X(t)\}$ with μ_X = 100 MPa and $\sigma_X = 25$ MPa, using the Gerber correction and:
(a) the S-N curve found in Example 10.1.
(b) the S-N curve found in Example 10.2.
(c) the S-N curve found in Exercise 10.1

****

10.4 A structural joint has been subjected to constant amplitude fatigue tests at the single level of $S_r = 200$ MPa, and the observed fatigue life was $N_f = 4 \times 10^5$ cycles. Based on experience with similar joints, it is estimated that the parameter m of the S-N curve is in the range of $3.5 \leq m \leq 4.5$. Based on this limited information, choose acceptable levels for the standard deviation σ_X of the narrowband, mean zero, Gaussian stress which will be applied to the joint in actual service. You need only consider the two limiting values of $m = 3.5$ and $m = 4.5$, not all the possible m values. Solve the problem for the design conditions of:
(a) $E[N(T)] = 10^6$
(b) $E[N(T)] = 10^8$

****

10.5 A machine part has been subjected to constant amplitude fatigue tests at the single level of $S_r = 100$ MPa, and the observed fatigue life was $N_f = 5 \times 10^6$ cycles. Based on experience with similar parts it is estimated that the parameter m of the S-N curve is in the range of $3.0 \leq m \leq 4.5$. Based on this limited information choose acceptable levels for the standard deviation σ_X of the narrowband, mean zero, Gaussian stress which will be applied to the joint in actual service. You need only consider the two limiting values of $m = 3.0$ and $m = 4.5$, not all the possible m values. Solve the problem for the design conditions of:
(a) $E[N(T)] = 10^7$
(b) $E[N(T)] = 10^9$.

****

10.6 The S-N curve for a particular connection has been found to be $N_f = (10,000 / S_r)^{3.5}$. In service it is subjected to a stress process $\{X(t)\}$ which is mean zero and has $\sigma_X = 125$ MPa. Furthermore, $\{X(t)\}$ can be modeled as the response of a SDF oscillator subjected to a Gaussian white noise excitation. The natural frequency and damping ratio of the oscillator are $\omega_0 = 3$ rad/sec and $\zeta = 0.02$. Use the Rayleigh approximation to estimate the expected hours of service prior to fatigue failure.

****

Single Moment Method

****

10.7 Let the mean zero, Gaussian stress process $\{X(t)\}$ have an autospectral density function of $S_{XX}(\omega) = S_0 e^{-\beta |\omega|}$. For a specimen with $m = 3$, compare the fatigue lives predicted by the Rayleigh approximation and the single moment method.

****

10.8 Let the mean zero, Gaussian stress process $\{X(t)\}$ have an autospectral density function of

$$S_{XX}(\omega) = S_0\left[1 - \frac{1}{5}\left(\frac{|\omega|}{\omega_0} - 1\right)\right]\left[U(|\omega| - \omega_0) - U(|\omega| - 6\omega_0)\right]$$

For a specimen with $m = 4$, compare the fatigue lives predicted by the Rayleigh approximation and the single moment method

****

10.9 Let the mean zero, Gaussian stress process $\{X(t)\}$ have an autospectral density function of

$$S_{XX}(\omega) = S_0 \qquad \text{for} \quad |\omega| \le \omega_0$$

$$= S_0\left(\frac{\omega_0}{|\omega|}\right)^4 \quad \text{for} \quad |\omega| > \omega_0$$

For a specimen with $m = 3$, compare the fatigue lives predicted by the Rayleigh approximation and the single moment method

****

10.10 Let the mean zero, Gaussian stress process $\{X(t)\}$ represent the response of a 2DF system with lightly damped modes. In particular, the fundamental mode has damping $\zeta_1 = 0.01$ and frequency $\omega_1 = 6\,\text{rad}/\text{sec}$, and the second mode has $\zeta_2 = 0.02$ and frequency $\omega_2 = 80\,\text{rad}/\text{sec}$. The standard deviations of the modal stress values are $\sigma_{X_1} = 100\,\text{MPa}$ and $\sigma_{X_2} = 7.5\,\text{MPa}$. Make an approximate comparison of the fatigue lives predicted by the Rayleigh approximation and the single moment method for a specimen with $m = 4$. Do this by considering the two modal responses to be independent and concentrated at their natural frequencies ω_1 and ω_2, so that the autospectral density can be approximated by the form used in Example 10.7.

****

Non-Gaussian Effects

****

10.11 Consider the fatigue life of a structural joint for which the S-N curve has been found to be $N = KS^{-3}$. In actual use the joint will be subjected to a non-Gaussian narrowband stochastic stress $\{X(t)\}$ with mean zero and standard deviation σ_X. This narrowband stress can be written as $X(t) = A(t)\cos[\omega_a t + \theta(t)]$, with $A(t)$ and $\theta(t)$ being independent, $\theta(t)$ being uniformly distributed on the set of possible values, and $A(t)$ having an exponential distribution:

$$p_{A(t)}(u) = \frac{1}{\sigma_X}\exp\left(\frac{-u}{\sigma_X}\right)U(u)$$

Compare the fatigue life predicted by:
(a) using the Rayleigh approximation.
(b) using $S_r = 2A(t)$ and $p_{A(t)}(u)$ in Eq. 10.13.
(c) using Eq. 10.28.

****

10.12 Consider the fatigue life of a structural joint for which the S-N curve has been found to be $N = KS^{-4}$. In actual use the joint will be subjected to a non-Gaussian narrowband stochastic stress $\{X(t)\}$ with mean zero and standard deviation σ_X. This narrowband stress can be written as $X(t) = A(t)\cos[\omega_a t + \theta(t)]$, with $A(t)$ and $\theta(t)$ being independent, $\theta(t)$ being uniformly distributed on the set of possible values, and $A(t)$ having a Weibull distribution of the form

$$p_{A(t)}(u) = \frac{\pi u^3}{4\sigma_X^4} \exp\left(\frac{-\pi u^4}{16\sigma_X^4}\right) U(u)$$

Compare the fatigue life predicted by:

(a) using the Rayleigh approximation.

(b) using $S_r = 2A(t)$ and $p_{A(t)}(u)$ in Eq. 10.13.

(c) using Eq. 10.28.

****

Appendix A

Analysis of Random Variables

A.1 Probability Distribution

A random variable is a mathematical tool which we can use to describe an entity which must take on some real value, but for which we are uncertain regarding what that value will be.[1] For example, in a structural dynamics problem, the uncertain quantity of interest might be the force which will occur on a structure at some specified future instant of time, or it might be some measure of response such as displacement, velocity or acceleration at the specified time. Since we are uncertain about the value of the uncertain entity, the best description we can hope to find is one which gives the probability of its taking on particular values, or values in any particular subset of the set of possible values. Thus, the description of a random variable is simply a description of its probabilities. It should perhaps be noted that there is nothing to be gained by debating whether a certain physical quantity *is* a random variable. The more pertinent question is whether our uncertainty about the value of the quantity can be usefully *modeled* by a random variable. As in all other areas of applied mathematics, it is safe to assume that our mathematical model is never identical to a physical quantity, but that does not necessarily preclude our using the model to obtain meaningful results.

Probabilities are always defined for sets of possible outcomes, called events. For a problem described by a single random variable X, any event of interest is always equivalent to the event of X belonging to some union (finite or infinite) of disjoint intervals of the real line. In some problems we are most interested in events of the type $X = u$, in which u is a particular real number. In order to include this situation within the idea of events corresponding to intervals, we can consider the event to be $X \in I_u$ with $I_u = [u,u]$ being an interval which includes only the single point u. In other problems, we are more interested in intervals of finite length. Probably the most general way to describe the probabilities associated with a given random variable is with the use of the **cumulative distribution function**, which will be written as $F_X(u)$. The argument of this function is always a real number, and the domain of definition of the function is the entire real line. That is, for any random variable the argument of the cumulative distribution function can be any real number. The definition of the $F_X(\cdot)$ function is in terms of a probability of X being smaller than (or equal to) a given number. Arbitrarily choosing u to denote the argument, we can write the definition as

$$F_X(u) \equiv P(X \leq u) \tag{A.1}$$

Thus, $F_X(u)$ is exactly the probability of X being within the infinite interval to the left of u on the real line: $F_X(u) = P\{X \in (-\infty,u]\} \equiv P(-\infty < X \leq u)$. Again, it should be kept in mind that u can be

---

[1] One can also define complex random variables, but in this Appendix we will restrict our attention to the slightly simpler case of real random variables.

any real number on the real line. Also, it should be kept in mind that u is not the random variable. It is a given real number, about which there is no uncertainty. The random variable is X and our uncertainty about its value is represented by the values of the function $F_X(u)$ for given values of u.

****

Example A.1: Consider X to be a random variable which is the numerical value of the result from a single roll of a standard gaming die. Thus, X has six possible values $\{1,2,3,4,5,6\}$, and is equally likely to have any one of these values: $P(X=1) = P(X=2) = P(X=3) = P(X=4) = P(X=5) = P(X=6) = 1/6$. Find the cumulative distribution function for X.

The cumulative distribution function for this random variable is found simply by summing outcomes which fall within the interval $(-\infty, u]$:

$$
\begin{aligned}
F_X(u) &= 0   && \text{for} && -\infty < u < 1 \\
       &= 1/6   && \text{for} && 1 \le u < 2 \\
       &= 2/6   && \text{for} && 2 \le u < 3 \\
       &= 3/6   && \text{for} && 3 \le u < 4 \\
       &= 4/6   && \text{for} && 4 \le u < 5 \\
       &= 5/6   && \text{for} && 5 \le u < 6 \\
       &= 6/6 = 1 && \text{for} && 6 \le u < \infty
\end{aligned}
$$

Note that this $F_X(u)$ function is defined over the entire real line $-\infty < u < \infty$, even though the random variable X has a very limited set of possible values. For example, $F_X(\pi) = 3/6 = 0.5$.

****

Example A.2: Let X denote a real number chosen "at random" from the interval $[0,10]$. Find the cumulative distribution function of X.

In this case there is a continuous set of possible values for the random variable, so there are infinitely many values which the random variable X might assume and it is equally likely that X will take on any one of these values. Obviously, this requires that the probability of X being equal to any particular one of the possible values must be zero, since the total probability assigned to the set of all possible values is always unity for any random variable (an axiom of probability theory, called "total probability"). Thus, it is not possible to define the probabilities of this random variable by giving the probability of events of the type $\{X = u\}$, but there is no difficulty in using the cumulative distribution function. It is given by

$$
\begin{aligned}
F_X(u) &= 0   && \text{for} && -\infty < u < 0 \\
       &= 0.1u   && \text{for} && 0 \le u < 10 \\
       &= 1   && \text{for} && 10 \le u < +\infty
\end{aligned}
$$

For example, $P(0 \le X \le 5) = F_X(5) = 0.5$.

****

In both of the examples, note that the function $F_X(u)$ starts at zero for $u \to -\infty$ and increases monotonically to unity for $u \to +\infty$. These limits of zero and unity, and the property of being monotonically increasing (or, more precisely, "monotonically nondecreasing") are characteristic of the cumulative distribution function of any random variable. They follow directly from axioms used in the definition of probability theory. In fact, one can define a random variable with a distribution function equal to $F(u)$ for any real function $F(u)$ which approaches zero for $u \to -\infty$, approaches unity for $u \to +\infty$, and is monotonically nondecreasing and continuous from the right for all finite u values.

Example A.1 is an illustration of the category of "discrete" random variables, whereas Example A.2 is a continuous random variable. Precisely, a random variable is discrete if it may only assume values within a discrete set. (It has zero probability of being outside the discrete set.) The $F_X(u)$ function for a discrete random variable is always of the "stairstep" form, with the magnitude of the discontinuity at any particular point u_j being the probability that $X = u_j$. In Example A.1 the number of possible values for X was finite, but this is not necessary for a discrete random variable. For example, $P(X = j) = 2^{-j}$ for $j \in$ {set of positive integers} gives a well-defined random variable X. In this case the steps in $F_X(u)$ become smaller and smaller as u becomes larger and larger, so that $F_X(u)$ does approach unity for $u \to +\infty$. A random variable is said to be continuous if its cumulative distribution function is continuous. Although the designations "discrete" and "continuous" are useful, they are not comprehensive. That is, there are also random variables which are neither discrete or continuous. These are sometimes called mixed random variables.

****

Example A.3: Let the random variable X represent the DC voltage output from some force transducer, and let the distribution be uniform on the set $[-4, 16]$ such that $F_X(u) = 0.05(u+4)$ for $-4 \leq u \leq 16$. Of course, $F_X(u)$ is zero to the left and unity to the right of the $[-4, 16]$ interval. Now let another (mixed) random variable Y represent the output from a voltmeter which reads only from zero to 10 volts, and which has X as the input. Whenever $X \in [0, 10]$ we will have $Y = X$, but we will get $Y = 0$ whenever $X < 0$ and $Y = 10$ whenever $X > 0$. Find the cumulative distribution function for Y.

From $P(Y \leq u)$ we get

$$\begin{aligned} F_Y(u) &= 0 & \text{for} \quad & u < 0 \\ &= 0.05(u+4) & \text{for} \quad & 0 \leq u < 10 \\ &= 1 & \text{for} \quad & u > 10 \end{aligned}$$

This $F_Y(u)$ function has discontinuities of 0.2 at $u = 0$ and 0.3 at $u = 10$, representing the finite probabilities that Y takes on these two particular values.

****

In describing cumulative distribution functions, it is often convenient to use the simple discontinuous function called the unit step function.[2] We will use the notation $U(\cdot)$ to denote this

---

[2]   This function is widely used in many engineering areas, and is sometimes called the Heaviside

function, and define it as

$$U(x) = 0 \qquad \text{for} \qquad x < 0$$
$$= 1 \qquad \text{for} \qquad x \geq 0 \tag{A.2}$$

Using the unit step function, it is possible to define a "stair-step" function as a summation. In particular, the cumulative distribution function for a discrete random variable such as that given in Example A.1 can be written as

$$F_X(u) = \sum_j p_j U(u - x_j) \tag{A.3}$$

in which p_j is the magnitude of the discontinuity at $u = x_j$. Note that for any u value the summation in Eq. A.3 is over all the possible values of j, whether that number be finite or infinite. However, for a given u value, some of the terms may contribute nothing. Specifically, there is no contribution to the summation for any j value for which $x_j > u$, since that corresponds to a $U(\cdot)$ function with a negative argument.

****

Example A.4: Use the unit step function to write a single expression for the cumulative distribution function of the discrete random variable X of Example A.1, having $P(X = 1) = P(X = 2) = P(X = 3) = P(X = 4) = P(X = 5) = P(X = 6) = 1/6$.

The summation in Eq. A.3 is simply

$$F_X(u) = \frac{1}{6}[U(u - 1) + U(u - 2) + U(u - 3) + U(u - 4) + U(u - 5) + U(u - 6)]$$

****

Example A.5: Use the unit step function to write a single expression for the cumulative distribution function for the mixed random variable Y of Example A.3, having

$$F_Y(u) = 0 \qquad \text{for} \qquad u < 0$$
$$= 0.05(u + 4) \quad \text{for} \qquad 0 \leq u < 10$$
$$= 1 \qquad \text{for} \qquad u > 10$$

The discontinuous $F_Y(u)$ is exactly given by

$$F_Y(u) = 0.05(u + 4)[U(u) - U(u - 10)] + U(u - 10)$$
$$= (0.2 + 0.05u)U(u) + (0.8 - 0.05u)U(u - 10)$$

****

These examples illustrate how the unit step function can be used to write the cumulative

function or the Heaviside step function. More detail on the unit step function and its relationship to the Dirac delta function are presented in Appendix C.

distribution function for a discrete or mixed random variable as a single equation which is valid everywhere on the real line, rather than using a piecewise description of the function, as was done in Examples A.1, A.2, and A.3. This is strictly a matter of convenience, but it will often be useful. Note that the fact that $U(\cdot)$ was defined to be continuous from the right assures that a cumulative distribution function written according to Eq. A.3 will have the proper value at any point of discontinuity.

A.2 Probability Density Functions

For many of our calculations, it will be more convenient to describe the probability distribution of a random variable X by using what is called the probability density function $p_X(\cdot)$ rather than the cumulative distribution function $F_X(\cdot)$. The $p_X(\cdot)$ function gives the probability per unit length along the line of possible values for a continuous random variable. Using $p_X(u)$ is analogous to describing the mass distribution of a nonuniform rod by giving the mass per unit length at each location u, whereas using $F_X(u)$ is like describing the rod by telling how much mass is located to the left of u for any particular u value. If $F_X(u)$ is continuous and differentiable everywhere, then one can define the probability density function as

$$p_X(u) = \frac{d}{du} F_X(u) \tag{A.4}$$

The inverse of this relationship is

$$F_X(u) = \int_{-\infty}^{u} p_X(v)\, dv \tag{A.5}$$

since this gives both Eq. A.4 and the limiting value of $F_X(-\infty) = 0$. Equation A.5 illustrates the fundamental nature of probability density functions — the integral of $p_X(u)$ over an interval gives the probability that X lies within that interval. In Eq. A.5, of course, the interval is $(-\infty, u]$. A special case of Eq. A.5 is that

$$\int_{-\infty}^{\infty} p_X(u)\, du = 1 \tag{A.6}$$

This is equivalent to $F_X(\infty) = 1$, and is another form of the axiom of total probability. The probability of X being within any finite interval $[a,b]$ is given by

$$P(a \le X \le b) = \int_{a}^{b} p_X(u)\, du = F_X(b) - F_X(a) \tag{A.7}$$

One may note that the final expression $F_X(b) - F_X(a)$ in Eq. A.7 actually gives $P(a < X \le b)$, excluding $P(X = a)$. This is identical to the probability sought in Eq. A.7 for a continuous random variable, though, since $P(X = a)$ is the magnitude of the discontinuity in $F_X(u)$ at $u = a$, and this must be zero for a continuous $F_X(u)$ function. Equation A.7 can also be used to write an infinitesimal expression which may help to illustrate the meaning of the probability density function. In particular, if $p_X(u)$ is continuous on the infinitesimal interval $(u \le X \le u + du)$, then it can be considered to be constant across the interval so that we have

$$P(u \leq X \leq u + du) = \int_u^{u+du} p_X(w)\,dw = p_X(u)\,du$$

Thus, at any point u for which $p_X(u)$ is continuous, the probability density gives the probability of X being in the neighborhood of u, in the sense that $p_X(u)\,du$ is the probability of being in the infinitesimal increment of length du.

The fact that any cumulative distribution function is nondecreasing tells us that any probability density function determined from Eq. A.5 will be nonnegative. In fact, nonnegativity and Eq. A.6 are the only restrictions on a probability density function. For any function $p(\cdot)$ which satisfies these two conditions, one can define a random variable having $p(\cdot)$ as its probability density function. Note, in particular, that there is no requirement that $p(\cdot)$ be continuous. A discontinuity in $p_X(u)$ corresponds to an instantaneous change in the *slope* of $F_X(u)$. So long as the number of points of discontinuity is countable, one can show that probabilities of X are uniquely defined by $p_X(u)$ even if that function is not uniquely defined at the points of discontinuity.

****
Example A.6: Find the probability density function for the random variable X of Example A.2, for which

$$\begin{aligned} F_X(u) &= 0 & \text{for} & \quad -\infty < u < 0 \\ &= 0.1u & \text{for} & \quad 0 \leq u < 10 \\ &= 1 & \text{for} & \quad 10 \leq u < +\infty \end{aligned}$$

Differentiating the cumulative distribution function, the probability density function is found to be

$$\begin{aligned} p_X(u) &= 0.1 & \text{for} & \quad 0 \leq u < 1 \\ &= 1 & \text{otherwise} \end{aligned}$$

or, equivalently,

$$p_X(u) = 0.1[U(u) - U(u-10)]$$
****

Any random variable X, such as that in the example just given, for which $p_X(u)$ has a constant value over all the possible values of X, is said to have a uniform distribution. Note that for any random variable, including one with a uniform distribution, the $p_X(u)$ function is defined on the entire real line (except possibly at points of discontinuity), even though it will be exactly zero in the vicinity of any u value which is not a possible value of X. Note that Example A.6 illustrates the use of the unit step function to simplify the form of the probability density function for a random variable which would otherwise require the use of a piecewise description.

The definition of the probability density function according to Eqs. A.4 and A.5 is only applicable to continuous random variables, unless one extends the boundaries of ordinary calculus. In particular, no bounded function $p_X(u)$ can satisfy Eq. A.5 if $F_X(u)$ contains discontinuities, as occur for a discrete or mixed random variable. This limitation of the probability density function can be removed by using the Dirac delta function $\delta(\cdot)$, which is defined by the following properties:[3]

$$\delta(x) = 0 \quad \text{for} \quad x \neq 0$$
$$= \infty \quad \text{for} \quad x = 0$$

and

$$\int_{-\infty}^{\infty} \delta(x - x_0) f(x) dx = f(x_0) \tag{A.8}$$

for any function $f(\cdot)$ which is finite and continuous at the point $x = x_0$. The δ function can also be thought of as the formal derivative of the unit step function (see Appendix C):

$$\delta(x) = \frac{d}{dx} U(x) \tag{A.9}$$

so it allows one to formulate the derivative of a cumulative distribution function $F_X(u)$ which contains discontinuities.

Using the Dirac delta function one can formally describe any random variable by using a probability density function. For a discrete random variable described by Eq. A.3, in particular, one obtains

$$p_X(u) = \sum_j p_j \delta(u - x_j) \tag{A.10}$$

The p_j multiplier of a term $\delta(u - x_j)$ in such a description of $p_X(u)$ for a discrete or mixed random variable always gives the finite probability that $X = x_j$.

****

Example A.7: Use the Dirac delta function to write an expression for the probability density function of the discrete random variable X of Example A.1, having $P(X = 1) = P(X = 2) = P(X = 3) = P(X = 4) = P(X = 5) = P(X = 6) = 1/6$.

This probability density function can be written as

$$p_X(u) = \frac{1}{6}[\delta(u-1) + \delta(u-2) + \delta(u-3) + \delta(u-4) + \delta(u-5) + \delta(u-6)]$$

This expression is exactly the formal derivative of the cumulative distribution function in Example A.4.
****

---

[3] Strictly speaking, $\delta(\cdot)$ is not a function, but it can be thought of as a limit of a sequence of functions, as explored in Appendix C.

Example A.8: Find a probability density function for the mixed random variable Y of Examples A.3 and A.5, having

$$F_Y(u) = (0.2 + 0.05u)U(u) + (0.8 - 0.05u)U(u - 10)$$

Differentiating this cumulative distribution function gives

$$p_Y(u) = 0.05[U(u) - U(u - 10)] + 0.2\delta(u) + 0.3\delta(u - 10)$$

but this does require noting that $(0.8 - 0.05u)\delta(u - 10) \equiv 0.3\delta(u - 10)$, since the terms are equal at $u = 10$ and are zero everywhere else.

****

This example illustrates an important feature of Dirac delta functions. Any term $g(x)\delta(x - x_0)$ can be considered to be identical to $g(x_0)\delta(x - x_0)$, since both are zero for $x \neq x_0$. A related property is that

$$g(x)\delta(x - x_0) \equiv 0 \qquad \text{if} \qquad g(x_0) = 0 \qquad\qquad \text{(A.12)}$$

That is, $g(x)\delta(x - x_0)$ can be considered to be identically zero for all x values if $g(x_0) = 0$. Clearly $g(x)\delta(x - x_0)$ is zero for $x \neq 0$, but its value seems to be indeterminate at $x = 0$, since it is then the product of zero and infinity. However the integral of the expression is zero, by the definition of the Dirac delta function in Eq. A.8. Since Dirac delta functions are useful only as they contribute to integrals, this demonstrates that $g(x)\delta(x - x_0)$ with $g(x_0) = 0$ can never contribute to any finite expression. Thus, it can be considered as identically zero and dropped from calculations. A particular example of this is the term $x\delta(x)$, which can always be considered to be zero.

A.3 Joint and Marginal Distributions of Random Variables

In many problems, we must use more than one random variable to describe the uncertainty that exists about various outcomes of a given activity. Example A.3 represents one very simple situation involving two random variables. In that particular case, Y is a function of X, so that if one knows the value of X then one knows exactly the value of Y. If one thinks of the (X,Y) plane, for this example, then all the possible outcomes lie on a simple (piecewise linear) curve. In other problems, there may be a less direct connection between the random variables of interest. For example, the set of possible values and/or the probability of any particular value for one random variable Y may depend on the value of another random variable X.

As with a single random variable, one can always describe the probabilities of two or more random variables by using a cumulative distribution function. For two random variables X and Y, this can be written as

$$F_{XY}(u,v) \equiv P[X \leq u, Y \leq v] \qquad\qquad \text{(A.11)}$$

in which the comma within the brackets on the right-hand side of the expression represents the intersection operation. That is, the probability denoted is for the joint event that $X \le u$ and $Y \le v$. The function $F_{XY}(\cdot)$ is defined on the two-dimensional space of all possible (X,Y) values, and it is called the joint cumulative distribution function. When we generalize to more than two or three random variables, it will often be more convenient to use a vector notation. In particular, we will use an arrow over a symbol to indicate that the quantity involved is a vector (which may also be viewed as a matrix with only one column). Thus, we will write

$$\vec{X} = \begin{bmatrix} X_1 \\ X_2 \\ \vdots \\ X_n \end{bmatrix}, \qquad \vec{u} = \begin{bmatrix} u_1 \\ u_2 \\ \vdots \\ u_n \end{bmatrix}$$

and use the notation

$$F_{\vec{X}}(\vec{u}) \equiv F_{X_1 X_2 \cdots X_n}(u_1, u_2, \cdots, u_n) \equiv P\left[\bigcap_{j=1}^{n} (X_j \le u_j) \right]$$

for the general joint cumulative distribution function of n random components, in which the intersection symbol $\bigcap$ describes the event of the joint occurrence of all the terms in the set for $j = 1$ to n.

As with a single random variable, one can convert from a cumulative distribution function to a probability density function by differentiation. The situation is slightly more complicated now, though, since we must use multiple partial derivatives:

$$p_{XY}(u,v) = \frac{\partial^2}{\partial u \partial v} F_{XY}(u,v), \qquad p_{\vec{X}}(\vec{u}) \equiv \frac{\partial^n}{\partial u_1 \partial u_2 \cdots \partial u_n} F_{\vec{X}}(\vec{u}) \qquad (A.13)$$

In order to obtain probabilities from an n-dimensional joint probability density function, one must use an n-fold integration such as

$$F_{\vec{X}}(\vec{u}) = \int_{-\infty}^{u_n} \cdots \int_{-\infty}^{u_1} p_{\vec{X}}(\vec{v}) \, dv_1 \cdots dv_n \qquad (A.14)$$

for the probability defined as the joint cumulative distribution function, or

$$P[(X,Y) \in A] = \iint_A p_{XY}(u,v) \, du \, dv \qquad (A.15)$$

for the probability of (X,Y) falling within any arbitrary region A on the plane of possible values for the pair.

****

Example A.9: Let the joint distribution of (X,Y) be such that the possible outcomes (u,v) are the points in the rectangle $(-1 \leq u \leq 2, -1 \leq v \leq 1)$. Furthermore, let every one of these possible outcomes be "equally likely." Note that we are not dealing with discrete random variables in this example, since the set of possible values is not discrete. As with a single random variable on a continuous set of possible values, the term equally likely denotes a constant value of the probability density function. Thus, we say that

$$p_{XY}(u,v) = C \qquad \text{for} \qquad -1 \leq u \leq 2, -1 \leq v \leq 1$$
$$= 0 \qquad \text{otherwise}$$

in which C is some constant. Find the value of the constant C and the joint cumulative distribution function $F_{XY}(u,v)$ for all values of (u,v).

We find the value of C by using the "total probability" property that (X,Y) must fall somewhere within the space of possible values, and that the probability of being within a set is the double integral over that set, as in Eq. A.15. Thus, we obtain $C = 1/6$, since the rectangle of possible values has an area of 6, and the joint probability density function has the constant value C everywhere in the rectangle:

$$1 = P[-1 \leq X \leq 2, -1 \leq Y \leq 1)]$$
$$= P[(X,Y) \in A] = \iint_A p_{XY}(u,v)\,dudv = \int_{-1}^{1}\int_{-1}^{2} C\,dudv = 6C$$

We may now calculate the joint cumulative distribution of these two random variables by integrating as in Eq. A.14, with the result that

$$F_{XY}(u,v) = C(u+1)(v+1) = (u+1)(v+1)/6 \qquad \text{for} \qquad -1 \leq u \leq 2, -1 \leq v \leq 1$$
$$= 3C(v+1) = (v+1)/2 \qquad \text{for} \qquad u > 2, -1 \leq v \leq 1$$
$$= 2C(u+1) = (u+1)/3 \qquad \text{for} \qquad -1 \leq u \leq 2, v > 1$$
$$= 6C = 1 \qquad \text{for} \qquad u > 2, v > 1$$
$$= 0 \qquad \text{otherwise}$$

One can verify that this cumulative distribution function is continuous. Thus, the random variables X and Y are said to have a continuous joint distribution. Clearly for this particular problem the description given by the density function is much simpler than that given by the cumulative distribution function.

****

Example A.10: Let the joint probability density function of X and Y be equal to the constant C on the set of possible values, as in Example A.4, but let the space of possible values be a triangle in the (u,v) plane such that

$$p_{XY}(u,v) = C \qquad \text{for} \quad 1 \leq v \leq u \leq 5$$
$$= 0 \qquad \text{otherwise}$$

A sketch of this joint probability density function is shown. Find the value of the constant C and the joint cumulative distribution function $F_{XY}(u,v)$ for all values of (u,v).

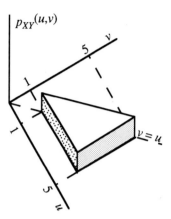

Again we use total probability to find that $C = 1/8$, since the triangle of possible values has an area of 8 and the density function is constant on that triangle. Integration gives the joint cumulative distribution function as the continuous function

$$
\begin{aligned}
F_{XY}(u,v) &= (2u-v-1)(v-1)/16 && \text{for} && 1 \le v \le u \le 5 \\
&= (u-1)^2/16 && \text{for} && v>u,\ 1 \le u \le 5 \\
&= (9-v)(v-1)/16 && \text{for} && 1 \le v \le 5,\ u>5 \\
&= 1 && \text{for} && u>5,\ v>5 \\
&= 0 && \text{otherwise}
\end{aligned}
$$

****

Example A.11: Consider two random variables X and Y with the joint cumulative distribution given by

$$
F_{XY}(u,v) = \left[\frac{(1-e^{-4u})(1-e^{-4v})}{2} + \frac{(1-e^{-4u})(1-e^{-3v})}{4} + \frac{(1-e^{-3u})(1-e^{-4v})}{4}\right]U(u)U(v)
$$

in which the unit step function has been used to convey the information that the nonzero function applies only in the first quadrant of the (u,v) plane.

Find the joint probability density function for X and Y.

Taking the mixed partial derivative according to Eq. A.13 gives

$$
p_{XY}(u,v) = \left(8e^{-4u-4v} + 3e^{-4u-3v} + 3e^{-3u-4v}\right)U(u)U(v)
$$

Note that no $\delta(u)$ or $\delta(v)$ Dirac delta functions have been included in the derivative, even though such terms do appear as derivatives of the $U(u)$ and $U(v)$ unit step functions. However, it is found that each Dirac delta function is multiplied by an expression which is zero at the single point where the Dirac delta function is nonzero, so that they can be ignored, as in Eq. A.12.

****

Some Properties of Joint Probability Distributions: It is left to the reader to verify that for any problem involving two random variables X and Y, the joint cumulative distribution function and joint probability distribution function must satisfy the following properties:

$$
F_{XY}(-\infty,-\infty) \equiv \lim_{\substack{u \to -\infty \\ v \to -\infty}} F_{XY}(u,v) = 0 \tag{A.16}
$$

$$F_{XY}(u, -\infty) \equiv \lim_{v \to -\infty} F_{XY}(u, v) = 0 \tag{A.17}$$

$$F_{XY}(-\infty, v) \equiv \lim_{u \to -\infty} F_{XY}(u, v) = 0 \tag{A.18}$$

$$F_{XY}(\infty, \infty) \equiv \lim_{\substack{u \to \infty \\ v \to \infty}} F_{XY}(u, v) = 1 \tag{A.19}$$

$$F_{XY}(\infty, v) \equiv \lim_{u \to \infty} F_{XY}(u, v) = F_Y(v) \tag{A.20}$$

$$F_{XY}(u, \infty) \equiv \lim_{v \to \infty} F_{XY}(u, v) = F_X(u) \tag{A.21}$$

$$p_{XY}(u, v) \geq 0 \quad \text{for all } (u, v) \text{ values} \tag{A.22}$$

$$\int_{-\infty}^{\infty} p_{XY}(u, v) dv = p_X(u) \tag{A.23}$$

$$\int_{-\infty}^{\infty} p_{XY}(u, v) du = p_Y(v) \tag{A.24}$$

$$\int_{-\infty}^{\infty} \int_{-\infty}^{\infty} p_{XY}(u, v) \, du dv = 1 \tag{A.25}$$

These properties can also be extended to problems with any number of random variables. A few of these extensions are:

$$F_{\vec{X}}(-\infty, u_2 \cdots, u_n) = 0 \tag{A.26}$$

$$F_{\vec{X}}(\infty, \cdots, \infty) = 1 \tag{A.27}$$

$$F_{\vec{X}}(u_1, \infty, \cdots, \infty) = F_{X_1}(u_1) \tag{A.28}$$

$$F_{\vec{X}}(u_1, u_2, \infty, \cdots, \infty) = F_{X_1 X_2}(u_1, u_2) \tag{A.29}$$

$$\int_{-\infty}^{\infty} \cdots \int_{-\infty}^{\infty} p_{\vec{X}}(\vec{u}) \, du_1 \cdots du_n = 1 \tag{A.30}$$

$$\int_{-\infty}^{\infty} \cdots \int_{-\infty}^{\infty} p_{\vec{X}}(\vec{u}) \, du_2 \cdots du_n = p_{X_1}(u_1) \tag{A.31}$$

$$\int_{-\infty}^{\infty} \cdots \int_{-\infty}^{\infty} p_{\vec{X}}(\vec{u}) \, du_3 \cdots du_n = p_{X_1 X_2}(u_1, u_2) \tag{A.32}$$

In problems involving more than one random variable, the one-dimensional distributions of single random variables are commonly referred to as the marginal distributions. Using this terminology, Eqs. A.20, A.21, A.23, A.24, A.28, and A.31 give formulas for deriving marginal distributions from joint distributions. Similarly Eqs. A.29 and A.32 can be considered to give two-dimensional marginal distributions for problems involving more than two random variables.

****

Example A.12: Find the marginal distributions for X and Y for the joint probability distribution of Example A.9 with

$$p_{XY}(u,v) = 1/6 \quad \text{for} \quad -1 \le u \le 2, -1 \le v \le 1$$
$$= 0 \qquad\qquad \text{otherwise}$$

Note that Eq. A.23 gives $p_X(u) \equiv 0$ for $u < -1$ or $u \ge 2$, since the integrand of that formula is zero everywhere on the specified interval. For the nontrivial situation of $-1 \le u \le 2$, the integral gives

$$p_X(u) \equiv \int_{-1}^{1} \left(\frac{1}{6}\right) dv = \frac{1}{3}$$

Thus, this marginal probability density function can be written as

$$p_X(u) = \frac{1}{3}[U(u+1) - U(u-2)]$$

We can now obtain the marginal cumulative distribution function of X either by integrating $p_X(u)$ or by applying Eq. A.21 to the joint cumulative distribution function given in Example A.9. Either way, we obtain

$$F_X(u) = \frac{u+1}{3}[U(u+1) - U(u-2)] + U(u-2)$$

Proceeding in the same way to find the marginal distribution of Y gives

$$p_Y(v) = \frac{1}{2}[U(v+1) - U(v-1)]$$

and

$$F_Y(v) = \frac{v+1}{2}[U(v+1) - U(v-1)] + U(v-1)$$

****

Example A.13: Find the marginal distributions for X and Y for the probability distribution of Example A.10 with

$$p_{XY}(u,v) = [U(u-1) - U(u-5)][U(v-1) - U(v-u)]/8$$

First integrating $p_{XY}(u,v)$ with respect to v, in order to apply Eq. A.23, gives

$$p_X(u) = \frac{u-1}{8}[U(u-1) - U(u-5)]$$

which gives a cumulative distribution of

$$F_X(u) = \frac{(u-1)^2}{16}[U(u-1) - U(u-5)] + U(u-5)$$

which is exactly the same as is obtained by taking $v = \infty$ in $F_{XY}(u,v)$, confirming Eq. A.21. Similarly

$$p_Y(v) = \frac{5-v}{8}[U(v-1) - U(v-5)]$$

and

$$F_Y(v) = \left(\frac{5(v-1)}{8} - \frac{(v^2-1)}{16}\right)[U(v-1) - U(v-5)] + U(v-5)$$

****

Example A.14: Find the marginal probability distribution for X for the random variables of Example A.11 with

$$F_{XY}(u,v) = \left[\frac{(1-e^{-4u})(1-e^{-4v})}{2} + \frac{(1-e^{-4u})(1-e^{-3v})}{4} + \frac{(1-e^{-3u})(1-e^{-4v})}{4}\right]U(u)U(v)$$

Since $F_{XY}(u,v)$ is given in the original statement of the problem, it is convenient to set $v = \infty$ in that equation, and obtain $F_X(u)$ from Eq. A.21 as

$$F_X(u) = \left(1 - \frac{3e^{-4u} + e^{-3u}}{4}\right)U(u)$$

Differentiating this expression gives

$$p_X(u) = 3\left(e^{-4u} + e^{-3u}/4\right)U(u)$$

It is left to the reader to verify that this agrees with the result of integrating $p_{XY}(u,v)$ from Example A.11 with respect to v. We will not derive the marginal distribution for Y, in this example, since one can see from symmetry that it will have the same form as the marginal distribution for X.

****

A.4 Probability Distribution of a Function of a Random Variable

For any reasonably smooth function $g(\cdot)$, we can define a new random variable $Y = g(X)$ in terms of a given random variable X. We now want to consider how the probability distribution of Y relates to that of X. First let us consider the simplest case of $g(\cdot)$ being a monotonically increasing function. Then $g(\cdot)$ has a unique inverse $g^{-1}(\cdot)$ and the event $\{Y \leq v\}$ is identical to the event $\{X \leq g^{-1}(v)\}$. Thus, $F_Y(v) = F_X[g^{-1}(v)]$, and taking the derivative with respect to v of both sides of this equation gives the probability density function as

$$p_Y(v) = p_X[g^{-1}(v)]\frac{d}{dv}g^{-1}(v) = \frac{p_X[g^{-1}(v)]}{\left[\dfrac{dg(u)}{du}\right]_{u=g^{-1}(v)}} \qquad \text{for} \quad g(\cdot) \text{ monotonically increasing}$$

Similarly, if $g(\cdot)$ is decreasing then $\{Y \le v\} = \{X \ge g^{-1}(v)\}$ so that $F_Y(v) = 1 - F_X[g^{-1}(v)]$ and

$$p_Y(v) = -p_X[g^{-1}(v)]\frac{d}{dv}g^{-1}(v) = \frac{-p_X[g^{-1}(v)]}{\left[\dfrac{dg(u)}{du}\right]_{u=g^{-1}(v)}} \qquad \text{for} \quad g(\cdot) \text{ monotonically decreasing}$$

Noting that the derivative of g is positive if $g(\cdot)$ is monotonically increasing and negative if $g(\cdot)$ is monotonically decreasing allows these two probability density function results to be combined as

$$p_Y(v) = \frac{p_X[g^{-1}(v)]}{\left|\dfrac{dg(u)}{du}\right|_{u=g^{-1}(v)}} \qquad \text{if} \quad g(\cdot) \text{ is monotonic} \qquad (A.33)$$

Note that the point $X = g^{-1}(v)$ maps into $Y = v$. Thus, it is surely logical that the probability of Y being in the neighborhood of v depends only on the probability of X being in the neighborhood of $g^{-1}(v)$. The derivative of g appearing in the probability density function reflects the fact that an increment of length du does not map into an equal length increment dv. That is, rather than having $p_Y(v)$ equal to $p_X[g^{-1}(v)]$, we must have $p_Y(v)dv = p_X[g^{-1}(v)]du$ and it is the du / dv ratio which introduces the derivative of the g function. An important special case of the monotonic function is the linear relationship $Y = c + bX$, for which $p_Y(v) = p_X[(v-c)/b]/|b|$.

For a general $g(\cdot)$ function, there may be many inverse points $u = g_j^{-1}(v)$. In this situation, Eq. A.33 becomes

$$p_Y(v) = \sum_j \frac{p_X[g_j^{-1}(v)]}{\left|\dfrac{dg(u)}{du}\right|_{u=g_j^{-1}(v)}} \qquad (A.34)$$

with the summation being over all points $X = g_j^{-1}(v)$ which map into $Y = v$.

The results given here can also be generalized to situations involving vectors $\vec{X}$ and $\vec{Y} = \vec{g}(\vec{X})$. The situation corresponding to the monotonic scalar g function is when $\vec{g}$ has a unique inverse. This can only happen if the dimension of $\vec{X}$ and $\vec{Y}$ is the same, and in that case one can obtain a result which resembles Eq. A.34:

$$p_{\vec{Y}}(\vec{v}) = \frac{p_{\vec{X}}[\vec{g}^{-1}(\vec{v})]}{\left\| \dfrac{d\vec{g}(u)}{d\vec{u}} \right\|_{\vec{u}=\vec{g}^{-1}(\vec{v})}} \qquad \text{if the inverse is unique} \qquad (A.35)$$

in which the derivative term in the denominator denotes a matrix

$$\frac{d\vec{g}(u)}{d\vec{u}} = \begin{bmatrix} \dfrac{dg_1}{du_1} & \dfrac{dg_1}{du_2} & \cdots & \dfrac{dg_1}{du_n} \\ \dfrac{dg_2}{du_1} & \dfrac{dg_2}{du_2} & \cdots & \dfrac{dg_2}{du_n} \\ \vdots & \vdots & \ddots & \vdots \\ \dfrac{dg_n}{du_1} & \dfrac{dg_n}{du_2} & \cdots & \dfrac{dg_n}{du_n} \end{bmatrix}$$

and the double bars around the matrix in Eq. A.35 denote the absolute value of the determinant of the matrix.[4] As with the scalar situation, the linear transformation is an important special case. This can be written for the vectors as $\vec{Y} = \vec{C} + \mathbf{B}\vec{X}$ in which $\mathbf{B}$ is a square matrix. It is then easily shown that the matrix in the denominator of Eq. A.35 is exactly $\mathbf{B}$, so that

$$p_{\vec{Y}}(\vec{v}) = \frac{p_{\vec{X}}[\vec{g}^{-1}(\vec{v})]}{\|\mathbf{B}\|} \qquad \text{for} \qquad \vec{Y} = \vec{C} + \mathbf{B}\vec{X}$$

A.5   Conditional Probability Distributions

For any two events, the conditional probability of one event given the other event is defined to be the ratio of two probabilities. Precisely, it is the probability of the intersection divided by the probability of the conditioning event. Thus, for the two events A and B, the conditional probability of A given B is

$$P(A|B) = \frac{P(AB)}{P(B)} \qquad (A.36)$$

with the vertical bar being read as "given." The conditional probability of Eq. A.36 is generally not defined if $P(B) = 0$. Intuitively, $P(A|B)$ refers to the likelihood that A will occur simultaneously with a known occurrence of B. Note that $P(A|B)$ must be exactly zero if it is impossible for A and B to occur simultaneously, since the probability of their intersection is then zero.

For random variable problems, we will usually be interested in conditioning the probability distribution of one random variable based on certain events involving the outcome for that same

---

[4]   It may also be noted that the determinant of the matrix is called the Jacobian of the transformation, so that the denominator in Eq. A.35 is the absolute value of the Jacobian.

random variable or a different random variable. If this conditioning event has a nonzero probability, then Eq. A.36 provides an adequate tool for defining the conditional distribution. For example, for a random variable X having the entire real line as possible values, we can define a conditional cumulative distribution function given $X \leq 10$ as

$$F_X(u|X \leq 10) \equiv P(X \leq u|X \leq 10) = \frac{P[X \leq \min(u,10)]}{P(X \leq 10)} = \frac{F_X[\min(u,10)]}{F_X(10)}$$

$$= F_X(u) / F_X(10) \qquad \text{for} \qquad u \leq 10$$

$$= 1 \qquad \text{for} \qquad u > 10$$

The $P(X \leq u)$ and $P(X \leq 10)$ terms can be evaluated directly from the unconditional cumulative distribution function $F_X(u)$ if that is known, or from integration of the unconditional probability density function $p_X(u)$. One can then define a conditional probability density function for X as

$$p_X(u|X \leq 10) \equiv \frac{\partial}{\partial u} F_X(u|X \leq 10)$$

$$= p_X(u) / F_X(10) \qquad \text{for} \qquad u \leq 10$$

$$= 0 \qquad \text{for} \qquad u > 10$$

Similarly, in a problem involving the two random variables X and Y, one might need to compute a conditional distribution for X given some range of outcomes for Y. For example, one particular conditional cumulative distribution function is

$$F_X(u|Y \leq 10) \equiv P(X \leq u|Y \leq 10) = \frac{P[X \leq u, Y \leq 10]}{P(Y \leq 10)} = \frac{F_{XY}(u,10)}{F_Y(10)}$$

and the corresponding conditional probability density function is

$$p_X(u|Y \leq 10) \equiv \frac{\partial}{\partial u} F_X(u|Y \leq 10) = \frac{1}{F_Y(10)} \frac{\partial F_{XY}(u,10)}{\partial u}$$

These procedures can easily be extended to situations with more than two random variables. In summary, whenever the conditioning event has a nonzero probability, the recommended procedure for finding the conditional probability distribution for some random variable X is first to write the conditional cumulative distribution according to Eq. A.36, as was done in the two examples. After that one can obtain the conditional probability density function by taking a derivative:

$$F_X(u|B) = \frac{P(\{X \leq u\} \cap B)}{P(B)} \qquad \text{(A.37)}$$

$$p_X(u|B) \equiv \frac{\partial}{\partial u} F_X(u|B) \qquad \text{for} \qquad P(B) > 0 \qquad \text{(A.38)}$$

****

Example A.15: Reconsider the random variables X and Y of Examples A.9 and A.12, with joint probability distribution of

$$p_{XY}(u,v) = 1/6 \qquad \text{for} \qquad -1 \le u \le 2, \ -1 \le v \le 1$$
$$= 0 \qquad \text{otherwise}$$

Find the conditional cumulative distribution function and conditional density function for the random variable X given that $Y > 0.5$.

To calculate the conditional probability of any event A given the event $Y > 0.5$, we need the probability of $Y > 0.5$ and the probability of the intersection of A with $Y > 0.5$. Thus, to compute the conditional cumulative distribution function for X we need

$$P(Y > 0.5) = \int_{0.5}^{\infty}\int_{-\infty}^{\infty} p_{XY}(u,v)\,dudv = \int_{0.5}^{1}\int_{-1}^{2}\left(\frac{1}{6}\right)dudv = 0.25$$

and

$$P(X \le u, Y > 0.5) = \int_{0.5}^{\infty}\int_{-\infty}^{u} p_{XY}(u,v)\,dudv$$

which gives

$$P(X \le u, Y > 0.5) = 0 \qquad\qquad\qquad \text{for} \quad u < -1$$
$$= \int_{0.5}^{1}\int_{-1}^{u}\left(\frac{1}{6}\right)dudv = \frac{u+1}{12} \quad \text{for} \quad -1 \le u \le 2$$
$$= \int_{0.5}^{1}\int_{-1}^{2}\left(\frac{1}{6}\right)dudv = \frac{1}{4} \qquad \text{for} \quad u > 2$$

Taking the ratio of probabilities, as in Eq. A.37, gives the conditional cumulative distribution function as

$$F_X(u|Y > 0.5) = 0 \qquad \text{for} \quad u < -1$$
$$= \frac{u+1}{3} \quad \text{for} \quad -1 \le u \le 2$$
$$= 1 \qquad \text{for} \quad u > 2$$

****
Example A.16: For the random variables of Example A.10 with

$$p_{XY}(u,v) = [U(u-1) - U(u-5)][U(v-1) - U(v-u)]/8$$

find the conditional cumulative distribution function and conditional density function for the random variable Y given that $X \le 3$.

Proceeding as in the previous example:

$$P(X \le 3) = \int_{1}^{3}\int_{v}^{3}\left(\frac{1}{8}\right)dudv = \frac{1}{4}$$

and

$$P(X \le 3, Y \le v) = 0 \qquad \text{for} \qquad v < 1$$
$$= \int_1^v \int_w^3 \left(\frac{1}{8}\right) du dw = \frac{(v-1)(5-v)}{16} \qquad \text{for} \qquad 1 \le v \le 3$$
$$= 1/4 \qquad \text{for} \qquad v > 3$$

so that

$$F_Y(v|X \le 3) = 0 \qquad \text{for} \qquad v < 1$$
$$= (v-1)(5-v)/4 \qquad \text{for} \qquad 1 \le v \le 3$$
$$= 1 \qquad \text{for} \qquad v > 3$$

Differentiating this expression gives the conditional probability density function as

$$p_Y(v|X \le 3) = [(3-v)/2] [U(v-1)] - U(v-3)]$$

****

A slightly different sort of conditioning arises in many random variable problems. In particular, one may wish to condition by an event with zero probability, and this requires a new definition, since Eq. A.36 no longer applies. In particular, one often is interested in the conditional distribution of one random variable given a precise value of another random variable. If the conditioning random variable has a continuous distribution, then this gives a conditioning event with zero probability. The definition of the conditional distribution for the situation in which X and Y have a continuous joint distribution is in terms of the conditional probability density function:

$$p_X(u|Y = v) = \frac{p_{XY}(u, v)}{p_Y(v)} \qquad (A.39)$$

which is defined only for v values giving $p_Y(v) \ne 0$. Note that Eq. A.39 is very similar in form to Eq. A.36, but the terms in Eq. A.39 are all probability density functions, while those in A.36 are probabilities. Of course, Eq. A.39 is consistent with a limit of Eq. A.36, and one can convert Eq. A.39 into infinitesimal increments of probability by multiplying the density terms by infinitesimal increments of length:

$$p_X(u|Y = v) du = \frac{p_{XY}(u, v) du dv}{p_Y(v) dv}$$

which is essentially a statement that

$$P(u < X < u + du | Y = v) = \frac{P(u < X < u + du, v < Y < v + dv)}{P(v < Y < v + dv)}$$

which is of the form of Eq. A.36 for the infinitesimal probabilities. Since the conditional distribution in this situation is defined by a conditional probability density function, one must integrate if the conditional cumulative distribution function is needed:

$$F_X(u|Y=v) = \int_{-\infty}^{u} p_X(w|Y=v)dw \qquad\qquad (A.40)$$

****

Example A.17: Find the conditional probability distribution of X given the event $Y=v$ and the conditional distribution of Y given $X=u$ for the joint probability distribution of Examples A.9, A.12, and A.15 with

$$p_{XY}(u,v)=1/6 \qquad \text{for} \qquad -1\le u\le 2, -1\le v\le 1$$
$$=0 \qquad \text{otherwise}$$

We note that $p_X(u|Y=v)$ is defined only for $-1\le v\le 1$, since $p_Y(v)$ is zero otherwise. For $-1\le v\le 1$ we use Eq. A.39 and the marginal probability density function derived in Example A.12 to obtain

$$p_X(u|Y=v) = \frac{1/6}{1/2}[U(u+1)-U(u-2)] = \frac{1}{3}[U(u+1)-U(u-2)]$$

Integrating this expression gives the conditional cumulative distribution function as

$$F_X(u|Y=v) = \frac{u+1}{3}[U(u+1)-U(u-2)] + U(u-2)$$

Similarly,

$$p_Y(v|X=u) = \frac{1}{2}[U(v+1)-U(v-1)]$$

and

$$F_Y(v|X=u) = \frac{u+1}{2}[U(v+1)-U(v-1)] + U(v-1)$$

for $-1\le u\le 2$, and the conditional distribution is undefined for other values of u.

****

Example A.18: For the random variables of Examples A.10 and A.13 with

$$p_{XY}(u,v) = [U(u-1)-U(u-5)][U(v-1)-U(v-u)]/8$$

find the conditional distribution of X given $Y=v$ and the conditional distribution of Y given $X=u$.

In Example A.13, we found that

$$p_X(u) = (u-1)[U(u-1)-U(u-5)]/8, \qquad p_Y(v) = (5-v)[U(v-1)-U(v-5)]/8$$

Thus, we can now take the ratio of joint and marginal probability density functions, according to Eq. A.39, to obtain

$$p_Y(v|X=u) = \frac{p_{XY}(u,v)}{p_X(u)} = \frac{1}{u-1}[U(v-1)-U(v-u)] \quad \text{for } 1 \le u \le 5$$

and

$$p_X(u|Y=v) = \frac{p_{XY}(u,v)}{p_Y(v)} = \frac{1}{5-v}[U(u-v)-U(u-5)] \quad \text{for } 1 \le v \le 5$$

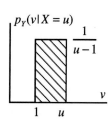

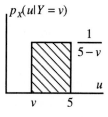

These conditional density functions are shown in the sketchs. They show that when $X=u$ is known, then the set of possible values of Y is limited to the interval $[1,u]$, and that $[v,5]$ gives the set of possible values for X when $Y=v$ is known. On these sets of possible values, both of the conditional distributions are uniform (each of the conditional density functions is a constant).

Integrating these conditional probability density functions gives the conditional cumulative distribution functions as

$$F_Y(v|X=u) = \frac{v-1}{u-1}[U(v-1)-U(v-u)] \quad \text{for } 1 \le u \le 5$$

and

$$F_X(u|Y=v) = \frac{u-v}{5-v}[U(u-v)-U(u-5)] \quad \text{for } 1 \le v \le 5$$

These are also sketched.

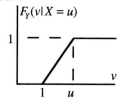

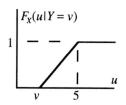

****

Example A.19: Find the conditional distribution of X given $Y=v$ for the probability distribution of Examples A.11 and A.14 with

$$p_{XY}(u,v) = \left(8e^{-4u-4v} + 3e^{-4u-3v} + 3e^{-3u-4v}\right)U(u)U(v)$$

We can use symmetry and the marginal probability density function derived in Example A.14 to write

$$p_Y(v) = 3\left(e^{-4v} + e^{-3v}/4\right)U(v)$$

then Eq. A.39 gives

$$p_X(u|Y=v) = \frac{p_{XY}(u,v)}{p_Y(v)} = \frac{\left(8e^{-4u-4v} + 3e^{-4u-3v} + 3e^{-3u-4v}\right)}{3\left(e^{-4v} + e^{-3v}/4\right)}U(u) \quad \text{for } v \ge 0$$

As a more specific example,

$$p_X(u|Y=4) = (3.91e^{-4u} + 0.0683e^{-3u})U(u)$$

****

It is important to remember that any conditional cumulative distribution function or conditional probability density function has the same mathematical characteristics as any other cumulative distribution function or probability density function. Specifically, any conditional cumulative distribution function, such as those given by Eqs. A.37 and A.40, satisfies the condition of being monotonically increasing from zero to unity as one considers all possible arguments of the function (that is, as u is increased from negative infinity to positive infinity). Similarly, any conditional probability density function, such as those given by Eqs. A.38 and A.39, satisfies the conditions of nonnegativity and having a unit integral (Eq. A.6) which are necessary for a probability density function.

In many situations, it is convenient to give the initial definition of a problem in terms of conditional distributions, rather than in terms of joint distributions. We can always rewrite Eq. A.39 as

$$p_{XY}(u,v) = p_X(u|Y=v)p_Y(v) \tag{A.41}$$

or

$$p_{XY}(u,v) = p_Y(v|X=u)p_X(u) \tag{A.42}$$

so that if we know the marginal distribution for one random variable and the conditional distribution for a second random variable given the value of the first, then we also know the joint distribution.

****

Example A.20: Let a random variable X be uniform on the set $[0,10]$, so that its probability density function is given by

$$p_X(u) = 0.1[U(u) - U(u-10)]$$

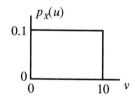

Let another random variable Y be uniform on the set $[0,X]$. That is, if we are given the information that $X=u$, then Y is uniform on $[0,u]$:

$$p_Y(v|X=u) = \frac{1}{u}[U(v) - U(v-u)]$$

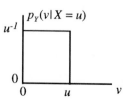

Find the joint probability density function of X and Y, and identify the domain on which this density function is nonzero.

Substituting into Eq. A.26 gives

$$p_{XY}(u,v) = (0.1/u)[U(u) - U(u-10)][U(v) - U(v-u)]$$

and this function is nonzero on the triangular region described by $0 \le v \le u \le 10$. This region is shown on a sketch of the (u,v) plane.

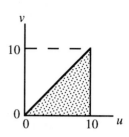

****

Example A.21: Let the random variable X again be uniform with probability density function of

$$p_X(u) = 0.1[U(u) - U(u-10)]$$

Let Y be a biased estimate of X such that it is always greater than X. In particular, let the conditional distribution be of the exponential form

$$p_Y(v|X=u) = \lambda\, e^{-\lambda(v-u)} U(v-u)$$

in which $\lambda > 0$ is a known constant.

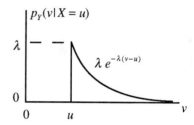

Find the joint probability density function of X and Y, and identify the domain on which this density function is nonzero.

Multiplying the marginal and conditional probability densities gives

$$p_{XY}(u,v) = 0.1\lambda\, e^{-\lambda(v-u)}[U(u) - U(u-10)]U(v-u)$$

and this is nonzero on the infinite strip shaded in the sketch.

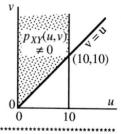

****

It is also possible to use the idea of conditional and joint probability distributions to describe a problem in which one random variable is a function of another random variable. However, such functional relationships always give degenerate conditional and joint distributions. For example, let Y be a function of X, say $Y = g(X)$. Now if we are given the event $X = u$, then the only possible value of Y is $g(u)$. By using the Dirac delta function, though, we can write a conditional probability density function as $p_Y(v|X=u) = \delta[v - g(u)]$. This function is an acceptable probability density function, since it does have a unit integral when integrated over the real line, and it is nonzero only when v is equal to the possible value for Y. By using Eq. A.42 we can now obtain a joint probability density function as $p_{XY}(u,v) = p_X(u)\delta[v - g(u)]$. Clearly this joint probability density function is nonzero only on a one-dimensional subset of the two dimensional (u,v) plane, and it is infinite on this one-dimensional subset. This degeneracy is typical of what one obtains when a joint distribution is used to describe the relationship between one random variable and a function of that random variable.

A.6   Independence of Random Variables

The intuitive idea of a random variable Y being independent of another random variable X is that knowledge of the value of X gives absolutely no information about the possible values of Y, or about the likelihood that Y will take on any of those possible values. Conditional probability has been defined precisely for the purpose of conveying this type of informational connection between random variables. Thus, it would seem natural to use conditional probabilities in defining the concept of independence. There are minor mathematical difficulties in using this approach for the definition, though, so we will use an alternative statement as the definition of independence, then show that the definition is consistent with the intuitive concept. Specifically, we will say that two random variables X and Y are defined to be **independent** if and only if

$$p_{XY}(u,v) = p_X(u)p_Y(v) \quad \text{for all } u \text{ and } v \qquad (A.43)$$

or, equivalently,

$$F_{XY}(u,v) = F_X(u)F_Y(v) \quad \text{for all } u \text{ and } v \qquad (A.44)$$

Thus, the definition is in terms of joint and marginal probability distributions rather than conditional distributions.

Using Eq. A.43 along with Eq. A.39 to compute the conditional distributions for independent X and Y gives

$$p_Y(v|X = u) = p_Y(v) \quad \text{provided that } p_X(u) \neq 0 \qquad (A.45)$$

and

$$p_X(u|Y = v) = p_X(u) \quad \text{provided that } p_Y(v) \neq 0 \qquad (A.46)$$

One can also show that

$$F_Y(v|X = u) = F_Y(v) \quad \text{provided that } p_X(u) \neq 0 \qquad (A.47)$$

and

$$F_X(u|Y = v) = F_X(u) \quad \text{provided that } p_Y(v) \neq 0 \qquad (A.48)$$

and various other statements such as

$$p_Y(v|X \leq u) = p_Y(v) \quad \text{provided that } P(X \leq u) \neq 0 \qquad (A.49)$$

or

$$F_X(u|Y > v) = F_X(u) \quad \text{provided that } P(Y > v) \neq 0 \tag{A.50}$$

Equations A.45 - A.50 are examples of the intuitive idea of independence discussed in the first paragraph of this section. Specifically, independence of X and Y implies that knowledge of the value of X gives absolutely no information about the probability distribution of Y, and knowledge of Y gives absolutely no information about the distribution of X. Expressions such as Eqs. A.45 - A.48 have not been used as the definition of independence simply because, for any given problem, each of these equations may be defined only for a subset of all possible (u,v) values, whereas Eqs. A.43 and A.44 hold for all (u,v) values.

****

Example A.22: Reconsider the probability distribution of Example A.9 with

$$p_{XY}(u,v) = \frac{1}{6}[U(u+1) - U(u-2)][U(v+1) - U(v-1)]$$

Are X and Y independent?

We have already found the marginal distributions for this problem in Example A.12, so they can easily be used to check for independence. In particular, we can use the previously evaluated functions $p_X(u) = [U(u+1) - U(u-2)]/3$ and $p_Y(v) = [U(v+1) - U(v-1)]/2$ to verify that for every choice of u and v we get $p_{XY}(u,v) = p_X(u)p_Y(v)$. Thus, Eq. A.43 is satisfied, and X and Y are independent.

We could, equally well, have demonstrated independence by using the joint and marginal cumulative distribution functions to show that Eq. A.44 was satisfied. We can also use the conditional probabilities that have been derived for this problem to verify the intuitive idea of independence — that knowledge of one of the random variables gives no information about the distribution of the other random variable. For example, we found $F_X(u|Y > 0.5)$ in Example A.15 and it is easily verified that this is identical to the marginal cumulative distribution function $F_X(u)$ for all values of u. Also, we found $p_X(u|Y = v)$ and this is identical to $p_X(u)$, provided that $-1 \leq v \leq 1$ so that the conditional distribution is defined. These are two of many equations that we could write to confirm that knowledge of Y gives no information about either the possible values of X, or the probability distribution on those possible values.

****

Example A.23: Determine whether X and Y are independent for the probability distribution of Example A.10 with

$$p_{XY}(u,v) = [U(u-1) - U(u-5)][U(v-1) - U(v-u)]/8$$

Using the marginal distributions derived for this problem in Example A.13, we have

$$p_X(u)p_Y(v) = \left(\frac{u-1}{8}\right)\left(\frac{5-v}{8}\right)[U(u-1) - U(u-5)][U(v-1) - U(v-5)]$$

and this is clearly not the same as $p_{XY}(u,v)$. Thus, X and Y are not independent.

In comparing $p_X(u)p_Y(v)$ with $p_{XY}(u,v)$, it is worthwhile to note two types of differences. Probably the first discrepancy that the reader will note is that on the domain where they are nonzero the functions are $(u-1)(5-v)/64$ and $1/8$, and these clearly are not the same functions of u and v. One can also note, though, that $p_X(u)p_Y(v)$ is nonzero on the square defined by $1 \le u \le 5$, $1 \le v \le 5$, whereas $p_{XY}(u,v)$ is nonzero on only the lower right half of this square. This difference of the domain of nonzero values of the functions is sufficient to prove that the functions are not the same. This latter comparison is sometimes an easy way to show that $p_X(u)p_Y(v) \ne p_{XY}(u,v)$ in a problem in which both the joint and marginal probability density functions are complicated.

Recall that in Examples A.16 and A.18 we derived conditional probability density functions for this problem:

$$p_Y(v|X \le 3) = \left(\frac{3-v}{2}\right)[U(v-1)] - U(v-3)]$$

and

$$p_Y(v|X = u) = \frac{1}{u-1}[U(v-1) - U(v-u)]$$

If X and Y were independent then both of these conditional probability density functions for Y would be the same as its marginal probability density function, which we found as

$$p_Y(v) = \frac{5-v}{8}[U(v-1) - U(v-5)]$$

Clearly neither of these conditional functions is the same as the marginal. Showing that any one conditional function for Y given information about X is not the same as the corresponding marginal function for Y is sufficient to prove that X and Y are not independent. Knowing something about the value of X (such as $X = u$, or $X \le 3$) does give information about the distribution of Y. In this example, the given information about X restricts the range of possible values of Y, as well as the probability distribution on those possible values. The following example will illustrate a case in which the probabilities for one random variable are modified by knowledge of another random variable, even though the domain is unchanged.
****

Example A.24: Determine whether X and Y are independent for the probability distribution of Examples A.11 and A.14 with
$$p_{XY}(u,v) = \left(8e^{-4u-4v} + 3e^{-4u-3v} + 3e^{-3u-4v}\right)U(u)U(v)$$

Using the marginal distribution from Example A.14, we have

$$p_X(u)p_Y(v) = 3\left(e^{-4u} + e^{-3u}/4\right)\left(e^{-4v} + e^{-3v}/4\right)U(u)U(v)$$

This time we do find that the domain of nonzero values is the same for $p_X(u)p_Y(v)$ as for $p_{XY}(u,v)$ (that is, the first quadrant), but the functions of u and v are not the same. Thus, X and Y are not independent.

Looking at conditional probabilities, we found in Examples A.14 and A.19 that the marginal distribution for X could be described by

$$p_X(u) = 3\left(e^{-4u} + e^{-3u}/4\right)U(u)$$

and the conditional distribution had

$$p_X(u|Y=v) = \frac{\left(8e^{-4u-4v} + 3e^{-4u-3v} + 3e^{-3u-4v}\right)}{3\left(e^{-4v} + e^{-3v}/4\right)}U(u)$$

Thus, the set of possible values of X is the infinite interval $[0,\infty)$, and knowledge of the value of Y does not change this set. Knowledge of the value of Y does change the probability distribution on this set of possible values, though, confirming that X and Y are not independent.

A.7 Expected Values

For any random variable X, the expected value is defined as

$$E(X) = \int_{-\infty}^{\infty} u\,p_X(u)\,du \qquad (A.51)$$

Note that this expression gives a weighted average over all the possible values of X. The weighting function is the probability density function. Thus, u is a particular value of X, $p_X(u)\,du$ is the probability of X being in the vicinity of u, and the integration gives us a "sum" of all such terms. In a weighted integral we usually expect to find an integral such as that in Eq. A.51 normalized by an integral of the weighting function. When the weighting function is a probability density, though, that normalization integral is exactly unity, so can be omitted. Thus, the expected value $E(X)$ is the probability weighted average of all the possible values of X. It is also called the expectation or, most commonly, the mean of X.

One must be careful to remember the meaning of $E(X)$. The name "expected value" could easily lead to a misinterpretation of $E(X)$ as being something like the most likely value of X, but this is generally not true, and in some situations we will find that $E(X)$ is not even a possible value of X.

Example A.25: Find the mean value of the random variable X having an exponential distribution

$$p_X(u) = 2e^{-2u}U(u)$$

Applying Eq. A.51 gives

$$E(X) = \int_{-\infty}^{\infty} u\, p_X(u)\, du = 2 \int_{0}^{\infty} u\, e^{-2u}\, du = 0.5$$

One would generally say that the most likely value of this X random variable is zero, since $p_X(u)$ is larger for $u = 0$ than for any other u value, indicating that there is higher probability of X being in the neighborhood of zero than in the neighborhood of any other possible value. The fact that $E(X) = 0.5$, though, reinforces the idea that $E(X)$ is an average of all the possible outcomes, and need not coincide with a large value of the probability density function.

****

Example A.26: Find the expected value of a random variable X which is uniformly distributed on the two intervals $[-2,-1]$ and $[1,2]$, so that the probability density function is as shown on the sketch.

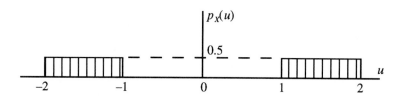

Integration of Eq. A.51 gives

$$E(X) = \int_{-\infty}^{\infty} u\, p_X(u)\, du = \int_{-2}^{-1} u(0.5)\, du + \int_{1}^{2} u(0.5)\, du = 0$$

In this case, $E(X) = 0$ is not within the set of possible values for X. It is, however, the average of all the equally likely possible values. Similarly, for the discrete distribution of Example A.1 we find that $E(X) = 3.5$, which is surely not a possible value for the outcome of a die roll, but is the average of the six equally likely outcomes.

****

Similar to Eq. A.51, we now define the expected value of any function of the random variable X as

$$E[g(X)] = \int_{-\infty}^{\infty} g(u)\, p_X(u)\, du \tag{A.52}$$

As before, we have $E[g(X)]$ as a weighted average of all the possible values of $g(X)$, since $p_X(u)\, du$ gives the probability of $g(X)$ being in the neighborhood of $g(u)$ due to X being in the neighborhood of u. Integrating over all the possible values of X gives $g(u)$ varying over all the possible values of $g(X)$, and $E[g(X)]$ is the probability weighted average of $g(X)$.

****

Example A.27: Find $E(X^2)$ and $E[\sin(3X)]$ for the random variable X of Examples A.2 and A.6, for which

$$p_X(u) = 0.1[U(u) - U(u-10)]$$

Using the probability density function in the appropriate integrals, we have

$$E[X^2] = \int_{-\infty}^{\infty} u^2 \, p_X(u) \, du = \int_0^{10} u^2 (0.1) \, du = 33.3$$

and

$$E[\sin(3X)] = \int_{-\infty}^{\infty} \sin(3u) \, p_X(u) \, du = \int_0^{10} \sin(3u)(0.1) \, du = \frac{1 - \cos(30)}{30}$$

∗∗∗

Since any function of a random variable is a random variable, we could write $Y = g(X)$ for the random quantity considered in Eq. A.52 and derive the probability distribution of Y from that of X. In particular, Eq. A.34 gives $p_Y(v)$ in terms of $p_X(u)$. After having found $p_Y(v)$, we could use Eq. A.51 to write

$$E(Y) = \int_{-\infty}^{\infty} v \, p_Y(v) \, dv \tag{A.53}$$

It can be shown that Eq. A.52 will always give the same result as Eq. A.53. The difference between the two is that Eq. A.53 prescribes calculation of the weighted average by summing over all possible values of $Y = g(X)$ in the order of increasing Y, whereas Eq. A.52 involves summing the same terms in the order of increasing X. Both Eq. A.52 and A.53 are exactly correct for calculating $E[g(X)]$, but in many situations we will find it easier to apply Eq. A.52.

∗∗∗

Example A.28: For the random variables of Example A.3 with

$$F_X(u) = 0.05(u+4)[U(u+4) - U(u-16)] + U(u-16)$$

and $Y = X[U(X) - U(X-10)] + 10U(X-10)$, find $E(Y)$ both by using Eq. A.52 and by using Eq. A.53.

In Eq. A.52 we need $p_X(u)$, which is

$$p_X(u) = 0.05[U(u+4) - U(u-16)]$$

along with the function $g(u)$ given by $g(u) = u[U(u) - U(u-10)] + 10U(u-10)$. The integral then gives

$$E[Y] = \int_{-\infty}^{\infty} g(u) \, p_X(u) \, du = \int_{-4}^{0} (0) \, (0.05) \, du + \int_0^{10} u \, (0.05) \, du + \int_{10}^{16} (10) \, (0.05) \, du$$
$$= 0 + 2.5 + 3 = 5.5$$

In order to use Eq. A.53, we use the probability density obtained for $Y = g(X)$ in Example A.8 to write

$$E(Y) = \int_{-\infty}^{\infty} v\, p_Y(v)\, dv = \int_{-\infty}^{\infty} v\left(0.05[U(v) - U(v-10)] + 0.2\delta(v) + 0.3\delta(v-10)\right) dv$$

$$= 0.05\int_{0}^{10} v\, dv + 0.2\int_{-\infty}^{\infty} v\,\delta(v)\, dv + 0.3\int_{-\infty}^{\infty} v\,\delta(v-10)\, dv$$

$$= 2.5 + 0 + 3 = 5.5$$

Of course, the two results are identical.
****

An important generalization of Eq. A.52 is needed for the situation where the new random quantity is a function of more than one random variable, such as $g(X,Y)$. The idea of a probability weighted average over all possible values of the function is maintained and the joint probability density of X and Y is used to write

$$E[g(X,Y)] = \int_{-\infty}^{\infty}\int_{-\infty}^{\infty} g(u,v)\, p_{XY}(u,v)\, du\, dv \qquad (A.54)$$

For a function of many random variables, we can use the corresponding vector notation

$$E\left[g(\vec{X})\right] = \int_{-\infty}^{\infty}\cdots\int_{-\infty}^{\infty} g(\vec{u})\, p_{\vec{X}}(\vec{u})\, du_1\cdots du_n \qquad (A.55)$$

in which n is the number of components of the random vector $\vec{X}$ and of the deterministic vector $\vec{u}$ of possible values of $\vec{X}$.

****
Example A.29: Find $E(X^2)$ and $E(XY)$ for the distribution of Example A.10 with

$$p_{XY}(u,v) = [U(u-1) - U(u-5)][U(v-1) - U(v-u)]/8$$

We have two choices of how to calculate $E(X^2)$. We can use Eq. A.54 with $g(u,v) = u^2$ and the given joint density function $p_{XY}(u,v)$, or we can use Eq. A.51 with $g(u) = u^2$ and the marginal probability density function $p_X(u)$ which we derived in Example A.13. First using Eq. A.54 we have

$$E(X^2) = \int_{-\infty}^{\infty}\int_{-\infty}^{\infty} u^2\, p_{XY}(u,v)\, du\, dv = \frac{1}{8}\int_{1}^{5}\int_{v}^{5} u^2\, du\, dv = \frac{1}{8}\int_{1}^{5}\frac{5^3 - v^3}{3}\, dv$$

$$= \frac{5^3}{6} - \frac{5^4 - 1}{96} = 14.33$$

then Eq. A.51 gives

$$E(X^2) = \int_{-\infty}^{\infty} u^2 \, p_X(u) \, du = \int_1^5 u^2 \left(\frac{u-1}{8} \right) du = \frac{5^4-1}{32} - \frac{5^3-1}{24} = 14.33$$

The result from Eq. A.51 is simpler when the marginal probability density is known. Equation A.54 involves a double integration and one can consider that one of these integrations (the one with respect to v) has already been performed in going from $p_{XY}(u,v)$ to $p_X(u)$. Thus, there is only a single integration (with respect to u) remaining in applying Eq. A.51.

For the function $g(X,Y) = XY$ we naturally use Eq. A.54 and obtain

$$E(XY) = \int_{-\infty}^{\infty} \int_{-\infty}^{\infty} uv \, p_{XY}(u,v) \, du dv = \frac{1}{8} \int_1^5 \int_v^5 uv \, du dv = \frac{1}{8} \int_1^5 \left(\frac{5^2 - v^2}{2} \right) v dv$$

$$= \frac{5^2}{16} \frac{5^2-1}{2} - \frac{5^4-1}{64} = 9$$

****

One very important property of expectation is **linearity**. This follows directly from the fact that the expected value is defined as an integral, and integration is a linear operation. The consequences of this are that $E(X + Y) = E(X) + E(Y)$ and $E(bX) = b \, E(X)$ for any random variables X and Y.

****

Example A.30: Find $E[X(X - Y)]$ for the distribution of Example A.10 with

$$p_{XY}(u,v) = [U(u-1) - U(u-5)][U(v-1) - U(v-u)]/8$$

We could do this directly from Eq. A.54 by using $g(u,v) = u(u-v)$, but we can find the result more simply by noting that

$$E[X(X - Y)] = E(X^2 - XY) = E(X^2) - E(XY)$$

and using the results obtained in Example A.29 to give

$$E[X(X - Y)] = 14.33 - 9 = 5.33$$

****

It should be noted that we cannot reverse the order of the expectation operation and nonlinear functions such as squaring or multiplication of random variables. That is, $E(X^2) \neq [E(X)]^2$ and $E(XY)$ generally does not equal $E(X)E(Y)$. We are only allowed to reverse the order of expectation and other linear operations.

Note also that there is no assurance that $E[g(X)]$ will exist in general. In particular, if $|X|$ is not bounded and $p_X(u)$ does not decay sufficiently rapidly then the integral of $g(u)p_X(u)$ in Eq. A.52 may not exist. We do know that $p_X(u)$ is integrable, since it cannot be a probability density function unless its integral is unity, but this does not ensure that $g(u)p_X(u)$ will be integrable unless $g(u)$ is bounded. For example, $p_X(u) = U(u-1)/u^2$ is an acceptable probability density function but it gives an infinite value for $E(X)$ since the integrand does not decay fast enough to give a finite value for that integral. Also, $E(X^2)$ and many other expected values are infinite for this random variable.

A.8 Moments of Random Variables

The quantities called moments of a random variable are simply special cases of the expectation of a function of the random variable. In particular, the jth moment of X is defined to be $E(X^j)$, corresponding to a special case of $E[g(X)]$. We will find it convenient to use a special name (mean) and notation (μ_X) for the first moment, which is simply the expected value of the random variable

$$\textbf{mean:} \quad \mu_X = E(X) = \int_{-\infty}^{\infty} u \, p_X(u) \, du \qquad (A.56)$$

The reason for this special treatment of the first moment is that it plays a special role in many of our calculations. Similarly, the second moment $E(X^2)$ is commonly called the **mean squared value** of the random variable, but we will not use any special notation for it.

One can also define cross moments of any pair of random variables (X,Y). The term $E(X^j Y^k)$ is the cross moment of order (j,k). The most commonly studied of these cross moments is $E(XY)$. This cross moment of order $(1,1)$ is called the **cross-product** of the random variables.

In some situations it is more convenient to use a special form of the moments, in which one first subtracts the mean of each random variable. These moments of the form $E[(X-\mu_X)^j]$ and $E[(X-\mu_X)^j (Y-\mu_Y)^k]$ are called **central moments**. Two of these central moments are given special names and notations. They are

$$\textbf{variance:} \quad \sigma_X^2 = E[(X-\mu_X)^2] = E[X^2 - 2\mu_X E(X) + \mu_X^2] = E(X^2) - \mu_X^2 \qquad (A.57)$$

and

$$\textbf{covariance:} \quad K_{XY} = E[(X-\mu_X)(Y-\mu_Y)] = E(XY) - \mu_X \mu_Y \qquad (A.58)$$

The square root of the second moments of a random variable are also given special names. In particular, $[E(X^2)]^{1/2}$ is called the **root-mean-square value** or **rms** of X, and the **standard deviation** of X is $\sigma_X \equiv \left(E[(X-\mu_X)^2] \right)^{1/2}$.

Note that Eqs. A.57 and A.58, in addition to giving the definitions of variance and covariance as central moments, also show how they can be evaluated in terms of ordinary non-central moments. It is easily shown that this sort of expansion can be found for any central moments. Specifically, the central moment of order j can be written as a linear combination of non-central moments of order 1 to j, and $E[(X-\mu_X)^j(Y-\mu_Y)^k]$ can be written as a linear combination of non-central cross moments involving orders 1 to j of X and orders 1 to k of Y. In some situations, it is useful to realize that the converse is also true — that non-central moments can be written as linear combinations of central moments of lower and equal order. For the moments of X, the explicit relationships can be written as

$$E[(X-\mu_X)^j] = \sum_{i=0}^{j} \frac{(-1)^i j!}{i!(j-i)!} \mu_X^i E(X^{j-i}) = \sum_{i=0}^{j-2} \frac{(-1)^i j!}{i!(j-i)!} \mu_X^i E(X^{j-i}) - (-1)^j (j-1)\mu_X^j \quad (A.59)$$

and

$$E(X^j) = E\left([(X-\mu_X)+\mu_X]^j\right) = \sum_{i=0}^{j} \frac{j!}{i!(j-i)!} \mu_X^i E[(X-\mu_X)^{j-i}]$$

$$= \sum_{i=0}^{j-2} \frac{j!}{i!(j-i)!} \mu_X^i E[(X-\mu_X)^{j-i}] + \mu_X^j \qquad (A.60)$$

One other form of the covariance deserves special attention. This form is called the correlation coefficient and can be viewed either as a normalized form of covariance or as a cross-product of a modified form of the original random variables:

$$\textbf{correlation coefficient:} \quad \rho_{XY} = \frac{K_{XY}}{\sigma_X \sigma_Y} = E\left[\left(\frac{X-\mu_X}{\sigma_X}\right)\left(\frac{Y-\mu_Y}{\sigma_Y}\right)\right] \qquad (A.61)$$

This form of covariance information is very commonly used, and its significance will be investigated somewhat following Example A.32. The random variables $(X-\mu_X)/\sigma_X$ and $(Y-\mu_Y)/\sigma_Y$ appearing in Eq. A.61 are sometimes called **standardized forms** of X and Y. The standardized form of any random variable, obtained by subtracting the mean value then dividing by the standard deviation, is a new random variable which is a linear function of the original random variable. The linear function has been chosen such that the standardized random variable always has zero mean and unit variance. The next two moments of the standardized random variable are given special names

$$\textbf{skewness:} \quad E\left[\left(\frac{X-\mu_X}{\sigma_X}\right)^3\right] = \frac{E[(X-\mu_X)^3]}{\sigma_X^3} \qquad (A.62)$$

and

$$\textbf{kurtosis:} \quad E\left[\left(\frac{X-\mu_X}{\sigma_X}\right)^4\right] = \frac{E[(X-\mu_X)^4]}{\sigma_X^4} \qquad (A.63)$$

but these quantities are used much less frequently than the first and second moments.

Example A.31: Let the random variable X have the exponential distribution with probability density function

$$p_X(u) = \lambda\, e^{-\lambda u}\, U(u)$$

in which λ is a positive constant. Find a general formula for $E(X^j)$.

We can simplify the integral

$$E(X^j) = \int_{-\infty}^{\infty} u^j\, p_X(u)\, du = \int_{0}^{\infty} u^j\, \lambda e^{-\lambda u}\, du$$

by using the change of variables $v = \lambda u$, giving

$$E(X^j) = \frac{1}{\lambda^j} \int_{0}^{\infty} v^j e^{-v}\, dv$$

This gives us an integral which does not depend on the parameter λ. Furthermore it is a quite common integral, called the gamma function and commonly written as

$$\Gamma(a) = \int_{0}^{\infty} v^{a-1} e^{-v}\, dv$$

The gamma function is a continuous function of a on the range $a > 0$, and at integer values of a it is equal to the factorial for a shifted argument: $\Gamma(a) = (a-1)!$. Thus, we have

$$E(X^j) = \frac{1}{\lambda^j}\Gamma(j+1) \quad \text{and} \quad E(X^j) = \frac{j!}{\lambda^j} \quad \text{for } j = \text{an integer.}$$

Note that there is no difficulty with existence of $E(X^j)$ for any j value. The exponential decay of the probability density function assures the existence of all moments of X.[5]
****

Example A.32: Let the random variable X have the Gaussian distribution with probability density function

$$p_X(u) = \frac{1}{\sqrt{2\pi}\,\sigma}\exp\left[-\frac{1}{2}\left(\frac{u-\mu}{\sigma}\right)^2\right]$$

in which μ and σ are constants. Find general formulas for the central moments $E[(X-\mu_X)^j]$.

First we note that the probability density function is symmetric about the line $u = \mu$, so that we immediately

---

[5] It should also be noted that the gamma function has a recursive property similar to that of the factorial even if a is not an integer. In particular, $\Gamma(a+1) = a\Gamma(a)$ for any a value except zero and negative integers, for which the gamma function does not exist. The gamma function arises fairly frequently in applied probability because many phenomena can be modeled using a probability density function containing an exponential term of some sort.

know that $\mu_X = \mu$ and

$$E[(X - \mu_X)^j] = \int_{-\infty}^{\infty} (u - \mu)^j p_X(u)\, du$$

If j is an odd integer then $(u - \mu)^j$ is antisymmetric about $u = \mu$ so that the integral must be zero, giving $E[(X - \mu_X)^j] = 0$ for j odd. For j even, $(u - \mu)^j$ is symmetric about $u = \mu$ so we can use only the integral on one side of this point of symmetry, then double it:

$$E[(X - \mu_X)^j] = 2 \int_{\mu}^{\infty} (u - \mu)^j p_X(u)\, du \quad \text{for} \quad j \text{ even}$$

Using the change of variables $v = (u - \mu)^2 / (2\sigma^2)$ gives

$$E[(X - \mu_X)^j] = \frac{1}{\sqrt{\pi}} \int_0^{\infty} (2\sigma^2 v)^{j/2} e^{-v} \frac{dv}{v^{1/2}} = \frac{1}{\sqrt{\pi}} (2\sigma^2)^{j/2} \Gamma\left(\frac{j+1}{2}\right) \quad \text{for} \quad j \text{ even}$$

in terms of the gamma function introduced in Example A.31. Applying this relationship will require knowledge of the gamma function for arguments which are half-integers. All such terms can be obtained from the fact that $\Gamma(1/2) = \sqrt{\pi}$ and then using the recursive property of the gamma function. Thus,

$$\Gamma\left(\frac{3}{2}\right) = \frac{1}{2}\Gamma\left(\frac{1}{2}\right) = \frac{\sqrt{\pi}}{2}$$

$$\Gamma\left(\frac{5}{2}\right) = \frac{3}{2}\Gamma\left(\frac{3}{2}\right) = \frac{3\sqrt{\pi}}{4}$$

and

$$\Gamma\left(\frac{j+1}{2}\right) = \frac{j-1}{2}\Gamma\left(\frac{j-1}{2}\right) = \frac{j-1}{2}\frac{j-3}{2}\cdots\frac{1}{2}\sqrt{\pi} = \frac{(1)(3)\cdots(j-1)}{2^{j/2}}\sqrt{\pi}$$

and

$$E[(X - \mu_X)^j] = (1)(3)\cdots(j-1)\sigma^j \quad \text{for} \quad j \text{ even}$$

Thus, $j = 2$ shows us that the parameter σ^2 is the variance $\sigma_X^2 \equiv E[(X - \mu_X)^2]$. Similarly $j = 4$ gives $E[(X - \mu)^4] = 3\sigma^4$ which shows that $kurtosis = 3$ for a Gaussian random variable.[6]
****

To illustrate the significance of covariance and correlation coefficient, we will now investigate some results called Schwarz inequalities. Consider any two random variables W and Z for which the joint probability distribution is known. Now let b and c be two real numbers and we will investigate the following mean squared value of a linear combination of the random variables

$$E[(bW - cZ)^2] = b^2 E(W^2) - 2bc E(WZ) + c^2 E(Z^2)$$

---

[6] Appendix B gives more information on the properties of Gaussian random variables.

Our purpose is to choose the relationship between b and c such as to minimize this expectation. If $E(W^2) \neq 0$, then we can rewrite the equation as

$$E[(bW - cZ)^2] = \left[b\sqrt{E(W^2)} - c\frac{E(WZ)}{\sqrt{E(W^2)}} \right]^2 + c^2\left[E(Z^2) - \frac{[E(WZ)]^2}{E(W^2)} \right]$$

Note that b appears only in the first term on the right-hand side of this expression, and that this term is nonnegative. Thus, for any joint distribution of W and Z and any choice of c we will minimize the expression on the left-hand side of the equation by choosing b/c such that the first term on the right-hand side is zero:

$$\frac{b}{c} = \frac{E(WZ)}{E(W^2)}$$

giving

$$E[(bW - cZ)^2] = c^2\left[E(Z^2) - \frac{[E(WZ)]^2}{E(W^2)} \right]$$

Since the left-hand side must be nonnegative for all values of b and c it is required that the right-hand side also be nonnegative. Thus, we have proved that $[E(WZ)]^2$ must be less than or equal to $E(W^2)E(Z^2)$. Furthermore, our choice of b/c makes $E[(bW - cZ)^2] = 0$ if and only if $[E(WZ)]^2 = E(W^2)E(Z^2)$. For the situation with $E(W^2) \neq 0$ this is a proof of one form of the Schwarz inequality:

Schwarz Inequality I: For any two random variables W and Z

$$[E(WZ)]^2 \leq E(W^2)E(Z^2) \tag{A.64}$$

and equality holds if and only if there exist constants b and c, not both zero, such that

$$E[(bW - cZ)^2] = 0 \tag{A.65}$$

Note that Eq. A.65 is simply a condition that all the probability of the joint distribution of W and Z lies on the straight line described by $(bW - cZ) = 0$, so that there is a simple linear functional relationship between W and Z. If $E(W^2) = 0$ but $E(Z^2) \neq 0$ then we prove the inequality by simply modifying this procedure to choose $c/b = E(WZ)/E(Z^2)$. If both $E(W^2) = 0$ and $E(Z^2) = 0$, then the result is trivial.

A more general form of the Schwarz inequality is found by letting W and Z be standardized random variables. In particular, let

$$W = \frac{X - \mu_X}{\sigma_X}, \qquad\qquad Z = \frac{Y - \mu_Y}{\sigma_Y}$$

In terms of X and Y, the Schwarz inequality can be written as

Schwarz Inequality II: For any two random variables X and Y

$$\rho_{XY}^2 \le 1 \tag{A.66}$$

and equality holds if and only if there exist constants $\tilde{a}$, $\tilde{b}$ and $\tilde{c}$, not all zero, such that

$$E[(\tilde{b}X - \tilde{c}Y - \tilde{a})^2] = 0 \tag{A.67}$$

indicating a linear relationship between X and Y.

Note that the values of the coefficients in Eq. A.67 can be related to those in Eq. A.65 as $\tilde{b} = b / \sigma_X$, $\tilde{c} = c / \sigma_Y$, and $\tilde{a} = \tilde{b}\mu_X - \tilde{c}\mu_Y$.

The correlation coefficient ρ_{XY} (or perhaps ρ_{XY}^2) can be considered to give the extent to which there is a linear relationship between the random variables X and Y. The limits of the correlation coefficient are $\rho_{XY} = \pm 1$ (so that $\rho_{XY}^2 = 1$), and in these limits there is a perfect linear functional relationship between X and Y of the form $\tilde{b}X - \tilde{c}Y = \tilde{a}$. That is, in these limiting cases one can find a linear function of X which is exactly equal to Y and/or a linear function of Y which is exactly equal to X. The other extreme is when $\rho_{XY} = 0$, which requires that $K_{XY} = 0$, and is called **uncorrelated random variables.** One can say that there is no linear relationship between uncorrelated random variables. For other values of ρ_{XY} (i.e., $0 < \rho_{XY}^2 < 1$) one may say that there is some linear relationship between X and Y, but any linear function of X is an imperfect approximation of Y and any linear function of Y is an imperfect approximation of X. Linear regression presents a slightly different way of looking at this matter. In linear regression of Y on X, one compares Y with a linear function $(a + bX)$, with a and b chosen to minimize $E[(a + bX - Y)^2]$. When this is done, it is found that the minimum value of $E[(a + bX - Y)^2]$ is $\sigma_Y^2(1 - \rho_{XY}^2)$, indicating that the mean squared error in this best linear fit is directly proportional to $(1 - \rho_{XY}^2)$. Again we see that ρ_{XY}^2 gives the extent to which there is a linear relationship between X and Y.

The preceding paragraph notes that $\rho_{XY} = 1$ and $\rho_{XY} = -1$ both give a perfect linear relationship between X and Y, but it says nothing about the significance of the sign of ρ_{XY}. Reviewing the proof of the Schwarz inequality, it is easily seen that the slope of the linear approximate relationship between Y and X or Z and W has the same sign as $E(WZ)$, which is the same sign as ρ_{XY}. Thus, $\rho_{XY} > 0$ indicates a positive slope for the linear approximation, while $\rho_{XY} < 0$ indicates a negative slope. The slope of the linear regression of Y on X is $(\rho_{XY}\sigma_Y / \sigma_X)$.

It should be kept in mind that having random variables X and Y uncorrelated does not generally mean that they are independent. Rather, $\rho_{XY} = 0$ (or $K_{XY} = 0$) only implies that there is not a linear relationship between X and Y. On the other hand, if X and Y are independent then they must also be uncorrelated. This is easily shown as follows

$$E(XY) = \int_{-\infty}^{\infty} \int_{-\infty}^{\infty} uv\, p_{XY}(u,v)\, du\, dv$$

$$= \int_{-\infty}^{\infty} \int_{-\infty}^{\infty} uv\, p_X(u) p_Y(v)\, du\, dv \quad \text{if} \quad X \text{ and } Y \text{ are independent}$$

$$= \int_{-\infty}^{\infty} u\, p_X(u)\, du \int_{-\infty}^{\infty} v\, p_Y(v)\, dv = E(X)E(Y)$$

so that the covariance of X and Y is

$$K_{XY} = E(XY) - E(X)E(Y) = 0 \quad \text{if} \quad X \text{ and } Y \text{ are independent.}$$

Independence of X and Y is generally a much stronger condition than being uncorrelated, implying no relationship of any kind (linear or nonlinear) between X and Y.

****

Example A.33: Let the probability density of X be uniform on $[-2,-1]$ and $[1,2]$, as shown in Example A.26, and let Y be a simple square of X: $Y = X^2$. Consider the correlation and dependence between X and Y.

Clearly X and Y are not independent. In fact, the joint probability density of X and Y is degenerate with $p_{XY}(u,v) \neq 0$ only on the curve $v = u^2$. Nonetheless, $E(X) = 0$ and $E(XY) \equiv E(X^3) = 0$ so that $K_{XY} = 0$, proving that X and Y are uncorrelated even though functionally dependent. They are uncorrelated because there is no linear relationship between them which is even approximately correct. Linear regression of Y on X for this problem can be shown to give a line of zero slope, again affirming that Y cannot be approximated (to any degree of accuracy at all) by a linear function of X.
****

The engineering reader may find it useful to relate the concept of moments of random variables to a more familiar topic of moments of areas or of rigid bodies. In particular, the expression for finding the mean of a random variable is exactly the same as that for finding the centroid of the area under the curve describing the probability density function. In general, to find the centroid one takes this first moment integral, then divides by the total area. For a probability density function, though, the area is unity, so that the first moment is exactly the centroidal location. This will often prove to be an easy way to detect errors in the evaluation of a mean value. If it is obvious that the value you have computed could not be the centroid of the area under the probability density curve, then you have made a mistake. Also, note that if the probability density is symmetric about some particular value, then that point of symmetry must be the mean value, just as it is the centroid of the area.

Similarly for the second moments, $E(X^2)$ is the same as the moment of inertia of the area under the probability density curve about the line at $u = 0$, and the variance σ_X^2 is the same as the centroidal moment of inertia. Extending these mechanics analogies to joint distributions shows that the point in two-dimensional space $u = \mu_X$ and $v = \mu_Y$ is at the centroid of the volume under the joint probability density function $p_{XY}(u,v)$, and the cross-product $E(XY)$ and covariance K_{XY} are the same as products of inertia of the volume under the probability density function.

We previously noted that it is convenient to use vector notation when we are dealing with many random variables. Consistent with this, we can use a matrix to organize all the information related to various cross-products and covariances of n random variables. In particular, the expression

$$E[\vec{X}\vec{X}^T] = E\left(\begin{bmatrix} X_1 \\ X_2 \\ \vdots \\ X_n \end{bmatrix} [X_1, X_2, \cdots, X_n]\right)$$

gives a square symmetric matrix in which the (j,k) component is the cross-product of X_j and X_k. Similarly,

$$\mathbf{K}_{XX} = E\left[(\vec{X} - \vec{\mu}_X)(\vec{X} - \vec{\mu}_X)^T\right] \qquad (A.68)$$

defines a square symmetric covariance matrix $\mathbf{K}_{XX}$ in which the (j,k) component of $\mathbf{K}_{XX}$ is the covariance of X_j and X_k. Note that the diagonal elements of $\mathbf{K}_{XX}$ are simply the variances of the components.[7] The reason for the double subscript on the covariance matrix is that we will sometimes also want to consider a matrix of covariances of different vectors defined by

$$\mathbf{K}_{XY} = E\left[(\vec{X} - \vec{\mu}_X)(\vec{Y} - \vec{\mu}_Y)^T\right] \qquad (A.69)$$

Note that $\mathbf{K}_{XY}$ generally need not be symmetric, or even square.

A.9 Conditional Expectation

For any given event A we will define the conditional expected value of any random variable X given A, $E(X|A)$, to be exactly the expectation defined in Eq. A.51, but using the conditional probability density function $p_X(u|A)$ in place of $p_X(u)$. Thus,

---

[7]   One can also show that the matrices $E[\vec{X}\vec{X}^T]$ and $\mathbf{K}_{XX}$ have the property of being nonnegative definite. This property says that none of the eigenvalues of either matrix is negative. More directly relevant to our purposes is the fact that $\mathbf{K}_{XX}$, for example, is nonnegative definite if $\vec{v}^T\mathbf{K}_{XX}\vec{v} \geq 0$ for all vectors $\vec{v}$. It is easy to prove that $\mathbf{K}_{XX}$ has this property since $\vec{v}^T\mathbf{K}_{XX}\vec{v}$ is the variance for a new scalar random variable $Y = \vec{v}^T\vec{X}$.

$$E(X|A) = \int_{-\infty}^{\infty} u \, p_X(u|A) \, du \tag{A.70}$$

Similarly

$$E[g(X)|A] = \int_{-\infty}^{\infty} g(u) \, p_X(u|A) \, du \tag{A.71}$$

and

$$E[g(\vec{X})|A] = \int_{-\infty}^{\infty} \cdots \int_{-\infty}^{\infty} g(\vec{u}) \, p_{\vec{X}}(\vec{u}|A) \, du_1 \cdots du_n \tag{A.72}$$

The key idea is that a conditional probability density function truly is a probability density function. It can be used in all the same ways as any other probability density function, including as a weighting function for calculating expected values. The conditional probability density function given an event A gives the revised probabilities of all possible outcomes based on the knowledge of A. Similarly the expected value calculated using this conditional probability density function gives the revised probability weighted average of all possible outcomes based on the knowledge of A.

****

Example A.34: For the distribution of Example A.10 with

$$p_{XY}(u,v) = [U(u-1) - U(u-5)][U(v-1) - U(v-u)]/8$$

find $E(Y|X \leq 3)$, $E(Y^2|X \leq 3)$, and $E(Y|X = u)$ for all u for which it is defined.

In Example A.16 we found that $p_Y(v|X \leq 3) = [(3-v)/2] [U(v-1)] - U(v-3)]$, so we can now integrate to obtain

$$E(Y|X \leq 3) = \int_1^3 v \left(\frac{3-v}{2} \right) dv = \frac{5}{3}$$

and

$$E(Y^2|X \leq 3) = \int_1^3 v^2 \left(\frac{3-v}{2} \right) dv = 3$$

We also know from Example A.18 that the conditional distribution of Y given $X = u$ is defined for $1 \leq u \leq 5$ and is given by

$$p_Y(v|X = u) = \frac{1}{u-1}[U(v-1) - U(v-u)]$$

so that we now obtain

$$E(Y|X = u) = \int_1^u \left(\frac{v}{u-1} \right) dv = \frac{1}{u-1} \frac{u^2-1}{2} = \frac{u+1}{2} \quad \text{for} \quad 1 \leq u \leq 5$$

Actually we could have used symmetry to write this last expected value without performing integration. The $p_Y(v|X = u)$ conditional probability density function is uniform on the set $[1,u]$ so the conditional mean must be at the center of this range.

****

Example A.35: Find $E(X|Y=v)$ and $E(X^2|Y=v)$ for all v values for which they are defined for the random variables of Example A.11 with

$$p_{XY}(u,v) = \left(8e^{-4u-4v} + 3e^{-4u-3v} + 3e^{-3u-4v}\right)U(u)U(v)$$

In Example A.19 we found that

$$p_X(u|Y=v) = \frac{\left(8e^{-4u-4v} + 3e^{-4u-3v} + 3e^{-3u-4v}\right)}{3\left(e^{-4v} + e^{-3v}/4\right)}U(u) \quad \text{for} \quad v \geq 0$$

so we now obtain

$$E(X|Y=v) = \int_0^\infty u\left[\frac{\left(8e^{-4u-4v} + 3e^{-4u-3v} + 3e^{-3u-4v}\right)}{3\left(e^{-4v} + e^{-3v}/4\right)}\right]du = \frac{9 + 40e^{-v}}{36(1+4e^{-v})} \quad \text{for} \quad v \geq 0$$

and

$$E(X^2|Y=v) = \int_0^\infty u^2 \frac{\left(8e^{-4u-4v} + 3e^{-4u-3v} + 3e^{-3u-4v}\right)}{3\left(e^{-4v} + e^{-3v}/4\right)}du = \frac{27 + 136e^{-v}}{216(1+4e^{-v})} \quad \text{for} \quad v \geq 0$$

One can perform a simple check on these results by looking at the limiting case for v tending to infinity. In this situation we will have $e^{-4v} \ll e^{-3v}$ so that we can neglect the e^{-4v} terms in the conditional probability density function, reducing it to $4e^{-4u}$. This is an exponential distribution for which the mean and mean squared values are $1/4$ and $2/4^2 = 1/8$. It is easy to verify that our conditional expected values do tend to these limits as v tends to infinity, so that the e^{-v} terms become negligible.
****

Recall the special case of independent random variables. We know that if X and Y are independent, then the conditional distribution of Y given any information about X is the same as the marginal distribution of X — that is, it is the same as if no information were given about X. Thus, if X and Y are independent, then the conditional expectation of Y given any information about X will be the same as if the information had not been given. It will be simply $E(Y)$.

****
Example A.36: For the distribution of Example A.9 with

$$p_{XY}(u,v) = \frac{1}{6}[U(u+1) - U(u-2)][U(v+1) - U(v-1)]$$

find the conditional expected values, $E(Y|X=u)$ and $E(X|Y=v)$ for all u and v for which they are defined, and also find $E(X|Y>0.5)$.

Since the random variables have been shown to be independent in Example A.22, we know that

$E(Y|X = u) = E(Y)$, and $E(X|Y = v) = E(X)$ for u and v values for which they are defined. Similarly, $E(X|Y > 0.5) = E(X)$. Using the marginal probability density functions derived in Example A.12 then gives $E(Y|X = u) = 0$ for $-1 \leq u \leq 2$, $E(X|Y = v) = 0.5$ for $-1 \leq v \leq 1$, and $E(X|Y > 0.5) = 0.5$.

****

A.10   Generalized Conditional Expectation

Our discussion of conditional expectation, to this point, has involved the expectation given some specific event. These conditional expectations are always deterministic, being derived from integrals of conditional probability density functions. Now we want to define a conditional expectation which is, itself, a random variable. The uncertainty about the conditional expectation will be introduced by allowing there to be uncertainty about the given event. In order to be more specific about this idea, let us consider the expected value of a random variable X given a value of a random variable Y: $E(Y|X = u)$. In general, this conditional expectation will be different for each value of the deterministic variable u. Thus, the conditional expectation can be considered to define a function of u. For the moment, let us write this as

$$f(u) \equiv E(Y|X = u)$$

Now that we have the definition of this function, we might also consider the same function with a random variable for an argument. In particular, we will be interested in taking the random variable X to be the argument of $f(\cdot)$. Now we can investigate the expected value of $f(X)$, just as we might for any other function of a random variable. We write

$$E[f(X)] = \int_{-\infty}^{\infty} f(u) p_X(u) du$$

but we also know from its definition that $f(u)$ is given by

$$f(u) = \int_{-\infty}^{\infty} v \, p_Y(v|X = u) dv$$

Substituting $f(u)$ into the integral for $E[f(X)]$ gives

$$E[f(X)] = \int_{-\infty}^{\infty} \int_{-\infty}^{\infty} v \, p_Y(v|X = u) \, p_X(u) \, dv \, du$$

but the product of probability density functions in this integrand is exactly the joint probability density function for X and Y:

$$p_{XY}(u,v) = p_X(u) \, p_Y(v|X = u)$$

so that the integral for $E[f(X)]$ becomes exactly the same as one for evaluating $E[g(X,Y)]$ with $g(X,Y) = Y$. Thus,

$$E[f(X)] = E(Y)$$

The usual notation for the conditional expectation random variable defined earlier as $f(X)$ is simply $E[Y|X]$. That is, $E[Y|X]$ is the function of u defined by $E[Y|X = u]$ when the deterministic quantity u is replaced by the random variable X. The result that we have just proved can then be written as

$$E[E(Y|X)] = E(Y) \tag{A.73}$$

which can be stated in words as follows: The expected value of the conditional expected value is the unconditional expected value. This statement may sound rather confusing, but the meaning of Eq. A.73 should be quite understandable. The left-hand side of Eq. A.73 can be thought of as depending on the joint distribution of X and Y. Evaluating $E(Y|X)$ corresponds to finding the probability weighted average over all possible values of Y for a given value of X, then the second expectation gives the probability weighted average over all possible values of X. Equation A.73 can also be generalized to a form

$$E\big(g(X)E[h(Y)|X]\big) = E[g(X)h(Y)] \tag{A.74}$$

for any functions $g(\cdot)$ and $h(\cdot)$ for which the expectations exist. The definition of $E[h(Y)|X]$ in Eq. A.74, of course, is $f(X)$ with the function $f(\cdot)$ defined by $f(u) = E[h(Y)|X = u]$. One can verify Eq. A.74 by substituting

$$E[h(Y)|X = u] = \int_{-\infty}^{\infty} h(v) p_Y(v|X = u) \, dv$$

into

$$E\big(g(X)E[h(Y)|X]\big) = \int_{-\infty}^{\infty} g(u) E[h(Y)|X = u] p_X(u) \, du$$

In some problems it will be found to be quite convenient to use Eqs. A.73 and A.74 to evaluate the expectations given on the right-hand sides of the equations. In particular, Eq. A.73 may be simpler than finding and using $p_Y(v)$ to find $E(Y)$, and A.74 may be simpler than finding and using $p_{XY}(u,v)$ to find $E[g(X)h(Y)]$. Most importantly, there are situations in which Eq. A.73 can be used when you are given insufficient information to describe $p_Y(v)$, and others in which A.74 can be used when you are given insufficient information to describe $p_{XY}(u,v)$.

****

Example A.37: Find $E(Y)$ for the probability distribution of Example A.20 with

$$p_X(u) = 0.1[U(u) - U(u - 10)]$$

and

$$p_Y(v|X = u) = \frac{1}{u}[U(v) - U(v - u)]$$

From the fact that the conditional distribution of Y given $X = u$ is uniform on the set $[0, u]$, we can immediately note that $E(Y|X = u) = u/2$. Thus, $E(Y|X) = X/2$ is the conditional expectation random variable obtained by substituting X for u in $E(Y|X = u)$. Now using Eq. A.73 gives $E(Y) = E[E(Y|X)] = E(X)/2$. Using the fact that the distribution of X is uniform on the set $[0,10]$ gives $E(X) = 5$, so that we have $E(Y) = 2.5$ without ever having evaluated $p_Y(v)$ or explicitly worked with $p_{XY}(u,v)$.

One can, of course, do this problem by first evaluating $p_Y(v)$, but that is less convenient. Using the joint probability density function we can write

$$p_Y(v) = \int_{-\infty}^{\infty} p_{XY}(u,v)\,du = \int_v^{10}\left(\frac{0.1}{u}\right)du[U(v)-U(v-10)]$$

giving

$$p_Y(v) = 0.1\log(10/v)[U(v)-U(v-10)]$$

The reader can verify that this probability density function does give $E(Y) = 2.5$, but it seems clear that the approach using Eq. A.73 is the simpler procedure for evaluating the expectation.
****
Example A.38: Let the random variable X have an exponential distribution

$$p_X(u) = 2\,e^{-2u}U(u)$$

and let the conditional mean and mean squared values of Y be given as

$$E(Y|X = u) = 3u \quad \text{and} \quad E(Y^2|X = u) = 10u^2 + 2u$$

Find $E(Y)$, $E(Y^2)$, and $E(X^2Y^2)$.

This is a situation in which there clearly is not enough given information to allow us to write out $p_Y(v)$ or $p_{XY}(u,v)$. It is not unusual to have practical problems in which one is given only partial information of this sort, particularly when the information about the Y random variable has been obtained strictly from statistical analysis of measured data. We can use Eqs. A.73 and A.74, though. We can say that $E(Y|X) = 3X$ so that Eq. A.73 gives $E(Y) = 3E(X)$. Now using the distribution of X we find that $E(X) = 0.5$, so that $E(Y) = 1.5$. Similarly, $E(Y^2|X) = 10X^2 + 2X$ so that $E(Y^2) = 10E(X^2) + 2E(X)$. From $p_X(u)$ we find that $E(X^2) = 0.5$, and finally $E(Y^2) = 6$. This result for $E(Y^2)$ can be considered either as an application of Eq. A.74 with $g(X) = 1$ and $h(Y) = Y^2$, or as an application of Eq. A.73 with the second random variable being taken as Y^2 rather than the Y that was written previously.

For the final expectation we need to use Eq. A.74 with $g(X) = X^2$ and $h(Y) = Y^2$. Substituting gives

$$E(X^2Y^2) = E[X^2E(Y^2|X)] = E[X^2(10X^2 + 2X)] = 10E(X^4) + 2E(X^3)$$

Integrals using $p_X(u)$ give $E(X^3) = 0.75$ and $E(X^4) = 1.5$, so that $E(X^2Y^2) = 16.5$.
****

A.11 Characteristic Function of a Random Variable

The characteristic function provides an alternative form for giving a complete description of a probability density function. Recall that we have previously noted that either the cumulative distribution function or the probability density function gives such a complete description. We will now show that the function called the characteristic function gives the all the information that is included within the probability density function. The characteristic function for a random variable X will be denoted by $M_X(\theta)$ and is defined as

$$M_X(\theta) = E\left(e^{-i\theta X}\right) = \int_{-\infty}^{\infty} e^{-i\theta u} p_X(u)\, du \qquad (A.75)$$

Thus, $M_X(\theta)$ is the expected value of a complex function of X and it can be evaluated from an integral with a complex integrand. One can, of course, convert this to the evaluation of two real integrals by using $e^{-i\theta u} = \cos(\theta u) - i\sin(\theta u)$. Those familiar with Fourier analysis will note that Eq. A.75 gives the characteristic function $M_X(\theta)$ as a form of the Fourier transform of the probability density function $p_X(u)$.[8] Noting that this is a Fourier transform allows us to use known results to write the corresponding inverse Fourier transform formula, giving

$$p_X(u) = \frac{1}{2\pi} \int_{-\infty}^{\infty} M_X(\theta) e^{i\theta u}\, d\theta \qquad (A.76)$$

Actually, this inverse formula is an equality only at points where $p_X(u)$ is continuous, but this is sufficient for our purposes. Changing the value of $p_X(u)$ only at points of discontinuity would not change integrals involving $p_X(u)$ in the integrand. Thus, knowledge of $p_X(u)$ everywhere except at points of discontinuity is sufficient to give the cumulative distribution function, which is a complete description of the probability distribution. This proves our earlier assertion that knowing $M_X(\theta)$ gives all the information necessary for a complete description of the probability distribution.

There are several reasons why one might choose to use the characteristic function of a random variable. Many times, the motivation for using $M_X(\theta)$ will be that it simplifies some analytical development or proof. There is one property of characteristic functions, though, which sometimes proves to be very useful for simple calculations. This is the so-called moment generating property and it is obtained by differentiating Eq. A.75 with respect to θ. In particular, if we take the jth derivative then we obtain

$$\frac{d^j}{d\theta^j} M_X(\theta) = (-i)^j E\left(X^j e^{-i\theta X}\right)$$

and letting $\theta = 0$ in this expression gives

---

[8] Appendix D gives a brief introduction to Fourier analysis.

$$E(X^j) = i^j \left[\frac{d^j}{d\theta^j} M_X(\theta) \right]_{\theta=0} \qquad (A.77)$$

Thus, if one knows the characteristic function for a random variable, then the moments of the random variable can be found by differentiating that characteristic function, then evaluating the derivatives at $\theta = 0$. If many moments are needed, then it may be easier to find them this way rather than be performing an integral for each moment according to Eq. A.52. Of course, if the probability distribution of X is given as $p_X(u)$, then the integration of Eq. A.75 must be performed before one can begin to evaluate moments according to Eq. A.77. This is only one integration, though, rather than one for each moment being calculated.

Some readers may be acquainted with the real function $E(e^{-rX})$ which is a Laplace transform of $p_X(u)$, and is called the moment generating function. The moment equations from this function are simpler than those from Eq. A.77 inasmuch as they are real, as is $E(e^{-rX})$ itself for real values of r. The disadvantage of $E(e^{-rX})$ is that it does not exist for all probability density functions, or for all values of r for other probability density functions. The condition for the existence of a Fourier transform, though, is simply that the original function be absolute value integrable. This means that the characteristic function $M_X(\theta)$ exists for all values of θ if

$$\int_{-\infty}^{\infty} |p_X(u)| \, du < \infty$$

but this condition is met for any random variable. In fact the integral shown is exactly unity. Thus, one never needs to be concerned about the existence of the characteristic function.

****

Example A.39: Find the characteristic function for the random variable with the general exponential distribution

$$p_X(u) = \lambda\, e^{-\lambda u} U(u)$$

and verify that the mean and mean-squared values obtained by taking derivatives of $M_X(\theta)$ agree with those obtained in Example A.31 by integration of the probability density function.

In Example A.31 we used a change of variables of $v = \lambda u$ to obtain

$$E(X) = \lambda \int_0^\infty u e^{-\lambda u} \, du = \frac{1}{\lambda} \int_0^\infty v e^{-v} \, dv = \frac{1}{\lambda}$$

and

$$E(X^2) = \lambda \int_0^\infty u^2 e^{-\lambda u} \, du = \frac{1}{\lambda^2} \int_0^\infty v^2 e^{-v} \, dv = \frac{2}{\lambda^2}$$

Similarly, using $v = (\lambda + i\theta)u$ gives the characteristic function as

$$M_X(\theta) = \lambda \int_0^\infty e^{-iu\theta} e^{-\lambda u}\, du = \lambda \int_0^\infty e^{-u(\lambda+i\theta)}\, du$$

$$= \left(\frac{\lambda}{\lambda+i\theta}\right) \int_{C_1} e^{-v}\, dv$$

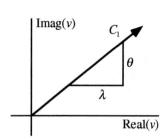

The contour C_1 in the final integral is the straight line shown in the sketch

The fact that the contour is made up of complex v values causes the evaluation of the integral to be nontrivial. In fact, though, it can be found from the study of complex functions that for this integrand the integral over C_1 is identical to the integral from zero to infinity along the real axis.[9] Thus, one obtains

$$M_X(\theta) = \left(\frac{\lambda}{\lambda+i\theta}\right)$$

Note that $M_X(0) = 1$, as is always the case. The first two derivatives are

$$M_X'(\theta) = \frac{-i\lambda}{(\lambda+i\theta)^2}, \qquad M_X''(\theta) = \frac{2i^2\lambda}{(\lambda+i\theta)^3}$$

giving $M_X'(0) = -i/\lambda$ and and $M_X''(0) = -2/\lambda^2$. Using Eq. A.77 then gives $E(X) = 1/\lambda$ and $E(X^2) = 2/\lambda^2$, confirming our results from integration.
****

Example A.40: Find the characteristic function and calculate the mean and mean-squared values by taking derivatives of $M_X(\theta)$ for the probability distribution of Examples A.1, A.4, and A.7 with

$$p_X(u) = \frac{1}{6}[\delta(u-1) + \delta(u-2) + \delta(u-3) + \delta(u-4) + \delta(u-5) + \delta(u-6)]$$

Using this probability density function, we obtain

$$M_X(\theta) = \int_0^\infty \frac{e^{-i\theta u}}{6}[\delta(u-1) + \delta(u-2) + \delta(u-3) + \delta(u-4) + \delta(u-5) + \delta(u-6)]du$$

which is easily integrated by the property of the Dirac delta function to give

$$M_X(\theta) = \frac{e^{-i\theta} + e^{-2i\theta} + e^{-3i\theta} + e^{-4i\theta} + e^{-5i\theta} + e^{-6i\theta}}{6}$$

From this one obtains

---

[9] The reader is cautioned that considerable care should be used in making changes of variables or changing contours of integration in problems involving complex functions. The complex analysis theorems governing these matters are not covered in this book.

$$M_X'(0) = \frac{-i - 2i - 3i - 4i - 5i - 6i}{6} = \frac{-7i}{2}$$

and

$$M_X''(0) = \frac{(-i)^2 + (-2i)^2 + (-3i)^2 + (-4i)^2 + (-5i)^2 + (-6i)^2}{6} = \frac{-91}{6}$$

Thus, Eq. A.77 gives $E(X) = 7/2$ and $E(X^2) = 91/6$.

****

One can easily generalize the characteristic function formulas to joint distributions of random variables. Using vector notation, one can write

$$M_{\vec{X}}(\vec{\theta}) = E(e^{-i\vec{\theta}^T\vec{X}}) = \int_{-\infty}^{\infty}\cdots\int_{-\infty}^{\infty} e^{-i\vec{\theta}^T\vec{u}} p_{\vec{X}}(\vec{u})\, du_1\cdots du_n \qquad (A.78)$$

in which the superscript ^{T} denotes the transpose of the vector so that $\vec{\theta}^T\vec{X} = (\theta_1 X_1 + \cdots + \theta_n X_n)$. The inverse formula is

$$p_{\vec{X}}(\vec{u}) = \frac{1}{(2\pi)^n} \int_{-\infty}^{\infty}\cdots\int_{-\infty}^{\infty} e^{i\vec{\theta}^T\vec{u}} M_{\vec{X}}(\vec{\theta})\, d\theta_1\cdots d\theta_n \qquad (A.79)$$

and moments are found from formulas such as

$$E\left(X_1^j X_2^k\right) = i^{(j+k)} \left[\frac{\partial^{j+k}}{\partial\theta_1^j \partial\theta_2^k} M_{\vec{X}}(\vec{\theta})\right]_{\vec{\theta}=\vec{0}}$$

or

$$E\left(X_1^{j_1}\cdots X_n^{j_n}\right) = i^{(j_1+\cdots+j_n)} \left[\frac{\partial^{j_1+\cdots+j_n}}{\partial\theta_1^{j_1}\cdots\partial\theta_n^{j_n}} M_{\vec{X}}(\vec{\theta})\right]_{\vec{\theta}=\vec{0}} \qquad (A.80)$$

A particularly convenient property of characteristic functions relates to the finding of marginal distributions from joint distributions. One finds that a marginal characteristic function is found from a joint characteristic function simply by setting one or more of the θ arguments equal to zero. For example, if $M_{XYZ}(\theta_1, \theta_2, \theta_3) \equiv E[\exp(-i\theta_1 X - i\theta_2 Y - i\theta_3 Z)]$ is known, then $M_{XY}(\theta_1, \theta_2) \equiv E[\exp(-i\theta_1 X - i\theta_2 Y)] = M_{XYZ}(\theta_1, \theta_2, 0)$, $M_X(\theta_1) = M_{XYZ}(\theta_1, 0, 0)$, etc.

A.12 Power Series for Characteristic Function

Recall that Eq. A.77 gave a simple relationship between the moments of a random variable and the derivatives of its characteristic function when evaluated at the origin of θ space. Thus, if all the moments of X exist, one can use them in a Taylor series expansion for $M_X(\theta)$ about the origin (i.e., a Maclaurin expansion):

$$M_X(\theta) = \sum_{j=0}^{\infty} \frac{\theta^j}{j!} \left[\frac{d^j M_X}{d\theta^j} \right]_{\theta=0} = \sum_{j=0}^{\infty} \frac{(i\theta)^j}{j!} E(X^j) \tag{A.81}$$

This equation represents a very important property of many probability distributions. Knowledge of all the moments of a random variable is usually sufficient information to provide a complete description of its probability distribution. In particular, from knowledge of all the $E(X^j)$ terms, one can write $M_X(\theta)$ according to Eq. A.81, then Eq. A.76 allows determination of the probability density function. It should be noted, though, that this procedure does not always work. The log-normal distribution is probably the most commonly used distribution for which knowledge of all the moments does not uniquely define the probability distribution (Kendall and Stuart, 1977). We are not suggesting that this is a practical method for finding $p_X(u)$ in an applied problem, even when there is no difficulty with uniqueness. Rather, it is an illustration of how much information is needed for a complete description of most probability distributions.

Sometimes it is convenient to work with the natural logarithm of $M_X(\theta)$ instead of with $M_X(\theta)$ itself. This function $\log[M_X(\theta)]$ is called the log-characteristic function. One can also expand the log-characteristic function as a Taylor series like the one in Eq. A.81, but the coefficients are different, of course. Specifically, we will write

$$\log[M_X(\theta)] = \sum_{j=0}^{\infty} \frac{(i\theta)^j}{j!} \kappa_j(X) \tag{A.82}$$

with

$$\kappa_j(X) = \left[\frac{d^j}{d\theta^j} \log[M_X(\theta)] \right]_{\theta=0} \tag{A.83}$$

The term $\kappa_j(X)$ is called the jth cumulant of X. We introduce these terms because it is sometimes very convenient to use cumulants rather than moments to describe a probability distribution. One can use Eq. A.83 to derive the relationship between any particular cumulant $\kappa_j(X)$ and the moments of X up the jth order. The general relationship is quite complicated, but the first few cumulants are given by

$$\kappa_1(X) = E(X) = \mu_X \tag{A.84}$$

$$\kappa_2(X) = E[(X - \mu_X)^2] = \sigma_X^2 \tag{A.85}$$

$$\kappa_3(X) = E[(X - \mu_X)^3] = \sigma_X^3 \times (skewness) \tag{A.86}$$

$$\kappa_4(X) = E[(X - \mu_X)^4] - 3\left(E[(X - \mu_X)^2] \right)^2 = \sigma_X^4 \times [kurtosis - 3] \tag{A.87}$$

For joint distributions, one can use joint cumulants of the form

$$\kappa_J(\underbrace{X_1,\cdots,X_1}_{j_1},\cdots,\underbrace{X_n,\cdots,X_n}_{j_n}) = \frac{1}{i^J}\left[\frac{\partial^J}{\partial\theta_1^{j_1}\cdots\partial\theta_n^{j_n}}\log[M_{\vec{X}}(\vec{\theta})]\right]_{\vec{\theta}=\vec{0}} \tag{A.88}$$

in which the order of the cumulant is $J \equiv j_1+\cdots+j_n$. This gives the series expansion

$$\log[M_{\vec{X}}(\vec{\theta})] = \sum_{j=0}^{\infty}\frac{(i\theta_1)^{j_1}\cdots(i\theta_n)^{j_n}}{j_1!\cdots j_n!}\kappa_J(\underbrace{X_1,\cdots,X_1}_{j_1},\cdots,\underbrace{X_n,\cdots,X_n}_{j_n}) \tag{A.89}$$

just as the joint characteristic function can be expanded in terms of the joint moments as

$$M_{\vec{X}}(\vec{\theta}) = \sum_{j=0}^{\infty}\frac{(i\theta_1)^{j_1}\cdots(i\theta_n)^{j_n}}{j_1!\cdots j_n!}E(X_1^{j_1}\cdots X_n^{j_n}) \tag{A.90}$$

One can also use the log-characteristic function to prove a property of joint cumulants which is sometimes of considerable importance in applications. In particular, we will prove that

$$\kappa_{n+1}\left(X_1,\cdots,X_n,\sum a_j Y_j\right) = \sum a_j\kappa_{n+1}(X_1,\cdots,X_n,Y_j) \tag{A.91}$$

which can be considered to be a property of linearity, and includes the distributive property. To show this, we use Eq. A.88 to write

$$\kappa_{n+1}\left(X_1,\cdots,X_n,\sum a_j Y_j\right) = \frac{1}{i^{n+1}}\frac{\partial^{n+1}}{\partial\theta_1\cdots\partial\theta_n\partial\phi}\log\left(E[e^{i\vec{\theta}^T\vec{X}+i\phi\sum a_j Y_j}]\right)\bigg|_{\vec{\theta}=\vec{0},\phi=0}$$

Performing the differentiation with respect to ϕ gives

$$\kappa_{n+1}\left(X_1,\cdots,X_n,\sum a_j Y_j\right) = \frac{1}{i^n}\frac{\partial^n}{\partial\theta_1\cdots\partial\theta_n}\frac{E[\sum a_j Y_j e^{i\vec{\theta}^T\vec{X}+i\phi\sum a_j Y_j}]}{E[e^{i\vec{\theta}^T\vec{X}+i\phi\sum a_j Y_j}]}\bigg|_{\vec{\theta}=\vec{0},\phi=0}$$

and setting $\phi = 0$, then rearranging terms allows this to be written as

$$\kappa_{n+1}\left(X_1,\cdots,X_n,\sum a_j Y_j\right) = \frac{1}{i^n}\sum a_j\frac{\partial^n}{\partial\theta_1\cdots\partial\theta_n}\frac{E[Y_j e^{i\vec{\theta}^T\vec{X}}]}{E[e^{i\vec{\theta}^T\vec{X}}]}\bigg|_{\vec{\theta}=\vec{0}}$$

This expression, though, is exactly what one would obtain from performing exactly the same operations on each of the cumulant terms appearing on the right-hand side of Eq. A.91. Thus, Eq. A.91 is proved to be true. One can also let some of the X_j arguments in Eq. A.91 be identical, so that the expression applies to any sort of joint cumulant term rather than being restricted to being first order in each of the X_j arguments.

A.13 Importance of Moment Analysis

In many random variable problems (and in much of the analysis of stochastic processes), one performs detailed analysis of only the first and second moments of the various quantities, with occasional consideration of skewness and/or kurtosis. One reason for this is surely the fact that analysis of mean, variance, or mean squared value is generally much easier than analysis of probability distributions. Furthermore, in many problems one has some idea of the shape of the probability density functions, so that knowledge of moment information may allow evaluation of the parameters in that shape, thereby giving an estimate of the complete probability distribution. If the shape has only two parameters to be chosen, in particular, then knowledge of mean and variance will generally suffice for this procedure. In addition to these pragmatic reasons, though, the results in Eqs. A.81, A.82, A.89, and A.90 give a theoretical justification for focusing attention on the lower-order moments. Specifically, mean, variance, skewness, kurtosis, etc., in that order, are the first items in an infinite sequence of information which would give a complete description of any problem. In most situations it is not possible for us to achieve the complete description, but it is certainly logical for us to focus our attention on the first items in the sequence.

Exercises

****

Distribution of One Random Variable

****

A.1. Consider a random variable X with cumulative distribution function

$$F_X(u) = 0 \qquad \text{for} \qquad u < 0$$
$$= u^3 \qquad \text{for} \qquad 0 \le u < 1$$
$$= 1 \qquad \text{for} \qquad u \ge 1$$

Find the probability density function for X.

****

A.2. Consider a random variable X with cumulative distribution function

$$F_X(u) = (1 - u^{-2}/2)U(u-1)$$

Find the probability density function for X.

****

A.3. Consider a random variable X with probability density function

$$p_X(u) = 2(u+1)^{-3}U(u)$$

Find the cumulative distribution function for X.

****

A.4. Consider a random variable X with probability density function

$$p_X(u) = \frac{3}{4}u(2-u)[U(u) - U(u-2)]$$

Find the cumulative distribution function for X.

****

Joint and Marginal Distributions

****

A.5. Let X and Y be two random variables with the joint cumulative distribution function

$$F_{XY}(u,v) = \left((1-e^{-2u})(1-e^{-3v}) + (6/5)[(u+v)e^{-2u-3v} - ue^{-2u} - ve^{-3v}]\right)U(u)U(v)$$

(a) Find the joint probability density function $p_{XY}(u,v)$.

(b) Find the marginal cumulative distribution functions $F_X(u)$ and $F_Y(v)$.

(c) Find the marginal probability density functions $p_X(u)$ and $p_Y(v)$, and verify that they satisfy Eqs. A.23 and A.24 as well as Eq. A.4.

****

A.6. Let X and Y be two random variables with joint cumulative distribution function

$$\begin{aligned}
F_{XY}(u,v) &= u^3(v-1)^2 &&\text{for}\quad 0\le u\le 1,\ 1\le v\le 2\\
&= 1 &&\text{for}\quad u>1,\ v>2\\
&= u^3 &&\text{for}\quad 0\le u\le 1,\ v>2\\
&= (v-1)^2 &&\text{for}\quad u>1,\ 1\le v\le 2\\
&= 0 &&\text{otherwise}
\end{aligned}$$

(a) Find the joint probability density function $p_{XY}(u,v)$.

(b) Find the marginal cumulative distribution functions $F_X(u)$ and $F_Y(v)$.

(c) Find the marginal probability density functions $p_X(u)$ and $p_Y(v)$, and verify that they satisfy Eqs. A.23 and A.24 as well as Eq. A.4.

****

A.7. Let the joint probability density function of two random variables X and Y be given by

$$\begin{aligned}
p_{XY}(u,v) &= C &&\text{for } (u,v) \text{ inside the shaded area on the sketch}\\
&= 0 &&\text{otherwise}
\end{aligned}$$

(a) Find the value of the constant C.

(b) Find both marginal probability density functions: $p_X(u)$ and $p_Y(v)$.

(c) Find $F_{XY}(0.5, 0.75)$. That is, find the joint cumulative distribution function $F_{XY}(u,v)$ only for arguments of $u = 0.5$, $v = 0.75$.

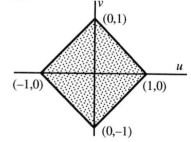

****

A.8 Let X and Y be two random variables with joint probability density function

$$p_{XY}(u,v) = Cve^{-2uv-5v}U(u)U(v)$$

(a) Find the value of the constant C.

(b) Find the marginal probability density function $p_X(u)$.

(c) Find the joint cumulative distribution function $F_{XY}(u,v)$. [Hint: It is easier to do the integration in the X direction first.]

(d) Find the marginal cumulative distribution function $F_X(u)$, and verify that it satisfies Eq. A.21 as well as Eqs. A.4 and A.5.

****

A.9. Let X and Y be two random variables with joint probability density function

$$p_{XY}(u,v) = Ce^{-u-4v}U(u)U(v-u)$$

(a) Find the value of the constant C.
(b) Find the marginal probability density function $p_X(u)$.
(c) Find the joint cumulative distribution function $F_{XY}(u,v)$.
(d) Find the marginal cumulative distribution function $F_X(u)$, and verify that it satisfies Eq. A.21 as well as Eqs. A.4 and A.5.

****

Conditional Distributions

****

A.10. Let X and Y be the two random variables described in Exercise A.5.
(a) Find the conditional cumulative distribution function $F_X(u|A)$ given the event $A = \{0 \le Y \le 2\}$.
(b) Find the conditional probability density function $p_X(u|A)$.

****

A.11. Let X and Y be the two random variables described in Exercise A.6.
(a) Find the conditional cumulative distribution function $F_Y(v|A)$ given the event
   $A = \{0 \le X \le 0.75\}$.
(b) Find the conditional probability density function $p_Y(v|A)$.

****

A.12. Let X and Y be the two random variables described in Exercise A.7.
(a) Find the conditional cumulative distribution function $F_Y(v|A)$ given the event $A = \{Y \le 0.5\}$.
(b) Find the conditional probability density function $p_Y(v|A)$.

****

A.13. Let X and Y be the two random variables described in Exercise A.8.
(a) Find the conditional cumulative distribution function $F_X(u|A)$ given the event $A = \{X \ge 3\}$.
(b) Find the conditional probability density function $p_X(u|A)$.

****

A.14. Let X and Y be the two random variables described in Exercise A.7.
(a) Find the conditional probability density function $p_Y(v|X = u)$.
(b) Find the cumulative distribution function $F_Y(v|X = u)$.

****

A.15. Let X and Y be the two random variables described in Exercise A.8.
(a) Find the conditional probability density functions $p_X(u|Y = v)$.
(b) Find the conditional cumulative distribution functions $F_X(u|Y = v)$.

****

A.16. Let X and Y be the two random variables described in Exercise A.9.
(a) Find the conditional probability density functions $p_Y(v|X = u)$.
(b) Find the conditional cumulative distribution functions $F_Y(v|X = u)$.

****

A.17. Let X and Y be the two random variables described in Exercise A.5.
(a) Find the conditional probability density functions $p_X(u|Y = v)$ and $p_Y(v|X = u)$.

(b) Find the conditional cumulative distribution functions $F_X(u|Y=v)$ and $F_Y(v|X=u)$.
****

A.18. Let X and Y be the two random variables described in Exercise A.6.
(a) Find the conditional probability density functions $p_X(u|Y=v)$ and $p_Y(v|X=u)$.
(b) Find the conditional cumulative distribution functions $F_X(u|Y=v)$ and $F_Y(v|X=u)$.
****

Independence
****

A.19. Are the random variables X and Y of Exercises A.5 and A.17 independent?
****

A.20. Are the random variables X and Y of Exercise A.6 and A.18 independent?
****

A.21. Are the random variables X and Y of Exercise A.7 and A.14 independent?
****

A.22. Are the random variables X and Y of Exercise A.8 and A.15 independent?
****

A.23. Are the random variables X and Y of Exercise A.9 and A.16 independent?
****

Mean and Variance of One Random Variable
****

A.24. Consider a random variable X with probability density function
$$p_X(u) = Cu^4[U(u+1) - U(u-2)]$$
(a) Find the value of the constant C.
(b) Find the mean value μ_X.
(c) Find the mean squared value $E(X^2)$.
(d) Find the variance σ_X^2.
****

A.25. Consider the random variable X of Exercise A.4.
(a) Find the mean value μ_X.
(b) Find the mean squared value $E(X^2)$.
(c) Find the variance σ_X^2.
****

A.26. Consider a random variable X with probability density function
$$p_X(u) = \frac{a}{b+u^6}$$
in which a and b are positive constants. For what positive integer j values does $E(X^j) < \infty$ exist?
****

Moments of Jointly Distributed Random Variables
****

A.27. Let X and Y be the two random variables of Exercise A.7.
(a) Find $E(X)$ and $E(Y)$.

(b) Find $E(X^2)$ and $E(Y^2)$.

(c) Find the variances and covariance σ_X^2, σ_Y^2, and K_{XY}.

(d) Are X and Y correlated?

****

A.28. Let X and Y be the two random variables of Exercise A.6.

(a) Find $E(X)$ and $E(Y)$.

(b) Find $E(X^2)$ and $E(Y^2)$.

(c) Find the variances and covariance σ_X^2, σ_Y^2, and K_{XY}.

(d) Are X and Y correlated?

****

A.29. Let X and Y be the two random variables of Exercise A.9.

(a) Find $E(X)$ and $E(Y)$.

(b) Find $E(X^2)$ and $E(Y^2)$.

(c) Find the variances and covariance σ_X^2, σ_Y^2, and K_{XY}.

(d) Are X and Y correlated?

****

A.30. The joint probability density function of two random variables is given by

$$p_{XY}(u,v) = 1 / \pi \qquad \text{for} \qquad u^2 + v^2 \le 1$$
$$= 0 \qquad \text{otherwise}$$

(a) Are X and Y uncorrelated?

(b) Are X and Y independent?

****

A.31. Let the probability density function of a random variable X be given by

$$p_X(u) = C[U(u) - U(u-5)]$$

and let the conditional probability density function of another random variable Y be given by

$$p_Y(v|X = u) = B(u)[U(v) - U(v-2)] \qquad \text{for} \qquad 0 \le u \le 5$$

(a) Find the value of the constant C and the function $B(u)$.

(b) Find the joint probability density function $p_{XY}(u,v)$ and indicate on a sketch of the (u,v) plane the region on which $p_{XY}(u,v) \ne 0$.

(c) Find $E(X^2)$.

(d) Find $E(XY)$.

(e) Are X and Y independent?

****

A.32. Consider a random variable X with probability density function

$$p_X(u) = ue^{-u}U(u)$$

Let another random variable Y have the conditional probability density function

$$p_Y(v|X = u) = u^{-1}[U(v) - U(v-u)] \qquad \text{for} \qquad u \ge 0$$

(a) Find the joint probability density function $p_{XY}(u,v)$ and indicate on a sketch of the (u,v) plane the region on which $p_{XY}(u,v) \ne 0$.

(b) Find $E(X^2)$.

(c) Find $E(Y)$.

(d) Find $E(X^2 Y)$.

****

A.33. Let the random variables $X(t)$ and $\dot{X}(t)$ denote the displacement and the velocity of response at a certain time t of a dynamic system. For a certain nonlinear oscillator the joint probability density function is $p_{X(t)\dot{X}(t)}(u,v) = C\exp(-\alpha|u^3|-\gamma v^2)$, in which C, α, and γ are positive constants.

(a) Find the probability density function $p_{X(t)}(u)$ for $X(t)$, including evaluation of C in terms of α and/or γ.

(b) Find $E[X^2(t)]$ in terms of α and/or γ.

[Methods developed in Chapter 9 will show that this problem corresponds to the response to Gaussian white noise of an oscillator with linear damping and a quadratic hardening spring.]

****

A.34. Consider a discrete element model of a beam in which the random sequence $\{U_1, U_2, U_3, \cdots\}$ represents normalized average slopes in unit length segments of the beam. The normalized deflections of the beam are then given by

$$X_1 = U_1, \quad X_2 = U_1 + U_2, \quad X_3 = U_1 + U_2 + U_3, \quad \text{etc.}$$

For $j \le 3$, $k \le 3$ it has been found that $E(U_j) = 0$, $E(U_j^2) = 1$, and $E(U_j U_k) = 0.8$ for $j \ne k$.

(a) Find $E(X_1^2)$, $E(X_2^2)$ and their correlation coefficient $\rho_{X_1 X_2}$.

(b) Let $W_1 = U_2 - U_1$ and $W_2 = U_3 - U_2$ (proportional to bending moments in the beam). Find $E(W_1^2)$, $E(W_2^2)$ and their correlation coefficient $\rho_{W_1 W_2}$.

(c) Let $V_1 = W_2 - W_1 = U_3 - 2U_2 + U_1$ (proportional to shear in the beam). Find $E(V_1^2)$.

****

Moments and Conditional Expectations

****

A.35. Consider two random variables X and Y with the joint probability density function

$$p_{XY}(u,v) = e^{-u}[U(v) - U(v-u)]U(u)$$

(a) Find both marginal probability density functions: $p_X(u)$ and $p_Y(v)$.

(b) Find the conditional probability density functions: $p_X(u|Y=v)$ and $p_Y(v|X=u)$.

(c) Find $E(Y|X=u)$.

****

A.36. Consider a random variable X with probability density function
$$p_X(u) = U(u) - U(u-1)$$

Let another random variable Y have the conditional probability density function
$$p_Y(v|X=u) = U(v-u) - U(v-u-1) \quad \text{for} \quad 0 \le u \le 1$$

(a) Find the joint probability density function $p_{XY}(u,v)$ and indicate on a sketch of the (u,v) plane the region on which $p_{XY}(u,v) \ne 0$.

(b) Find $p_Y(v)$ and sketch this probability density function versus v.

(c) Find the conditional expected value $E[Y|X=u]$ for $0 \le u \le 1$.

****

A.37. Consider a random variable X with probability density function
$$p_X(u) = 2e^{-2u}U(u)$$
Let another random variable Y have the conditional probability density function
$$p_Y(v|X=u) = e^u[U(v)-U(v-e^{-u})] \quad \text{for} \quad u \geq 0$$
(a) Find the joint probability density function $p_{XY}(u,v)$ and indicate on a sketch of the (u,v) plane the region on which $p_{XY}(u,v) \neq 0$.
(b) Find $p_Y(v)$ and sketch this probability density function versus v.
(c) Find the conditional expected value $E[Y|X=u]$ for $u \geq 0$.
(d) Find $E(Y)$.
****

A.38. Consider a random variable X with probability density function
$$p_X(u) = \frac{3}{4}u(2-u)[U(u)-U(u-2)]$$
Let another random variable Y have the conditional probability density function
$$p_Y(v|X=u) = u^{-1}[U(v)-U(v-u)] \quad \text{for} \quad 0 \leq u \leq 2$$
(a) Find the joint probability density function $p_{XY}(u,v)$ and indicate on a sketch of the (u,v) plane the region where $p_{XY}(u,v) \neq 0$.
(b) Find $p_Y(v)$ and sketch this function versus v.
(c) Find $E(Y|X=u)$ for $0 \leq u \leq 2$.
(d) Find $E(X^2Y)$.
****

A.39. Consider a random variable X with probability density function
$$p_X(u) = 3u^2[U(u)-U(u-1)]$$
Let another random variable Y have the conditional probability density function
$$p_Y(v|X=u) = [U(v)-U(v-u)]/u \quad \text{for} \quad 0 \leq u \leq 1$$
(a) Find the joint probability density function $p_{XY}(u,v)$ and indicate on a sketch of the (u,v) plane the region where $p_{XY}(u,v) \neq 0$.
(b) Find $p_Y(v)$ and sketch this function versus v.
(c) Find $E[Y|X=u]$ for $0 \leq u \leq 1$.
(d) Find $E[Y]$.
****

A.40. Consider a random variable X with probability density function
$$p_X(u) = 4u^3[U(u)-U(u-1)]$$
Let another random variable Y have the conditional probability density function
$$p_Y(v|X=u) = [U(v)-U(v-u^2)]/u^2 \quad \text{for} \quad 0 \leq u \leq 1$$
(a) Find the joint probability density function $p_{XY}(u,v)$ and indicate on a sketch of the (u,v) plane the region where $p_{XY}(u,v) \neq 0$.
(b) Find $p_Y(v)$ and sketch this function versus v.
(c) Find $E[Y|X=u]$ for $0 \leq u \leq 1$.
(d) Find $E[Y]$.
****

A.41. Consider a random variable X with probability density function:

$$p_X(u) = \frac{\pi}{4} \cos\left(\frac{\pi u}{2}\right) U(1-|u|)$$

Let another random variable Y have the conditional probability density function

$$p_Y(v|X=u) = \left[\cos\left(\frac{\pi u}{2}\right)\right]^{-1}\left(U(v) - U\left[v - \cos\left(\frac{\pi u}{2}\right)\right]\right) \quad \text{for} \quad |u| \leq 1$$

(a) Find the joint probability density function $p_{XY}(u,v)$, and indicate on a sketch of the (u,v) plane the region where $p_{XY}(u,v) \neq 0$.
(b) Find $E[Y|X=u]$ for $|u| \leq 1$.
(c) Find $E[Y]$.
(d) Find $p_Y(v)$ and sketch this probability density function versus v.
****

General Conditional Expectation
****

A.42. Let X be uniform on [-1,1]: $p_X(u) = 1/2$ for $|u| \leq 1$.
Let $E(Y|X=u) = u$, $E(Y^2|X=u) = 2u^2$ and $Z = XY$.
Find the mean and variance of Z.
[Hint: Begin with $E(Z|X=u)$ and $E(Z^2|X=u)$ then find the unconditional expectations.]
****

A.43. Let X be a random variable which is uniform on [-1,1]. That is, $p_X(u) = 1/2$ for $|u| \leq 1$.
Let $E(Y|X=u) = 2u^2$ and $E(Y^2|X=u) = 4u^4 + 2u^2$. Let $Z = XY$.
Find the mean and variance of Z.
****

Characteristic Function
****

A.44. Let the random variable X have a distribution which is uniform on $[c - h/2, c + h/2]$:
$$p_X(u) = [U(u-c+h/2) - U(u-c-h/2)]/h$$
(a) Determine the mean and variance of X from integration of $p_X(u)$.
(b) Find the characteristic function of X.
(c) Check the mean and variance of X by using derivatives of the characteristic function.
****

A.45. Let the random variable X have a discrete distribution with
$$p_X(u) = a\delta(u-1) + \frac{a}{4}\delta(u-2) + \frac{a}{9}\delta(u-3)$$
That is, $P(X = k) = a/k^2$ for $k = 1, 2,$ and 3.
(a) Determine the mean and variance of X by using $p_X(u)$.
(b) Find the characteristic function of X.
(c) Check the mean and variance of X by using derivatives of the characteristic function.
****

Appendix B

Gaussian Random Variables

B.1 Gaussian Distribution

The Gaussian distribution is emphasized here because it plays a key role in most areas of applied probability, including the analysis of stochastic dynamics. In the simplest sense, the term Gaussian simply refers to particular forms of the probability density function, the cumulative distribution function, and the characteristic function. It turns out, though, that random variables or stochastic processes having this distribution have a number of special properties which expedite numerical analysis. Furthermore, it seems that much of nature is approximately Gaussian, so that these models often give very acceptable estimates of important quantities. In addition, the Gaussian models form the basis for various procedures for studying non-Gaussian problems.

We will begin with the simplest Gaussian situation by letting W be a random variable with the probability density function

$$p_W(w) = \frac{e^{-w^2/2}}{\sqrt{2\pi}} \tag{B.1}$$

Of course, a necessary condition for this actually to be a probability density function is that

$$\int_{-\infty}^{\infty} e^{-w^2/2}\, dw = \sqrt{2\pi} \tag{B.2}$$

This very important integral can be found in mathematical reference books, or it can be derived by making a change of variables in order to use the gamma function (defined in Example A.31 of Appendix A) and the fact that $\Gamma(1/2) = \pi^{1/2}$.

Let us now investigate the characteristic function for W. From the fact that $M_W(\theta) \equiv E(e^{i\theta W})$ is the Fourier transform of the probability density function (see Eq. A.75) we have

$$M_W(\theta) = \frac{1}{\sqrt{2\pi}}\int_{-\infty}^{\infty} e^{-i\theta w} e^{-w^2/2}\, dw = \frac{1}{\sqrt{2\pi}}\int_{-\infty}^{\infty} \exp\left[-\frac{(w+i\theta)^2}{2}\right] e^{-\theta^2/2}\, dw = e^{-\theta^2/2}\int_{C_1} \frac{e^{-v^2/2}}{\sqrt{2\pi}}\, dv$$

The contour C_1 of the final integral is from $-\infty + i\theta$ to $+\infty + i\theta$ along a line parallel to the real axis in the space of complex v values. From the theory of complex variables one can show that the integral is the same as if it were along the real axis, though, and Eq. B.2 then gives[1]

---

[1] One could also note that we must have $M_W(0) = 1$ (see Eq. A.75) and use this as an alternative proof that the scaling of Eq. B.1 is correct.

$$M_W(\theta) = e^{-\theta^2/2} \tag{B.3}$$

Note the remarkable similarity between the probability density function in Eq. B.1 and the characteristic function in Eq. B.3. The function and its Fourier transform differ only by a constant. No other function has this property.

We can now use the characteristic function to calculate moments of W (see Eq. A.77). For the first two, we find $E(W) = 0$ and $E(W^2) = 1$. Thus, the mean and variance of W are zero and unity, respectively. Equation B.1 is sometimes called the Gaussian $(0,1)$ distribution to indicate these mean and variance values. It is extremely easy to use Eq. A.83 to find the cumulants of a Gaussian random variable. In fact, the log-characteristic function for W is simply $(-\theta^2/2)$, and this gives $\kappa_2(W) \equiv \sigma_W^2 = 1$ as the only nonzero cumulant of W.

The cumulative distribution function $P(W \le w)$ for the Gaussian $(0,1)$ random variable W is commonly denoted as $\Phi(w)$. Of course it can be written as an integral of the probability density function:

$$\Phi(w) = \frac{1}{\sqrt{2\pi}} \int_{-\infty}^{w} e^{-u^2/2}\, du = \frac{1 + \mathrm{erf}(w/\sqrt{2})}{2} \tag{B.4}$$

in which the error function defined by

$$\mathrm{erf}(u) = \frac{2}{\sqrt{\pi}} \int_0^u e^{-v^2}\, dv$$

is essentially a different form of the $\Phi(\cdot)$ function. A short table of the $\Phi(\cdot)$ function is given in Table B.1. More extensive tables of $\Phi(\cdot)$ are available in some probability books, and both it and the erf($\cdot$) function are tabulated in mathematical handbooks (for example, Abramowitz and Stegun, 1965). Many types of computer software to evaluate the functions are also readily available.

Note that the mean and variance values of W show that it is of the form of a so-called standardized random variable, having zero mean and unit variance. We can obtain a more general Gaussian random variable X by taking a linear function of W such that W is the standardized version of X. That is, if we take

$$X = \mu_X + \sigma_X W, \qquad W = \frac{X - \mu_X}{\sigma_X}$$

then we say that X is also a Gaussian random variable. One can relate the distribution of X to that of W by starting with the relationship between their cumulative distribution functions: $F_X(u) = F_W[(u - \mu_X)/\sigma_X]$. Taking the derivative of this equation with respect to u then gives the probability density relationship as

$$p_X(u) = \frac{1}{\sqrt{2\pi}\,\sigma_X}\exp\left[-\frac{1}{2}\left(\frac{u-\mu_X}{\sigma_X}\right)^2\right] \tag{B.5}$$

This is the most general form for the probability density function of a Gaussian random variable.

Similarly, we can use the relationship between X and W along with the definition of the characteristic function to write

$$M_X(\theta) = E\left(e^{-i\theta X}\right) = e^{-i\theta\mu_X}E\left(e^{-i\theta\sigma_X W}\right) = e^{-i\theta\mu_X}M_W(\sigma_X\theta)$$

which gives

$$M_X(\theta) = \exp\left(-i\theta\mu_X - \frac{\sigma_X^2\theta^2}{2}\right) \tag{B.6}$$

as the general form of the characteristic function of a Gaussian random variable. Note that the log-characteristic function for X is quadratic in θ, as it was for W. For X, though, the log-characteristic function contains a term which is linear in θ as well as the term involving θ^2, giving both the first two cumulants of X as being nonzero. This is as it must be, of course, since the first cumulant is the same as the mean for any random variable. Equation B.6 confirms the general relationships (see Eqs. A.84 and A.85) that the first two cumulants are the mean and the variance for the Gaussian random variable X. Furthermore, it shows that the cumulants higher than the second are exactly zero for this general Gaussian random variable. In fact, this property of the third- and higher-order cumulants being zero could be taken as the definition of a Gaussian random variable, since it is only true for a characteristic function of the form of Eq. B.6, and this must give a probability density function of the form of Eq. B.5.

Another alternative definition of the Gaussian distribution is that a random variable is Gaussian if the natural logarithm of its probability density function is quadratic. Thus, we might have a Gaussian random variable X with $p_X(u) = \exp(au^2 + bu + c)$. The fact that $p_X(u)$ must have a unit integral imposes the limitation that $a < 0$, as well as a relationship between the three constants a, b, and c. One can use this relationship along with the expressions relating μ_X and σ_X to a, b, and c to solve for a, b, and c, and the result is exactly Eq. B.5.

The central moments of a Gaussian random variable are derived in Example A.32, and the non-central moments can be derived by using the binomial expansion of Eq. A.60. Also, the cumulative distribution function can be shown to be

$$F_X(u) \equiv P(X \le u) = \Phi\left(\frac{u-\mu_X}{\sigma_X}\right) \tag{B.7}$$

Table B.1 Gaussian Cumulative Distribution Function

$$\Phi(u) = \frac{1}{\sqrt{2\pi}} \int_{-\infty}^{u} e^{-v^2/2}\, dv = \frac{1 + \mathrm{erf}(u/\sqrt{2})}{2}, \qquad \Phi(-u) = 1 - \Phi(u)$$

u	$\Phi(u)$	u	$\Phi(u)$	u	$\Phi(u)$	u	$\Phi(u)$
0.00	0.5000000	1.00	0.8413447	2.00	0.9772499	3.00	0.9986501
0.02	0.5079783	1.02	0.8461358	2.02	0.9783083	3.02	0.9987361
0.04	0.5159534	1.04	0.8508300	2.04	0.9793248	3.04	0.9988171
0.06	0.5239222	1.06	0.8554277	2.06	0.9803007	3.06	0.9988933
0.08	0.5318814	1.08	0.8599289	2.08	0.9812372	3.08	0.9989650
0.10	0.5398278	1.10	0.8643339	2.10	0.9821356	3.10	0.9990324
0.12	0.5477584	1.12	0.8686431	2.12	0.9829970	3.12	0.9990957
0.14	0.5556700	1.14	0.8728568	2.14	0.9838226	3.14	0.9991553
0.16	0.5635595	1.16	0.8769756	2.16	0.9846137	3.16	0.9992112
0.18	0.5714237	1.18	0.8809999	2.18	0.9853713	3.18	0.9992636
0.20	0.5792597	1.20	0.8849303	2.20	0.9860966	3.20	0.9993129
0.22	0.5870644	1.22	0.8887676	2.22	0.9867906	3.22	0.9993590
0.24	0.5948349	1.24	0.8925123	2.24	0.9874545	3.24	0.9994024
0.26	0.6025681	1.26	0.8961653	2.26	0.9880894	3.26	0.9994429
0.28	0.6102612	1.28	0.8997274	2.28	0.9886962	3.28	0.9994810
0.30	0.6179114	1.30	0.9031995	2.30	0.9892759	3.30	0.9995166
0.32	0.6255158	1.32	0.9065825	2.32	0.9898296	3.32	0.9995499
0.34	0.6330717	1.34	0.9098773	2.34	0.9903581	3.34	0.9995811
0.36	0.6405764	1.36	0.9130850	2.36	0.9908625	3.36	0.9996103
0.38	0.6480273	1.38	0.9162067	2.38	0.9913437	3.38	0.9996376
0.40	0.6554217	1.40	0.9192433	2.40	0.9918025	3.40	0.9996631
0.42	0.6627573	1.42	0.9221962	2.42	0.9922397	3.42	0.9996869
0.44	0.6700314	1.44	0.9250663	2.44	0.9926564	3.44	0.9997091
0.46	0.6772419	1.46	0.9278550	2.46	0.9930531	3.46	0.9997299
0.48	0.6843863	1.48	0.9305634	2.48	0.9934309	3.48	0.9997493
0.50	0.6914625	1.50	0.9331928	2.50	0.9937903	3.50	0.9997674
0.52	0.6984682	1.52	0.9357445	2.52	0.9941323	3.52	0.9997842
0.54	0.7054015	1.54	0.9382198	2.54	0.9944574	3.54	0.9997999
0.56	0.7122603	1.56	0.9406201	2.56	0.9947664	3.56	0.9998146
0.58	0.7190427	1.58	0.9429466	2.58	0.9950600	3.58	0.9998282
0.60	0.7257469	1.60	0.9452007	2.60	0.9953388	3.60	0.9998409
0.62	0.7323711	1.62	0.9473839	2.62	0.9956035	3.62	0.9998527
0.64	0.7389137	1.64	0.9494974	2.64	0.9958547	3.64	0.9998637
0.66	0.7453731	1.66	0.9515428	2.66	0.9960930	3.66	0.9998739
0.68	0.7517478	1.68	0.9535213	2.68	0.9963189	3.68	0.9998834
0.70	0.7580363	1.70	0.9554345	2.70	0.9965330	3.70	0.9998922
0.72	0.7642375	1.72	0.9572838	2.72	0.9967359	3.72	0.9999004
0.74	0.7703500	1.74	0.9590705	2.74	0.9969280	3.74	0.9999080
0.76	0.7763727	1.76	0.9607961	2.76	0.9971099	3.76	0.9999150
0.78	0.7823046	1.78	0.9624620	2.78	0.9972821	3.78	0.9999216
0.80	0.7881446	1.80	0.9640697	2.80	0.9974449	3.80	0.9999277
0.82	0.7938919	1.82	0.9656205	2.82	0.9975988	3.82	0.9999333
0.84	0.7995458	1.84	0.9671159	2.84	0.9977443	3.84	0.9999385
0.86	0.8051055	1.86	0.9685572	2.86	0.9978818	3.86	0.9999433
0.88	0.8105703	1.88	0.9699460	2.88	0.9980116	3.88	0.9999478
0.90	0.8159399	1.90	0.9712834	2.90	0.9981342	3.90	0.9999519
0.92	0.8212136	1.92	0.9725711	2.92	0.9982498	3.92	0.9999557
0.94	0.8263912	1.94	0.9738102	2.94	0.9983589	3.94	0.9999593
0.96	0.8314724	1.96	0.9750021	2.96	0.9984618	3.96	0.9999625
0.98	0.8364569	1.98	0.9761482	2.98	0.9985588	3.98	0.9999655
1.00	0.8413447	2.00	0.9772499	3.00	0.9986501	4.00	0.9999683

so that the $\Phi(\cdot)$ or the erf($\cdot$) function can be used to calculate the probabilities of most events in Gaussian problems.

****

Example B.1: Let $V(t)$ denote the wind velocity at time t at a given location. Given that $V(t)$ is Gaussian with $E[V(t)] = 10\,\mathrm{m/s}$ and mean squared value of $E[V^2(t)] = 125\,\mathrm{m^2/s^2}$, find the probability that $V(t)$ will exceed $25\,\mathrm{m/s}$.

Since the problem is Gaussian, we only need find the mean and variance as $\mu_{V(t)} = 10\,\mathrm{m/s}$ and $\sigma_{V(t)} = (125 - 10^2)^{1/2} = 5$ m/s and use Eq. B.7 to obtain

$$P[V(t) > 25] = 1 - P[V(t) \le 25] = 1 - \Phi[(25-10)/5] = 1 - \Phi(3) \approx 0.00135$$

****

B.2 Jointly Gaussian Random Variables

Next we wish to study the use of the Gaussian distribution for jointly distributed random variables. We will start with the simplest situation. Let $\vec{Z}$ be a vector random variable with n components, $\vec{Z} = (Z_1, Z_2, \cdots, Z_n)^T$, and let the Z_j components be independent and identically Gaussian with zero means and unit variances. That is, the probability density function of each Z_j is the same as Eq. B.1 and the characteristic function is the same as Eq. B.3. The fact that the Z_j components are independent allows us to write their joint probability density and characteristic functions as the products of the marginal functions, giving

$$p_{\vec{Z}}(\vec{z}) = \frac{1}{(2\pi)^{n/2}} \exp\left[-\frac{1}{2}\left(z_1^2 + z_2^2 + \cdots + z_n^2 \right) \right]$$

and

$$M_{\vec{Z}}(\vec{\theta}) = \exp\left[-\frac{1}{2}\left(\theta_1^2 + \theta_2^2 + \cdots + \theta_n^2 \right) \right]$$

It is more convenient, though, to use matrix algebra to rewrite the exponents in these expressions so that we have

$$p_{\vec{Z}}(\vec{z}) = \frac{1}{(2\pi)^{n/2}} \exp\left[-\frac{\vec{z}^T \vec{z}}{2} \right] \tag{B.8}$$

and

$$M_{\vec{Z}}(\vec{\theta}) = \exp\left[-\frac{\vec{\theta}^T \vec{\theta}}{2} \right] \tag{B.9}$$

in which the superscript ^{T} denotes the transpose of the vector.

Next consider the covariance matrix for this distribution. The independence of the Z_j

components tells us that the off-diagonal terms of the $\mathbf{K}_{ZZ} = E(\vec{Z}\vec{Z}^T)$ covariance matrix (see Eq. A.68) are all zero, and the unit variances, tells us that the diagonal elements are all unity. Thus, $\mathbf{K}_{ZZ}$ is the n-dimensional identity matrix:

$$\mathbf{K}_{ZZ} = \mathbf{I}_n = \begin{bmatrix} 1 & 0 & \cdots & 0 \\ 0 & 1 & \ddots & \vdots \\ \vdots & \ddots & \ddots & 0 \\ 0 & \cdots & 0 & 1 \end{bmatrix}$$

Now we will define a new random vector $\vec{X}$ as

$$\vec{X} = \mathbf{A}\vec{Z} + \vec{\mu}_X \tag{B.10}$$

so that each component of $\vec{X}$ is a linear combination of the components of $\vec{Z}$:

$$X_j = \sum_{k=1}^n A_{jk} Z_k + \mu_{X_j}$$

It is clear that $\vec{\mu}_X$ is the mean vector for $\vec{X}$ and the covariance matrix can be written as

$$\mathbf{K}_{XX} = E\left[(\vec{X} - \vec{\mu}_X)(\vec{X} - \vec{\mu}_X)^T\right] = \mathbf{A}\mathbf{K}_{ZZ}\mathbf{A}^T = \mathbf{A}\mathbf{A}^T \tag{B.11}$$

since $\mathbf{K}_{ZZ}$ is the identity matrix.

We can easily write the characteristic function for $\vec{X}$ as

$$M_X(\vec{\theta}) = E[\exp(-i\vec{\theta}^T\vec{X})] = E[\exp(-i\vec{\theta}^T\mathbf{A}\vec{Z} - i\vec{\theta}^T\vec{\mu}_X)] = \exp(-i\vec{\theta}^T\vec{\mu}_X)M_Z(\mathbf{A}^T\theta)$$

and Eq. B.9 now gives

$$M_X(\vec{\theta}) = \exp(-i\vec{\theta}^T\vec{\mu}_X - \frac{1}{2}\vec{\theta}^T\mathbf{A}\mathbf{A}^T\vec{\theta})$$

which can be rewritten according to Eq. B.11 as

$$M_X(\vec{\theta}) = \exp(-i\vec{\theta}^T\vec{\mu}_X - \frac{1}{2}\vec{\theta}^T\mathbf{K}_{XX}\vec{\theta}) \tag{B.12}$$

Equation B.12 is the general characteristic function for **jointly Gaussian** random variables. In fact, Eq. B.12 can be considered to be the definition of the jointly Gaussian distribution of random variables.

Finding the joint probability density function for a set of linear combinations of random variables (like $\vec{X}$) is not quite as easy as finding the characteristic function. The procedure is basically the same as was used in deriving Eq. B.5, and it is helpful if we limit attention to the special case with $\mathbf{A}$ being square and nonsingular. This means that the inverse $\mathbf{A}^{-1}$ exists and we can invert Eq. B.10 to give

$$\vec{Z} = \mathbf{A}^{-1}(\vec{X} - \vec{\mu}_X) \tag{B.13}$$

The probability density function can then be shown to be (see Eq. A.35)

$$p_{\vec{X}}(\vec{u}) = \frac{1}{\|\mathbf{A}\|} p_{\vec{Z}}\left[\mathbf{A}^{-1}(\vec{u} - \vec{\mu}_X)\right]$$

in which the double bars on $\mathbf{A}$ in the denominator denote the absolute value of the determinant of the matrix. Using Eq. B.8, we can write this more explicitly as

$$p_{\vec{X}}(\vec{u}) = \frac{1}{(2\pi)^{n/2}\|\mathbf{A}\|} \exp\left[-\frac{1}{2}(\vec{u} - \vec{\mu}_X)^T \mathbf{A}^{T^{-1}} \mathbf{A}^{-1}(\vec{u} - \vec{\mu}_X)\right]$$

Next we put the formula into standard form by eliminating the matrix $\mathbf{A}$ by using the covariance matrix $\mathbf{K}_{XX}$ instead. In particular, Eq. B.11 gives $\mathbf{K}_{XX}^{-1} = \mathbf{A}^{T^{-1}} \mathbf{A}^{-1}$ and $|\mathbf{K}_{XX}| = |\mathbf{A}|^2$. Thus,

$$p_{\vec{X}}(\vec{u}) = \frac{1}{(2\pi)^{n/2}\sqrt{|\mathbf{K}_{XX}|}} \exp\left[-\frac{1}{2}(\vec{u} - \vec{\mu}_X)^T \mathbf{K}_{XX}^{-1}(\vec{u} - \vec{\mu}_X)\right] \tag{B.14}$$

which is the standard form of the jointly Gaussian probability density function. Equation B.14, or some special case of it, is sometimes used as a definition of the jointly Gaussian relationship, but it is not quite as general as Eq. B.12. In particular, Eq. B.14 can be written only if $\mathbf{K}_{XX}$ is not singular, whereas Eq. B.12 does not have that restriction. Henceforth, we will slightly simplify our terminology. Whenever we say that a vector $\vec{X}$ is Gaussian, we will mean that its components are jointly Gaussian.

We previously noted that a random variable X is Gaussian if and only if its probability density function and its characteristic function are each the exponential of a quadratic function of their arguments. We can now extend this comment to a random vector $\vec{X}$. The matrix product in the exponent of Eq. B.14 includes only first- and second-order terms in the u_j arguments, and the one in the exponent in Eq. B.11 includes only first- and second-order terms in the θ_j arguments. Furthermore, in any problem, if the logarithm of $p_{\vec{X}}(\vec{u})$ is a quadratic function of the u_j arguments, then $\vec{X}$ is Gaussian. Similarly, if the logarithm of $M_{\vec{X}}(\vec{\theta})$ is a quadratic function of the θ_j arguments, then $\vec{X}$ is Gaussian.

••

Example B.2: Explicitly write out the probability density function and the characteristic function for two jointly Gaussian random variables X and Y.

Using the notation $\rho \equiv \rho_{XY}$ for the correlation coefficient, we can write the covariance matrix as

$$\mathbf{K}_{XY} = \begin{bmatrix} \sigma_X^2 & \rho\sigma_X\sigma_Y \\ \rho\sigma_X\sigma_Y & \sigma_Y^2 \end{bmatrix}$$

for which the determinant is $|\mathbf{K}_{XY}| = \sigma_X^2\,\sigma_Y^2(1-\rho^2)$ and the inverse is

$$\mathbf{K}_{XY}^{-1} = \frac{1}{\sigma_X^2\sigma_Y^2(1-\rho^2)}\begin{bmatrix} \sigma_Y^2 & -\rho\sigma_X\sigma_Y \\ -\rho\sigma_X\sigma_Y & \sigma_X^2 \end{bmatrix} = \frac{1}{(1-\rho^2)}\begin{bmatrix} \sigma_X^{-2} & -\rho\sigma_X^{-1}\sigma_Y^{-1} \\ -\rho\sigma_X^{-1}\sigma_Y^{-1} & \sigma_Y^{-2} \end{bmatrix}$$

Substituting these expressions into Eq. B.14 gives

$$p_{XY}(u,v) = \frac{1}{2\pi\,\sigma_X\sigma_Y\sqrt{1-\rho^2}}\exp\left\{\frac{-1}{2(1-\rho^2)}\left[\left(\frac{u-\mu_X}{\sigma_X}\right)^2 - 2\rho\left(\frac{u-\mu_X}{\sigma_X}\right)\left(\frac{v-\mu_Y}{\sigma_Y}\right) + \left(\frac{v-\mu_Y}{\sigma_Y}\right)^2\right]\right\}$$

This is the form of the probability density for any two random variables which are jointly Gaussian.

Finding the joint characteristic function from Eq. B.12 is even easier, since it does not require the inverse of the covariance matrix. The result is

$$M_{XY}(\theta,\phi) = \exp\left[-i\mu_X\theta - i\mu_Y\phi - \frac{\sigma_X^2\theta^2 + 2\rho\sigma_X\sigma_Y\theta\phi - \sigma_Y^2\phi^2}{2}\right]$$

****

Example B.3: Consider two random variables X_1 and X_2 with

$$p_{X_1X_2}(u,v) = A\exp\left\{-\frac{8}{3}\left[13 + 2u + u^2 - 14v - 2uv + 4v^2\right]\right\}$$

Find A, the means and variances of both X_1 and X_2, and their correlation coefficient ρ_X.

There are various ways that we could solve this problem. For example, the most straightforward approach would be to write out expressions for $E(X_1)$, $E(X_2)$, $E(X_1^2)$, $E(X_2^2)$, and $E(XY)$ involving integrals of $p_{X_1X_2}(u,v)$. From these expected values we could then find the means and variances of X_1 and X_2, and their correlation coefficient ρ_X. We could also find A from an integral of $p_{X_1X_2}(u,v)$. We will illustrate a method that is somewhat easier, though. We note that X_1 and X_2 are jointly Gaussian, since the probability density function is an exponential of a quadratic function of u and v. Thus, we will evaluate the unknowns by comparing terms in $p_{X_1X_2}(u,v)$ with those in the general jointly Gaussian probability density function given in Example B.2. First, note that the coefficient of u^2 in the exponent of $p_{X_1X_2}(u,v)$ is $-8/3$, but

Example B.2 tells us that this coefficient must be $[-2(1-\rho_X^2)\sigma_{X_1}^2]^{-1}$. Thus, we have

$$(1-\rho_X^2)\sigma_{X_1}^2 = 3/16$$

Next the coefficients of v^2 of $-32/3$ and $[-2(1-\rho_X^2)\sigma_{X_2}^2]^{-1}$ in the exponents of $p_{X_1 X_2}(u,v)$ and Example B.2, respectively, give

$$(1-\rho_X^2)\sigma_{X_2}^2 = 3/64$$

To avoid directly solving nonlinear equations, we eliminate $(1-\rho_X^2)$ by dividing this equation by the previous one, giving $\sigma_{X_1} = 2\sigma_{X_2}$. Now we note that the coefficients of uv in the two exponents are $16/3$ and $\rho_X[(1-\rho_X^2)\sigma_{X_1}\sigma_{X_2}]^{-1}$, so that

$$(1-\rho_X^2)\sigma_{X_1}\sigma_{X_2}/\rho_X = 3/16$$

Dividing this by the preceding equation gives $\sigma_{X_1} = 4\rho_X\sigma_{X_2}$. Along with $\sigma_{X_1} = 2\sigma_{X_2}$ this gives $\rho_X = 0.5$, and the original equations now give $\sigma_{X_1} = 0.5$ and $\sigma_{X_2} = 0.25$. Thus, consideration of the coefficients of u^2, v^2, and uv has given us the values of the second moment terms for X_1 and X_2. To find the mean values (first moments), we now consider the coefficients of u and v in the exponents of $p_{X_1 X_2}(u,v)$ and Example B.2. From u we get

$$-\frac{16}{3} = \frac{-1}{2(1-\rho_X^2)}\left(\frac{-2\mu_{X_1}}{\sigma_{X_1}^2} + \frac{2\rho_X\mu_{X_2}}{\sigma_{X_1}\sigma_{X_2}}\right)$$

and v gives

$$\frac{112}{3} = \frac{-1}{2(1-\rho_X^2)}\left(\frac{-2\mu_{X_2}}{\sigma_{X_2}^2} + \frac{2\rho_X\mu_{X_1}}{\sigma_{X_1}\sigma_{X_2}}\right)$$

Using our known values of σ_{X_1}, σ_{X_2}, and ρ_X gives $\mu_{X_2} - \mu_{X_1} = 1$ and $4\mu_{X_2} - \mu_{X_1} = 7$. Thus, we only need to solve simultaneous linear equations to find that $\mu_{X_1} = 1$ and $\mu_{X_2} = 2$.
****

It is important to realize that the procedure used to derive Eqs. B.12 and B.14 from Eqs. B.5 and B.6 can also be reversed. That is, if a random vector $\vec{X}$ is known to be described by Eq. B.12 or B.14, then one can find a matrix $\mathbf{A}$ such that Eq. B.10 applies with the components of $\vec{Z}$ being independent and Gaussian and having zero means and unit variances.

One of the important properties of the jointly Gaussian distribution is that it implies that each component is Gaussian. Furthermore, if $\vec{X}$ is Gaussian, then any subset of the components of $\vec{X}$ is also a set of jointly Gaussian random variables, so that each X_j is Gaussian, each pair (X_j, X_k) is jointly Gaussian, etc. This is easily demonstrated from the form of the characteristic function given in Eq. B.12, since marginal characteristic functions are obtained simply by setting some components of $\vec{\theta}$ equal to zero. Another important property relates to linear combinations of jointly Gaussian random variables. Given that $\vec{X}$ is Gaussian, we can define a new set of random variables as the components of $\vec{W} = \mathbf{B}\vec{X} + \vec{b}$, with $\mathbf{B}$ being any rectangular matrix with n columns, and we can be sure that $\vec{W}$ is also Gaussian. In addition, the components of $\vec{W}$ and the components of $\vec{X}$ are jointly Gaussian. The proof of these statements follows directly from the development leading to Eq. B.12. In particular, in going from Eq. B.10 to B.12 we showed that any set of random variables which are linear combinations of given independent Gaussian random variables constitute a Gaussian vector. Since a Gaussian vector $\vec{Z}$ with independent components exists such that $\vec{X} = \mathbf{A}\vec{Z} + \vec{\mu}_X$ for any Gaussian vector $\vec{X}$, we can rewrite any vector given by $\vec{W} = \mathbf{B}\vec{X} + \vec{b}$ as $\vec{W} = \mathbf{B}\mathbf{A}\vec{Z} + \mathbf{B}\vec{\mu}_X + \vec{b}$. This shows that the components of $\vec{W}$, as well as the components of $\vec{X}$, are linear combinations of the set of independent Gaussian random variables included in $\vec{Z}$.

****

Example B.4: Consider the two random variables of Example B.3, with the joint probability density function

$$p_{X_1 X_2}(u,v) = A \exp\left\{ -\frac{8}{3}\left[13 + 2u + u^2 - 14v - 2uv + 4v^2 \right] \right\}$$

Let $Y_1 = X_1 + X_2$ and $Y_2 = X_1 + 3X_2$. Find the joint probability density function $p_{Y_1 Y_2}(u,v)$.

Since X_1 and X_2 are jointly Gaussian, we know that Y_1 and Y_2 are also jointly Gaussian. Thus, we can write out $p_{Y_1 Y_2}(u,v)$ if we find the means and variances of Y_1 and Y_2 and their correlation coefficient ρ_Y. We have already found, in the previous example, that $\mu_{X_1} = 1$, $\mu_{X_2} = 2$, $\sigma_{X_1} = 0.5$, $\sigma_{X_2} = 0.25$, and $\rho_X = 0.5$. Thus, we can take expectations of the given relationships between (Y_1, Y_2) and (X_1, X_2) to obtain

$$\mu_{Y_1} = \mu_{X_1} + \mu_{X_2} = 3$$

and

$$\mu_{Y_2} = \mu_{X_1} + 3\mu_{X_2} = 7$$

Similarly,

$$E(Y_1^2) = E(X_1^2) + 2E(X_1 X_2) + E(X_2^2)$$

so that

$$E(Y_1^2) = \mu_{X_1}^2 + \sigma_{X_1}^2 + 2(\mu_{X_1}\mu_{X_2} + \rho_X \sigma_{X_1}\sigma_{X_2}) + \mu_{X_2}^2 + \sigma_{X_2}^2 = 9.4375$$

and

$$E(Y_2^2) = \mu_{X_1}^2 + \sigma_{X_1}^2 + 6(\mu_{X_1}\mu_{X_2} + \rho_X \sigma_{X_1}\sigma_{X_2}) + 9(\mu_{X_2}^2 + \sigma_{X_2}^2) = 50.1875$$

Now $\sigma_{Y_1}^2 = E(Y_1^2) - \mu_{Y_1}^2 = 0.4375$ and $\sigma_{Y_2}^2 = E(Y_2^2) - \mu_{Y_2}^2 = 1.1875$. Similarly,

$$E(Y_1 Y_2) = \mu_{X_1}^2 + \sigma_{X_1}^2 + 4(\mu_{X_1}\mu_{X_2} + \rho_X \sigma_{X_1}\sigma_{X_2}) + 3(\mu_{X_2}^2 + \sigma_{X_2}^2) = 21.6875$$

so that $\rho_Y \sigma_{Y_1}\sigma_{Y_2} = E(Y_1 Y_2) - \mu_{Y_1}\mu_{Y_2} = 0.6875$, and $\rho_Y = 0.9538$.

Substituting these values into the general form gives

$$p_{Y_1 Y_2}(u,v) = \frac{\exp\left\{\dfrac{-1}{2(0.09023)}\left[\left(\dfrac{u-3}{0.6614}\right)^2 - 2(.9538)\left(\dfrac{u-3}{0.6614}\right)\left(\dfrac{v-7}{1.0897}\right) + \left(\dfrac{v-7}{1.0897}\right)^2\right]\right\}}{2\pi(0.6614)(1.0897)\sqrt{0.09023}}$$

which can be shown to be

$$p_{Y_1 Y_2}(u,v) = \frac{4}{\sqrt{3}\pi}\exp\left\{-\frac{2}{3}\left[52 + 40u + 19u^2 - 32v - 22uv + 7v^2\right]\right\}$$

****

An additional property of the Gaussian distribution which has great importance concerns the conditions for independence. In particular, uncorrelated Gaussian random variables are always independent, whereas this is not true for most distributions. Thus, Gaussian random variables are independent if and only if they are uncorrelated. For most random variables the condition of being independent is much stronger than the condition of being uncorrelated (see Example A.33), but for Gaussian random variables the two conditions are identical. To demonstrate that lack of correlation proves independence, one only needs to study the form of the Gaussian probability density function or characteristic function. For two random variables X and Y one can see from the probability density function in Example B.2 that if $\rho = 0$, then the probability density function can be factored as $p_{XY}(u,v) = p_X(u)p_Y(v)$, and this is exactly the condition for independence of two random variables. Alternatively, one can note that if $\rho = 0$ then the characteristic function from Example B.2 factors as $M_{XY}(\theta,\phi) = M_X(\theta)M_Y(\phi)$, and this is an equivalent statement of the independence condition. For more than two random variables, one can demonstrate similar results by using Eqs. B.12 and B.14. In this case, one must note that if all the components of the vector $\vec{X}$ are uncorrelated then $\mathbf{K}_{XX}$ is a diagonal matrix. This, in turn, gives $\mathbf{K}_{XX}^{-1}$ as a diagonal matrix. When these matrices are diagonal it is easy to show that Eq. B.12 gives $M_{\vec{X}}(\vec{\theta})$ as the product of the n marginal Gaussian characteristic functions, and Eq. B.14 gives the same result for $p_{\vec{X}}(\vec{u})$ and the marginal Gaussian probability density functions.

The reader is cautioned that the jointly Gaussian property is much more restrictive than is the property of having marginal Gaussian distributions. That is, it is quite possible for random variables X and Y each to be individually Gaussian without their being jointly Gaussian. Thus, while the jointly Gaussian property always implies that the random variables are marginally Gaussian, the converse is not generally true.

****

Example B.5: Let the random variables X and Y have the probability density function of

$$p_{XY}(u,v) = \frac{1}{\pi}e^{-u^2/2}e^{-v^2/2}U(uv)$$

Find the marginal probability density functions $p_X(u)$ and $p_Y(v)$, and determine whether X and Y are independent.

Note that the unit step in this probability density function makes it be exactly zero on the second and fourth quadrants of the (u,v) space, since the argument (uv) is negative there. To find $p_X(u)$ we must integrate with respect to v:

$$p_X(u) = \int_{-\infty}^{\infty} p_{XY}(u,v)\,dv$$

For $u < 0$ this gives

$$p_X(u) = \int_{-\infty}^{\infty} \frac{1}{\pi} e^{-u^2/2} e^{-v^2/2} U(-v)\,dv = \frac{e^{-u^2/2}}{\pi} \int_{-\infty}^{0} e^{-v^2/2}\,dv = \frac{e^{-u^2/2}}{\sqrt{2\pi}}$$

and for $u \geq 0$ it is

$$p_X(u) = \int_{-\infty}^{\infty} \frac{1}{\pi} e^{-u^2/2} e^{-v^2/2} U(v)\,dv = \frac{e^{-u^2/2}}{\pi} \int_{0}^{\infty} e^{-v^2/2}\,dv = \frac{e^{-u^2/2}}{\sqrt{2\pi}}$$

so that the marginal distribution of X is exactly the Gaussian distribution with mean zero and unit variance:

$$p_X(u) = \frac{e^{-u^2/2}}{\sqrt{2\pi}}$$

Similarly, one can show that Y also has this Gaussian marginal distribution:

$$p_Y(v) = \frac{e^{-v^2/2}}{\sqrt{2\pi}}$$

The product of the marginal probability density functions is seen to be different than the joint probability density function, even though they are similar in form. Specifically, $p_X(u)p_Y(v)$ is nonzero on all four quadrants of the (u,v) space, whereas $p_{XY}(u,v)$ is twice as large as $p_X(u)p_Y(v)$ on the first and third quadrants, and zero on the second and fourth quadrants. Clearly, X and Y are not independent. One could also have answered the question of independence without performing any calculations. For the $p_{XY}(u,v)$ given, it is clear that X and Y have the same sign. Thus, if we were given $X = 5$, for example, then this would tell us that $Y \geq 0$. A statement like this is always contradictory to independence. If X and Y were independent, then knowledge of the value of X could tell us nothing about the probabilities for Y.

****

B.3   Conditional Distribution of Gaussian Random Variables

Another important property of the Gaussian distribution is revealed by the form of the conditional probability density function for some random variables given values for other random variables. In particular, let the vector $\vec{X}^{(n)}$ consist of n jointly Gaussian components, $\vec{X}^{(l)}$ be the vector of the first l components of $\vec{X}^{(n)}$, and $\vec{X}^{(n-l)}$ be the vector of the other $n-l$ components of $\vec{X}^{(n)}$. We will now investigate the conditional probability distribution of the components of $\vec{X}^{(n-l)}$ given values of the components of $\vec{X}^{(l)}$. The general formula for this situation (a generalization of Eq. A.39) is

$$p_{\vec{X}^{(n-l)}}[\vec{u}^{(n-l)} | \vec{X}^{(l)} = \vec{u}^{(l)}] \equiv p_{X_{l+1},\cdots,X_n}(u_{l+1},\cdots,u_n | X_1 = u_1,\cdots,X_l = u_l) = \frac{p_{\vec{X}^{(n)}}(\vec{u}^{(n)})}{p_{\vec{X}^{(l)}}(\vec{u}^{(l)})}$$

Simplifying the notation of Eq. B.14 by writing $\mathbf{C}^{(n)}$ for the inverse of the covariance matrix for all n random variables and $\mathbf{C}^{(l)}$ for the inverse of the covariance matrix for the first l random variables gives

$$p_{\vec{X}^{(n-l)}}[\vec{u}^{(n-l)} | \vec{X}^{(l)} = \vec{u}^{(l)}] = \frac{1}{(2\pi)^{(n-l)/2}} \sqrt{\frac{|\mathbf{C}^{(n)}|}{|\mathbf{C}^{(l)}|}} \frac{\exp\left(-\frac{1}{2}[\vec{u}^{(n)} - \vec{\mu}_u^{(n)}]^T \mathbf{C}^{(n)}[\vec{u}^{(n)} - \vec{\mu}_u^{(n)}]\right)}{\exp\left(-\frac{1}{2}[\vec{u}^{(l)} - \vec{\mu}_u^{(l)}]^T \mathbf{C}^{(l)}[\vec{u}^{(l)} - \vec{\mu}_u^{(l)}]\right)}$$

(B.15)

The exponential in the numerator of this expression is a quadratic function of the variables $(u_1,\cdots,u_n)$ and the exponential in the denominator is quadratic in $(u_1,\cdots,u_l)$. The ratio of the two exponential functions, then, also has the form of an exponential function of a quadratic function of the variables. In particular, this means that the exponent is a quadratic function of the components of $\vec{u}^{(n-l)}$, and the fact that $p_{\vec{X}^{(n-l)}}[\vec{u}^{(n-l)} | \vec{X}^{(l)} = \vec{u}^{(l)}]$ is an exponential of a quadratic function of the components of $\vec{u}^{(n-l)}$ assures us that it must be exactly a Gaussian conditional probability density function. Thus, if one finds the values of the conditional means

$$\vec{r} = E\left(\vec{X}^{(n-l)} | \vec{X}^{(l)} = \vec{u}^{(l)}\right)$$

(B.16)

and covariances

$$\mathbf{Q} = E\left([\vec{X}^{(n-l)} - \vec{r}][\vec{X}^{(n-l)} - \vec{r}]^T | \vec{X}^{(l)} = \vec{u}^{(l)}\right)$$

(B.17)

of the components of $\vec{X}^{(n-l)}$, then the conditional probability density function can be written in the form of Eq. B.14:

$$p_{\vec{X}^{(n-l)}}[\vec{u}^{(n-l)} | \vec{X}^{(l)} = \vec{u}^{(l)}] = \frac{1}{(2\pi)^{(n-l)/2}\sqrt{|\mathbf{Q}|}} \exp\left(-\frac{1}{2}[\vec{u}^{(n-l)} - \vec{r}]^T \mathbf{Q}^{-1}[\vec{u}^{(n-l)} - \vec{r}]\right)$$

(B.18)

Calculations for Gaussian problems can often be significantly simplified by using this fact that the conditional distributions are also of Gaussian form.

Comparing the form of Eqs. B.15 and B.18 shows that for a set of jointly Gaussian random variables, the conditional mean must be a linear combination of the given values. That is, there is always some matrix $\mathbf{B}$ such that the vector $\vec{r}$ in Eq. B.16 is of the form $\vec{r} = \mathbf{B}\vec{u}^{(l)}$, for which a typical component can be written as

$$r_j \equiv E\left(X_j \mid \vec{X}^{(l)} = \vec{u}^{(l)}\right) = \sum_{k=1}^{l} B_{jk} u_k \qquad (B.19)$$

This follows from the fact if $\vec{r}$ were of any other form, then the exponent in Eq. B.18 would not be quadratic in the $(u_1, \cdots, u_l)$ components, so could not match the exponent in Eq. B.15.

Comparison of Eqs. B.15 and B.18 also reveals an important property of the conditional variances and covariances pertaining to a set of jointly Gaussian random variables. In particular, Eq. B.18 shows that the elements of the matrix $\mathbf{Q}^{-1}$ are the multipliers of the $(u_{l+1}, \cdots, u_n)$ elements in the exponent of the conditional probability density function. However, the multipliers of the $(u_{l+1}, \cdots, u_n)$ elements in Eq. B.15 come only from the numerator of that equation, and that numerator does not depend on the values of the $(u_1, \cdots, u_l)$ components. Thus, the conditional variance and covariance terms do not depend on the given values of the other components. This is not the same as saying that the conditional variance is the same as an unconditional variance. In general, the conditional variance of a Gaussian random variable is less than the unconditional variance of the same random variable, but the amount of reduction in the variance due to conditioning is the same regardless of the value that is found for any other random variable. The following example illustrates this idea for the special case of two jointly Gaussian random variables.

****

Example B.6: For jointly Gaussian random variables X and Y, find the conditional mean and conditional variance of Y given $X = u$.

We can use Eq. B.5 and Example B.2 to write the conditional probability density function as

$$p_Y[v \mid X = u] = \frac{p_{XY}(u,v)}{p_X(u)}$$

$$= \frac{1}{\sqrt{2\pi}\,\sigma_Y\sqrt{1-\rho^2}} \exp\left\{ -\frac{1}{2}\left(\frac{1}{1-\rho^2} - 1\right)\left(\frac{u-\mu_X}{\sigma_X}\right)^2 \right.$$

$$\left. -\frac{1}{2(1-\rho^2)}\left[-2\rho\left(\frac{u-\mu_X}{\sigma_X}\right)\left(\frac{v-\mu_Y}{\sigma_Y}\right) + \left(\frac{v-\mu_Y}{\sigma_Y}\right)^2 \right] \right\}$$

Since this conditional Gaussian distribution must also agree with the single-variable Gaussian form of Eq. B.5, we can see that the denominator of the v^2 term in the exponent must be two times the conditional variance of Y. Thus, we find that the conditional variance of Y is

$$E\left([Y - E(Y|X = u)]^2 \mid X = u \right) = \sigma_Y^2(1-\rho^2)$$

This conditional variance is the same whatever value u is given for X. The value of the conditional mean can be found by identifying the terms which are linear in v in the exponent, and comparing them with the corresponding term in the single-variable Gaussian form. The result is

$$E(Y|X = u) = \mu_Y + \rho\sigma_Y\left(\frac{u - \mu_X}{\sigma_X}\right) = \mu_Y + \frac{K_{XY}}{\sigma_X^2}(u - \mu_X)$$

A little algebra confirms that use of this conditional variance and mean in the single variable Gaussian form does give exactly the preceding expression for the conditional probability density function.

The same procedures can be used to determine the conditional variance and mean of a random variable given the values of two or more other variables, and/or the covariance of two random variables given the value of one or more other random variables. Obviously , though, the expressions are more complicated in these situations.
****

Exercises
****

B.1   Let $\{X_j(t)\}$ denote the east-west displacement at the top of the jth story of a four-story building subjected to an earthquake. Each displacement is measured relative to the moving base of the building. Presume that these story motions are jointly Gaussian and mean zero, and that the covariance matrix at any particular instant of time t_0 is

$$\mathbf{K} = E[\vec{X}(t_0)\vec{X}^T(t_0)] = \begin{bmatrix} 100 & 180 & 223 & 231 \\ 180 & 400 & 540 & 594 \\ 223 & 540 & 900 & 1080 \\ 231 & 594 & 1080 & 1600 \end{bmatrix} \text{mm}^2$$

(a) Find the standard deviation $\sigma_{Y_j(t_0)}$ of each story shear deformation, $Y_j(t_0) = X_j(t_0) - X_{j-1}(t_0)$, in which $X_0(t_0) = 0$.
(b) Find the probability that the fourth story shear deformation will have $|Y_4(t_0)| > 50$ mm.
****

B.2   Consider two random variables X_1 and X_2 with the joint probability density function

$$p_{X_1X_2}(u,v) = \frac{1}{12\pi}\exp\left[-\frac{1}{2}\left(\frac{u-1}{2}\right)^2 - \frac{1}{2}\left(\frac{v-2}{3}\right)^2\right]$$

(a) Find the means and variances of both X_1 and X_2, and find their correlation coefficient ρ_X.
(b) Find the conditional probability density functions: $p_{X_1}(u|X_2 = v)$ and $p_{X_2}(v|X_1 = u)$.
****

B.3   Consider two random variables X_1 and X_2 with the joint probability density function

$$p_{X_1X_2}(u,v) = \frac{1}{12\pi}\exp\left[-\frac{1}{2}\left(\frac{u-1}{2}\right)^2 - \frac{1}{2}\left(\frac{v-2}{3}\right)^2\right]$$

and two other random variables defined as $Y_1 = X_1 - X_2$ and $Y_2 = X_1 + 2X_2$.

(a) Find the means and variances of both Y_1 and Y_2, and find their correlation coefficient ρ_Y.

(b) Give the joint probability density function of Y_1 and Y_2.

****

B.4 Given two random variables with a joint probability density function of

$$p_{XY}(u,v) = \frac{1}{2\pi\sigma_X\sigma_Y}\exp\left(-\frac{u^2}{2\sigma_X^2}-\frac{v^2}{2\sigma_Y^2}\right)\left[1+uv\exp\left(-\frac{u^2v^2}{2}\right)\right]$$

(a) Find the marginal probability density functions: $p_X(u)$ and $p_Y(v)$.

(b) Are X and Y independent?

(c) Are X and Y jointly Gaussian?

[Symmetry and antisymmetry can be used to simplify integrals.]

****

B.5 Consider two random variables X and Y with joint probability density function

$$p_{XY}(u,v) = \frac{1}{\sqrt{2\pi}}\exp\left(-\frac{u^2}{2}-|v|\right)U(uv)$$

(a) Find the marginal probability density functions: $p_X(u)$ and $p_Y(v)$.

(b) Find the conditional probability density functions: $p_X(u|Y=v)$ and $p_Y(v|X=u)$.

(c) Are X and Y independent?

(c) Find $E(Y|X=u)$.

Give all answers for all possible values of u and v.

****

Appendix C

Dirac Delta Functions

The fundamental properties of the so-called Dirac delta function $\delta(x)$ are:

$$\delta(x) = 0 \text{ for } x \neq 0, \quad \delta(0) = \infty \tag{C.1}$$

and

$$\int_{-\infty}^{\infty} \delta(x - x_0) f(x) dx = f(x_0) \tag{C.2}$$

for any function $f(\cdot)$ which is finite and continuous at the point $x = x_0$. Strictly speaking, $\delta(x)$ is not a function, since it is not finite at one point on the real line. Of course, saying that $\delta(0) = \infty$ is not an adequate definition of the behavior of the function at the origin, since infinity is not a number. For example, ∞ / ∞ can have any value from zero to infinity. The definition of the critical property of $\delta(\cdot)$ at the origin is given by Eq. C.2. We will only use Dirac delta functions in situations where the quantity of real interest is to be obtained from an integral involving the $\delta(\cdot)$ function, rather than from the value of $\delta(\cdot)$ at any single argument.

Another way of viewing the Dirac delta function is as the formal derivative of the unit step function. This interpretation follows directly from the fundamental properties of $\delta(\cdot)$. In particular, let a function $U(x)$ be defined as the integral of $\delta(x)$:[1]

$$U(x) = \int_{-\infty}^{x} \delta(u) du \tag{C.3}$$

with an initial condition of $U(-\infty) = 0$. Based on Eqs. C.1 and C.2, we then obtain

$$U(x) = 0 \text{ for } x < 0, \quad U(x) = 1 \text{ for } x > 0$$

This specification of $U(x)$, though, is identical to that given in Section A.1 for the unit step function, except for uncertainty about the value of $U(x)$ at the point $x = 0$.[2] If the derivative of Eq. C.3 existed, then it would be given by

$$\frac{dU}{dx} = \delta(x) \tag{C.4}$$

The difficulty with this procedure, of course, is that the unit step function is not truly differentiable at

---

[1] The reader is reminded of the equivalence of the indefinite integral of $f(x)$ and the definite integral of $f(\cdot)$ with x appearing only as the upper limit of the integral.

[2] In Section A.1 we defined $U(x)$ to be continuous from the right since this simplifies its usage in the description of cumulative distribution functions. This choice, though, is somewhat arbitrary.

the point $x = 0$, since it is discontinuous at that point.

In order to precisely define $\delta(x)$ it is necessary to consider a sequence of functions, since $\delta(x)$ is not truly a function. One way to do this is to consider a sequence of functions, each of which satisfies Eq. C.2, and which asymptotically approach the condition of Eq. C.1. For example,

$$\delta_j(x) = 0 \text{ for } x \le -2^{-j}, \quad \delta_j(x) = 0 \text{ for } x \ge 2^{-j}, \quad \delta_j(x) = 2^{j-1} \text{ for } |x| < 2^{-j} \quad \text{(C.5)}$$

or

$$\delta_j(x) = 0 \text{ for } x \le -2^{-j}, \quad \delta_j(x) = 0 \text{ for } x \ge 2^{-j}, \quad \delta_j(x) = 2^j(1 - 2^j|x|) \text{ for } |x| < 2^{-j} \quad \text{(C.6)}$$

Clearly, each member of these sequences of rectangles and triangles does exactly satisfy Eq. C.2, and as j increases toward infinity, either sequence comes closer and closer to meeting the condition of Eq. C.1. The term "generalized function" is sometimes used for a relationship like $\delta(x)$ which is singular, but can be approached asymptotically by a sequence of functions.

It may be noted that each sequence which asymptotically approaches $\delta(x)$ can be integrated, term by term, to give a sequence which converges to $U(x)$, except possibly at the point $x = 0$. For the two sequences given above the $U_j(x)$ integrals are exactly 1/2 at $x = 0$. This can be remedied by shifting the "pulses" to the left of the origin, such as replacing Eq. C.5 by

$$\delta_j(x) = 0 \text{ for } x \le -2^{-j+1}, \quad \delta_j(x) = 0 \text{ for } x \ge 0, \quad \delta_j(x) = 2^{j-1} \text{ for } -2^{-j+1} < x < 0 \quad \text{(C.7)}$$

but this also has certain disadvantages. In particular, the Dirac delta function is inherently symmetric $[\delta(-x) = \delta(x)]$ while the sequence in Eq. C.7 is always asymmetric for any finite j. The integral of Eq. C.5 or C.7 gives a sequence of piecewise linear functions which tend to the unit step function, while integration of Eq. C.6 gives a corresponding sequence which is piecewise quadratic. If needed, one can go so far as to use a sequence of analytic (i.e., infinitely differentiable) functions which asymptotically approaches $\delta(x)$. One example is

$$\delta_j(x) = \frac{2^j}{\sqrt{2\pi}} \exp\left(-2^{(2j-1)}x^2\right) \quad \text{(C.8)}$$

Term-by-term differentiation of such a sequence can be used also to provide a precise generalized function definition of the derivatives of a Dirac delta function. It is good to know that such a definition is possible, since there are situations in which it is convenient to use this derivative concept.

Appendix D

Fourier Analysis

The basic idea of Fourier analysis is to represent a function $x(t)$ as a sum, or linear combination, of harmonic components, in order to simplify its analysis. The simplest such situation is when $x(t)$ is defined only on a finite region. We will take this region to be of length T, and we will take it to be symmetric about the origin, $-T/2 \leq t \leq +T/2$, since this symmetry will lead to some simplifications later. In this case one can write

$$x(t) = \sum_{j=0}^{\infty} a_j \cos\left(\frac{2\pi j t}{T}\right) + \sum_{j=1}^{\infty} b_j \sin\left(\frac{2\pi j t}{T}\right) \tag{D.1}$$

Note that the frequencies of the harmonic terms have been taken such that each term contains an integer number of cycles of oscillations during the interval $-T/2 \leq t \leq +T/2$. Also, note that the a_0 term is simply a constant, so that the equation could equally well be written as

$$x(t) = a_0 + \sum_{j=1}^{\infty} a_j \cos\left(\frac{2\pi j t}{T}\right) + \sum_{j=1}^{\infty} b_j \sin\left(\frac{2\pi j t}{T}\right)$$

The problem, now, is to evaluate all the a_j and b_j coefficients in Eq. D.1. This task is made quite easy, though, by the orthogonality of the harmonic terms included. In particular,

$$\int_{-T/2}^{T/2} \cos\left(\frac{2\pi j t}{T}\right) \cos\left(\frac{2\pi k t}{T}\right) dt = 0 \qquad \text{for} \qquad j \neq k$$
$$= T/2 \qquad \text{for} \qquad j = k \neq 0$$
$$= T \qquad \text{for} \qquad j = k = 0$$

and

$$\int_{-T/2}^{T/2} \cos\left(\frac{2\pi j t}{T}\right) \sin\left(\frac{2\pi k t}{T}\right) dt = 0$$

First we multiply Eq. D.1 by $\cos(2\pi k t / T)$, for any integer k, then integrate both sides of the equation from $t = -T/2$ to $t = T/2$:

$$\int_{-T/2}^{T/2} x(t) \cos\left(\frac{2\pi k t}{T}\right) dt = \int_{-T/2}^{T/2} \cos\left(\frac{2\pi k t}{T}\right) \left[\sum_{j=0}^{\infty} a_j \cos\left(\frac{2\pi j t}{T}\right) + \sum_{j=1}^{\infty} b_j \sin\left(\frac{2\pi j t}{T}\right)\right] dt$$

The order of summation and integration can be reversed on the right hand side of this equation, and the orthogonality relationships cause all but one of these integral terms to be zero. In particular, the only non-zero term is the one $\cos(2\pi jt/T)$ term with $j = k$. Thus, one finds that

$$\int_{-T/2}^{T/2} x(t)\cos\left(\frac{2\pi kt}{T}\right) dt = a_k\left(\frac{T}{2}\right) \qquad \text{for} \quad k \neq 0$$

$$\int_{-T/2}^{T/2} x(t)\, dt = a_0 T$$

Rewriting these expressions with an index variable of j, instead of k, gives

$$a_0 = \frac{1}{T}\int_{-T/2}^{T/2} x(t)\, dt \tag{D.2}$$

$$a_j = \frac{2}{T}\int_{-T/2}^{T/2} x(t)\cos\left(\frac{2\pi jt}{T}\right) dt \qquad \text{for} \quad j \neq 0 \tag{D.3}$$

This gives the values for all the a_j coefficients of the $\cos(2\pi jt/T)$ terms in Eq. D.1. Performing the same operations using a multiplier of $\sin(2\pi kt/T)$ gives

$$b_j = \frac{2}{T}\int_{-T/2}^{T/2} x(t)\sin\left(\frac{2\pi jt}{T}\right) dt \qquad \text{for} \quad j \neq 0 \tag{D.4}$$

for the coefficients of the $\sin(2\pi jt/T)$ terms in Eq. D.1.

The usefulness of the Fourier series represented in Eqs. D.1 - D.4 depends on the fact that the series does converge, so that a truncated series of the form

$$x_N(t) = \sum_{j=0}^{N} a_j\cos\left(\frac{2\pi jt}{T}\right) + \sum_{j=1}^{N} b_j\sin\left(\frac{2\pi jt}{T}\right) \tag{D.5}$$

can be used as an approximation of $x(t)$. In particular, the truncated series converges to $x(t)$ as N goes to infinity at every point t where $x(t)$ is continuous. If $x(t)$ is discontinuous at the value t, then the truncated series converges to the average of the limits from the right and the left. The fact that Eq. D.1 can be used to represent any continuous function on $0 \leq t \leq T$ can be viewed as a statement that the harmonic functions form a complete basis for the space of these continuous functions. Thus, Eqs. D.1 - D.4 give a complete representation of $x(t)$ as a sum of harmonic components. Alternatively, one can say that Eq. D.1 is the harmonic decomposition of $x(t)$.

Sometimes it is more convenient to use an alternate form of Eq. D.1, based on the complex exponential representation of the harmonic functions. In particular, if one uses the identities

$$\cos\left(\frac{2\pi jt}{T}\right) = \frac{e^{2\pi ijt/T} + e^{-2\pi ijt/T}}{2}, \qquad \sin\left(\frac{2\pi jt}{T}\right) = \frac{e^{2\pi ijt/T} - e^{-2\pi ijt/T}}{2i}$$

then Eqs. D.1 - D.4 become

$$x(t) = \sum_{j=-\infty}^{\infty} c_j e^{2\pi i jt/T} \qquad\qquad (D.6)$$

with

$$c_j = \frac{1}{T} \int_{-T/2}^{T/2} x(t) e^{-2\pi i jt/T} \, dt \qquad\qquad (D.7)$$

in which $c_0 = a_0$ and

$$c_j = \frac{a_j}{2} + \frac{b_j}{2i} = \frac{a_j - ib_j}{2}, \qquad c_{(-j)} = \frac{a_j}{2} - \frac{b_j}{2i} = \frac{a_j + ib_j}{2} \qquad \text{for} \quad j > 0$$

The validity of this exponential form is also easily confirmed by direct evaluation of the coefficients in Eq. D.6. That is, rather than using Eqs. D.2 - D.4 to find the coefficients in Eq. D.6, one can use the orthogonality of the complex exponential functions, which can be written as

$$\int_{-T/2}^{T/2} e^{2\pi i jt/T} e^{2\pi i kt/T} \, dt = 0 \qquad \text{for} \quad j + k \neq 0$$

$$= T \qquad \text{for} \quad j + k = 0$$

Thus, multiplying Eq. D.6 by $\exp(2\pi i kt / T)$ and integrating directly gives the coefficient values given in Eq. D.7.

It should be noted that the Fourier series representation converges to $x(t)$ only within the interval $0 \leq t \leq T$. In fact, it was assumed in Eqs. D.1, D.5, and D.6 that $x(t)$ was defined only on that finite region. In some situations $x(t)$ is actually defined on a broader domain, but the series converges only within the $[-T/2, T/2]$ interval used in evaluating the coefficients according to Eqs. D.2 - D.4 or Eq. D.7. In fact, the functions of Eqs. D.1, D.5, and D.6 are periodic. For example, $x_N(t \pm T) = x_N(t)$, so that the series repeats itself with period T. This finite period representation is not adequate for many problems in which we wish to consider $x(t)$ to be aperiodic and to extend from $-\infty$ to ∞. A direct way of extending our Fourier analysis to include this situation is to let the period T tend to infinity. In investigating this limiting situation, we will use the exponential form of Eqs. D.6 and D.7, since it is somewhat simpler than the form using sine and cosine representations of the harmonic components.

First, we introduce a notation for the frequency of the j term in Eqs. D.6 and D.7:

$$\omega_j = j \Delta \omega, \qquad \Delta \omega = 2\pi / T$$

so that the equations can be written as

$$x(t) = \sum_{j=-\infty}^{\infty} c_j e^{i\omega_j t}, \qquad c_j = \frac{\Delta\omega}{2\pi} \int_{-T/2}^{T/2} x(t) e^{-i\omega_j t} \, dt$$

Letting T tend to infinity now gives $\Delta\omega \to d\omega$ and the summation tending to an integral, so that one can introduce a renormalized function $\tilde{x}(\omega_j) = c_j / \Delta\omega$ and write

$$x(t) = \int_{-\infty}^{\infty} \tilde{x}(\omega) e^{i\omega t} \, d\omega \tag{D.8}$$

with

$$\tilde{x}(\omega) = \frac{1}{2\pi} \int_{-\infty}^{\infty} x(t) e^{-i\omega t} \, dt \tag{D.9}$$

We will call the function $\tilde{x}(\omega)$ given in Eq. D.9 the Fourier transform of $x(t)$, and Eq. D.8 is then the inverse Fourier transform formula. Other terms such as integral Fourier transform, exponential Fourier transform, etc. are also used for Eq. D.9. Note that the frequency decomposition of the aperiodic, infinite period $x(t)$ function generally contains all frequencies. That is, the Fourier transform $\tilde{x}(\omega)$ is defined for all ω values, and Eq. D.8 is a superposition of $\tilde{x}(\omega)\exp(i\omega t)$ terms for all ω values.

One also has the option of deriving Eq. D.9 directly, without any consideration of the Fourier series introduced earlier. In particular, if one assumes that an $\tilde{x}(\omega)$ transform exists such that it is possible to write Eq. D.8, then one can use the orthogonality of the complex exponential functions to derive Eq. D.9. In particular,

$$\int_{-\infty}^{\infty} e^{i\psi t} \, dt = 2\pi \, \delta(\psi) \tag{D.10}$$

so that multiplying Eq. D.8 by $\exp(i\eta t)$ and integrating gives

$$\int_{-\infty}^{\infty} x(t) e^{i\eta t} \, dt = \int_{-\infty}^{\infty} \tilde{x}(\omega) \int_{-\infty}^{\infty} e^{i\eta t} e^{i\omega t} \, dt \, d\omega = 2\pi \int_{-\infty}^{\infty} \tilde{x}(\omega) \delta(\eta + \omega) \, d\omega = 2\pi \, \tilde{x}(-\eta)$$

which is easily rewritten as Eq. D.9. The orthogonality relationship of Eq. D.10 appears quite frequently in applications of Fourier analysis. It can be viewed as the Fourier transform of the function $f(t) = 1$. Written in the usual notation with ω, rather than η, representing frequency, this gives $\tilde{f}(\omega) = 2\pi \, \delta(\omega)$, showing that the only harmonic component of a constant is the term with zero frequency.

References

Abramowitz, M. and and Stegun, I. A. (1965). *Handbook of Mathematical Functions*, Dover Publications, Inc., New York.

Asano, K. and Iwan, W. D. (1984). "An alternative approach to the random response of bilinear hysteretic systems," *Earthquake Engineering and Structural Dynamics*, vol. 12, pp. 229-236.

Atalik, T. S. and Utku, S. (1976). "Stochastic linearization of multi-degree-of-freedom non-linear systems," *Earthquake Engineering and Structural Dynamics*, vol. 4, pp. 411-420.

Bendat, J. S. and Piersol, A. G. (1966). *Measurement and Analysis of Random Data*, Wiley, New York.

Bogdanoff, J. L. and Kozin, F. (1985) *Probabilistic Models of Cumulative Damage*, Wiley, New York.

Bouc, R. (1968). "Forced vibration of a mechanical system with hysteresis," Abstract, in *Proc. of the Fourth Conference on Nonlinear Oscillations*, p. 315, Academia Publishing, Prague, Czechoslovakia.

Cai, G. Q. and Lin, Y. K. (1988). "A new approximate solution technique for randomly excited nonlinear oscillators," *International Journal of Nonlinear Mechanics*, vol. 23, pp. 409-420.

Caughey, T. K. (1960a). "Classical normal modes in damped linear dynamic systems," *Journal of Applied Mechanics*, ASME, vol. 27, pp. 269-271.

Caughey, T. K. (1960b). "Random excitation of a system with bilinear hysteresis," *Journal of Applied Mechanics*, vol. 27, pp. 649-652.

Caughey, T. K. (1965). "On the response of a class of nonlinear oscillators to stochastic excitation," *Les Vibrations forceés dans les systèmes non-linéaires*, (Éditions du Centre National de la Recherche Scientifique), Paris, France, pp. 393-405.

Caughey, T. K. (1986). "On response of non-linear oscillators to stochastic excitation," *Probabilistic Engineering Mechanics*, vol. 1. no. 1, pp. 2-4.

Caughey, T. K. and O'Kelly, M. E. J. (1965). "Classical normal modes in damped linear dynamic systems," *Journal of Applied Mechanics*, ASME, vol. 32, pp. 583-588.

Caughey, T. K. and Stumpf, H. S. (1961). "Transient response of a dynamic system under random excitation," *Journal of Applied Mechanics*, ASME, vol. 28, pp. 563-566.

Chen, D. C-K. and Lutes, L. D. (1994). "First-passage time of secondary system mounted on a yielding structure," *Journal of Engineering Mechanics*, ASCE, vol. 120, pp. 814-834.

Choi, D-W., Miksad, R. W., Powers, E. J., and Fischer, F. J. (1985). "Application of digital cross-bispectral analysis techniques to model the nonlinear response of a moored vessel system in random seas," *Journal of Sound and Vibration*, vol. 99, no. 3, pp. 309-326.

Corotis, R. B., Vanmarcke, E. H., and Cornell, C. A. (1972). "First passage of nonstationary random processes," *Journal of the Engineering Mechanics Division*, ASCE, vol. 98, no. EM2, pp. 401-414.

Cramer, H. (1966). "On the intersection between the trajectories of a normal stationary stochastic process and a high level," *Arkiv för Matematik*, vol. 6. no. 2, pp. 337-349.

Cramer, H. and Leadbetter, M. R. (1967). *Stationary and Related Stochastic Processes*, Wiley, New York.

Crandall, S. H. (1970). "First crossing probabilities of the linear oscillator," *Journal of Sound and Vibration*, London, vol. 12, no. 3, pp. 285-299.

Crandall, S. H. (1980). "Non-Gaussian closure for random vibration excitation," *International Journal of Non-Linear Mechanics*, vol. 15, pp. 303-313.

Crandall, S. H., Chandaramani, K. L., and Cook, R. G. (1966). "Some first passage problems in random vibration," *Journal of Applied Mechanics*, ASME, vol. 33, no. 3, pp. 532-538.

Crandall, S. H. and Mark, W. D. (1963). *Random Vibrations in Mechanical Systems*, Academic Press, New York.

Der Kiureghian, A. (1980). "Structural response to stationary excitation," *Journal of the Engineering Mechanics Division*, ASCE, vol. 106, no. EM6, pp. 1195-1213.

Di Paola, M. and Falsone, G. (1993). "Stochastic dynamics of nonlinear systems driven by non-normal delta-correlated processes, *Journal of Applied Mechanics*, ASME, vol. 60, pp. 141-148.

Di Paola, M., Falsone, G., and Pirotta, A. (1992). "Stochastic response analysis of nonlinear systems under Gaussian inputs," *Probabilistic Engineering Mechanics*, vol. 7, pp. 15-21.

Di Paola, M. and Muscolino, G. (1990). "Differential moment equations of FE modeled structures with geometrical non-linearities," *International Journal of Non-Linear Mechanics*, vol. 25, pp. 363-373.

Ditlevsen, O. (1986). "Duration of Gaussian process visit to critical set," *Probabilistic Engineering Mechanics*, vol. 1, pp. 82-93

Dowling, N. E. (1972). "Fatigue failure predictions for complicated stress-strain histories," *Journal of Materials*, vol. 7, no. 1, pp. 71-87.

Downing, S. D. and Socie, D. F. (1982). "Simple rainflow counting algorithms," *International Journal of Fatigue*, vol. 4, no. 1, pp. 31-40.

Elishakoff, I. (1995). "Random vibration of structures: A personal perspective," *Applied Mechanics Reviews*, vol. 48, no. 12, Part 1, pp. 809-825.

Fuchs, H. O. and Stephens, R. I. (1980). *Metal Fatigue in Engineering*, Wiley, New York.

Grigoriu, M. (1995). *Applied Non-Gaussian Processes*, PTR Prentice Hall, Englewood Cliffs, NJ.

Ibrahim, R. A. (1985). *Parametric Random Vibration*, Wiley, New York.

Ibrahim, R. A. (1995). "Recent results in random vibrations of nonlinear mechanical systems," *50th Anniversary of the Design Engineering Division*, Transactions of the ASME, pp. 222-233.

Igusa, T., Der Kiureghian, A., and Sackman, J. L. (1984). "Modal Decomposition Method for Stationary Response of Non-Classically Damped Systems," *Earthquake Engineering and Structural Dynamics*, vol. 12, no. 1, pp.121-136.

Iwan, W. D. and Lutes, L. D. (1968). "Response of the bilinear hysteretic system to stationary random excitation," *Journal of the Acoustical Society of America*, vol. 43, pp. 545-552.

Iyengar, R. N. and Dash, P. K. (1978). "Study of the random vibration of nonlinear systems by the Gaussian closure technique," *Journal of Applied Mechanics*, ASME, vol. 45, pp. 393-399.

Jennings, P. C. (1964). "Periodic response of a general yielding structure," *Journal of the Engineering Mechanics Division*, ASCE, vol. 90, no. EM2, pp. 131-166.

Kareem, A., Zhao, J., and Tognarelli, M. A. (1995). "Surge response statistics of tension leg platforms under wind and wave loads: A statistical quadratization approach," *Probabilistic Engineering Mechanics*, vol. 10, pp. 225-240.

Kaul, M. K. and Penzien, J. (1974). "Stochastic analysis of yielding offshore towers," *Journal of the Engineering Mechanics Division*, ASCE, vol. 100, no. EM5, pp. 1025-1038.

Kazakov, I. E. (1965). "Generalization of the method of statistical linearization to multidimensional systems," *Automation and Remote Control*, vol. 26, pp. 1201-1206.

Kendall, M. and Stuart, A. (1977). *The Advanced Theory of Statistics*, Macmillan Publishing Co., Inc., New York.

Kutt, T. V. and Bienick, M. P. (1985). "Cumulative damage and fatigue life prediction," presented at the *26th Structures, Structural Dynamics, and Materials Conference*, Orlando, FL.

Larsen, C. E. and Lutes, L. D. (1991). "Predicting the fatigue life of offshore structures by the single-moment spectral method," *Probabilistic Engineering Mechanics*, vol. 6, no. 2, pp. 96-108.

Li, X-M., Quek, S-T., and Koh, C-G. (1995). "Stochastic response of offshore platforms by statistical cubicization," *Journal of Engineering Mechanics* , ASCE, vol. 121, pp. 1056-1068.

Lin, Y. K. (1967). *Probabilistic Theory of Structural Dynamics*, McGraw-Hill Book Company, New York. Reprinted in 1976 by Krieger Publishing Co., Melbourne, FL.

Lin, Y. K. and Cai, G. Q. (1995). *Probabilistic Structural Dynamics: Advanced Theory and Applications*, McGraw-Hill Book Company, New York.

Lindgren, G., and Rychlik, I. (1987). "Rain flow cycle distribution for fatigue life prediction under Gaussian load processes," *Fatigue Fracture Engineering Materials and Structures*, vol. 10, no. 3, pp. 251-260.

Longuet-Higgins, M. S. (1963). "The effect of non-linearities on statistical distributions in the theory of sea waves," *Journal of Fluid Mechanics*, vol. 17, no. 3, pp. 459-480.

Lutes, L. D. and Chen, D. C. K. (1992). "Stochastic response moments for linear systems," *Probabilistic Engineering Mechanics*, vol. 7, no. 3, pp. 165-173.

Lutes, L. D., Corazao, M., Hu, S-L. J., and Zimmerman, J. J. (1984). "Stochastic fatigue damage accumulation," *Journal of Structural Engineering*, ASCE, vol. 110, no. 11, pp. 2585-2601.

Lutes, L. D. and Larsen, C. E. (1990). "Improved spectral method for variable amplitude fatigue," *Journal of Structural Engineering*, ASCE, vol. 116, no. 4, pp. 1149-1164.

Lutes, L. D. and Wang, J. (1993). "Kurtosis effects on stochastic structural fatigue," in *Proc. of the International Conference on Structural Safety and Reliability* (ICOSSAR '93), Innsbruck, Austria, pp. 1091-1097.

Madsen, H. O., Krenk, S., and Lind, N. C. (1986). *Methods of Structural Safety*, Prentice-Hall, Inc., Englewood Cliffs, NJ.

Marcinkiewicz, J. (1938). "Sur une propriété de la loi de Gauss," *Mathematische Zeitschrift*, vol. 44, pp. 612-618.

Marple, Jr., S. L. (1987). *Digital Spectral Analysis: With Applications*, Prentice-Hall, Inc., Englewood Cliffs, NJ.

Matsuishi, M. and Endo, T. (1968). "Fatigue of metals subject to varying stress," presesented to the *Japan Society of Mechanical Engineers*, Fukuoka, Japan.

Miles, J. W. (1954). "On structural fatigue under random loading," *Journal of the Aeronautical Society*, vol. 21, pp. 753-762.

Moyal, J. E. (1949). "Stochastic processes and statistical physics," *Journal of the Royal Statistical Society (London)* , vol. B11, pp. 150-210.

Naess, A. and Ness, G. M. (1992). "The statistical distribution of second-order, sum-frequency response statistics of tethered platforms in random waves," *Applied Ocean Research*, vol. 14, no. 2, pp. 23-32.

Nigam, N. C. (1983). *Introduction to Random Vibrations*, M.I.T. Press, Cambridge, MA.

Ochi, M. K. (1986). "Non-Gaussian random processes in ocean engineering," *Probabilistic Engineering Mechanics*, vol. 1, no. 1, pp. 28-39.

Ortiz, K. and Chen, N. K. (1987). "Fatigue damage prediction for stationary wideband random stresses," in *Proc. of theFifth International Conference on Applications of Statistics and Probability in Soil and Structural Engineering (ICASP 5)*, Vancouver, B.C., pp. 309-316.

Papadimitriou, C. (1995). "Stochastic response cumulants of MDOF linear systems," *Journal of Engineering Mechanics*, ASCE, vol. 121, no. 11, pp. 1181-1192.

Papadimitriou, C. and Lutes, L. D. (1994). "Approximate analysis of higher cumulants for MDF random vibration," *Probabilistic Engineering Mechanics*, vol. 9, no. 1, pp. 71-82.

Papadimitriou, C. and Lutes, L. D. (1996). "Stochastic cumulant analysis of MDOF systems with polynomial-type nonlinearities," *Probabilistic Engineering Mechanics*, vol. 11, no. 1, pp. 1-13.

Priestly, M. B. (1988). *Non-Linear and Non-Stationary Time Series Analysis*, Academic Press, London.

Rice, S. O. (1944, 1945). "Mathematical analysis of random noise," *Bell System Technical Journal*, Vols. 23 and 24. Reprinted in *Selected Papers on Noise and Stochastic Processes*, edited by N. Wax (1954), Dover Publications, Inc., New York.

Roberts, J. B., (1976a). "First passage time for the envelope of a randomly excited linear oscillator," *Journal of Sound and Vibration*, London, vol. 46, no. 1, pp. 1-14.

Roberts, J. B., (1976b). "First passage probability for nonlinear oscillators," *Journal of the Engineering Mechanics Division*, ASCE, vol. 102, EM5, pp. 851-866.

Roberts, J. B. and Spanos P. D. (1986). "Stochastic averaging: An approximate method for solving random vibration problems," *International Journal of Non-linear Mechanics*, vol. 21, pp. 111-134.

Roberts, J. B. and Spanos, P. D. (1990). *Random Vibration and Statistical Linearization*, Wiley, New York.

Rugh, W. J. (1981). *Nonlinear System Theory: The Volterra/Wiener Approach*, Johns Hopkins University Press, Baltimore, MD.

Rychlik, I. (1989). "Simple approximations of the rain-flow-cycle distribution for discretized random loads," *Probabilistic Engineering Mechanics*, vol. 4, no. 1, pp. 40-48.

Sarkani, S., Kihl, D.P., and Beach, J.E. (1994). "Fatigue of welded joints under narrowband non-Gaussian loadings," *Probabilistic Engineering Mechanics*, vol. 9, no.3, pp.179-190.

Senthilnathan, A. and Lutes, L. D. (1991). "Nonstationary maximum response statistics for linear structures," *Journal of Engineering Mechanics*, ASCE, vol. 117, no. 2, pp. 294-311.

Schetzen, M. (1980). *The Volterra and Wiener Theories of Nonlinear Systems*, Wiley Interscience, New York.

Sobczyk, K. and Spencer, Jr., B. F. (1992). *Random Fatigue: From Data to Theory*, Academic Press, Inc., Boston, MA.

Soong, T. T. and Grigoriu, M. (1993). *Random Vibration of Mechanical and Structural Systems*, PTR Prentice Hall, Englewood Cliffs, NJ.

Spanos, P-T. D. (1983). "Spectral moments calculation of linear system output," *Journal of Applied Mechanics*, ASME, vol. 50, pp. 901-903.

Stratonovich, R. L. (1963). *Topics in the Theory of Random Noise*, Gordon and Breach, New York. Translated from the Russian by Silverman, R. A.

Tanabashi, R. and Kaneta, K. (1962). "On the relation between the restoring force characteristics of structures and the pattern of earthquake ground motion," in *Proc. of Japan National Symposium on Earthquake Engineering*, Tokyo, Japan.

To, C. W. S. and Li, D. M. (1991). "Equivalent nonlinearization of nonlinear systems to random excitations," *Probabilistic Engineering Mechanics*, vol. 6, pp. 184-192.

Vanmarcke, E. H. (1972). "Properties of spectral moments with applications to random vibration," *Journal of the Engineering Mechanics Division*, ASCE, vol. 98, no. EM2, pp. 425-446.

Vanmarcke, E. H. (1975). "On the distribution of the first-passage time for normal stationary random processes," *Journal of Applied Mechanics*, ASME, vol. 42, pp. 215-220.

Veletsos, A. S. and Ventura, C. E. (1986). "Modal analysis of non-classically damped linear systems," *Earthquake Engineering and Structural Dynamics*, vol. 14, pp. 217-243.

Wang, J. and Lutes, L. D. (1993). "Current induced skewness of dynamic response for offshore structures," in *Proc. of the International Conference on Structural Safety and Reliability* (ICOSSAR '93), Innsbruck, Austria, pp. 545-549.

Wen, Y. K. (1976). "Method for random vibration of hysteretic systems," *Journal of the Engineering Mechanics Division*, ASCE, vol. 102, no. EM2, pp. 249-263.

Wilson, E. L., Der Kiureghian, A., and Bayo, E. P. (1981). "A replacement for the SRSS method in seismic analysis," *Earthquake Engineering and Structural Dynamics*, vol. 9, pp. 187-192.

Winterstein, S. R. (1985). "Non-normal responses and fatigue damage," *Journal of Engineering Mechanics*, ASCE, vol. 111, no. 10, pp. 1291-1295.

Winterstein, S. R. (1988). "Nonlinear vibration models for extremes and fatigue," *Journal of Engineering Mechanics*, ASCE, vol. 114, no. 10, pp. 1772-1790.

Winterstein, S. R. and Cornell, C. A. (1985). "Energy fluctuation scale and diffusion models," *Journal of Engineering Mechanics*, ASCE, vol. 111, no. 2, pp. 125-142.

Winterstein, S. R., Ude, T. C., and Marthinsen, T. (1994). "Volterra models of ocean structures: Extreme and fatigue reliability," *Journal of Engineering Mechanics*, ASCE, vol. 120, pp. 1369-1385.

Wirsching, P. H. and Light, M. C. (1980). "Fatigue under wide band random stress," *Journal of the Structural Division*, ASCE, vol. 106, no. ST7, pp. 1593-1607.

Wirsching, P. H., Paez, T. L., and Ortiz, K. (1995). *Random Vibrations: Theory and Practice*, Wiley Interscience, New York.

Wu, W. F. and Lin, Y. K. (1984). "Cumulant-neglect closure for non-linear oscillators under random parametric and external excitations," *International Journal of Non-Linear Mechanics*, vol. 19, pp. 349-362.

Zhu, W. Q. and Yu, J. S. (1989). "The equivalent non-linear system method," *Journal of Sound and Vibration*, vol. 129, pp. 385-395

Author Index

Subject Index